Bioactive Molecules and Medicinal Plants

Kishan Gopal Ramawat
Jean-Michel Mérillon
(Eds.)

Bioactive Molecules
and Medicinal Plants

Professor Kishan Gopal Ramawat
Laboratory of Biomolecular
Technology, Department of Botany
M.L. Sukhadia University
Udaipur 313001
India
kg_ramawat@yahoo.com

Professor Jean-Michel Mérillon
Groupe d'Etude des Substances
Végétales à Activité Biologique
Laboratoire de Sciences Végétales,
Mycologie et Biotechnologie,
EA 3675 – UFR Pharmacie
Universite Victor Segalen
Bordeaux 2
146 Rue Leo-Saignat
F-33076 Bordeaux cedex
France
jean-michel.merillon@u-bordeaux2.fr

ISBN 978-3-540-74600-3   e-ISBN 978-3-540-74603-4

DOI 10.1007/978-3-540-74603-4

Library of Congress Control Number: 2007933308

© Springer-Verlag Berlin Heidelberg 2008

This work is subject to copyright. All rights are reserved, wether the whole or part of the material is concerned, specifically the rights of translation, reprinting, reuse of illustrations, recitation, broad-casting, reproduction on microfilm or any other way, and storage in data banks. Duplication of this publication or parts thereof is permitted only under the provisions of the German Copyright Law of September 9, 1965, in its current version, and permission for use must always be obtained from Springer. Violations are liable to prosecution under the German Copyright Law.

The use of general descriptive names, registered names, trademarks, etc. in this publication does not imply, even in the absence of a specific statement, that such names are exempt from the relevant protective laws and regulations and therefore free for general use.

Product liability: the publishers cannot guarantee the accuracy of any information about dosage and application contained in this book. In every individual case the user must check such information by consulting the relevant literature.

Reproduction, typesetting and production: LE-TEX Jelonek, Schmidt & Vöckler GbR, Leipzig
Cover design: WMXDesign GmbH, Heidelberg

Printed on acid-free paper   21/3180/YL   5 4 3 2 1 0   SPIN 11816423

springer.com

Preface

Use of medicinal plants is as old as human civilization and continuous efforts are being made to improve medicinal plants or produce their products in high amounts through various technologies. About 200,000 natural products of plant origin are known and many more are being identified from higher plants and microorganisms. Some plant-based drugs have been used for centuries and there is no alternative medicine for many drugs, such as cardiac glycosides. However, natural products research was sidelined to pave the way for combinatorial chemistry, which was expected to produce large numbers of synthetic compounds for high-throughput screening (HTS). This line of work has failed to deliver desirable results. Moreover, it is not possible for all pharmaceutical companies and institutions to adopt costly HTS technology. Therefore, medicinal plants and their bioactive molecules are always in demand and are a central point of research. While planning this book, we endeavored to incorporate articles that cover the entire gamut of current medicinal plants research.

The aim of this book was to review the current status of bioactive molecules and medicinal plants research in light of the surge in the demand for herbal medicine. The chapters focus on bioactive molecules (e.g., stilbenes and phytoestrogens) and on medicinal plants as a whole (e.g., *Bacopa monnierie*). We hope that this book will be useful for researchers in academia, industry, and agriculture planning.

Finally, we would like to acknowledge our contributors, who have made serious efforts to ensure the high scientific quality of this book. We also would like to thank our colleagues at Springer.

June 2007 *K.G. Ramawat and J.M. Mérillon*

About the editors

Professor K.G. Ramawat (born in 1952) received his M.Sc. (1974) and Ph.D. (1978, Plant Biotechnology) from the University of Jodhpur, Jodhpur, India and became a faculty member in January of 1979. He joined M.L. Sukhadia University as an Associate Professor in 1991 and became a Professor in 2001. He served as Head of the Department of Botany (2001–2004), was in charge of the Department of Biotechnology (2003–2004), was a member of the task force on medicinal and aromatic plants at the Department of Biotechnology (Government of India, New Delhi; 2002–2005), and was a coordinator of the UGC-DRS and DST-FIST programs (2002–2007). He did his postdoctoral study at the University of Tours, France (1983–85) and subsequently worked as visiting professor at the University of Tours (1991) and University of Bordeaux 2, France (1995, 1999, 2003, 2006). He visited Poland under the auspices of an INSA-PAN academic exchange program (2005). He has published more than 100 research papers and review articles in reputed journals and books. He has edited two books on the biotechnology of secondary metabolites and of medicinal plants (Scientific Publishers, Enfield, USA). Professor Ramawat has completed several major research projects from UGC, CSIR, ICAR, DBT, and DST, and has supervised the doctoral theses of 16 students. He has been a member of the Plant Tissue Culture Association of India since 1991.

Professor J.M. Mérillon (born in 1953) received his M.Pharma. (1979) and Ph.D. (1984) from the University of Tours, Tours, France. He joined the University of Tours as a faculty member in 1982, became associate professor in 1987, and a full Professor in 1993 at the faculty of Pharmacy, University of Bordeaux 2, Bordeaux, France. He is currently group leader of a "study group on biologically active plant substances" at the Institute of Vine and Wine Sciences, which comprises 20 scientists and research students. He has published more than 80 research papers in internationally recognized journals. He is involved in developing courses and research on phytochemistry and biological properties of compounds from vine and wine in France and has traveled widely as a senior professor. Scientists from several countries are working in his laboratory and his research is supported by funds from the Vinegrowers Association, Ministry of Higher Education and Research, and various private enterprises.

Contents

1 Drug Discovery from Plants
A.A. Salim, Y.-W. Chin and A.D. Kinghorn

1.1	The Role of Plants in Human History	1
1.2	The Role of Plant-Derived Compounds in Drug Development	4
1.2.1	Plant Secondary Metabolites as Drug Precursors	4
1.2.2	Plant Secondary Metabolites as Drug Prototypes	6
1.2.3	Plant Secondary Metabolites as Pharmacological Probes	8
1.3	Recent Developments in Drug Discovery from Plants	9
1.3.1	New Plant-Derived Drugs Launched Since 2001	9
1.3.2	Examples of Plant-Derived Compounds Currently Involved in Clinical Trials	11
1.3.3	Plant Extracts Currently Involved in Clinical Trials	15
1.4	Recent Trends and Future Directions	18

2 Grapevine Stilbenes and Their Biological Effects
P. Waffo-Teguo, S. Krisa, T. Richard and J.-M. Mérillon

2.1	Introduction	25
2.2	Epidemiology	26
2.3	Chemistry of Stilbenes	27
2.3.1	Characterisation	27
2.3.1.1	Monomers	27
2.3.1.2	Oligomers	29
2.3.2	Biosynthetic Pathway	32
2.3.3	Distribution in *Vitis vinifera*	33
2.3.4	Determination of Stilbenes in Wine	33
2.4	Biological and Pharmacological Activities	35
2.4.1	Bioavailability and Metabolism	35
2.4.2	Cardiovascular Protection	39
2.4.2.1	Antioxidant Activity	39
2.4.2.2	Antithrombotic and Vasoprotective Properties	41
2.4.2.3	Biological Activities after Ingestion of Polyphenols or Wine	42
2.4.3	Cancer Chemoprevention	43
2.4.4	Neurodegenerative Diseases	46
2.5	Conclusion	49

3	**Research into Isoflavonoid Phyto-oestrogens in Plant Cell Cultures** *M.T. Luczkiewicz*	
3.1	Introduction	56
3.2	The Influence of the Basic Experimental Media on the Biosynthesis of Isoflavones in In Vitro Cultures	57
3.3	The Influence of Physical Factors on the Biosynthesis and Accumulation of Isoflavonoids in In Vitro Cultures	59
3.4	The Effect of Technological Procedures on the Biosynthesis and Accumulation of Isoflavonoids in In Vitro Cultures	60
3.4.1	Elicitation	61
3.4.2	Supplementation with Biosynthesis Precursors	70
3.4.3	Biotransformation	71
3.4.4	Genetic Modifications	73
3.5	In Vitro Cultures of Legume Plants Oriented for Selective Production of Phyto-oestrogens	77

4	**Secondary Metabolite Production from Plant Cell Cultures: the Success Stories of Rosmarinic Acid and Taxol** *S. Kintzios*	
4.1	Introduction: Cell Factories at the Cross Point	86
4.2	Rosmarinic Acid	87
4.2.1	General Information	87
4.2.2	Historical Development of In Vitro RA Production – a Brief Overview	87
4.2.3	Stimulation of Biosynthetic Pathways Leads to Enhanced RA Accumulation In Vitro	88
4.2.4	Is RA Biosynthesis Growth Dependent?	89
4.2.5	Is RA Accumulation Related to Culture Differentiation?	90
4.2.6	Recent Attempts to Scale Up RA Production	91
4.2.7	RA Production in Immobilized Cell Cultures	92
4.3	Taxol	92
4.3.1	General Information	92
4.3.2	Historical Development of In Vitro Taxol Production – a Brief Overview	93
4.3.3	Stimulation of Biosynthetic Pathways Leads to Enhanced Taxol Accumulation In Vitro	94
4.3.4	Is Taxol Biosynthesis Growth and Differentiation Dependent?	96
4.3.5	Recent Attempts to Scale Up Taxol Production	97
4.3.6	Taxol Production in Immobilized Cell Cultures	97
4.4	Conclusions	98

Contents   XI

5   **Guggulsterone: a Potent Natural Hypolipidemic Agent from**
          ***Commiphora wightii* – Problems, Perseverance, and Prospects**
          K.G. Ramawat, M. Mathur, S. Dass and S. Suthar

5.1   Introduction 102
5.2   Distribution 102
5.3   Biology 102
5.4   Gum-Resin Production 103
5.5   Chemistry 104
5.6   Methods of Analysis 104
5.6.1   Thin-Layer Chromatography 104
5.6.2   High-Performance Liquid Chromatography 109
5.7   Traditional Therapeutic Uses 109
5.8   Pharmacology 110
5.8.1   Animal and Clinical Trials 110
5.8.2   Mechanism of Action 111
5.8.3   Other Potential Activities 112
5.8.4   Toxicity 113
5.9   Biotechnological Approaches 113
5.9.1   Micropropagation 113
5.9.2   Somatic Embryogenesis 114
5.9.3   Resin Canal Formation 115
5.9.4   Guggulsterone Production 115
5.10   Future Prospects 119

6   ***Silybum marianum* (L.) Gaertn: the Source of Silymarin**
          P. Corchete

6.1   Introduction 124
6.2   Botany 125
6.3   Chemical Composition of *S. marianum* Fruits 126
6.4   Pharmacology of Silymarin 130
6.4.1   Mechanisms of Action 131
6.4.1.1   Antioxidant Activity 131
6.4.1.2   Effects on Hepatocyte Membranes and Cellular Permeability .. 131
6.4.1.3   Effects on Receptor Binding of Toxins and Drugs 131
6.4.1.4   Stimulation of Protein Synthesis 132
6.4.1.5   Inhibition of Cell Proliferation in Hepatic Fibrosis .. 132
6.4.1.6   Anti-inflammatory Activity 132
6.4.2   Pharmacological Applications 132
6.4.2.1   Hepatoprotective Action 132
6.4.2.2   Hypocholesterolemic Action 134
6.4.2.3   Chemopreventive and Anticarcinogenic Effects 134
6.4.2.4   Anti-inflammatory Action 136
6.4.2.5   Other Actions 137

6.4.3	Bioavailability	138
6.4.4	Toxicology	139
6.4.5	Therapeutics	139
6.5	Biotechnology	140

7 The Production of Dianthrones and Phloroglucinol Derivatives in St. John's Wort
A. Kirakosyan, D.M. Gibson, and P.B. Kaufman

7.1	Introduction	149
7.2	Dianthrone and Phloroglucinol Derivatives Family of Compounds in *Hypericum perforatum*	150
7.2.1	Botany of *Hypericum*	152
7.2.2	Medicinal Uses of Hypericin and Hyperforin	152
7.3	Biotechnology for the Production of Hypericin and Hyperforin	154
7.3.1	Biosynthesis of Hypericin and Hyperforin in Mature Plants	154
7.3.2	Plant Cell Biotechnology	156
7.3.3	Influences on Hypericin and Hyperforin Productivity by Other Factors	159
7.3.4	New Directions for Hypericin and Hyperforin Production	161
7.4	Conclusions	162

8 Production of Alkaloids in Plant Cell and Tissue Cultures
D. Laurain-Mattar

8.1	Introduction	165
8.2	Correlation Between Organogenesis, Somatic Embryogenesis and Isoquinoline Alkaloid Accumulation	168
8.3	Hairy Roots and Tropane and Morphinan Alkaloid Accumulation	170
8.4	Conclusion and Perspective	171

9 *Bacopa monnieri*, a Nootropic Drug
M. Rajani

9.1	Introduction	175
9.2	Chemical Constituents	176
9.3	Analysis of Saponins of *B. monnieri*	184
9.3.1	Pharmacological Studies	184
9.3.2	Clinical Studies	188
9.3.3	Concluding Remarks	189
9.4	Biotechnology and Tissue Culture Studies on *B. monniera*	190

10	Chemical Profiling of *Nothapodytes nimmoniana* for Camptothecin, an Important Anticancer Alkaloid: Towards the Development of a Sustainable Production System *R. Uma Shaanker, B.T. Ramesha, G. Ravikanth, R.P. Gunaga, R. Vasudeva and K.N. Ganeshaiah*	
10.1	Introduction	198
10.2	*N. nimmoniana:* Ecology and Distribution	201
10.3	Basic Patterns of Accumulation of CPT in *N. nimmoniana*	203
10.4	Chemical Profiling of Populations of *N. nimmoniana* for CPT	204
10.5	Modeling Habitat Suitability for CPT Production	207
10.6	Development of a Sustainable Extraction Approach	209
10.7	Conclusions	210

11	Colchicine – an Overview for Plant Biotechnologists *S. Ghosh and S. Jha*	
11.1	Introduction	216
11.2	The Alkaloid Colchicine	218
11.3	Toxicity of Colchicine	218
11.4	Biological Effects of Colchicine	219
11.5	Colchicine as a Medicine	222
11.6	Botanical Use of Colchicine	222
11.7	Chemistry of Colchicine	223
11.8	Occurrence	225
11.9	Biotechnological Approaches for the Production of Colchicine	225
11.10	Conclusion	228

12	In Vitro Azadirachtin Production *S. Srivastava and A.K. Srivastava*	
12.1	Introduction	234
12.2	Chemistry of Azadirachtin	236
12.3	Mode of Action of Azadirachtin	237
12.4	Biosynthetic Pathway for Azadirachtin	238
12.5	Qualitative and Quantitative Analysis of Azadirachtin	238
12.6	Availability of Azadirachtin	242
12.7	Plant Cell/Tissue Culture: an Alternative for Azadirachtin Production	243
12.7.1	Azadirachtin Production from Plant Cell/Tissue Cultures of *Azadirachta indica*	243
12.7.2	Yield Improvement Strategies	244
12.7.2.1	Strain Improvement and Selection	245
12.7.2.2	Media Compositions and Culture Conditions	245

12.7.2.3	Application of Elicitors, Precursors and Permeabilising Agents	246
12.7.2.4	Genetic Engineering Approach	247
12.7.2.5	Somatic Embryogenesis and Regeneration	248
12.7.2.6	Two-Phase (Stage) Systems	248
12.7.2.7	Immobilisation	248
12.8	Stability of Azadirachtin	249
12.9	Scale-up of In Vitro Azadirachtin Production	249
12.10	Conclusion	250

13 Arabinogalactan Protein and Arabinogalactan: Biomolecules with Biotechnological and Therapeutic Potential
A. Pal

13.1	Introduction	256
13.2	Biological Activities of AGP	259
13.3	Role of AGPs in Reproductive Organ Development	261
13.4	Signaling Role of AGP	262
13.5	Abiotic Stress Tolerance Conferred by AGP	263
13.6	Probable Role in PCD	263
13.7	Commercial Uses of Gum Arabic	264
13.8	AG as Dietary Fiber and Prebiotics	264
13.9	AG as Immunomodulators and Immunity Enhancers	265
13.10	Echinacea-AG as a Nutraceutical	266
13.11	Other Uses of AG	266
13.12	Scope of Exploiting the Potentials of AGP and AG in Plant Biotechnology and Therapeutics	267
13.13	Concluding Remarks	267

14 Hairy Roots: a Powerful Tool for Plant Biotechnological Advances
S. Guillon, J. Trémouillaux-Guiller, P.K. Pati, and P. Gantet

14.1	Introduction	272
14.2	Hairy Roots Are on the Way to towards an Experimental Model	272
14.3	Improvement in the Productivity of Hairy Roots: Biotic and Abiotic Treatments or Metabolite Trapping	275
14.4	Potential Discovery of Metabolic Genes from Transcriptome Analysis of T-DNA Activation Tagging or Elicited Hairy Roots	277
14.5	RNA Silencing via Hairy Root: a Powerful Tool for Loss-of-Function Analyses of Genes	277
14.6	Metabolic Engineering of the Hairy Root System	278
14.7	Hairy Roots: A Novel System for Molecular Farming	279
14.8	Phytoremediation Process for Cleaning up the Environment and More Knowledge on Root Adsorption	279
14.9	Scale up and Technological Integration into Industry	280

| 14.10 | Perspectives | 281 |

15 **Hairy Roots of *Catharanthus roseus*: Efficient Routes to Monomeric Indole Alkaloid Production**
S. Guillon, P. Gantet, M. Thiersault, M. Rideau and J. Trémouillaux-Guiller

15.1	Introduction	286
15.2	Materials and Methods	286
15.2.1	Bacterial Strain	286
15.2.2	Plant Material	287
15.2.3	Hairy Root Induction	287
15.2.4	Liquid Hairy Root Culture	287
15.2.5	Methyl Jasmonate Treatment	287
15.2.6	Alkaloid Identification by Ceric Ammonium Sulphate Reagent	288
15.2.7	Serpentine and Ajmalicine Content Determination by Spectrofluorometry	288
15.2.8	Catharanthine Content Determination by High-Performance Liquid Chromatography Analysis	288
15.2.9	Statistical Analysis	289
15.3	Results and Discussion	289
15.3.1	Genetic Transformation of *Catharanthus* Leaves	289
15.3.2	Alkaloid Profiles	291
15.3.3	Alkaloid Contents	292
15.4	Conclusion	294

16 **Roseroot (*Rhodiola rosea* L.): Effect of Internal and External Factors on Accumulation of Biologically Active Compounds**
Z. Węglarz, J.L. Przybył and A. Geszprych

16.1	Introduction	298
16.2	Plant Characteristics	298
16.3	Intraspecific Variability	300
16.3.1	Distribution of Phenolic Compounds in Rhizomes and Roots	300
16.3.2	Quality of Raw Material of Different Origin	303
16.3.3	Individual Variation	304
16.4	Accumulation of Biomass and Biologically Active Compounds in the Underground Organs of Roseroot During Plant Development	305
16.5	Effect of Ecological Factors on the Accumulation of Biomass and Biologically Active Compounds in the Underground Organs of Roseroot	308
16.6	Effect of Post-harvest Treatment on the Quality of Raw Material and Extracts	310

17 Apoptosis and Plant-Derived Pharmaceuticals
L.F. Brisson

17.1	Introduction	317
17.1.1	Molecular Regulation of Apoptosis	317
17.2	Plant Antitumoral Substances	318
17.2.1	Plant Substances	318
17.2.2	Chemotherapeutic Drugs	319
17.2.2.1	Taxanes	319
17.2.2.2	Vinca Alkaloids	320
17.2.3	Chemopreventive Agents: Catechin and its Derivatives	321
17.3	Conclusion	322

18 The Indian Herbal Drugs Scenario in Global Perspectives
K.G. Ramawat and S. Goyal

18.1	Introduction	325
18.2	Indian System of Medicine	328
18.3	World-Wide Use of Medicinal Aromatic Plants	336
18.4	Supply and Demand of Medicinal Plants	337
18.5	Medicinal Plant Biodiversity	338
18.6	Traditional Medicine in Healthcare	340
18.7	Indian Pharmaceutical Industries	341
18.8	Quality of Herbal Drugs	342
18.9	Concluding Remarks	343

19 Phytochemical Standardization of Herbal Drugs and Polyherbal Formulations
M. Rajani and N.S. Kanaki

19.1	Introduction	349
19.1.1	Herbal drugs	349
19.1.2	Trade Scenario	350
19.1.3	Bottlenecks and Steps to be Taken	350
19.1.4	Strategy	351
19.1.5	Status of Herbal Drugs in Pharmacopoeias	351
19.1.5.1	Official	352
19.1.5.2	Others	352
19.2	Phytochemical Standardization	352
19.2.1	Sample Preparation	353
19.2.2	Preliminary Screening for Chemical Groups and Quantification of Chemical Groups	354
19.2.3	Phytochemical Profiles – Fingerprinting	354
19.2.3.1	Multiple Marker-Based Fingerprinting	355

19.2.4	Marker Compound Analysis	357
19.2.4.1	Marker Compounds	357
19.2.4.2	Quantification of Marker Compounds	357
19.2.5	Multiple Marker-Based Evaluation	358
19.2.6	Polyherbal Formulations	359
19.2.7	Hyphenated Techniques	361
19.2.8	Reference Compounds	364
19.3	Some Examples	364
19.3.1	Raw Material	364
19.3.1.1	Ammoniacum Gum	364
19.3.1.2	*Cinchona officinalis* Stem Bark	365
19.3.2	Formulation	366
19.3.2.1	Chandraprabhavati	366
19.4	Conclusion	366

Subject Index 371

List of Contributors

Editors

K.G. Ramawat
Laboratory of Biomolecular
Technology, Department of Botany
M.L. Sukhadia University
Udaipur 313001
India
kg_ramawat@yahoo.com

Jean-Michel Mérillon
Groupe d'Etude des Substances
Végétales à Activité Biologique
Laboratoire de Sciences Végétales
Mycologie et Biotechnologie
EA 3675 – UFR Pharmacie
Universite Victor Segalen
Bordeaux 2
146 Rue Leo-Saignat
F-33076 Bordeaux cedex
France
jean-michel.merillon@u-bordeaux2.fr

Authors

Louise F. Brisson
Department of Biochemistry
and Microbiology, Research in Heath
and Life Science Building
Laval University Quebec
Quebec
Canada
G1K 7P4
louise.brisson@bcm.ulaval.ca

S. Dass
Laboratory of Bio-Molecular
Technology, Department of Botany
M.L. Sukhadia University
Udaipur-313001
India

A. Douglas Kinghorn
College of Pharmacy
The Ohio State University
500 West 12th Avenue
Columbus
OH 43210-1291
USA
kinghorn.4@osu.edu

Young-Won Chin
College of Pharmacy
The Ohio State University
500 West 12th Avenue
Columbus
OH 43210-1291
USA

Purificación Corchete
Department of Plant Physiology
Faculty of Pharmacy
University of Salamanca
37007-Salamanca
Spain
corchpu@usal.es

K.N. Ganeshaiah
Department of Genetics and Plant
Breeding
University of Agricultural Sciences
GKVK Campus
Bangalore 560065
India

Pascal Gantet
Université Montpellier 2
UMR PIA 1096
Laboratoire de Biochimie
et de Physiologie Végétales
Place Eugène Bataillon
Bat 15 CC 002
34095 Montpellier cedex 5
France

A. Geszprych
Department of Vegetable
and Medicinal Plants
Warsaw Agricultural University
Nowoursynowska 159
02-776 Warsaw
Poland

Seemanti Ghosh
Centre of Advanced Study in
Cell and Chromosome Research
Department of Botany
University of Calcutta
35 Ballygunge Circular Road
Calcutta 700019
India

Donna M. Gibson
USDA Agricultural Research Service
Plant Protection Research Unit
U.S. Plant, Soil, and Nutrition
Laboratory
Tower Road
Ithaca
NY 14583
USA

Shaily Goyal
Laboratory of Biomolecular
Technology, Department of Botany
M.L. Sukhadia University
Udaipur 313001
India

Stéphanie Guillon
UPRES EA 2106 "Biomolécules
et Biotechnologies Végétales"
Université François Rabelais
UFR des Sciences Pharmaceutiques
Parc de Grandmont
37200 Tours
France

R. Gunaga
Department of Forest Biology
College of Forestry
Sirsi 581401
India

Sumita Jha
Centre of Advanced Study in
Cell and Chromosome Research
Department of Botany
University of Calcutta
35 Ballygunge Circular Road
Calcutta 700019
India
sjbot@caluniv.ac.in

Peter B. Kaufman
Department of Cardiac Surgery
University of Michigan
1150 West
Medical Center Dr, Ann Arbor
Michigan 48109-0686
USA

Niranjan S. Kanaki
B.V. Patel Pharmaceutical Education
and Research Development (PERD)
Centre
Thaltej
Ahmedabad – 380 054
Gujarat
India

List of Contributors

Spiridon Kintzios
Laboratory of Plant Physiology
Faculty of Agricultural
Biotechnology
Agricultural University of Athens
Iera Odos 75
11855 Athens
Greece
spiroskintzios@usa.net

Ara Kirakosyan
Department of Cardiac Surgery
University of Michigan
1150 West
Medical Center Dr, Ann Arbor
Michigan 48109-0686
USA
akirakos@umich.edu

Pratap Kumar Pati
Department of Botanical
and Environmental Sciences
Guru Nanak Dev University
Amritsar-143 005
India

Stéphanie Krisa
Groupe d'Etude des Substances
Végétales à Activité Biologique
Laboratoire de Sciences Végétales
Mycologie et Biotechnologie
EA 3675 – UFR Pharmacie
Universite Victor Segalen
Bordeaux 2
146 Rue Leo-Saignat
F-33076 Bordeaux cedex
France

Dominique Laurain-Mattar
Groupe S.U.C.R.E.S.
U.M.R. 7565 C.N.R.S.
Nancy-Université
BP 239
54506 Nancy-Vandoeuvre
France
dominique.laurain-mattar@pharma.uhp-nancy.fr

M. Mathur
Laboratory of Biomolecular
Technology, Department of Botany
M.L. Sukhadia University
Udaipur-313001
India

Professor Jean-Michel Mérillon
Groupe d'Etude des Substances
Végétales à Activité Biologique
Laboratoire de Sciences Végétales
Mycologie et Biotechnologie
EA 3675 – UFR Pharmacie
Universite Victor Segalen
Bordeaux 2
146 Rue Leo-Saignat
F-33076 Bordeaux cedex
France
jean-michel.merillon@u-bordeaux2.fr

Amita Pal
Plant Molecular and Cellular
Genetics
P 1/12 CIT Scheme VIIM
Kolkata – 700054
India
amita_pal@yahoo.com

J.L. Przybył
Department of Vegetable
and Medicinal Plants
Warsaw Agricultural University
Nowoursynowska 159
02-776 Warsaw
Poland

M. Rajani
B.V. Patel Pharmaceutical Education
and Research Development (PERD)
Centre
Thaltej
Ahmedabad – 380 054
Gujarat
India
rajanivenkat@hotmail.com

Professor K.G. Ramawat
Laboratory of Biomolecular
Technology, Department of Botany
M.L. Sukhadia University
Udaipur 313001
India
kg_ramawat@yahoo.com

B.T. Ramesha
Department of Crop Physiology
University of Agricultural Sciences
GKVK Campus
Bangalore 560065
India

G. Ravikanth
School of Ecology and Conservation
University of Agricultural Sciences
GKVK Campus
Bangalore 560065
India

Tristan Richard
Groupe d'Etude des Substances
Végétales à Activité Biologique
Laboratoire de Biophysique
EA 3675 – UFR Pharmacie
Universite Victor Segalen
Bordeaux 2
146 Rue Leo-Saignat
F-33076 Bordeaux cedex
France

Marc Rideau
UPRES EA 2106 "Biomolécules
et Biotechnologies Végétales"
Université François Rabelais
UFR des Sciences Pharmaceutiques
Parc de Grandmont
37200 Tours
France

Angela A. Salim
College of Pharmacy
The Ohio State University
500 West 12th Avenue
Columbus
OH 43210-1291
USA

S. Suthar
Laboratory of Biomolecular
Technology, Department of Botany
M.L. Sukhadia University
Udaipur-313001
India

R. Uma Shaanker
Department of Crop Physiology
University of Agricultural Sciences
GKVK Campus
Bangalore 560065
India
rus@vsnl.com

Martine Thiersault
UPRES EA 2106 "Biomolécules
et Biotechnologies Végétales"
Université François Rabelais
UFR des Sciences Pharmaceutiques
Parc de Grandmont
37200 Tours
France

Jocelyne Trémouillaux-Guiller
UPRES EA 2106 " Biomolécules
et Biotechnologies Végétales"
Université François Rabelais
UFR des Sciences Pharmaceutiques
Parc de Grandmont
37200 Tours
France
j.tremouillaux@telez.fr

List of Contributors

R. Vasudeva
Ashoka Trust for Research in
Ecology and the Environment #659
5th A Main
Hebbal
Bangalore 560024
India

Pierre Waffo-Teguo
Groupe d'Etude des Substances
Végétales à Activité Biologique
Laboratoire de Sciences Végétales
Mycologie et Biotechnologie
EA 3675 – UFR Pharmacie
Universite Victor Segalen
Bordeaux 2
146 Rue Leo-Saignat
F-33076 Bordeaux cedex
France

Z. Węglarz
Department of Vegetable
and Medicinal Plants
Warsaw Agricultural University
Nowoursynowska 159
02-776 Warsaw
Poland
weglarz@alpha.sggw.waw.pl

Chapter 1
Drug Discovery from Plants

A.A. Salim, Y.-W. Chin and A.D. Kinghorn (✉)

Division of Medicinal Chemistry and Pharmacognosy, College of Pharmacy,
The Ohio State University, Columbus, OH 43210, USA, e-mail: kinghorn.4@osu.edu

Abstract Many plant-derived compounds have been used as drugs, either in their original or semi-synthetic form. Plant secondary metabolites can also serve as drug precursors, drug prototypes, and pharmacological probes. Recent developments in drug discovery from plants, including information on approved drugs and compounds now in clinical trials, are presented. There are also several plant extracts or "phytomedicines" in clinical trials for the treatment of various diseases. In the future, plant-derived compounds will still be an essential aspect of the therapeutic array of medicines available to the physician, particularly with the availability of new hyphenated analytical methods such as LC-NMR-MS and LC-SPE-NMR to accelerate their future discovery.

Keywords Natural products, Plant-derived drugs, Drug discovery, Drug development, Drug precursors, Drug prototypes, Pharmacological probes, New therapeutic agents, Clinical trials, Accelerated discovery techniques

1.1 The Role of Plants in Human History

Over the centuries humans have relied on plants for basic needs such as food, clothing, and shelter, all produced or manufactured from plant matrices (leaves, woods, fibers) and storage parts (fruits, tubers). Plants have also been utilized for additional purposes, namely as arrow and dart poisons for hunting, poisons for murder, hallucinogens used for ritualistic purposes, stimulants for endurance, and hunger suppression, as well as inebriants and medicines. The plant chemicals used for these latter purposes are largely the secondary metabolites, which are derived biosynthetically from plant primary metabolites (e.g., carbohydrates, amino acids, and lipids) and are not directly involved in the growth, development, or reproduction of plants. These secondary metabolites can be classified into several groups according to their chemical classes, such alkaloids, terpenoids, and phenolics [1].

Ramawat KG, Mérillon JM (eds.), In: *Bioactive Molecules and Medicinal Plants*
Chapter DOI: 10.1007/978-3-540-74603-4_1, © Springer 2008

Arrow and dart poisons have been used by indigenous people in certain parts of the world with the principal ingredients derived from the genera *Aconitum* (Ranunculaceae), *Akocanthera* (Apocynaceae), *Antiaris* (Moraceae), *Chondrodendron* (Menispermaceae), *Strophanthus* (Apocynaceae), and *Strychnos* (Loganiaceae) [2]. Most compounds responsible for the potency of arrow and dart poisons belong to three plant chemical groups, namely the alkaloids (e.g., strychnine from *Strychnos* species), cardiac glycosides (e.g., ouabain from *Strophanthus* species), and saponins (e.g., a monodesmoside glucoside from *Clematis* species) [2].

In some cultures, toxic plant extracts were also used for murder and "trials by ordeal," where a person accused of a crime was given a noxious brew, and it was believed that if innocent, this suspect would survive this ordeal. Some of the plants well-documented for murder are henbane (*Hyoscyamus niger* L.), mandrake (*Mandragora officinarum* L.), deadly nightshade (*Atropa belladonna* L.), and some *Datura* species, all of which belong to the family Solanaceae [3]. Calabar bean (*Physostigma venenosum* Balf.) was famous for its use in trials by ordeal by people who lived on the Calabar Coast, West Africa [3]. Certain plants formerly used for arrow poisons, such as several *Aconitum* species, have also been used as medicines at lower dosages, for their analgesic and anti-inflammatory properties [4]. In fact, many compounds isolated from poisonous plants were later developed as therapeutic drugs, due to their desirable pharmacological actions [5, 6].

The use of hallucinogens in the past was usually associated with magic and ritual. However, these hallucinogens have been exploited as recreational drugs and accordingly may lead to habituation problems. Several well-recognized plants that contain hallucinogenic or psychoactive substances (the compound names are given in parentheses) include *Banisteriopsis caapi* (Spruce ex Griseb.) Morton (N,N-dimethyltryptamine), *Cannabis sativa* L. (Δ^9-*trans*-tetrahydrocannabinol), *Datura* species (scopolamine), *Erythroxylum coca* Lam. (cocaine), *Lophophora williamsii* (Salm-Dyck) J.M. Coult. (mescaline), *Papaver somniferum* L. (morphine), and *Salvia divinorum* Epling & Játiva (salvinorin A) [7, 8]. Several of these plants are also used as drugs due to their desired pharmacological activities, and some of the constituents of these plants have been developed into modern medicines, either in the natural form or as lead compounds subjected to optimization by synthetic organic chemistry [5, 6].

Plants have also been used in the production of stimulant beverages (e.g., tea, coffee, cocoa, and cola) and inebriants or intoxicants (e.g., wine, beer, kava) in many cultures since ancient times, and this trend continues till today. Tea (*Camellia sinensis* Kuntze) was first consumed in ancient China (the earliest reference is around CE 350), while coffee (*Coffea arabica* L.) was initially cultivated in Yemen for commercial purposes in the 9th century [3]. The Aztec nobility used to consume bitter beverages containing raw cocoa beans (*Theobroma cacao* L.), red peppers, and various herbs [3]. Nowadays, tea, coffee, and cocoa are important commodities and their consumption has spread worldwide. The active components of these stimulants are methylated xanthine derivatives, namely caffeine, theophylline, and theobromine, which are the main constituents of coffee, tea, and cocoa, respectively [9].

The most popular inebriants in society today are wine, beer, and liquor made from the fermentation of fruits and cereals. Wine was first fermented about 6000–8000 years ago in the Middle East, while the first beer was brewed around 5000–6000 BCE by the Babylonians [3]. The intoxicating ingredient of these drinks is ethanol, a by-product of bacterial fermentation, rather than secondary plant metabolites. Recent studies have shown that a low to moderate consumption of red wine is associated with reduction of mortality due to cardiovascular disease and cancer [10]. This health benefit has been suggested to be due to the presence of resveratrol, a hydroxylated stilbenoid found in the skin of grapes [11]. Kava, a beverage made from the root of *Piper methysticum* Roxb., has been a popular intoxicating beverage in Polynesia for centuries [3]. Kava is not normally consumed in this manner in the Western world, but has gained popularity as a botanical dietary supplement to ease the symptoms of stress, anxiety, and depression [12]. A study has shown that the anxiolytic activity of kava extract may be mediated in part by the kavalactone, dihydrokavain [13]. The consumption of kava has been associated with liver toxicity, although this is somewhat controversial. Recently, a study has shown that the alkaloid pipermethystine, found mostly in the leaves and stems of *Piper methysticum*, may be responsible for this toxicity [14].

Plants have formed the basis of sophisticated traditional medicine (TM) practices that have been used for thousands of years by people in China, India, and many other countries [9]. Some of the earliest records of the usage of plants as drugs are found in the Artharvaveda, which is the basis for Ayurvedic medicine in India (dating back to 2000 BCE), the clay tablets in Mesopotamia (1700 BCE), and the Eber Papyrus in Egypt (1550 BCE) [9]. Other famous literature sources on medicinal plant include "De Materia Medica," written by Dioscorides between CE 60 and 78, and "Pen Ts'ao Ching Classic of Materia Medica" (written around 200 CE) [9].

Before the realization that pharmacologically active compounds present in medicinal plants are responsible for their efficacy, the "doctrine of signatures" was often used to identify plants for treating diseases. For example, goldenrod with a yellow hue was used to cure jaundice, red-colored herbs were used to treat blood diseases, liverworts were used for liver diseases, pileworts for hemorrhoids, and toothworts for toothache [9]. In 1805, morphine became the first pharmacologically active compound to be isolated in pure form from a plant, although its structure was not elucidated until 1923 [9]. The 19th century marked the isolation of numerous alkaloids from plants (species in parentheses) used as drugs, namely, atropine (*Atropa belladonna*), caffeine (*Coffea arabica*), cocaine (*Erythroxylum coca*), ephedrine (*Ephedra* species), morphine and codeine (*Papaver somniferum*), pilocarpine (*Pilocarpus jaborandi* Holmes), physostigmine (*Physostigma venenosum*), quinine (*Cinchona cordifolia* Mutis ex Humb.), salicin (*Salix* species), theobromine (*Theobroma cacao*), theophylline (*Camellia sinensis*), and (+)-tubocurarine (*Chondodendron tomentosum* Ruiz & Pav.) [9]. Following these discoveries, bioactive secondary metabolites from plants were later utilized more widely as medicines, both in their original and modified forms [5, 6].

The correlation between the ethnomedical usage of medicinal plants and modern medicines discovered from those plants has been studied by Fabricant and Farnsworth [15]. Based on their analysis, 88 single chemical entities isolated from 72 medicinal plants have been introduced into modern therapy, many of which have the same or a similar therapeutic purpose as their original ethnomedical use [15]. Some of these plant-derived compounds, such as atropine (anticholinergic), codeine (cough suppressant), colchicine (antigout), ephedrine (bronchodilator), morphine (analgesic), pilocarpine (parasympathomimetic), and physostigmine (cholinesterase inhibitor) are still being used widely as single-agent or combination formulations in prescription drugs [5].

Nowadays, plants are still important sources of medicines, especially in developing countries that still use plant-based TM for their healthcare. In 1985, it was estimated in the Bulletin of the World Health Organization (WHO) that around 80% of the world's population relied on medicinal plants as their primary healthcare source [16]. Even though a more recent figure is not available, the WHO has estimated that up to 80% of the population in Africa and the majority of the populations in Asia and Latin America still use TM for their primary healthcare needs [17]. In industrialized countries, plant-based traditional medicines or phytotherapeuticals are often termed complementary or alternative medicine (CAM), and their use has increased steadily over the last 10 years. In the USA alone, the total estimated "herbal" sales for 2005 was $4.4 billion, a significant increase from $2.5 billion in 1995 [18]. However, such "botanical dietary supplements" are regulated as foods rather than drugs by the United States Food and Drug Administration (US FDA).

1.2 The Role of Plant-Derived Compounds in Drug Development

Despite the recent interest in drug discovery by molecular modeling, combinatorial chemistry, and other synthetic chemistry methods, natural-product-derived compounds are still proving to be an invaluable source of medicines for humans. The importance of plants in modern medicine has been discussed in recent reviews and reports [19–22]. Other than the direct usage of plant secondary metabolites in their original forms as drugs, these compounds can also be used as drug precursors, templates for synthetic modification, and pharmacological probes, all of which will be discussed briefly in turn in this section.

1.2.1 Plant Secondary Metabolites as Drug Precursors

Some natural products obtained from plants can be used as small-molecule drug precursors, which can be converted into the compound of interest by chemical

Chapter 1 Drug Discovery from Plants

modification or fermentation methods. The semisynthetic approach is usually used to resolve the shortage of supply due to the low yield of compounds from plants and/or the high cost of total synthesis. For compounds with complex structures and many chiral centers, protracted methods may be needed for their synthesis, and thus these methods would not be feasible economically. The following examples indicate that some secondary metabolites from plants are useful drug precursors, although they are not necessarily pharmacologically active in their original naturally occurring forms.

Cropping of the bark of the slow-growing Pacific yew tree, *Taxus brevifolia* Nutt., is not a feasible method to provide sufficient amounts of the antitumor drug paclitaxel (1, Taxol) to meet the market demand (paclitaxel was originally isolated in only 0.014% w/w yield from the bark of *Taxus brevifolia*) [23]. Even though this compound can be produced by total synthesis, this has proven to be inefficient in affording large quantities of paclitaxel [24, 25]. Fortunately, 10-deacetylbaccatin III (2) can be isolated in relatively large amounts from the needles of other related yew species, such as *Taxus baccata* L. (a renewable resource), and can be converted chemically in several steps into paclitaxel [26, 27]. During the period 1993–2002, the main pharmaceutical manufacturer, Bristol-Myers Squibb, adopted the semisynthetic method developed by the Holton research group to produce paclitaxel from 10-deacetylbaccatin III [27, 28]. Since 2002, Bristol-Myers Squibb has produced paclitaxel using a plant cell culture method, which will be mentioned in section 1.4 of this chapter [29].

Diosgenin (3), a steroidal sapogenin obtained from the tubers of various *Dioscorea* species that grow in Mexico and Central America, can be converted chemically in several steps into progesterone (4), a hormone that can be used as a female oral contraceptive [30]. Originally, progesterone was isolated from sow ovaries with a very low yield (20 mg from 625 kg of ovaries), and later was synthesized from cholesterol with very low efficiency [31]. Progesterone is also a key intermediate for the production of cortisone (5), an important anti-inflammatory drug. Progesterone can be converted into 11α-hydroxyprogesterone (6) by microbial hydroxylation at C-11, followed by chemical reactions, to produce cortisone (**5**) [32, 33]. Until now, diosgenin (**3**) is still an important starting material for the production of various steroid hormones.

Oseltamivir phosphate (**7**, Tamiflu) is an orally active neuraminidase inhibitor developed for the treatment and prophylaxis of influenza viruses A and B [34, 35]. The starting material for the oseltamivir synthesis is (−)-shikimic acid (**8**), an important biochemical intermediate in plants and microorganisms [36]. Previously, shikimic acid was extracted solely from the fruits of the shikimi tree (*Illicium verum* Hook.f.), also known as star anise, which contains a large amount of this compound [37]. Later on, shikimic acid was obtained from the fermentation of genetically engineered *Escherichia coli* strains, which are deficient in the shikimate kinase gene [38]. Currently, Roche, the drug manufacturer, still relies on both extraction and fermentation methods to obtain ton quantities of shikimic acid [37]. Several routes for the production of oseltamivir independent of shikimic acid have been developed [36, 39], but these alternatives are still not cost efficient [37].

1.2.2 Plant Secondary Metabolites as Drug Prototypes

Sneader has defined a drug prototype as "the first compound discovered in a series of chemically related therapeutic agents" [5]. As of 1996, from a total of 244 drug prototypes identified in one analysis from minerals, plants, animals, microbes, and chemical sources, plant secondary metabolites contributed 56 of these (23%) [5]. With advances in organic chemistry, medicinal chemists started preparing analogs from these drug prototypes to provide safer and more potent drugs. Sometimes, new compounds with novel pharmacological properties have

Chapter 1 Drug Discovery from Plants

been developed in the process of developing such derivatives. In the following examples, podophyllotoxin, camptothecin, and guanidine have been selected as drug prototypes with analogs having the same pharmacological action as the parent compound, while atropine is a drug prototype that has furnished many analogs that have additional pharmacological properties.

Several antineoplastic compounds isolated from plants, such as podophyllotoxin (**9**) and camptothecin (**10**), are too toxic or not water soluble enough for clinical application, and analogs with higher therapeutic indices such as etoposide (**11**, Vepesid) and topotecan (**12**, Hycamtin) have been developed in consequence [40, 41]. Due to their unique modes of anticancer activities, there is much interest in the clinical development of further derivatives of paclitaxel (**1**) and camptothecin (**10**) as anticancer therapeutic drugs [28, 41–43]. According to a recent review, of the 2255 cancer clinical trials recorded as of August 2003, 310 (or 13.7%) and 120 (or 5.4%) of the trials involved taxane- and camptothecin-derived drugs, respectively [43]. In 2002, it was estimated that the combined sales of paclitaxel and docetaxel (both taxanes), and topotecan and irinotecan (both based on the parent molecule camptothecin) constituted over 30% of the total global sales of cytotoxic drugs [44].

Guanidine (**13**) is a natural product with good hypoglycemic activity isolated from *Galega officinalis* L., but is too toxic for clinical use [45]. Many derivatives of guanidine have been synthesized, and metformin (dimethylbiguanide) (**14**) was later found to be clinically suitable for treatment of type II diabetes [46].

Atropine (**15**) is an artifact of the tropane alkaloid (−)-hyoscyamine, which racemizes during the extraction process from its plant of origin (*Atropa belladonna*). Atropine is a competitive antagonist of muscarinic acetylcholine receptors (antimuscarinic agent). Atropine is sometimes used in the ophthalmology area as a mydriatic agent, and has additional therapeutic uses as an antispasmodic. It is also used as a premedication for anesthesia, to decrease bronchial and salivary secretions, and to block the bradycardia (low heart rate) associated with the administration of anesthetic drugs [5]. Biological and physiological studies of a large number of synthetic atropine analogs have led to the introduction of new drugs with different therapeutic applications than the parent compound. Examples of drugs derived from the basic atropine skeleton include droperidol (**16**, antipsychotic), ipratropium bromide (**17**, bronchodilator for the treatment of asthma), loperamide (**18**, antidiarrheal), methadone (**19**, a morphine substitute for addicts), and pethidine (**20**, analgesic) [5].

1.2.3 Plant Secondary Metabolites as Pharmacological Probes

In addition to their direct contribution as drugs or drug protopyes to cure human disease, secondary metabolites of plant origin, such as phorbol esters and genistein, can be used as "pharmacological probes." Pharmacological probes help researchers to understand the mechanism of action of intracellular signal transductions and biological mechanisms related to human disease, which can aid the design of better drugs.

Genistein (**21**), an isoflavone found naturally in soybean (*Glycine max* Merr.), is an inhibitor of various protein tyrosine kinases (PTK), which are essential enzymes involved in intracellular signal transduction [47]. Genistein has been used to probe the interaction between PTK and cyclic nucleotide-gated (CNG) channels, which are important in mammalian olfactory and visual systems [48, 49]. By observing the effect of genistein on the CNG channels containing either homomeric or heteromeric subunits, specific subunits containing binding sites for PTKs can be identified [48]. Furthermore, the mechanism of inhibition of the CNG channels by PTKs has been studied with the aid of genistein as a probe [49].

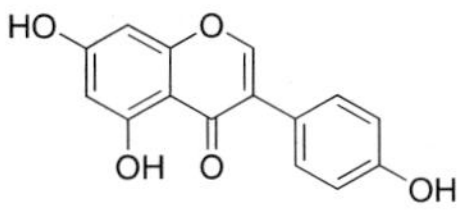

21

Phorbol is a tetracyclic diterpenoid plant secondary metabolite isolated as a hydrolysis product of croton oil from the seeds of *Croton tiglium* L. [50]. Various 12,13-diesters of phorbol have the capacity to act as tumor promoters, due in part to their role as protein kinase C (PKC) activators [51–53]. The most abundant phorbol ester derivative of croton oil, 12-*O*-tetradecanoylphorbol-13-acetate (TPA) (**22**), has been used in biomedical research in standard laboratory models of carcinogenesis promotion [54–56].

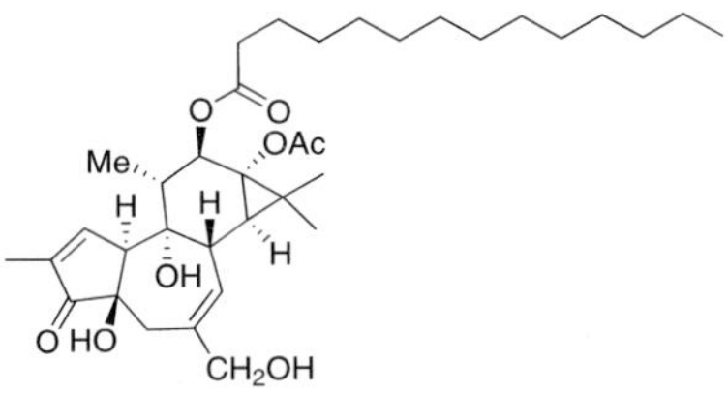

22

1.3 Recent Developments in Drug Discovery from Plants

Despite the large number of drugs derived from total synthesis, plant-derived natural products still contribute to the overall total number of new chemical entities (NCE) that continue to be launched to the market. Several reviews on drug discovery and development from natural sources (plants, marine fauna, microbes) have been published recently [42, 57–59]. The following sections will cover specifically the plant-derived drugs newly launched since 2001 and examples of some plant-derived compounds currently in clinical trials.

1.3.1 *New Plant-Derived Drugs Launched Since 2001*

In the past 6 years, five new drugs derived from natural products, namely, apomorphine hydrochloride, galanthamine hydrobromide, nitisinone, tiotropium bromide, and varenicline, have been approved by the US FDA. The following is a brief description of each drug and their therapeutic use.

Galantamine (**23**, Razadyne, Reminyl, Nivalin) was first marketed in 2001 in the USA for the symptomatic treatment of patients with early-onset Alzheimer's disease [58]. Galantamine (also known as galanthamine) is an alkaloid that was initially isolated from the snowdrop (*Galanthus woronowii* Losinsk.) in the early 1950s, and has since been found in other plants in the family Amaryllidaceae [60]. Galantamine slows the process of neurological degeneration by inhibiting acetylcholinesterase as well as binding to and modulating the nicotinic acetylcholine receptor [60, 61]. Due to the limited availability of the plants of origin of this compound, galantamine is now produced by total synthesis.

Nitisinone (**24**, Orfadin) was approved by the FDA in 2002 for the treatment of hereditary tyrosinemia type 1 (HT-1) [58]. HT-1 is a rare pediatric disease caused by a deficiency of fumaryl acetoacetate hydrolase (FAH), an enzyme essential in the tyrosine catabolism pathway. FAH deficiency leads to the accumulation of toxic substances in the body, resulting in liver and kidney damage [62]. Nitisinone is a derivative of leptospermone (**25**), a new class of herbicide from the bottlebrush plant [*Callistemon citrinus* (Curtis) Skeels]. Both nitisinone and leptospermone inhibit 4-hydroxyphenyl pyruvate dioxygenase (HPPD), the enzyme involved in plastoquinone and tocopherol biosynthesis in plants [63]. In humans, inhibition of HPPD prevents tyrosine catabolism, leading to the accumulation of tyrosine metabolites, 4-hydroxyphenyl pyruvic acid and 4-hydroxyphenyl lactic acid, which can be excreted through the urine [64].

Apomorphine (**26**, Apokyn) was approved by the FDA in 2004 as an injectable drug for the symptomatic treatment for Parkinson's disease patients during episodes of "hypomobility" (e.g., persons unable to move or to perform daily activities) [65]. Apomorphine is a synthetic derivative of morphine (**27**), but unlike morphine, apomorphine does not have opioid analgesic properties, and instead is a short-acting dopamine D_1 and D_2 receptor agonist [66].

Tiotropium bromide (**28**, Spiriva), an atropine analog, was approved by the FDA in 2005 for the treatment of bronchospasm associated with chronic obstructive pulmonary disease (COPD), including chronic bronchitis and emphysema [67].

Varenicline (**29**, Chantix), based on the plant quinolizidine alkaloid, cytisine (**30**), has been approved by the FDA since 2006 as an aid to smoking cessation [68–70]. Cytisine (**30**), an alkaloid isolated from *Cytisus laburnum* L., has been used to treat tobacco dependence in Eastern Europe (Bulgaria, Germany, Poland, and Russia) for the last 40 years [71]. Cigarette smoking has been linked to several diseases including cardiovascular disease, COPD, many cancers (particularly lung, mouth, and esophageal), and pregnancy-related complications. Varenicline (**29**) is a partial agonist with a high affinity for the $\alpha_4\beta_2$ nicotinic acetylcholine receptor, and is a full agonist at α_7 neuronal nicotinic receptors [70].

1.3.2 Examples of Plant-Derived Compounds Currently Involved in Clinical Trials

Although relatively few plant-derived drugs have been launched onto the market the last 6 years, many plant-derived compounds are currently undergoing clinical trial for the potential treatment of various diseases. The majority of such drugs under clinical development are in the oncological area, including new analogs of known anticancer drugs based on the camptothecin-, taxane-, podophyllotoxin-, or vinblastine-type skeletons [42]. Examples of compounds with carbon skeletons different from the existing plant-derived drugs used in cancer chemotherapy will be discussed below, namely, betulinic acid, ceflatonine (homoharringtonine), combretastatin A4 phosphate, ingenol-3-angelate, phenoxodiol, and protopanaxadiol. In the antiviral area, bevirimat and celgosivir are currently undergoing clinical trials for the treatment of human immunodeficiency viral (HIV) and hepatitis C viral (HCV) infections, respectively. Capsaicin is in clinical trial for the treatment of severe postoperative pain, while huperzine is being developed for the treatment of Alzheimer's disease.

Betulinic acid (**31**) is a lupane-type triterpene that is widely distributed in the plant kingdom, and this compound, along with various derivatives, has been shown to have anticancer, antibacterial, antimalarial, anti-HIV, anthelminthic, anti-inflammatory, and antioxidant properties [72, 73]. In 1995, a research group from the University of Illinois at Chicago reported that betulinic acid selectively inhibited human melanoma in both *in vitro* and *in vivo* model systems, and induced apoptosis in Mel-2 human melanoma cells [74]. This compound was further developed under the Rapid Access to Intervention Development program of the United States National Cancer Institute [75], and is currently undergoing phase I/II clinical trials for treatment of dysplastic melanocytic nevi, a preliminary symptom that may lead to melanomas of the skin [76].

Bevirimat (**32**, PA-457), a semisynthetic compound derived from betulinic acid, is being developed by Panacos Pharmaceuticals (Watertown, MA, USA) as a new class of antiretroviral drug [77]. Bevirimat blocks HIV-1 maturation by disrupting a late step in the *Gag* processing pathway, causing the virions released to be noninfectious, thus terminating the viral replication [78]. This compound is currently undergoing phase II clinical trials, and phase III trials are expected to commence in 2007 [77, 79].

Capsaicin (**33**) is a capsaicinoid-type amide that causes the burning sensation in the mouth associated with eating chilli peppers [80]. Upon topical application, capsaicin desensitizes the neurons and lowers the threshold for thermal, chemical, and mechanical nociception by direct activation of the transient receptor potential channel, vanilloid subfamily member 1 [80, 81]. Low-concentration capsaicin (0.025–0.075%) creams and dermal patches are now available without prescription to relieve the pain associated with osteoarthritis, rheumatoid arthritis, postherpetic neuralgia, psoriasis, and diabetic neuropathy [82]. Anesiva (San Fransisco, CA, USA) has developed a capsaicin formulation for internal use, for the treatment of severe postoperative pain, post-traumatic neuropathic pain, and musculoskeletal diseases, which is currently undergoing various phase II clinical trials [83].

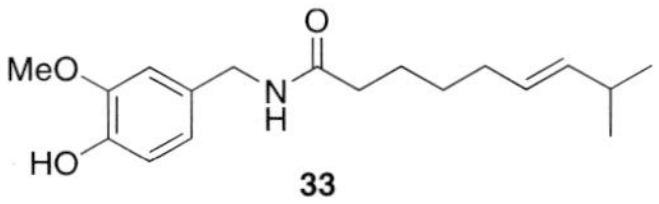

Chapter 1 Drug Discovery from Plants

33

Ceflatonine (**34**), a synthetic version of homoharringtonine produced by ChemGenex Pharmaceuticals (Menlo Park, CA, USA), is currently undergoing phase II/III clinical trials for the treatment of patients with chronic myeloid leukemia that is resistant to the first-line therapy, Gleevec [84]. Homoharringtonine is an alkaloid isolated from the Chinese evergreen tree *Cephalotaxus harringtonia* K. Koch. [84]. Homoharringtonine affects several cellular pathways, including the regulation of genes associated with apoptosis and angiogenesis [85].

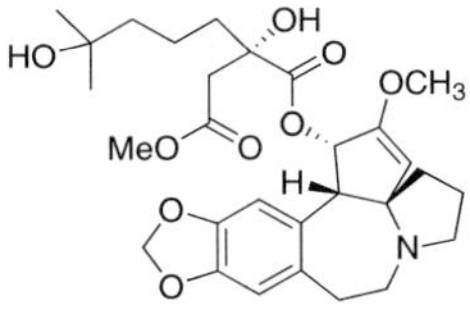

34

Celgosivir (**35**, MX-3253), developed by MIGENIX (Vancouver, Canada), is a semisynthetic derivative of the alkaloid castanospermine (**36**), which is isolated from the Australian tree *Castanospermum australe* A. Cunningham ex R. Mudie [86, 87]. Celgosivir is an α-glucosidase I inhibitor and has shown in vitro synergy with various interferons [88]. Celgosivir is currently undergoing phase IIb clinical trials as a combination therapy with peginterferon α2b and ribavirin for the treatment of patients with chronic HCV infection [89].

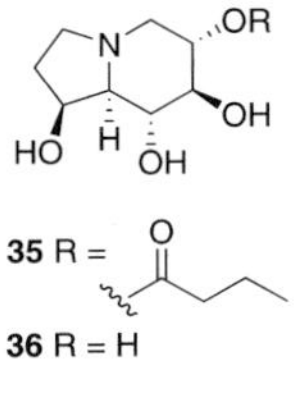

35 R =

36 R = H

Combretastatin A4 phosphate (**37**, CA4P) is a disodium phosphate prodrug of the natural stilbene combretastatin A4 (**38**) isolated from the South African tree *Combretum caffrum* Kuntze [90]. CA4P is being developed by OXiGENE (Waltham, MA, USA) to treat anaplastic thyroid cancer in combination with other anticancer drugs and also for myopic macular degeneration, both in phase II clinical trials [91]. Combretastatin is a vascular targeting agent that functions by destroying existing tumor vasculature by inducing morphological changes within the endothelial cells [90].

37 R = OPO$_3$Na$_2$
38 R = OH

Huperzine A (**39**) is an alkaloid with a potent acetylcholinesterase inhibitory activity isolated from the Chinese club moss *Huperzia serrata* (Thunb. ex Murray) Trevis. [92]. Huperzine A is currently available in the USA as "nutraceutical" or "functional food". The National Institute on Aging, at the National Institutes of Health (Bethesda, USA) [93], in collaboration with Neuro-Hitech (New York, NY, USA) [94] are currently evaluating the safety and efficacy of huperzine A in a phase II clinical trial [95]. A prodrug of huperzine A, ZT-1 (**40**), is currently being evaluated by Debiopharm (Lausanne, Switzerland) in phase II clinical trials for the potential treatment of Alzheimer's disease, and has shown efficacy in patients with mild to moderate symptoms [96].

Ingenol 3-angelate (**41**, PEP005) is a diterpene ester isolated from the medicinal plant *Euphorbia peplus* L., a species used traditionally to treat skin conditions such as warts and actinic keratoses [97]. PEP005 kills tumor cells via two mechanisms: (1) by inducing primary necrosis of tumor cells, and (2) by potently activating PKC. This is also associated with an acute T-cell-independent inflammatory response that is characterized by a pronounced neutrophil infiltration [98]. PEP005, developed by Peplin (Brisbane, Australia), is currently undergoing phase II clinical trials as a topical formulation for the treatment of actinic keratosis and basal cell carcinoma [99].

Morphine (**27**), an opiate analgesic alkaloid isolated from *Papaver somniferum*, is a drug that is still used widely today for the alleviation of severe pain [5]. Morphine is metabolized into morphine-3-glucuronide and morphine-

6-glucuronide (**42**, M6G) in the human body; but of these two metabolites, only M6G possesses analgesic activity [100, 101]. M6G is being developed by CeNeS (Cambridge, UK) as a treatment for postoperative pain, and is currently undergoing phase III trials in Europe, with phase III clinical trials in the USA expected to commence in 2007 [102]. The results of clinical testing to date have shown that M6G gives the same postoperative pain relief as morphine, but causes less postoperative nausea and vomiting [102].

Phenoxodiol (**43**), a synthetic analog of daidzein (**44**), an isoflavone from soybean (*Glycine max* Merr.), is being developed by Marshall Edwards (North Ryde, Australia) for the treatment of cervical, ovarian, prostate, renal, and vaginal cancers [103]. Phenoxodiol is a broad-spectrum anticancer drug that induces cancer cell death through inhibition of antiapoptotic proteins including XIAP and FLIP [104]. Phase III clinical trials of phenoxodiol as a treatment for ovarian cancer has started in Australia, with phase II trials currently underway in the USA [105].

Protopanaxadiol (**45**), a triterpene aglycone hydrolyzed from various Korean ginseng (*Panax ginseng* C. A. Mey.) saponins [106], has been shown to exhibit apoptotic effects on cancer cells through various signaling pathways, and has also been reported to be cytotoxic against multidrug-resistant tumors [106–108]. PanaGin Pharmaceuticals (British Columbia, Canada) is developing protopanaxadiol (Pandimex) for the treatment of lung cancer and other solid tumors, and is currently undergoing phase I clinical study in the USA [109]. Pandimex has been marketed in the People's Republic of China under conditional approval for the treatment of advanced cancers of the lung, breast, pancreas, stomach, colon, and rectum [110].

1.3.3 Plant Extracts Currently Involved in Clinical Trials

There are new forms of registered plant-derived medicines (phytomedicines) that are not single chemical entities. These more complex drugs are subjected to quality control via extract standardization procedures involving either or both compounds with known biological activity or inactive "marker compounds" present in high concentration [111]. The following are examples of standardized plant extracts that have undergone clinical trial for the treatment of several diseases, including osteoarthritis and cancer, and as a pain reliever.

Devil's Claw (*Harpagophytum procumbens* DC.) has been used for thousands of years in Africa for the treatment of fever, rheumatoid arthritis, and skin conditions, and is currently available as an alternative treatment for pain and osteoarthritis [112]. Harpagoside (**46**), one of the major components of the plant, has been shown to suppress lipopolysaccharide-induced inducible nitric oxide synthase and cyclooxygenase (COX)-2 expression through inhibition of nuclear factor-κB activation [113]. Several clinical trials have shown a *Harpagophytum procumbens* extract containing 50–60 mg of harpagoside to be effective in treating pain [114]. This *Harpagophytum procumbens* extract is currently undergoing phase II clinical trials in the USA for the treatment of hip and knee osteoarthritis [115].

Flavocoxid (Limbrel), a proprietary blend of natural flavonoids from *Scutellaria baicalensis* Georgi and *Acacia catechu* Willd., is being marketed in the USA by Primus Pharmaceuticals (Scottsdale, AZ, USA) under prescription as a "medical food" therapy for osteoarthritis [116]. A medical food is not a drug, nor a dietary supplement, and is defined by the FDA as a "formulated food that is consumed under the supervision of a physician and is intended for the specific management of a disease" [117]. Flavocoxid is currently undergoing a phase I clinical trial in the USA for the treatment of knee osteoarthritis. The active components of flavocoxid include baicalin (**47**) and cathechin (**48**), two flavonoids with anti-inflammatory and antioxidant properties [118]. This product works by inhibiting the cyclooxygenase (COX-1 and COX-2) and lipoxygenase (5-LOX) enzyme systems, two major inflammatory pathways involved in osteoarthritis that process arachidonic acid into inflammatory metabolites [119].

Ginkgo extracts are produced from the dried leaves of the *Ginkgo biloba* L. tree, a unique species with no close living relatives, that has been described as a "living fossil." The standardized ginkgo extract (EGb 761) contains approximately 24% flavone glycosides (primarily quercetin, kaempferol, and isorhamnetin) and 6% terpenoid lactones [2.8–3.4% ginkgolides A (**49**), B (**50**), and C (**51**), and 2.6–3.2% bilobalide (**52**)] [120]. Ginkgo extract is used for the treatment of early-stage Alzheimer's disease (AD), vascular dementia, peripheral claudication, and tinnitus of vascular origin [121]. Several reviews on the studies of ginkgo extract for the treatment of patients with Alzheimer's disease and dementia have been published [122–124]. In the USA, *Ginkgo biloba* extract (240 mg/day) is currently undergoing phase III clinical trials to prevent dementia and the onset of Alzheimer's disease in older individuals [125].

49 R$_1$ = H, R$_2$ = H
50 R$_1$ = OH, R$_2$ = H
51 R$_1$ = OH, R$_2$ = OH

52

Mistletoe (*Viscum album* L.) extract (Iscador) has been used as a complementary treatment in cancer patients in various European countries (e.g., Austria, Germany, Switzerland, and the United Kingdom) [126]. In the USA, mistletoe extract is currently undergoing phase II clinical trials as a supplemental treatment for lung cancer patients receiving conventional chemotherapy [127], and in phase I trials as a combination drug with gemcitabine (a synthetic antitumor drug) for patients with advanced solid tumors [128]. Mistletoe extract has been shown to have cytotoxicity against tumor cells and immunomodulatory activity, but the mechanism of action is poorly understood [126]. Mistletoe contains a cytotoxic lectin (viscumin) and several cytotoxic proteins and polypeptides (viscotoxins) that have been shown to induce tumor necrosis, increase natural killer cell activity, increase the production of interleukins 1 and 6, activate macrophages, induce programmed cell death (apoptosis), and protect DNA in normal cells during chemotherapy [129–132].

Sativex, developed by GW Pharmaceuticals (Wiltshire, UK), is a standardized extractive of *Cannabis sativa* L. with an almost 1:1 ratio of the cannabinoids, Δ^9-tetrahydrocannabinol (**53**) and cannabidiol (**54**), for the treatment of neuropathic pain in patients with multiple sclerosis [133]. In 2005, Sativex oromucosal spray was approved by Health Canada as an adjunctive treatment for the symptomatic relief of neuropathic pain in multiple sclerosis patients [134]. In the USA, Sativex began phase III clinical trials for multiple sclerosis patients in 2006 [135].

1.4 Recent Trends and Future Directions

Plant-derived and other natural product secondary metabolites have provided many novel prototype bioactive molecules, some of which have led to important drugs that are available on the market today. In spite of this, in the last 10 years or so, most large pharmaceutical companies have either terminated or scaled down their natural products drug-discovery programs, largely in favor of performing combinatorial chemistry, which can generate libraries consisting of millions of compounds [58, 136]. The roles of large pharmaceutical companies in the field of natural products have now been taken over to some extent by small biotechnology companies, which are specializing in lead identification from natural product extracts and the development of these leads into drugs [58, 137]. Many of the plant-derived drugs currently undergoing clinical trials were obtained and promoted by these emerging "biotech" companies, some of which were mentioned in the previous section.

In the past, drug discovery of bioactive compounds from plants was time-consuming, and the process of identifying the structures of active compounds from an extract could take weeks, months, or even years, depending on the complexity of the problem. Nowadays, the speed of bioassay-guided fractionation has been improved significantly by improvements in instrumentation such as high-performance liquid chromatography (HPLC) coupled to mass spectrometry (MS)/MS (liquid chromatography, LC-MS), higher magnetic field-strength nuclear magnetic resonance (NMR) instruments, and robotics to automate high-throughput bioassays. The introduction of capillary NMR (cap-NMR) spectroscopy is a recent major breakthrough for the characterization of compounds that are extremely limited in quantity in their organisms of origin [138, 139].

The high sensitivity of the cap-NMR probe has allowed for the combination of NMR spectroscopy with other analytical "hyphenated" techniques, such as LC-NMR-MS and LC-solid phase extraction (SPE)-NMR [140, 141]. The LC-NMR-MS technique generally requires deuterated solvents during the chromatographic separation, or alternatively, solvent suppression can be used for nondeuterated solvents [141, 142]. In contrast, the LC-SPE-NMR technique does not require deuterated solvents during the chromatographic separation, and, furthermore, it allows for sample enrichment through repeated chromatographic runs using SPE before NMR analysis [140]. A state-of-the-art integrated system for LC-NMR-MS and LC-SPE-NMR-MS has been developed and the hardware can be switched from LC-NMR-MS to LC-SPE-NMR-

Chapter 1   Drug Discovery from Plants   19

MS with minimal reconfiguration [140]. LC-SPE-NMR in combination with HPLC-electrospray ionization mass spectrometry (ESIMS) has been used for the rapid identification of compounds present in crude extracts of plants, as exemplified by the identification of sesquiterpene lactones and esterified phenylpropanoids in *Thapsia garganica* L. [143], and the characterization of constituents of *Harpagophytum procumbens* [144].

The development of automated high-throughput techniques has allowed for rapid screening of plant extracts; thus, the biological assay is no longer the rate-limiting step in the drug-discovery process. With advances in data handling systems and robotics, 100,000 samples can be assayed in just over 1 week when using a 384-well format [42]. Screening of plant extract libraries can be problematic due to the presence of compounds that may either autofluoresce or have UV absorptions that interfere with the screen readout, but prefractionation of extracts can be used to alleviate some of these types of problems. Also, most high-throughput screening assay methods have been developed with computational filtering methods to identify and remove potentially problematic compounds that can give false-positive results [145].

In the future, the routine use of NMR "hyphenated" techniques will allow for quick "dereplication" (a process of eliminating known and active compounds in the plant extracts that have been studied previously), and high-throughput screening will permit the rapid identification of the active compounds [140]. For example, duplicate SPE plates can be generated during the HPLC separation, with one plate used to prepare samples for high-throughput screening, while the other plate is kept as a reference. The structure(s) of compounds in wells of these plates that show(s) activity can be determined by cap-NMR and MS, and known compounds can be ruled out quickly based on their NMR spectroscopic and MS information. In instances where the active compound has a new structure, further isolation can be carried out from the plant material, provided there is enough sample. Alternatively, the compound can be synthesized for further bioassay, and combinatorial chemistry can be used to design new analogs based on the parent molecules.

Adequate and continuous supplies of plant-derived drugs are essential to meet the market demand. For compounds that are uneconomical to synthesize, and only available in a small quantities from plants, the use of plant cell cultures is an alternative production method. Plants accumulate secondary metabolites at specific developmental stages, and by manipulating the environmental conditions and medium, many natural products have been synthesized in cell cultures in larger percentage yields than those evident in whole plants [146, 147]. Paclitaxel (**1**) has been produced successfully by plant cell fermentation (PCF) technology, and, as mentioned earlier, the supply of the important anticancer drug, paclitaxel, by its manufacturer, Bristol-Myers Squibb, is now obtained by PCF technology [148–150]. Other plant-derived compounds that can currently be produced by cell cultures include the *Catharanthus* alkaloids [151], diosgenin from *Dioscorea* [152], and the *Panax ginseng* ginsenosides [153].

In conclusion, plants have provided humans with many of their essential needs, including life-saving pharmaceutical agents. In the last 6 years, five new plant-derived drugs have been launched onto the market, and many more are currently undergoing clinical trials. As a vast proportion of the available higher plant species have not yet been screened for biologically active compounds, drug discovery from plants should remain an essential component in the search for new medicines, particularly with the development of highly sensitive and versatile analytical methods.

References

1. Harborne JB (1984) Phytochemical Methods: A Guide to Modern Techniques of Plant Analysis, 2nd edn. Chapman and Hall, New York
2. Bisset NG (1989) J Ethnopharmacol 25:1
3. Mann J (2000) Murder, Magic and Medicine, 2nd edn. Oxford University Press, Oxford, UK
4. Bisset NG (1991) J Ethnopharmacol 32:71
5. Sneader W (1996) Drug Prototypes and their Exploitation, Wiley, Chichester, UK
6. Samuelsson G (2004) Drugs of Natural Origin, 5th edn, Apotekarsocieteten, Stockholm
7. Halpern JH, Sewell RA (2005) Life Sci 78:519
8. McCurdy CR, Scully SS (2005) Life Sci 78:476
9. Sneader W (2005) Drug Discovery: a History, Wiley, Chichester, UK
10. King RE, Bomser JA, Min DB (2006) Comprehen Rev Food Sci Food Safety 5:65–70
11. Fulda S, Debatin K-M (2006) Cancer Detect Prev 30:217
12. Bilia AR, Gallori S, Vincieri FF (2002) Life Sci 70:3077
13. Smith KK, Dharmaratne HRW, Feltenstein MW, Broom SL, Roach JT, Nanayakkara NPD, Khan IA, Sufka KJ (2001) Psychopharmacology 155:86
14. Nerurkar PV, Dragull K, Tang C-S (2004) Toxicol Sci 79:106
15. Fabricant DS, Farnsworth NR (2001) Environ Health Perspect 109:69
16. Farnsworth NR, Akerele O, Bingel AS, Soejarto DD, Guo Z (1985) Bull WHO 63:965
17. Anon (2003) World Health Organization fact sheet No. 134, revised May 2003. Available at http://www.who.int/mediacentre/factsheets/fs134/en/
18. Blumenthal M, Ferrier GKL, Cavaliere C (2006) HerbalGram 71:64
19. Fowler MW (2006) J Sci Food Agric 86:1797
20. Gurib-Fakin A (2006) Mol Aspects Med 27:1
21. Jones WP, Chin YW, Kinghorn AD (2006) Curr Drug Targets 7:247
22. Balunas MJ, Kinghorn AD (2006) Life Sci 78:431
23. Kingston DGI (2000) J Nat Prod 63:726
24. Holton RA, Somoza C, Kim H-B, Liang F, Biediger RJ, Boatman PD, Shindo M, Smith CC, Kim S, Nadizadeh H, Suzuki Y, Tao C, Vu P, Tang S, Zhang P, Murthi KK, Gentile LN, Liu JH (1995) The total synthesis of paclitaxel starting with camphor. In: Georg GI, Chen TT, Ojima I, Vyas DM (eds) ACS Symposium Series 583 (Taxane Anticancer Agents, Basic Science and Current Status). American Chemical Society, Washington DC, p 288
25. Nicolau KC, Guy RK (1995) The total synthesis of paclitaxel by assembly of the ring system. In: Georg GI, Chen TT, Ojima I, Vyas DM (eds) ACS Symposium Series 583 (Taxane Anticancer Agents, Basic Science and Current Status). American Chemical Society, Washington DC, p 302
26. Denis J-N, Greene AE (1988) J Am Chem Soc 110:5917
27. Holton RA, Biediger RJ, Boatman PD (1995) Semisynthesis of taxol and taxotere. In: Suffness M (ed) Taxol, Science and Applications. CRC Press, Boca Raton, FL, p 97

28. Kingston DGI (2006) Taxol and its analogs. In: Cragg GM, Kingston DGI, Newman DJ (eds) Anticancer Agents from Natural Products. Taylor and Francis, Boca Raton, FL, p 89
29. Ritter SK (2004) Chem Eng News 82:25
30. Wall ME (1960) Am Perfumer Aromatics 75:63
31. Applezweig N (1962) Steroid Drugs. McGraw-Hill, New York
32. Mancera O, Zaffaroni A, Rubin BA, Sondheimer F, Rosenkranz G, Djerassi C (1952) J Am Chem Soc 74:3711
33. Mancera O, Ringold HJ, Djerassi C, Rosenkranz G, Sondheimer F (1953) J Am Chem Soc 75:1286
34. Ward P, Small I, Smith J, Suter P, Dutkowski R (2005) J Antimicrob Chemother 55:i5
35. Graeme L (2006) Future Virol 1:577
36. Abrecht S, Harrington P, Iding H, Karpf M, Trussadi R, Wirz B, Zutter U (2004) Chimia 58:621
37. Yarnell, A (2005) Chem Eng News 83:22
38. Kramer M, Bongaerts J, Bovenberg R, Kremer S, Muller U, Orf S, Wubbolts M, Raeven L (2003) Metab Eng 5:277
39. Yeung YY, Hong S, Corey EJ (2006) J Am Chem Soc 128:6310
40. Lee K-H, Xiao Z (2005) Podophyllotoxins and analogs. In: Cragg GM, Kingston DGI, Newman DJ (eds) Anticancer Agents from Natural Products. Taylor and Francis, Boca Raton, FL, p 71
41. Rahier NJ, Thomas CJ, Hecht SM (2005) Camptothecin and its analogs. In: Cragg GM, Kingston DGI, Newman DJ (eds) Anticancer Agents from Natural Products. Taylor and Francis, Boca Raton, FL, p 5
42. Butler MS (2005) Nat Prod Rep 22:162
43. Cragg GM, Newman DJ (2004) J Nat Prod 67:232
44. Oberlies NH, Kroll DJ (2004) J Nat Prod 67:129
45. Bailey CJ, Day C (2004) Pract Diab Int 21:115
46. Krentz AJ, Bailey CJ (2005) Drugs 65:385
47. Grynkiewicz G, Achmatowicz O, Pucko W (2000) Herba Pol 46:151
48. Molokanova E, Savchenko A, Kramer RH (2000) J Gen Physiol 115:685
49. Molokanova E, Kramer RH (2001) J Gen Physiol 117:219
50. Hecker E (1968) Cancer Res 28:2338
51. Castagna M (1987) Biol Cell 59:3
52. Perchellet JP, Perchellet EM (1988) Pharmacology 2:325
53. Kazanietz, MG (2005) Biochim Biophys Acta 1754:296
54. Estensen RD (1984) J Exp Pathol 1:71
55. Montesano R, Orci L (1985) Cell 42:469
56. Droms KA, Malkinson AM (1991) Mol Carcinog 4:1
57. Newman DJ, Cragg GM, Snader KM (2003) J Nat Prod 66:1022
58. Butler MS (2004) J Nat Prod 67:2141
59. Chin YW, Balunas MJ, Chai HB, Kinghorn AD (2006) AAPS J 8:E239
60. Howes M-JR, Perry NSL, Houghton PJ (2003) Phytother Res 17:1
61. Heinrich M, Teoh HL (2004) J Ethnopharmacol 92:147
62. McKiernan PJ (2006) Drugs 66:743
63. Duke SO, Dayan FE, Romagni JG, Rimando AM (2000) Weed Res 40:99
64. Hall MG, Wilks MF, Provan WM, Eksborg S, Lumholtz B (2001) Br J Clin Pharmacol 52:169
65. U.S. Food and Drug Administration. CDER new molecular Entity (NME) drug and new biologic approvals in calendar year 2004. Available at http://www.fda.gov/cder/rdmt/nmecy2004.htm
66. Deleu D, Hanssens Y, Northway MG (2004) Drugs Aging 21:687
67. Koumis T, Samuel S (2005) Clin Ther 27:377
68. Niaura R, Jones C, Kirkpatrick P (2006) Nature Rev Drug Discov 5:537
69. Feret B, Orr K (2006) Formulary 41:265

70. Mihalak KB, Caroll FI, Luerje CW (2006) Mol Pharmacol 70:801
71. Etter J-F (2006) Arch Intern Med 166:1553
72. Cichewicz RH, Kouzi SA (2004) Med Res Rev 24:90
73. Yogeeswari P, Sriram D (2005) Cur Med Chem 12:657
74. Pisha E, Chai H, Lee I-S, Chagwedera TE, Farnsworth NR, Cordell GA, Beecher CWW, Fong HHS, Kinghorn AD, Brown DM, Wani MC, Wall ME, Hieken TJ, Das Gupta TK, Pezzuto JM (1995) Nat Med 1:1046
75. Further information available at http://nihroadmap.nih.gov/raid/
76. U.S. National Institutes of Health. Evaluation of 20% betulinic acid ointment for treatment of dysplastic nevi (moderate to severe dysplasia) Available at http://clinicaltrials.gov/ct/show/NCT00346502
77. Temesgen Z, Feinberg JE (2006) Curr Opin Investig Drugs 7:759
78. Li F, Goila-Gaur R. Salzwedel K, Kilgore NR, Reddick M, Matallana C, Castillo A, Zoumplis D, Martin DE, Orenstein JM, Allaway GP, Freed EO, Wild CT (2003) Proc Natl Acad Sci U S A 100:13555
79. Panacos: Press release August 22, 2005. Press release and further information available at http://www.panacos.com
80. Bevan S (1999) Capsaicin and pain mechanisms. In: Brain SD, Moore PK (eds) Pain and neurogenic inflammation. Birkhaeuser, Basel, p 61
81. Szallasi A, Appendino G (2004) J Med Chem 47:2717
82. Minami T, Bakoshi S, Nakano H, Mine O, Muratani T, Mori H, Ito S (2001) Anesth Analg 93:419
83. Anesiva, Inc. Further information available at http://www.anesiva.com/wt/page/pipeline
84. ChemGenex Pharmaceuticals: Press release September 20, 2006. Press release and further information available at http://www.chemgenex.com
85. Itokawa H, Wang X, Lee KH (2005) Homoharringtonine and related compounds. In: Cragg GM, Kingston DGI, Newman DJ (eds) Anticancer Agents from Natural Products. Taylor and Francis, Boca Raton, FL, p 47
86. Hohenschutz LD, Bell EA, Jewess PJ, Leworthy DP, Pryce RJ, Arnold E, Clardy J (1981) Phythochemistry 20:811
87. Sorbera LA, Castaner J, Garcia-Capdevila L (2005) Drugs Future 30:545
88. Whitby K, Taylor D, Patel D, Ahmed P, Tyms AS (2004) Antivir Chem Chemother 15:141
89. MIGENIX Inc: Press release November 6, 2006. Press release and further information available at http://www.migenix.com
90. Pinney KG, Jelinek C, Edvardsen K, Chaplin DJ, Pettit GR (2005) The discovery and development of the combrestatatins. In: Cragg GM, Kingston DGI, Newman DJ (eds) Anticancer Agents from Natural Products. Taylor and Francis, Boca Raton, FL, p 23
91. OXiGENE, Inc: Press releases June 5, 2006 and September 12, 2006. Press release and further information available at http://www.oxigene.com
92. Zhu D-Y, Tan C-H, Li Y-M (2006) The overview of studies on huperzine A: a natural drug for the treatment of Alzheimer's disease. In: Liang X-T, Fang W-S (eds) Medicinal Chemistry of Bioactive Natural Products. John Wiley Sons, Hoboken, NJ, p 143
93. Further information available at http://www.nia.nih.gov/
94. Further information available at http://www.neurohitech.com
95. U.S. National Institutes of Health. Huperzine A in Alzheimer's disease. Available at http://www.clinicaltrials.gov/ct/show/NCT00083590
96. DebioPharm. Further information available at http://www.debiopharm.com
97. Hampson P, Wang K, Lord JM (2005) Drugs Future 30:1003
98. Challacombe JM, Suhrbier A, Parsons PG, Jones B, Hampson P, Kavanagh D, Rainger GE, Morris M, Lord JM, Le TTT, Diem H-L, Ogbourne SM (2006) J Immunol 177:8123
99. Peplin, Ltd.: ASX release August 1, 2006. Press release and further information available at http://www.peplin.com

100. Lotsch J, Geisslinger G (2001) Clin Pharmacokinet 40:485
101. Yamada H, Ishii K, Ishii Y, Ieiri I, Nishio S, Morioka T, Oguri K (2003) J Toxicol Sci 28:395
102. CeNeS Pharmaceuticals: News updates September 28, 2006 and October 19, 2006. News release and further information available at http://www.cenes.com
103. Gamble JR, Xia P, Hahn CN, Drew JJ, Drogemuller CJ, Brown D, Vadas MA (2006) Int J Cancer 118:2412
104. Kamsteeg M, Rutherford T, Sapi E, Hanczaruk B, Shahabi S, Flick M, Brown D, Mor G (2003) Oncogene 22:2611
105. Marshall Edwards, Inc.: Current news November 27, 2006. Current news and further information available at http://www.marshalledwardsinc.com
106. Nagai M, Tanaka O, Shibata S (1966) Tetrahedron Lett 7:4797
107. Jia W, Yan H, Bu X, Liu G, Zhao Y (2004) J Clin Oncol 22:9663
108. Popovich DG, Kitts DD (2004) Can J Physiol Pharmacol 82:183
109. Li G, Wang Z, Sun Y, Liu K, Wang Z (2006) Basic Clin Pharm Toxicol 98:588
110. PanaGin Pharmaceuticals, Inc. Further information available at http://www.panagin.com
111. Gaedcke F, Stenhoff B, Blasius H (2000) Herbal Medicinal Products, Scientific and Regulatory Basis for Development, Quality Assurance, and Marketing Authorization, CRC Press, Boca Raton, FL
112. McGregor G, Fiebich, B, Wartenberg A, Brien S, Lewith G, Wegener T (2005) Phytochemistry Rev 4:47
113. Huang TH-W, Tran VH, Duke RK, Tan T, Chrubasik S, Roufogalis, BD, Duke CC (2006) J Ethnopharmacol 104:149
114. Chrubasik S, Conradt C, Roufogalis BD (2004) Phytother Res 18:187
115. U.S. National Institute of Health. Trial evaluating devil's claw for the treatment of hip and knee. Available at http://www.clinicaltrials.gov/ct/gui/show/NCT00295490
116. Primus Pharmaceuticals: Further information available at http://www.primusrx.com
117. Anon. Center for food safety and applied nutrition. Available at http://www.cfsan.fda.gov/~dms/ds–medfd.html
118. U.S. National Institute of Health. Flavocoxid, a plant-derived therapy for the treatment of knee osteoarthritis. Available at http://www.clinicaltrials.gov/ct/show/NCT00294125
119. Further information available at http://www.limbrel.com
120. Anon (2003) Drugs R&D 4:188
121. Gertz H-J, Kiefer M (2004) Curr Pharm Design 10:261
122. Christen Y, Mathiex-Fortunet H (2003) The effects of Ginkgo biloba extract EGb 761 in Alzheimer's disease: from the mechanism of action to clinical data. In: Vellas BJ (ed) Research and Practice in Alzheimer's Disease, Vol 7. Springer, Berlin, Heidelberg, New York, p 276
123. Christen Y (2006) *Ginkgo biloba* extract and Alzheimer's disease: is the neuroprotection explained merely by the antioxidant action? In: Luo Y, Packer L (eds) Oxidation Stress and Disease, 22 (Oxidative Stress and Age-Related Neurodegeneration). CRC Press, Boca Raton, FL, p 43
124. Smith JV, Luo Y (2006) *Ginkgo biloba* extract EGb 761 extends life span and attenuates H_2O_2 levels in *Caenorhabditis elegans* model of Azheimer's disease. In: Luo Y, Packer L (eds) Oxidation Stress and Disease, 22 (Oxidative Stress and Age-Related Neurodegeneration), CRC Press, Boca Raton, FL, p 301
125. U.S. National Institutes of Health. *Ginkgo biloba* prevention trial in older individuals. Available at http://clinicaltrials.gov/ct/show/NCT00010803
126. Maldacker J (2006) Arzneimittelforschung 56:497
127. U.S. National Institute of Health. Iscar for supplemental care in Stage IV lung cancer. Available at http://clinicaltrials.gov/ct/show/NCT00079794

128. U.S. National Institute of Health. Gemcitabine combined with mistletoe in treating patients with advanced solid tumors. Available at http://clinicaltrials.gov/ct/show/NCT00049608
129. Elluru S, Van Huyen J-PD, Delignat S, Prost F, Bayry J, Kazatchkine MD, Kaveri SV (2006) Arzneimittelforschung 56:461
130. Kovacs E, Link S, Toffol-Schmidt U (2006) Arzneimittelforschung 56:467
131. Harmsma M, Ummelen M, Dignef W, Tusenius KJ, Ramaekers FCS (2006) Arzneimittelforschung 56:474
132. Buessing A (2006) Arzneimittelforschung 56:508
133. Guy GW, Stott CG (2005) The development of Sativex®, a natural cannabis-based medicine. In: Mechoulam R (ed) Cannabinoids as Therapeutics (Milestones in Drug Therapy). Birkhaeuser, Basel, p 231
134. GW Pharmaceuticals: Press release April 19, 2005. Press release and further information available at http://www.gwpharm.com
135. U.S. National Institute of Health. Sativex® versus placebo when added to existing treatment for central neuropathic pain in MS. Available at http://www.clinicaltrials.gov/ct/gui/show/NCT00391079
136. Rouhi AM (2003) Chem Eng News 81(41):77
137. Rouhi AM (2003) Chem Eng News 81(41):93
138. Martin GE (2005) Annu Rep NMR Spectrosc 56:1
139. Schroeder FC, Gronquist M (2006) Angew Chem Int Ed Engl 45:7122
140. Corcoran O, Spraul M (2003) Drug Discov Today 8:624
141. Lewis RJ, Bernstein MA, Duncan SJ, Sleigh CJ (2005) Magn Reson Chem 43:783
142. Wann M-H (2005) Application of LC-NMR in pharmaceutical analysis. In: Ahuja S, Dong MW (eds) Handbook of Pharmaceutical Analysis by HPLC. Elsevier, Amsterdam, p 569
143. Lambert M, Wolfender J-L, Stærk D, Christensen SB, Hostettmann K, Jaroszewski JW (2007) Anal Chem 79:727
144. Clarkson C, Stærk D, Hansen SH, Smith PJ, Jaroszewski JW (2006) J Nat Prod 69:1280
145. Walters WP, Namchuk M (2003) Nature Rev Drug Discov 2:259
146. Zhong JJ (2001) Biochemical engineering of the production of plant-specific secondary metabolites by cell suspension cultures. In: Zhong JJ (ed) Advances in Biochemical Engineering/Biotechnology, 72 (Plant Cells). Springer, Berlin, Heidelberg, New York, p 1
147. Kirakosyan A (2006) Plant biotechnology for the production of natural products. In: Cseke LJ, Kirakosyan A, Kaufman BP, Warber SL, Duke JA, Brielmann HL (eds) Natural Products from Plants, 2nd edn. CRC Press, Boca Raton, FL, p 221
148. Tabata H (2004) Paclitaxel production by plant-cell-culture technology. In: Zhong JJ (ed) Advances in Biochemical Engineering/Biotechnology, 87 (Biomanufacturing). Springer, Berlin, Heidelberg, New York, p 1
149. Zhong JJ, Yue C-J (2005) Plant cells: secondary metabolite heterogeneity and its manipulation. In: Nielsen J (ed) Advances in Biochemical Engineering/Biotechnology, 100 (Biotechnology for the Future). Springer, Berlin, Heidelberg, New York, p 53
150. Tabata H (2006) Curr Drug Targets 7:453
151. Van Der Heijden R, Jacobs DI, Snoeijer W, Hallard D, Verpoorte R (2004) Curr Med Chem 11:607
152. Rokem JS, Tal B, Goldberg I (1985) J Nat Prod 48:210
153. Moyano E, Osuna L, Bonfill M, Cusido RM, Palazon J, Tortoriello J, Pinol MT (2005) Bioproduction of triterpenes on plant cultures of *Panax ginseng* and *Gaphimia glauca*. In: Pandalai SG (ed) Recent Research Developments in Plant Science, Vol. 3. Research Signpost, Kerala, India, p 195

Chapter 2
Grapevine Stilbenes and Their Biological Effects

P. Waffo-Teguo, S. Krisa, T. Richard and J.-M. Mérillon (✉)

Groupe d'Etudes des Substances Végétales à Activité Biologique, EA 3675,
Institut des Sciences de la Vigne et du Vin – Université
Bordeaux2 – 33076 Bordeaux Cedex, France,
e-mail: jean-michel.merillon@u-bordeaux2.fr

Abstract Vine and wine are an abundant source of polyphenolic compounds, including mainly flavonoids and stilbenes. The latter appear to constitute a large class of compounds, monomers and oligomers (dimers, trimers, tetramers) resulting from different oxidative condensations of the resveratrol monomer. Stilbenes exhibit potent biological activities in vitro on several targets that might be able to influence favourably several physiological and pathological processes, and to provide a protective effect against cardiovascular diseases and cancer, as suggested by epidemiological studies. Trans-resveratrol, the most studied stilbene, shows great promise in the treatment of leading diseases. Resveratrol acts through multiple pathways on the same pathology such as cancer or cardiovascular diseases. Its absorption appears to be high, but the oral bioavailability of unchanged resveratrol is very low due to rapid and extensive metabolism.

Keywords Polyphenols, Stilbenes, Resveratrol, *Vitis vinifera*, Cardiovascular protection, cancer chemoprevention and neuroprotection

2.1 Introduction

Vitis vinifera L. is a perennial woody vine belonging to the Vitaceae family. It is a productive plant that is considered to be the world's premier fruit, with nearly 9 million hectares of viticultural land in 1990. It is used for wine, juice, fresh consumption (table grapes), dried fruit and distilled liquor.

Polyphenolics are important constituents of grapes in determining the colour, taste and body of wines. Unlike other alcoholic beverages, red wine, which is obtained after maceration, contains phenolic compounds in high concentrations of up to 4 g/l; relatively low quantities are present in white and rosé wines (i.e. about one-tenth of those in red wines) [1]. Among these compounds, stilbenes constitute an important subclass with respect to the diversity of molecules, and its levels can reach 50 mg/l in red wine.

According to several epidemiological studies, polyphenols from grapes and wines have significant health benefits, since they possess cancer chemopreventive, cardioprotective, and neuroprotective activities. Among these phenolic compounds, *trans*-resveratrol, which belongs to the stilbene family, is a major active ingredient and can prevent or slow the progression of major diseases, as well as extend the lifespan of various organisms from yeast to vertebrates [2].

2.2 Epidemiology

Some epidemiologic studies in the United States have examined the relationship between wine consumption and the risk of cancer. Contrary to other alcoholic beverages, moderate wine consumption was associated with a decrease (or no increase) in the risk of oral and pharyngeal cancer [3, 4] and breast cancer [5]. Farchi et al. [6] using small cohorts in Italy found a minimum risk of cancer and cardiovascular disease associated with moderate alcohol consumption (wine being the main beverage). The inverse association between moderate alcohol consumption and coronary heart disease is well established by several reports, and a portion of this benefit seems to be from the alcohol itself [7]. According to World Health Organization statistics (1995), the reduction in the mortality rate from coronary heart disease in France as compared to the USA was 61% for men and 69% for women. As compared to the UK, the reduction was 68% and 71%, respectively [8, 9]. This finding constitutes the "French paradox" because saturated fat intakes and serum cholesterol concentrations are similar in the three countries. This paradox could be attributable in part to high wine consumption [8]. Renaud's group have evaluated prospectively the health risk of grape wine and beer drinking in 36,250 middle-aged men, from eastern France [10]. Compared to abstainers, a moderate daily intake of wine (22–32 g of alcohol, i.e. 2–3 glasses) was associated with a lower risk of death due to cardiovascular diseases (40%), cancer (22%), other causes (42%), and all causes (33%). In the case of beer drinkers, the risk for cardiovascular diseases was less significant (30%) and there was no reduction in the mortality from cancer and all causes. Moreover, only moderate wine drinkers have a lower hypertension-related mortality [11].

Similar results have been found in several large Danish cohort studies in which it was found that wine intake has a beneficial effect on all-cause mortality, and this was attributable to a reduction in death from both coronary heart disease and cancer [12, 13]. Other alcoholic beverages (beer, spirits) are positively related to death from cancer. Concerning the risk of upper digestive tract cancer, a considerable increase in the risk was observed from a moderate intake of the latter, but not for wine [14]. Moreover, during the past 15 years, mortality from coronary heart disease has declined by 30% in Denmark, which may be due to the increase in wine drinking during that period (from 17 to 30% of the total alcohol intake) that accompanied the opening up of the European market [12].

These studies have established non-linear relationships between alcohol consumption and mortality, which are either U- or J-shaped. This indicates that only moderately drinking wine may have beneficial effects on health, whereas excessive consumption is harmful and associated with increased mortality.

A protective effect of wine against the occurrence of dementia was also found in Aquitaine (France) in a prospective study that involved approximately 4000 subjects aged 65 years and over [15]. The risk of developing a dementia is divided by about a factor of two among moderate drinkers (0.25–0.5 l), compared to non-drinkers and light drinkers. The most frequent age-related degenerative disease is Alzheimer's disease (AD), which accounts for approximately 70% of cases of dementia. Similar results were found in a 2-year follow-up study of elderly people in China [16]. Flavonoid intake appears to be inversely related to the risk of incident dementia [17].

However, some studies that involved self reporting have found that wine drinkers have a healthier diet than people who drink beer or spirits, which may explain why wine has an additional beneficial effect on health. In Danish supermarkets, wine buyers made more purchases of healthy food items such as olives, fruit and vegetables than people who buy beer [18].

2.3 Chemistry of Stilbenes

Stilbenes naturally occur in several plant families, such as the Cyperaceae, Dipterocarpaceae, Gnetaceae and Vitaceae [19, 20]. Grapes (Vitaceae) and products manufactured from grapes are viewed as the most important dietary sources of these substances [21, 22]. They are a family of molecules belonging to the non-flavonoid polyphenol group. The essential structural skeleton comprises two aromatic rings joined by an ethylene bridge (C_6-C_2-C_6). From this relatively simple structure, there is a large array of compounds: monomers, for which the number and position of hydroxyl groups, the substitution with sugars, methyl, methoxy and other residues and the steric configuration of the molecules vary (Fig. 2.1), and oligomers, which are the result of different oxidative condensations of the resveratrol monomer (e.g. dimers, trimers, tetramers).

2.3.1 Characterisation

2.3.1.1 Monomers

Among stilbene monomers (Fig. 2.1), resveratrol (3,5,4'-trihydroxystilbene) has been identified as the major biologically active compound, and most of the studies have focussed on it. The two isomeric forms of resveratrol (*cis*- and

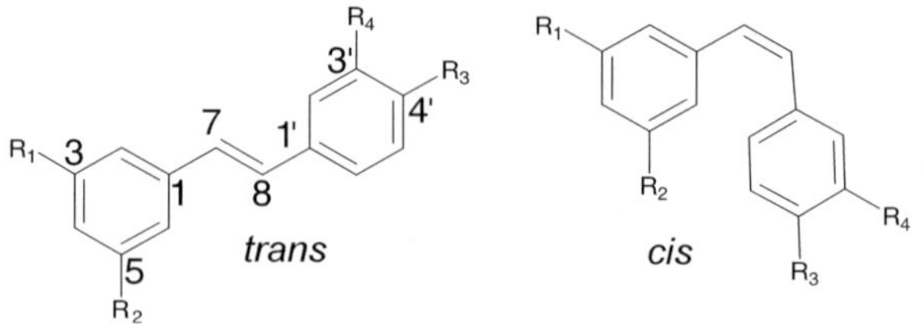

names	R₁	R₂	R₃	R₄
cis- and trans-resveratrol	OH	OH	OH	H
trans-pterostilbene	OCH$_3$	OCH$_3$	OH	H
cis- and trans-piceid	OGlc	OH	OH	H
cis- and trans-resveratroloside	OH	OH	OGlc	H
cis- and trans-resveratrol 3,5-O-β-diglucoside	OGlc	OGlc	OH	H
cis- and trans-resveratrol 3,4'-O-β-diglucoside	OGlc	OH	OGlc	H
trans-resveratrol 3,5,4'-O-β-triglucoside	OGlc	OGlc	OGlc	H
trans-piceatannol	OH	OH	OH	OH
cis- and trans-astringin	OGlc	OH	OH	OH

Fig. 2.1 Structure of the main stilbene monomer derivatives from *Vitis vinifera*

trans-) have different chemical characteristics and biological activities. The *trans-*isomer is usually the more stable, and *cis-trans-*interconversions can occur in the presence of heat or ultraviolet (UV) light. These two compounds, having maximal absorbance at 286 (*cis-*isomer) and 306 nm (*trans-*isomer) [23], are highly fluorescent at 374 nm when excited at 330 nm [24].

Other simple stilbenes have been isolated in *Vitis vinifera*: *trans-*pterostilbene [25] and piceatannol. Besides the aglycone of resveratrol cited above, some resveratrol glucoside derivatives have been identified, such as piceid and resveratroloside, two β-glucosides of resveratrol [26–28], together with astringin (piceatannol 3-O-β-glucoside). These compounds exist in their two isomeric forms, *cis* and *trans* [28]. Furthermore, resveratrol di- and triglucoside derivatives have been recently isolated from *Vitis vinifera* [29, 30].

2.3.1.2 Oligomers

In addition to stilbene monomers, some oligomers have been isolated from *Vitis vinifera*; these are the dimers, trimers and tetramers. These oligomers result from the different oxidative condensations of the resveratrol monomer.

2.3.1.2.1 Dimers

The dimers are divided into two major groups. One group (A) contains one, five-membered oxygen heterocyclic ring bearing an aromatic ring (benzofuran

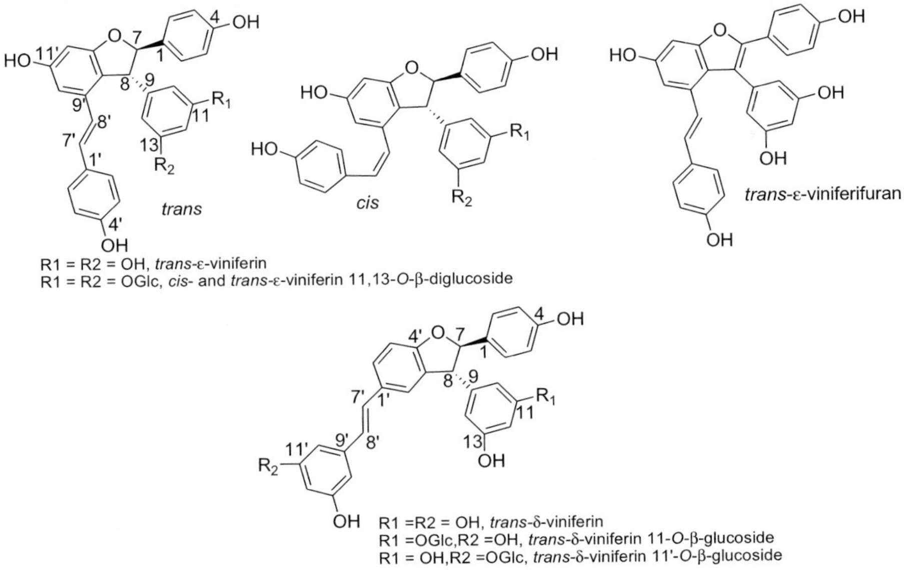
trans
cis
trans-ε-viniferifuran
R1 = R2 = OH, trans-ε-viniferin
R1 = R2 = OGlc, cis- and trans-ε-viniferin 11,13-O-β-diglucoside
R1 =R2 = OH, trans-δ-viniferin
R1 =OGlc,R2 =OH, trans-δ-viniferin 11-O-β-glucoside
R1 = OH,R2 =OGlc, trans-δ-viniferin 11'-O-β-glucoside

Fig. 2.2 Structure of the main stilbene dimer derivatives (group A) from *Vitis vinifera*

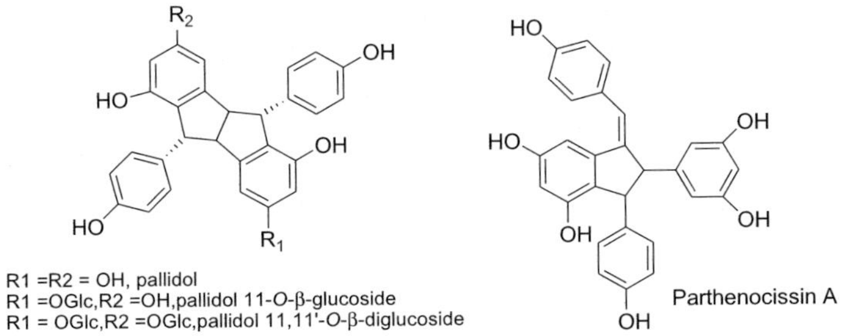
R1 =R2 = OH, pallidol
R1 =OGlc,R2 =OH,pallidol 11-O-β-glucoside
R1 = OGlc,R2 =OGlc,pallidol 11,11'-O-β-diglucoside
Parthenocissin A

Fig. 2.3 Structure of the main stilbene dimer derivatives (group B) from *Vitis vinifera*

ring; Fig. 2.2). Three members of this group are ε-viniferin, which can be either substituted or not with sugars [31, 32], δ-viniferin (also named resveratrol dehydrodimer), which can be either glucosylated or not [33], and ε-viniferifuran [34].

Dimers in the other group (group B) do not contain an oxygen heterocyclic ring. Among the dimers in this group, pallidol has been isolated in *Vitis vinifera*, as well as its mono- and di-glucoside [33, 32], and parthenocissin A (Fig. 2.3).

2.3.1.2.2 Trimers

Recently, a resveratrol trimer was detected in grapevine infected by downy mildew using high performance liquid chromatography (HPLC) coupled to atmospheric pressure photoionisation mass spectrometry. The structure was thought to be α-viniferin [35] (Fig. 2.4).

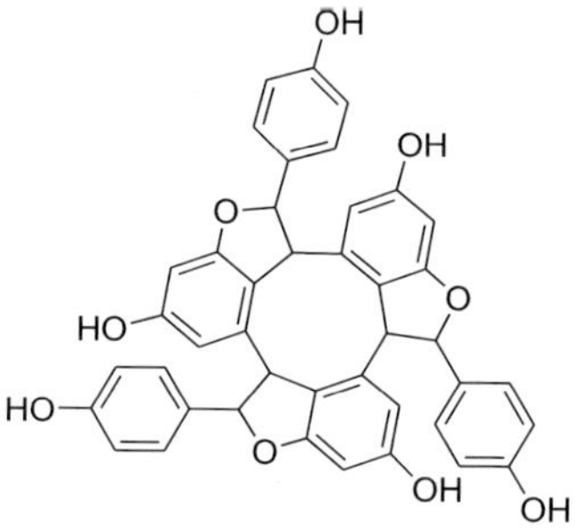

Fig. 2.4 Structure of the stilbene trimer from *Vitis vinifera*

2.3.1.2.3 Tetramers

In addition to resveratrol dimers and trimer, stilbene tetramers were isolated from *Vitis vinifera*. These can be divided into three groups:
1. Tetramers in the first group contain a bicyclo[6.3.0]undecane ring system; viniferol A (Fig. 2.5).
2. The second tetramer group has a bicyclo[5.3.0]decane ring system. Resveratrol tetramers belonging to this group are: viniferol B and C, vaticanol B, and vaticaphenol A (Fig. 2.5).
3. The third group contains a benzofuran system, usually *trans*-2-aryl-2,3-benzofuran moiety (Fig. 2.6). Tetramers belonging this group are: vitisifuran A and B and iso- and hopeaphenol (two cyclic symmetric tetramers) [36].

Chapter 2 Grapevine Stilbenes and Their Biological Effects

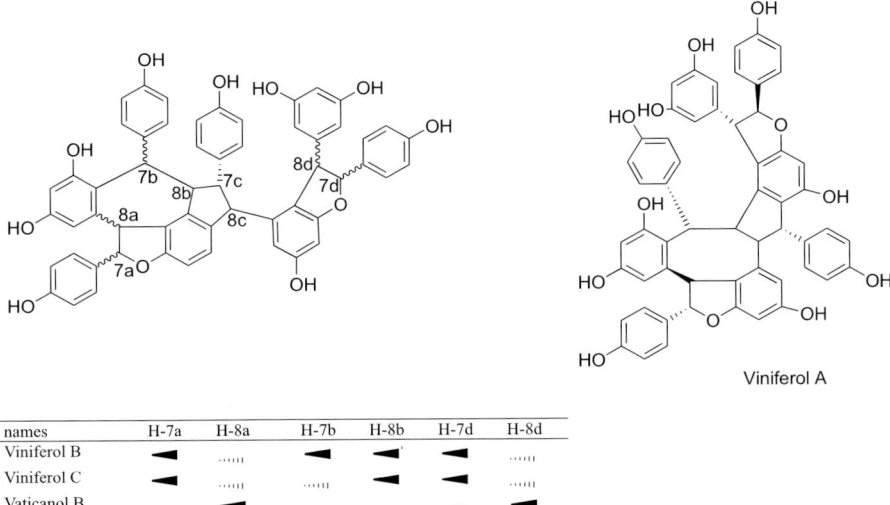

Fig. 2.5 Structure of the stilbene tetramer from *Vitis vinifera*

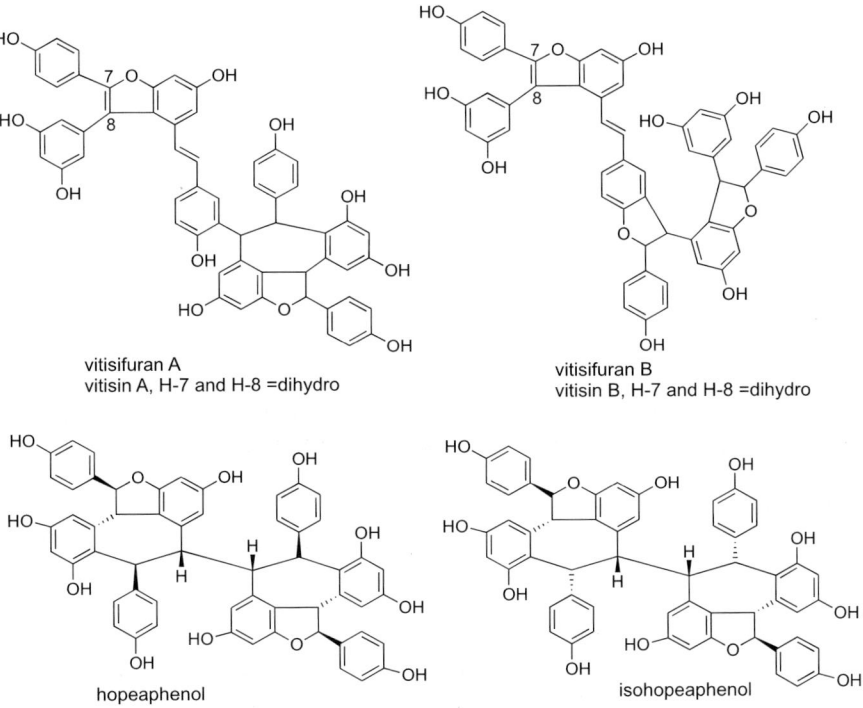

Fig. 2.6 Structure of the stilbene tetramer from *Vitis vinifera*

2.3.2 Biosynthetic Pathway

The immediate precursors of resveratrol are p-coumaroyl coenzyme A (CoA) and malonyl CoA in a molar ratio of 1:3. The latter is derived from elongation of acetyl CoA units and the former from phenylalanine, which can be synthesised from sugars via the shikimate pathway. Following oxidative deamination catalysed by phenylalanine ammonia lyase, phenylalanine is converted to cinnamic acid, which in turn is hydroxylated enzymatically to *p*-coumaric acid. In the final step, *p*-coumaroyl CoA is generated from the free coenzyme by a specific CoA ligase. The condensation of p-coumaroyl CoA with three molecules of malonyl CoA is accomplished through the activity of stilbene synthase, which leads to *trans*-resveratrol (Fig. 2.7). However, the exact biosynthetic formation of these derivatives is unknown.

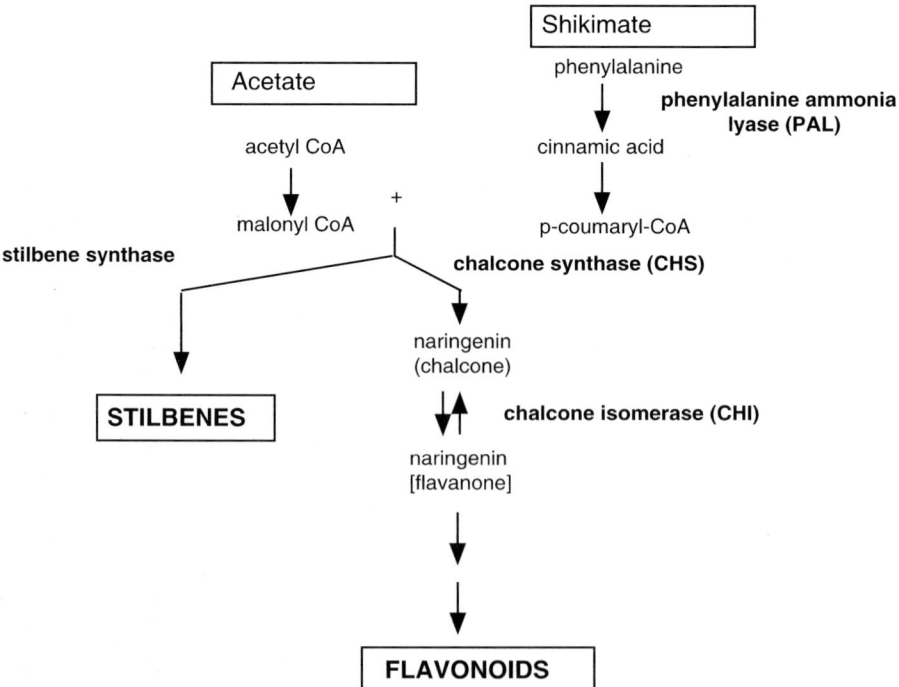

Fig. 2.7 General biosynthesis pathway of stilbenes and flavonoids

2.3.3 Distribution in Vitis vinifera

In *Vitis vinifera* varieties, several different hydroxystilbenes are present in several parts of the grape plant as constitutive compounds of the lignified organs (roots, canes, seeds, stems, ripe cluster stems), and as induced substances (in leaves and berries), probably acting as phytoalexins in the mechanisms of grape resistance against pathogens. Both resveratrol and piceid can be found in grape products, with the concentration of the glucoside usually being significantly higher than the aglycone [1].

In grape berries, stilbene synthesis is located primarily at the skin cells and it is absent or low in the fruit flesh [37, 38]. The greater part of the resveratrol in the skins is in both glycosidic forms (piceid isomers) [39], whereas pterostilbene is detected at very low levels in healthy and immature grape berries [40]. Romero-Perez et al. [41] observed that piceids are the major components of grape juices, averaging a total of 4 mg/l in red grape juices and 0.5 mg/l in white ones. Only *trans-* and *cis-* resveratrol were detected in the seeds.

In leaves infected by *Botrytis cinerea*, the main stilbenes detected [25] were *trans*-resveratrol (90 µg/g fresh weight), ε-viniferin (30 µg/g) and α-viniferin (20 µg/g). Resveratrol dehydrodimer and piceids were detected in leaves, induced by UV irradiation [42, 43].

The stilbenes usually found in *Vitis vinifera* stems are *trans-* and *cis-* resveratrol, piceatannol ε-viniferin [44], viniferol A, hopeaphenol and isohopeaphenol [45], viniferol B and C, vaticanol B and vaticaphenol A [46]. A tetramer, *cis-* and *trans*-vitisin B, has been isolated from the corks of *Vitis vinifera* [47].

Several stilbenes have been isolated from grape cell suspension cultures: resveratrol, piceid, resveratroloside and astringin, all in the two isomeric forms *cis-* and *trans* (Fig. 2.1) Recently, three new resveratrol diglucosides, *cis-* and *trans*-resveratrol 3,5-O-β-diglucoside [30] and *trans*-resveratrol 3,4'-O-β-diglucoside [29] (Fig. 2.1) have been isolated together with a new resveratrol triglucoside, *trans*-resveratrol 3,5,4'-O-β-triglucoside [30] (Fig. 2.1). Furthermore, resveratrol dimers have been identified, among them, *trans*-δ-viniferin together with *trans*-δ-viniferin 11- and 11'-O-β-glucoside and pallidol [33].

2.3.4 Determination of Stilbenes in Wine

In the first paper describing the occurrence of *trans*-resveratrol in wine, Siemann and Creasy (1992) [48] found very low concentrations in white and red wines (<1 mg/l). The highest concentrations were found by Mattivi (1993) [49] in Italian red wines: he recorded the highest concentrations in Cabernet Sauvignon (1.33–7.17 mg/l), Merlot (3.14–6.03 mg/l) and Pinot Noir (3.22–5.93 mg/l).

Resveratrol is now known to occur in wine in both free (*cis* and *trans*) and glycosidically bound forms. Free *trans-* and *cis*-resveratrol are present in a concentration range of 0.2–3 mg/l in red wines and 0.1–0.8 mg/l in white wines [50, 51]. Our study corroborates these studies, which show that red wines from various countries and regions have a low mean concentration of *trans-* and *cis-*resveratrol, less than 5 mg/l. Concerning *trans-* and *cis-*piceid, we found that their levels exceeded those of the free isomers and reached maximal concentrations of 26 mg/l and 24 mg/l in red wines, respectively [52]. Ribeiro de Lima et al. (1999) [53] even determined piceids in Portuguese red wines in concentrations up to 68 mg/l. The maximal value for *cis-* piceid in white wines was found to be 0.9 mg/l, that of the *trans-* isomer being 2.9 mg/l. *Trans-*astringin is sometimes found at higher concentrations than piceids in wines, some wines having concentrations over 30 mg/l. However, there were wide variations between the wine samples, and *trans-*astringin was not detected in all of them.

In addition to the monomers of stilbenes, some resveratrol dimers have been characterised from the wines: from German commercial white wines (Riesling), ε-viniferin diglucosides and pallidol mono- and di-glucosides have been identified at very low levels (<0.05 mg/l) [32]; from French commercial red wines, low levels (0.5–4.8 mg/l) of *trans-*ε-viniferin, parthenocissin A and pallidol have been isolated [54]; from commercial Brazilian red wines, *trans-*δ-viniferin has been found, but only in the youngest vintage (2002), with an average level of 11.7 mg/l [55].

As already reported [56], we found that total stilbene levels reached mean concentrations of up to 20 mg/l, often with a predominance of the glucoside isomers, depending on multiple factors such as grape cultivar, fungal pressure and climate.

For the first time, hopeaphenol (Fig. 2.6), a resveratrol tetramer, was identified in wines from North Africa. Furthermore, this molecule was quantified in ten commercial wines from North Africa by HPLC coupled to diode array detection (Fig. 2.8). The concentration was found to ranging between 0.3 and 3.8 mg/l [57]. Besides hopeaphenol, *trans-*resveratrol, pallidol and *trans-*ε-viniferin were also present in these wines [57].

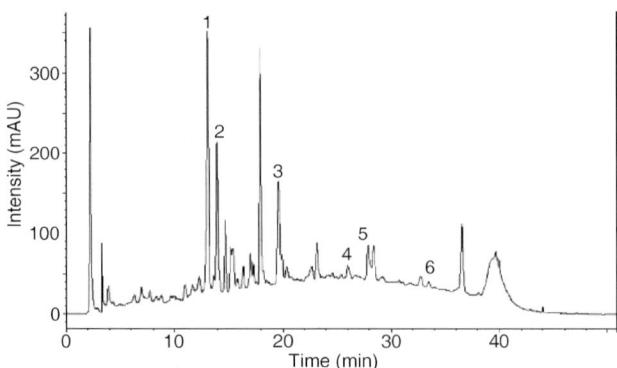

Fig. 2.8 Chromatographic profiles using diode array detection for a directly injected sample of merlot red wine. **1**, *trans-*piceid; **2**, astilbin; **3**, pallidol; **4**, *trans-*resveratrol; **5**, hopeaphenol; **6**, *trans-*ε-viniferin [57]

It has been shown that pallidol and some other resveratrol oligomers may be fungal metabolites of resveratrol [58]. Thus, the occurrence of this compound in wine is certainly due to the oxidation of resveratrol by fungus in infected berries used for vinification.

Red wines contain larger amounts of stilbenes than white wines, regardless of the oenological technology applied. The extent of maceration with skins and seeds during fermentation is the main factor determining the concentration of stilbenes in wines. They generally require long maceration on the skins to be extracted efficiently [38, 59].

2.4 Biological and Pharmacological Activities

Phenolic compounds like stilbenes have a wide range of pharmacological and biological actions in three major domains: cancer, cardiovascular disease and neurodegeneration. The response in humans, however, depends on their absorption and in vivo metabolism.

2.4.1 Bioavailability and Metabolism

Stilbenes are naturally occurring polyphenolic compounds that have been reported to have potential preventive activities in human diseases. Among these stilbenes, *trans*-resveratrol, which is found mainly in peanuts, grapes and red wine, is one of the most important in terms of biological activity, since it has been reported to exert anticarcinogenic, antioxidant and cardioprotective effects [2]. Resveratrol is used as traditional medicine in East Asian countries. In wine, it is present in small quantities compared to one of its glucosides, *trans*-piceid. However, little is known about the absorption and the bioavailability of these two stilbenes in humans. Indeed, their potential biological activities in vivo are dependent upon their absorption and subsequent access to the target tissues. Numerous studies have shown that dietary polyphenols are subjected to metabolic conversion not only in the liver, but also during their absorption in the intestine before reaching the systemic circulation.

Andlauer and co-workers [60] investigated the absorption of resveratrol using an isolated preparation of luminally and vascularly perfused rat small intestine. They showed that 46% of the luminally administered resveratrol was extracted by the small intestine and 21% appeared on the vascular side. The majority of the absorbed resveratrol were two glucuronides (80%). However, they did not identify the glucuronidation sites of this stilbene. In our laboratory, we identified two glucuronides of resveratrol using human liver microsomes, which corresponded to glucuronidation at positions 3 and 4' [61].

In order to investigate the transport and the metabolism of stilbenes in intestine, we used the human epithelial cell line Caco-2, which possesses intestinal enterocyte-like properties in vitro [62].

First, we examined the mechanisms of transport of *trans*-resveratrol and *trans*-piceid. The results demonstrated the uptake of these polyphenols across the apical membrane of Caco-2 cells, with a higher cellular accumulation in the cells for resveratrol than for piceid. Thus, we investigated a possible interaction of *trans*-piceid with the sodium-dependent glucose transporter (SGLT-1). The cellular accumulation of the stilbene was measured in the presence of different inhibitors of SGLT-1 (glucose, phlorizin or ouabain) and a decrease of uptake rate was observed. From these findings, the involvement of SGLT-1 in the absorption of *trans*-piceid has been deduced, whereas *trans*-resveratrol seems to use passive diffusion.

Second, we investigated the possibility that the multidrug-related protein (MRP)-2, an efflux pump present on the apical membrane, is involved in *trans*-resveratrol and *trans*-piceid efflux by Caco-2 cells. In the presence of MK-571 and indomethacin (MRP inhibitors), the rate of cellular accumulation of these compounds increased. This result was not observed in the presence of verapamil, a P-glycoprotein inhibitor. Thus, the effect in the presence of MK-571 seems to implicate MRP-2 as being responsible for the transcellular efflux of the two stilbenes [63].

The transepithelial transport of *trans*-piceid (apical to basolateral transport) was also measured and the apparent permeability coefficient during the 6 h of the experiment declined rapidly. This observation is likely to be attributable to the formation of metabolites.

Deglycosylation of some polyphenols has been observed during their absorption in rat small intestine [64]. In our study, after incubation of Caco-2 cells with *trans*-piceid, we not only detected *trans*-resveratrol on both the apical and basolateral sides, but also inside the cells. These results show that *trans*-piceid can be deglycosylated into *trans*-resveratrol. Moreover, by using protein extracts obtained from rat small intestine and inhibitors of the two glycosidases presents in the intestine, lactase phlorizin hydrolase (LPH), and cytosolic-ß-glucosidase (CBG), we confirmed that these two enzymes are implicated in the hydrolysis of *trans*-piceid.

This study shows clearly that the transepithelial transport of *trans*-piceid occurs at a high rate, and that this compound is deglycosylated by the two intestinal glucosidases LPH and CBG. There are two possible pathways by which *trans*-piceid might be hydrolysed in the intestine. The first is cleavage by CBG after passing the brush border membrane via the SGLT-1. The second is deglycosylation on the luminal surface of the intestinal epithelium by the membrane-bound enzyme LPH, followed by passive diffusion of the released aglycone, which is further metabolised in the cells in two glucuronides [65]. Figure 2.9 shows the proposed pathways for the intestinal absorption and metabolism of *trans*-piceid and its aglycone *trans*-resveratrol.

Some studies describe the bioavailability of labelled *trans*-resveratrol *in vivo*. First, Soleas and co-workers [66] used tritiated resveratrol and showed that

77–80% of this stilbene may be absorbed in the rat intestine. However, only trace amounts of radioactivity were detectable in liver, kidney, heart or spleen.

Second, in our laboratory we investigated the absorption and tissue distribution of ^{14}C-resveratrol following oral administration to mice [67]. The con-

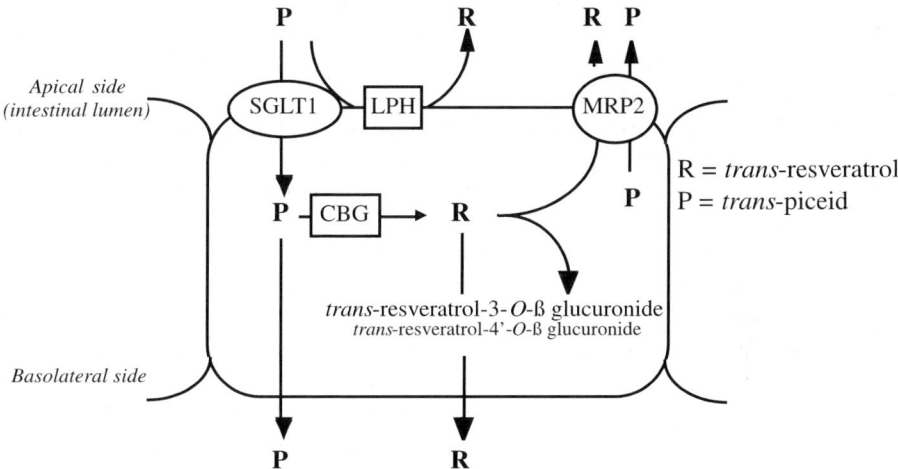

Fig. 2.9 The proposed pathways for intestinal absorption and metabolism of *trans*-piceid and its aglycone *trans*-resveratrol

Table 2.1 Radioactivity in digested organs of mice given a single oral dose of 7.4 kBq ^{14}C-resveratrol. Each value represents the mean ± SD ($n = 3$). Means within a row with different superscript letters (a or b) are significantly different from each other ($P < 0.05$). n.c. Not collected

Organ	Radioactivity (dpm/100 mg tissue)		
	1.5 h	3 h	6 h
Duodenum	2208 ± 1436[a]	1841 ± 183[a]	933 ± 201[a]
Colon	106 ± 80[a]	300 ± 83[b]	88 ± 26[a]
Liver	441 ± 90[a]	374 ± 48[a]	189 ± 28[b]
Kidney	342 ± 165[a]	552 ± 51[a]	263 ± 123[b]
Lung	n.c.	380 ± 148	n.c.
Spleen	n.c.	312 ± 62	n.c.
Heart	n.c.	210 ± 132	n.c.
Brain	n.c.	196 ± 47	n.c.
Testis	n.c.	193 ± 9	n.c.

[1] Means within a row with different superscript letters (a or b) are significantly different ($P < 0.05$).
[2] Each value represents the mean ± SDs (n = 3).
[3] Not collected.

centrations of drug-related radioactivity in the various organs are shown in Table 2.1. Three hours after administration, radioactivity was found in various organs, such as the brain, lung, heart, liver, kidney, spleen, duodenum, colon and testis. The highest concentration of radioactivity was found in the duodenum. The kidney was the next most heavily labelled organ, followed by the lung and the liver. Substantial activity was present in the colon and spleen, while moderate activity was present in the heart, testis and brain.

The low concentration of radioactivity in the colon suggests that the faecal route is a minor route for elimination. On the contrary, decreasing concentrations of radioactivity in the kidney over time indicates that renal excretion might be one of the major elimination routes for ^{14}C-labelled resveratrol. This observation is supported by the high concentrations recorded in urine.

We also found that the concentration of radioactivity in the whole blood was relatively low and constant during the experimental period (6 h). These observations are consistent with the results recorded earlier by Soleas et al. [66].

Marier and colleagues [68] studied the pharmacokinetics of *trans*-resveratrol following oral administration to rat (50 mg/kg). They determined resveratrol and glucuronide metabolite concentrations in plasma samples. They showed that resveratrol was bioavailable at 38% when administered in a solution of hydroxypropyl ß-cyclodextrin, and its systemic exposure was approximately 46-fold lower than that of these glucuronides. In the same year, Yu and co-workers [69] detected resveratrol glucuronide and resveratrol sulphate as the resveratrol metabolites in serum samples from mice after administration of 20 mg resveratrol/kg body weight. However, only traces of unconjugated resveratrol were observed.

Marier et al. [68] also assessed the enterohepatic recirculation of resveratrol by diverting the bile cannula from a bile-donor rat into the duodenum of a bile-recipient rat. The resveratrol was administered orally and plasma samples were collected. The plasma concentrations of resveratrol and glucuronides in bile-donor rats declined. Over the 4- to 8-h time period, significant plasma concentrations of resveratrol and glucuronides in bile-recipient rats were observed and coincided with the sudden peaks in plasma concentrations obtained in intact rats receiving intravenous or oral doses. The authors suggested that the enterohepatic recirculation contributes significantly to the overall systemic exposures of aglycone and glucuronide in rat.

More recently, some groups investigated the bioavailability of resveratrol in humans. Two of them evaluated the absorption after oral administration of 25 mg of unlabelled resveratrol [70] or ^{14}C-labelled resveratrol [71]. They obtained similar results for plasma concentrations. Golberg et al. [70] showed that plasma concentrations of conjugated resveratrol ranged from 416 to 471 µg/l, with no differences among the three matrices used (grape juice, white wine and V-8 homogenised vegetable cocktail), and peaked at 30 min after consumption. Walle et al. [71] determined a peak plasma resveratrol equivalent concentration (total radioactivity) of 491 µg/l at 1 h after the dose. In the two studies, the plasma concentration of free resveratrol aglycone was less than 2% of the total resveratrol concentration.

Walle et al. [71] also used an intravenous dose of 0.2 mg to assess the metabolism in the body plasma of the unchanged compound. They collected plasma samples at 10 min and 30 min after intravenous resveratrol infusion in three subjects. The 10-min samples contained a small amount of resveratrol (3.7–16.4 µg/l), and when examining the 30-min samples, two samples had no resveratrol, but sulphate conjugates were detected in these samples. This result suggests a rapid metabolism of this stilbene in the body. Most of the radioactivity after the oral doses was recovered from the urine (53–85%), but also in faeces (0.3–38%). For structure identification of resveratrol metabolites in the urine, a larger unlabelled dose (100 mg) was given to one subject. Five major metabolites were detected and identified by liquid chromatography/mass spectrometry: two resveratrol monoglucuronides, a dihydroresveratrol monoglucuronide, a resveratrol sulphate and a dihydroresveratrol sulphate. The sulphate conjugates excreted in the urine were the major metabolites.

It may be concluded from these studies that the absorption of resveratrol appears to be very high. However, the oral bioavailability of unchanged resveratrol is very low due to rapid and extensive metabolism. Indeed, in vitro and in vivo studies have demonstrated that resveratrol was metabolised to glucuronide and sulphate conjugates. These metabolites occur in the liver, but the rapid appearance of metabolites in plasma suggests that resveratrol is also partly metabolised in the small intestine [72]. Today, a question remains: how the resveratrol with its low bioavailability could present an undeniable in vivo efficacy?

2.4.2 Cardiovascular Protection

2.4.2.1 Antioxidant Activity

Free radicals derived from molecular oxygen, such as superoxide, hydroxyl, hydroperoxyl radicals and nitric oxide (NO), are constantly generated in vivo for specific metabolic purposes [73]. Free-radical concentrations are increased either by their overproduction or by a deficiency in antioxidant defence systems. The reactivity of radicals can cause severe damage to biological molecules, especially to DNA, lipids and proteins [73]. This damage probably contributes to the development of major chronic diseases including cancer, Parkinson's disease (PD), senile dementia and atherosclerosis.

There is increasing evidence that oxidised low-density lipoproteins (LDLs) may be responsible for promoting atherogenesis. Oxidised LDLs are a key component in endothelial injury. They may directly injure the endothelium and play an initial role in the increased adherence and migration of monocytes and lymphocytes into the subendothelial space [74]. Oxidised LDLs may bypass the normal tight control exercised by the classical LDL receptor in the macrophages and be endocytosed via non-regulated scavenger receptors. This rapid accumulation of cholesterol and cholesteryl ester leads to foam cell formation [75]. Hence, it is conceivable that oxidative stress accelerates atherogenesis by enhancing LDL oxidation and increasing its accumulation in foam cells. The presence of antioxidants can interfere with the peroxidation process by removing the alkoxyl or peroxyl radicals.

In in vitro studies with total phenolic compounds from red wine, Frankel et al. [76] recorded that red wine diluted 1000-fold, containing 10 µmol/l phenolics, inhibited human LDL oxidation significantly more than α-tocopherol. In our laboratory, we have studied the antioxidant activities of the pure stilbenes found in wine [28]. Co-existence of an antioxidant, A, and a free radical, R°, such as the reactive oxygen species generated by an oxidative stress, or 1,1 diphenyl-2-picryl-hydrazyl (DPPH) leads to the disappearance of this free radical and to the apparition of the free radical A°, according to the reaction: $A + R° ==> A° + R$. On Cu^{2+}-induced lipid peroxidation on the LDL, we showed that three stilbenes (*trans*-resveratrol, astringin and piceatannol) were more efficient than Trolox (the water-soluble vitamin E analogue). On the stilbenes studied, the conjugation between rings A and B via a planar C2 unsaturated structure allows an electron delocalisation across the molecules for stabilisation of the radical that appeared, which explains the relative antioxidant properties of all these compounds. No important difference was found between the *trans* and *cis* structures of each molecule, except for resveratrol, with a better activity for *trans*-resveratrol. The glycosylation of *trans*-stilbenes reduces their activity when compared to the corresponding aglycones; this difference is less important on *cis* structure. It is worthy to note the importance of the two hydroxyl groups in the orthodiphenolic arrangement in the B ring. Actually, astringin, which possesses this catechol structure and consequently a supplementary OH on the B ring as compared to piceid, has an activity six times higher for the *trans* and *cis* structures. Among these molecules, the most potent antioxidant is piceatannol, which possesses four hydroxyl groups, including the catechol structure in the B ring. Furthermore, piceatannol is two times more efficient than Trolox. On DPPH, no important difference is found between *trans* and *cis* structures of each molecule. The glycosylation reduces their activity when compared to the corresponding aglycones. *Trans*-astringin and piceatannol have the best activity.

Polyphenols such as resveratrol can also act by another mechanism (i.e. their complexation with metal ions, for example iron or copper), which are involved in the generation of free radicals and lipid peroxidation [77]. Moreover, α-tocopherol (the principal form of vitamin E), which functions as a major antioxidant in human LDLs, can be recycled from its free radical form (α-tocopheryl) by a phenolic compound. Resveratrol can also prevent the initial events of atherosclerosis in endothelial cells by inhibition of the enzymatic systems producing reactive oxygen species such as NADPH oxidase and hypoxanthine/xanthine oxidase, and by inhibition of both the expression of adhesion molecules and the monocyte adhesion to endothelial cells [78].

Red wine polyphenols are also effective in reducing oxidative damage on normal human red blood cells in vitro [79]. In fact, erythrocytes are particularly exposed to oxidative damage because of the oxygen carrier in the presence of high polyunsaturated fatty acid content on their membranes and high cellular concentration of haemoglobin.

2.4.2.2 Antithrombotic and Vasoprotective Properties

It was observed that de-alcoholised red wine containing total phenolics inhibits platelet aggregation induced by ADP and thrombin human [80]. This report suggested that *trans*-resveratrol is particularly active by inhibiting the synthesis of certain eicosanoids. We screened various stilbenes obtained from *Vitis vinifera* cell cultures for their antiplatelet properties on human platelet-rich plasma in which aggregation was induced either with arachidonic acid (AA), ADP, collagen or the thromboxane A2 mimetic U-46619 [81]. Significant inhibition of AA-induced aggregation was observed for the free hydroxylated compounds (*trans*- and *cis*-resveratrol, and piceatannol), with IC_{50} values ranging from 17 to 38 µM, similar to that of acetylsalicylic acid (25 µM). In contrast, the corresponding glycosylated derivatives were weakly active or devoid of significant activity. These results are in agreement with other reported data [82, 83]. On the other hand, only the free hydroxylated stilbenes showed an antiplatelet activity when collagen or ADP was used as a stimulating agent, but at much higher concentrations. In addition, all derivatives were found to be inactive when U-46619 was tested as an aggregation inducer. Based on the findings, the anti-aggregating profile observed for stilbenes clearly points to an interaction of the compounds with the platelet AA pathway through a mechanism consistent with an inhibition of cyclooxygenase. Moreover, *trans*-resveratrol was found to inhibit calcium influx in thrombin-stimulated human platelets [84]. Indeed, the increase in free cytosolic calcium concentration is the major intracellular stimulus involved in platelet aggregation. Moreover, other results [85] showed that the inhibition of platelet activation by resveratrol may correlate with partial modification of flux in the phosphoinositide cycle and a decrease in phosphatidylinositol 4,5-bisphosphate available for signalling in the cells.

Significant inhibition of AA-induced aggregation was also observed for a tetramer, vitisin A, with an IC_{50} value lower than that of acetylsalicylic acid, but ε-viniferin (a dimer) had no inhibitory effect [86].

Several studies have shown that the extent of coronary artery stenosis due to atherosclerotic plaque formation and expansion into the arterial lumen is not sufficient to explain the incidence of clinical events associated with atherosclerosis [87]. It appears that the generation of clinical events involves plaque rupture, resulting in thrombus formation and arterial occlusion. This rupture is induced by vasomotor disturbances in which oxidised low-density lipoproteins may be involved [87].

Resveratrol is able to regulate vasomotion, which is impaired in atherosclerosis. The key regulators of the vasomotor function are the vasodilatator NO and the vasoconstrictor endothelin-1 [78]. It was shown that red wine and other grape products that contain polyphenols can induce endothelium-dependent vasorelaxation in the rat isolated aorta, an action that appears to be mediated by the NO–cGMP pathway [88, 89]. Resveratrol enhances the expression and activity of endothelial NO synthase [90], and inhibits endothelin-1 secretion and endothelin-1 gene expression in human umbilical vein endothelial cells [91]. Moreover, red wine polyphenols, in particular *trans*-resveratrol, have been shown to inhibit the proliferation and the migration of vascular smooth muscle cells in the intima [92].

Taken together, these in vitro biological activities of polyphenols such as *trans*-resveratrol, (antioxidant, anti-atherogenic, anti-thrombotic, vasorelaxant and anti-hypertensive) could explain the beneficial effects of wine in the prevention of cardiovascular disease [93].

2.4.2.3 Biological Activities after Ingestion of Polyphenols or Wine

Some *in vivo* studies have been carried out in animals and human volunteers in order to show these protective effects after wine or pure compound consumption. Klurfeld and Kritchewsky [94] found that red wine notably reduced coronary atherosclerosis in rabbit. Using a hamster model of atherosclerosis, Auger and co-workers [95] showed that the aortic fatty streak area was significantly reduced (76%) in the group receiving resveratrol, at a level mimicking a moderate consumption of red wine. Intravenous and intragastric administration of red wine, grape juice and not white wine inhibited in vivo platelet activity and thrombosis in canine coronary arteries [96]. After 2–4 months of beverage consumption, rats exhibited a reduction in platelet aggregation at the same rate by alcohol, red wine and white wine [97]. By contrast, only red wine did not result in a rebound effect on platelets after deprivation of the alcoholic beverage. This protective effect appears to be essentially associated with polyphenols, which could counteract the known increased lipid peroxidation observed in association with alcohol drinking. Intragastric administration of resveratrol for 12 weeks to hypercholesterolaemic rabbits improved the endothelial function, reduced plasma endothelin-1 levels and induced a significant elevation in NO levels [93, 98]. Numerous data in animals suggest strongly that piceatannol and in particular resveratrol might protect against ischaemic damage during myocardial infarction and brain damage following cerebral ischaemia [2].

In human experimental studies, the ingestion of red wine led to an increase in the serum antioxidant activity, which peaked after 1–2 h [99, 100]. The daily consumption of red wine for 2 weeks resulted in a reduction in the susceptibility of LDLs to oxidation [101–103], as well as a decrease in platelet aggregation and an increase in high-density-lipoprotein-cholesterol [104, 105]. In these studies, white wine exhibited no (or negative) activity. However, these results could not be reproduced by other authors [106–109].

It was also reported that coronary flow-velocity reserve and flow-mediated dilatation of the brachial artery increased specifically after the intake of red wine by volunteers, certainly due to the improvement of endothelial function and the vasorelaxant effects of polyphenols [110, 111]. Similar results were obtained in patients with coronary heart disease who ingested a red grape polyphenol extract containing mainly flavanols and phenolic acids, but also *trans*-resveratrol and ε-viniferin [112].

2.4.3 Cancer Chemoprevention

The term "chemoprevention" can be defined as the ingestion of non-toxic quantities of chemical agents (dietary or pharmaceutical) that are capable of preventing, inhibiting or reversing the process of carcinogenesis [113]. This relatively new concept is based on epidemiological studies that suggest a strong link between the environment (excluding genetic susceptibility) and cancer. There is evidence that 50–80% of human cancer is potentially preventable. In the United States, approximately 35% of cancer deaths are attributable to variation in diet. Indeed, carcinogenesis is a multistage process, and tumour development can occupy an important portion of the life span of an individual following long exposure to exogenous factors (such as dietary factors), which largely determine the incidence of carcinogenesis [114, 115]. This offers numerous opportunities for intervention before malignant tumours develop.

The stages of carcinogenesis include [116, 117]:
1. Initiation: this results from the exposure of normal cells to carcinogenic agents, which cause genetic change. Spontaneous mutations also occur through endogenous mechanisms such as free radical production, or through DNA replication errors, even though there may be no exposure to carcinogenic agents.
2. Promotion: selective clonal expansion of initiated cells, which leads to the appearance of a benign tumour.
3. Progression: genetic change and conversion of the tumour from benign to malignant.

Chemopreventive agents can act by various mechanisms on this process [115, 116]:
1. Anti-initiating activities: inhibition of carcinogen formation in the body and of uptake; inhibition of the metabolic activation of carcinogens by phase I enzymes, such as cytochrome P450 enzymes, or increase in their detoxification by phase II enzymes, such as transferases, leading to an easier excretion; scavenging of free radicals and trapping ultimate carcinogens, preventing their interactions with DNA. These compounds are referred to as "blocking agents" due to their ability to prevent initiation. Other compounds that inhibit the carcinogenic process after initiation are classified as "suppressing agents".
2. Anti-promoting and anti-proliferative activities: antioxidant activities because free radicals are also involved in the phases of promotion and progression; anti-inflammatory effects by inhibition of AA metabolism, the inhibitors of cyclooxygenase and lipoxygenase being considered as inhibitors of

tumour promotion; inhibition of cell proliferation by modulation of signal transduction, inhibition of polyamine metabolism, inhibition of oncogene activities, induction of differentiation, increase in intercellular communications, increase in apoptosis or inhibition of angiogenesis.

Clifford and co-workers [118] reported that a diet with red wine solids rich in phenols delayed the onset of tumours in transgenic mice that spontaneously develop externally visible tumours without carcinogen pre-treatment. Red wine polyphenols administered to rats with the diet also inhibited colon carcinogenesis induced by chemical compounds, and a significant decrease in the basal level of DNA oxidative damage to the colon mucosa was observed in rats not treated with carcinogens [119]. The study of tumorigenesis in a two-stage mouse skin cancer model showed that topical application of resveratrol reduced the number of skin tumours per mouse by up 98% and lowered drastically the percentage of mice with tumours [120, 121]. Since Jang's famous paper [120], systemic administration of resveratrol has been shown to inhibit the initiation and the development of tumours in about 30 rodent cancer models, but there are a few exceptions in which no benefit has been found [2]. The other results [122] show the efficacy of low doses of resveratrol in rat model of colon carcinogenesis, suggesting that the concentration present in a food sources such as red wine, could be active [2].

In addition, several reports indicate that *trans*-resveratrol inhibits the proliferation of a wide variety of tumour cells (e.g., see [123–127]). Among the other stilbenes, piceatannol, α- and ε-viniferin, hopeaphenol, pallidol, ampelopsin-A, vaticanol B and C, and pterostilbene also exert cytotoxicity and/or anti-proliferative effects on different tumour cell lines [128–136]. We studied the effects of grapevine stilbenes on cultured human liver myofibroblasts [137]. Liver myofibroblasts are major actors in the development of liver fibrosis and cancer progression. *Trans*-resveratrol markedly reduced the proliferation and migration of myofibroblasts. It can also deactivate human liver myofibroblasts. Other stilbenes such as *trans*-piceid and *cis*-resveratrol were ineffective. Moreover, *trans*-resveratrol blocks hepatocyte growth-factor-induced invasion of hepatocellular carcinoma cells by an unidentified post-receptor mechanism [138].

The cancer chemopreventive activity of *trans*-resveratrol was established in various assays reflecting the three major stages of carcinogenesis [120, 139, 140].

To find new cancer chemopreventive agents in wine, we have evaluated the inhibitory effect of grape stilbenes on carcinogen-induced preneoplastic lesions in mouse mammary gland organ culture, which is a relevant model with which to evaluate the efficacy of potential chemopreventive agents [141]. Moreover, these polyphenols have been evaluated for their potential to inhibit cyclooxygenase. Of the two isoenzymes that lead to the formation of prostaglandins, cyclooxygenase-1, which is constitutively expressed in most tissues, is considered to be involved in physiological cell–cell signalling, whereas cyclooxygenase-2, which is induced by specific events in a limited number of cell types, appears to be involved in inflammation and mitogenesis [142]. There is now evidence for a strong link between chronic inflammation and cancer [143]. Indeed, several pro-inflammatory gene products have been identified that play a critical mediating role in the suppression of apoptosis, proliferation, angiogenesis, invasion and metastasis. Among these gene products are interleukins, chemokines, tumour necrosis factor, metalloproteinases, vascular endothelial growth factor, inducible NO synthase, lipoxygenase and cyclooxygenase-2. The expression of all of these genes is regulated mainly by the transcription factors nuclear factor-κB (NF-κB) and activation protein-1 (AP-1), which are constitutively active in most tumours and are induced by carcinogens [144]. In our study, *trans*-astringin, *trans*-resveratrol and *trans*-piceatannol exhibit significant inhibition of 7,12-dimethylbenz[a]anthracene-induced preneoplastic lesions in mouse mammary gland organ cultures [145]. We showed in this investigation that two structural criteria appear to be important for this activity of the stilbenoids (i.e. the presence of *trans* geometrical isomerism and the absence of glycosylation – except astringin). The mechanism by which polyphenolic compounds inhibit carcinogenesis has not been clearly established. *Trans*-resveratrol directly inhibits cyclooxygenase activity. However, *cis*-resveratrol is also active against this enzyme, but exhibited no discernable activity on a carcinogen-induced preneoplastic lesion model. Thus, it seems that activity against the cyclooxygenase-2 target is not sufficient to explain the action of *trans*-resveratrol, so other mechanisms might have involved. Moreover, our data show that *trans*-astringin and its aglycone *trans*-piceatannol inhibit the induction of preneoplastic lesions without any apparent activity against the cyclooxygenase-2 enzyme. These two stilbenoids could thus act by another mechanism. Interestingly, in contrast to *trans*-resveratrol, they are inactive against physiological cyclooxygenase-1, the inhibition of which may lead to side effects such as gastric lesions and renal toxicity [142]. These results suggest that *trans*-astringin and *trans*-piceatannol are attractive new candidates for cancer chemoprevention. Moreover, two stilbene dimers isolated from grape cell culture, *trans*-δ-viniferin and its 11-O-β-D-glucopyranoside, demonstrated strong cyclooxygenase-2 inhibitory activity, but specificity is lacking in both cases [33].

Stilbenes seem to act by several specific mechanisms. Indeed, resveratrol may inhibit carcinogenesis by affecting the molecular events in the three stages [117]:
1. The anti-initiation activity was demonstrated by its antioxidant and antimutagenic effects, inhibition of carcinogen bioactivation, induction of phase II drug-metabolising enzymes and the stimulation of DNA repair.
2. The anti-promotion activity was shown by its blocking action of the stimuli-mediated mitogen-activated protein kinase pathway activation, inhibition of polyamine synthesis and increase of polyamine catabolism, and inhibition of the production of pro-inflammatory mediators via cyclooxygenase-2 (activity and expression) and lipoxygenase pathways. Resveratrol also may inhibit NF-κB and AP-1 activation [146]. Resveratrol can arrest the cell cycle by a blockage, which varies with cell type, in G1, S, or G2/M phase, and induce apoptosis through p53-dependent mechanism in several cancer cell lines. We described a cell-surface resveratrol receptor on the extracellular domain of integrin αVβ3 in breast cancer cells. Binding of resveratrol to in-

tegrin, principally to the β3 monomer, was essential for transduction of the stilbene signal into p53-dependent apoptosis of these cells [147]. Evidence also suggests that resveratrol acts through a p53-independent mechanism in some cell types [146].
3. The inhibition of progression/invasion was also reported. Resveratrol can affect the expression of the inducible NO synthase gene, which is partly controlled by NF-κB, and thus reduce the abnormal level of NO that contributes to inflammation and angiogenesis. Moreover, resveratrol can inhibit angiogenesis through the inhibition of the necessary polyamines, vascular endothelial growth factor, and of the expression of adhesion molecules and matrix metalloproteinases (also involved in tumour invasion and metastasis). In vivo studies show the inhibition of tumour-induced neovascularisation and metastasis by resveratrol delivered systemically [2, 148, 149].

Resveratrol can be converted to piceatannol by cytochrome P450 enzymes present in liver [150] or overexpressed in a wide variety of human tumours [151]. Piceatannol has a known antileukaemic activity and is also a tyrosine kinase inhibitor [152]. Protein-tyrosine kinases are important mediators of a variety of mitogenic signalling pathways, including those associated with several growth factors. Aberrant or overexpressed protein-tyrosine kinases are associated with several cancers [153]. Piceatannol can also induce apoptosis in human tumour cell lines [136], inhibit the lipopolysaccharide-induced production of critical mediators of the inflammatory response such as interleukins and tumour necrosis factor in different models [154, 155], and suppress NF-κB activation induced by various inflammatory agents [156]. Moreover, picetannol has antimetastatic activities, which might be due to the inhibition of angiogenesis [129].

With regard to the other stilbenes, ε-viniferin displayed a more potent inhibitory effect than resveratrol on human cytochrome P450 enzymes involved in bioactivation of numerous carcinogens [157]. ε-Viniferin also possesses anti-inflammatory properties [158], as do vaticanol B, α-viniferin, vitisin A, vitisifuran A and hopeaphenol [134, 159]. An orally administered extract containing bergenin, hopeaphenol, vaticanol B and C, and ε-viniferin, exhibits an antitumoral effect against subcutaneously allografted sarcoma in mice [132]. Pterostilbene possesses the ability to induce apoptosis in leukaemia cells [135].

Concerning resveratrol, phase I and II clinical trials are currently underway in the USA for patients with colon cancer.

2.4.4 Neurodegenerative Diseases

For demographic reasons, the percentage of neurodegenerative diseases is on the increase. These disorders, which result from the deterioration of neurons, are classified into two classes: (1) movement disorder pathologies like in PD; (2) cognitive deterioration pathologies and dementia like in AD.

Resveratrol promotes anti-ageing effects in numerous organisms. It modulates pathomechanisms of debilitating neurological disorders, such as ischaemia, Huntington's disease (HD), PD and AD [160]. In rat hippocampal neurons, resveratrol inhibits voltage-activated potassium currents, suggesting that it is useful for treating ischaemic brain injury [161]. In midbrain dopaminergic neurons, resveratrol protects neuron cultures against several type of insults related to PD pathogenesis, like the cytotoxic effects induced by 1-methyl-4-phenyl pyrimidium, sodium azide, thrombin and DNA damage [162]. In HD, resveratrol rescued mutant polyglutamine-specific cell death in neuronal cells derived from HdhQ111 knock-in mice and from transgenic *Caenorhabditis elegans*, both of which are models for HD [163]. In a gerbil ischaemia model, administration of resveratrol during the early stage of cerebral ischaemia protected against neuronal death in the hippocampal CA1 area, and concomitantly inhibited glial cell activation [164]. Moreover, in this in vivo model, results showed that resveratrol, after formation of glucuronide conjugates, enters the bloodstream and can cross the blood–brain barrier. The concentration of resveratrol required to achieve neuroprotective actions was in range of 10–100 µM. Different mechanisms, such as antioxidative actions and regulation of gene transcription, may be involved in the protective actions of resveratrol.

Nevertheless, the major part of neurodegenerative pathological studies was focussed on AD. Indeed, AD is the most common type of neurodegenerative disorder, accounting for 65% of all dementias, with the prevalence estimated to be between 1 and 5% among people aged 65 years, which doubles every 4 years to reach about 30% at 80 years [165]. Histopathology reveals that one of the major characteristics of AD is the abundant protein deposit in selected neurons. These deposits are the result of the extracellular accumulation of amyloid-β (Aβ) peptide [166]. Aβ originates from proteolytic cleavages of the transmembrane amyloid precursor protein (APP) [167]. APP can be cleaved by different proteases, called α-, β- and γ-secretases. Whereas the α-secretase cleaves APP into a non-toxic amyloid form, β-secretase followed by γ-secretase cleave the APP to form the mature peptide. Aβ accumulation leads to formation and deposition of senile plaques and neurofibrillary tangles, which promote pro-inflammatory responses and activate neurotoxic pathways, leading to the dysfunction and death of brain cells [168].

Different studies have investigated the effects of resveratrol and some others stilbenes on AD, suggesting that stilbenes could modulate multiple mechanisms of AD pathology [169]. In PC12 cells, resveratrol has been shown to protect against the Aβ peptide-induced toxicity by influencing apoptotic signalling pathways, reducing changes in mitochondrial membrane potential and inhibiting the accumulation of intracellular reactive oxygen intermediates [170]. In APP695-transfected cell lines, resveratrol from 20 µM could reduce the secretion of Aβ. The treatment of cells with selective proteasome inhibitors, lactacystin, Z-GPFL-CGO or YU1001, significantly blocked the resveratrol-induced decrease of Aβ [171]. These findings demonstrate a proteasome-dependent anti-amylogenic activity of resveratrol. In hippocampal primary neurons, resveratrol significantly attenuated Aβ-induced cell death in a concentration-dependent manner. Results indicate that the protein kinase C pathway is involved in the neuroprotective action of resveratrol [172]. In this study, resveratrol was used as pre-, co- and post-treatment; in all cases, resveratrol exhibited its neuroprotective effects. Jeon et al. (2006) [173] reported that resveratrol, oxyresveratrol and scirpusin A have a potent inhibitory activity on β-secretase. Among the secretases, β-secretase is an attractive target for the inhibition of amyloid production. Taken together, these results suggest that resveratrol and other stilbenes protect neurons against Aβ-induced injuries.

We investigated the potentially inhibitory activity of various stilbenes on Aβ aggregation [174]. Indeed, it was demonstrated that Aβ aggregates have neurotoxic effects in cell culture and in vivo [167]. Thus, finding molecules to prevent the aggregation of Aβ could be of therapeutic value in AD pathology. We have compared the inhibitor Aβ polymerisation activity of various stilbenes like resveratrol, piceid, resveratrol diglucoside, piceatannol, astringin and ε-viniferin with curcumin, an anti-amyloidogenic polyphenol [175], as a control (Fig. 2.10). The anti-amyloidogenic activity of the molecules studied is in the following order: resveratrol ≈ piceid > curcumin > diglucoside≈astringin≈piceatannol≈ε-viniferin. These results showed that resveratrol and piceid could be effective anti-amyloidogenic molecules, whereas bulk structures like ε-viniferin might be less active. Moreover, piceatannol, which differs from resveratrol in only one hydroxyl group, exhibits less inhibitor activity. This could be due to a specific binding of resveratrol with free Aβ. Thus, resveratrol and piceid could be important molecules for therapeutic development.

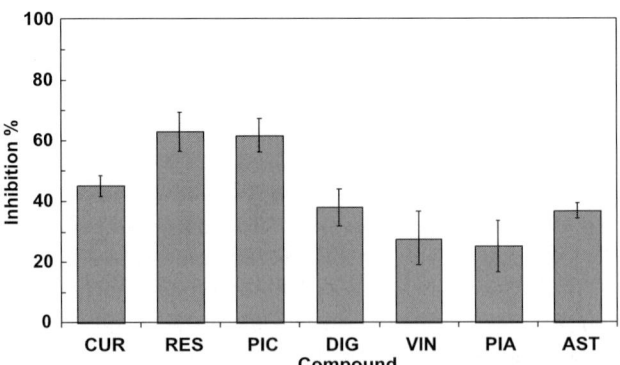

Fig. 2.10 Polyphenol amyloid-β fibril inhibition compared to that of curcumine. Means and standard deviation of three independent measurements are shown. *CUR* curcumin, *RES* resveratrol, *PIC* piceid, *DIG* diglucoside, *VIN* ε-viniferin, *PIA* piceatannol, *AST* astringin

2.5 Conclusion

Vine and wine are abundant sources of polyphenolic compounds, including mainly flavonoids and stilbenes. The latter appear to constitute a large class of compounds and exhibit potent biological activities in vitro on several targets. *Trans*-resveratrol, the most studied stilbene, shows great promise in the treatment of leading diseases. Resveratrol acts through multiple pathways on the same pathology, such as cancer or cardiovascular diseases. This seems to suggest a cooperative action. Baur and Sinclair (2006) [2] give two possible explanations: (1) the similarity of resveratrol with an endogenous signalling molecule like, for example, oestrogen, or (2) the "xenohormesis hypothesis", which proposes that organisms have evolved to respond to chemical cues in their diets. Could resveratrol and similar molecules form the next class of wonder-drugs? [2].

References

1. Waterhouse AL, Teissèdre PL (1997) Levels and phenolics in California varietal wines In: Watkins TR (ed) Wine Nutritional and Therapeutic Benefits. ACS Symp. Series 661, Washington, p 12
2. Baur JA, Sinclair DA (2006) Nat Rev Drug Discovery 5:493
3. Blot WJ, McLaughlin JK, Winn DM, Austin DF, Greenberg RS, Preston-Martin S, Bernstein L, Schoenberg JB, Stemhagen A, Fraumeni JJF (1988) Cancer Res 48:3282
4. Macfarlane GJ, Zheng T, Marshall JR, Boffetta P, Niu S, Brasure J, Merletti F, Boyle P (1995) Eur J Cancer 31B:181
5. Longnecker MP, Newcomb PA, Mittendorf R, Greenbreg ER, Clapp RW, Bogdan GF, Baron J, MacMahon B, Willett CW (1995) J Natl Cancer Inst 87:923
6. Farchi G, Fidanza F, Mariotti S, Menotti A (1992) Int J Epidemiol 21:74
7. Rimm EB, Klatsky A, Grobbee D, Stampfer MJ (1996) Br Med J 312:731
8. Renaud S, De Lorgeril M (1992) Lancet 339:1523
9. Renaud S, Guégen R, Schenker J, D'Houtaud A (1998) Epidemiology 9:184
10. Renaud S, Guéguen R, Siest G, Salomon R (1999) Arch Intern Med 159:1865
11. Renaud SC, Guéguen R, Conard P, Lanzmann-Petithory D, Orgogozo JM, Henry O (2004) Am J Clin Nutr 80:621
12. Gronbaek M, Deis A, Sorensen T, Becker U, Schnohr P, Jensen G (1995) Br Med J 310:1165
13. Gronbaek M, Becker U, Johansen D, Gottschau A, Schnohr P, Hein HO, Jensen G, Sorensen T (2000) Ann Intern Med 133:411
14. Gronbaek M, Becker U, Johansen D, Tonnesen H, Jensen G, Sorensen T (1998) Br Med J 317:844
15. Orgogozo JM, Dartigues JF, Lafont S, Letenneur L, Commenges D, Salamon R, Renaud S, Breteler MB (1997) Rev Neurol 153:185
16. Deng J, Zhou DHD, Li J, Wang YJ, Gao C, Chen M (2006) Clin Neurol Neurosurg 108:378
17. Commenges D, Scotet V, Renaud S, Jacqmin-Gadda H, Barberger-Gateau P, Dartigues JF (2000) Eur J Epidemiol 16:357
18. Johansen D, Friis K, Skovenborg E, Gronbaek M (2006) BMJ 332:519

19. Hart JH (1981) Rev Phytopathol 19:437
20. Sotheeswaran S, Pasupathy V (1993) Phytochemistry 32:1083
21. Mattivi F, Reniero F, Korhammer S (1995) J Agric Food Chem 43:1820
22. Goldberg DM, Ng E, Karumanchiri A, Diamandis EP, Soleas GJ (1996) Am J Enol Vitic 47:415
23. Trela BC, Waterhouse AL (1996) J Agric Food Chem 44:1253
24. Pezet R, Pont V, Cuenat P (1994) J Chromatogr A 663:19
25. Langcake P, Comford CA, Pryce RJ (1979) Phytochemistry 18:1025
26. Waterhouse AL, Lamuela-Raventos RM (1994) Phytochemistry 37: 571
27. Waffo-Téguo P, Decendit A, Vercauteren J, Deffieux G, Mérillon JM (1996) Phytochemistry 42:1591
28. Waffo-Téguo P, Fauconneau B, Deffieux G, Huguet F, Vercauteren J, Mérillon JM (1998) J Nat Prod 61:655
29. Decendit A, Waffo-Téguo P, Richard T, Krisa S, Vercauteren J, Monti JP, Deffieux G, Mérillon JM (2002) Phytochemistry 60:795
30. Larronde F, Richard T, Delaunay JC, Decendit A, Monti JP, Krisa S, Mérillon JM (2005) Planta Med 71:888
31. Langcake P, Pryce RJ (1977) Experientia 13:151
32. Baderschneider B, Winterhalter P (2000) J Agric Food Chem 48:2681
33. Waffo-Téguo P, Lee D, Cuendet M, Mérillon JM, Pezzuto JM, Kinghorn AD (2001) J Nat Prod 64:136
34. Ito J, Tayaka Y, Niwa M, Oshima YA (1999) Tetrahedron 55:2529
35. Jean-Denis JB, Pezet R, Tabacchi R (2006) J Chromatogr A 1112:263
36. Ito J, Niwa M, Oshima YA (1997) Heterocycles 45:1809
37. Creasy LL, Coffee M (1988) J Am Soc Hort Sci 113:230
38. Jeandet P, Bessis R, Maume BF, Meunier P, Peyron D, Trollat P (1995) J Agric Food Chem 43:316
39. Roggero JP (1997) Biofactors 6:441
40. Pezet R, Pont V (1988) Plant Physiol Biochem 26:603
41. Romero-Pérez AI, Ibern-Gomez M, Lamuela-Raventos RM, De la Torre-Boronat MC (1999) J Agric Food Chem 47:1533
42. Jeandet P, Breuil AC, Adrian M, Weston LA, Debord S, Meunier P, Maune G, Bessis R (1997) Anal Chem 69:5172
43. Douillet-Breuil AC, Jeandet P, Adrian M, Bessis R (1999) J Agric Food Chem 47:4456
44. Bourhis M, Théodore N, Weber JF, Vercauteren J (1996) Isolation and identification of (Z) and (E)-e-viniferins from stalks of Vitis vinifera. In: Vercauteren J, Chèze C, Dumon MC, Weber JF (eds) Polyphenols Communications 96. INRA, Groupe Polyphénols, Bordeaux, France, p 43
45. Yan KX, Terashima K, Takaya Y, Niwa (2001) Tetrahedron 57:2711
46. Yan KX, Terashima K, Takaya Y, Niwa (2002) Tetrahedron 58:6931
47. Ito J, Niwa M (1996) Tetrahedron 30:9991
48. Siemann EH, Creasy LL (1992) Am J Enol Vitic 43:49
49. Mattivi F (1993) Z Lebensm Unters Forsch 196:522
50. Lamuela-Raventos RM, Romero-Pérez AN, Waterhouse AL, de la Torre-Boronat MC (1995) J Agric Food Chem 43:281
51. Goldberg DM, Ng. E, Karumanchiri A, Yan J, Diamandis EP, Soleas GJ (1995) J Chromatogr A 708:89
52. Vitrac X, Castagnino C, Delaunay JC, Waffo-Téguo P, Vercauteren J, Deffieux G, Mérillon JM (2001) J Agric Food Chem 49:59
53. Ribeiro de Lima MT, Waffo-Téguo P, Teissèdre PL, Pujolas A, Vercauteren J, Cabanis JC, Mérillon JM (1999) J Agric Food Chem 47:2666
54. Vitrac X, Monti JP, Vercauteren J, Deffieux G, Mérillon JM (2002) Anal Chim Acta 458:103

55. Vitrac X, Bornet A, Vanderlinde R, Valls Josep, Richard T, Delaunay JC, Mérillon JM, Teissèdre PL (2005) J Agric Food Chem 53:5664
56. Carando S, Teissedre PL, Waffo-Téguo P, Cabanis JC, Deffieux G, Mérillon JM (1999) J Chromatogr A 64:748
57. Guebailia HA, Chira K, Richard T, Mabrouk T, Furiga A, Monti JP, Delaunay JC, Mérillon JM (2006) J Agric Food Chem 54:9559
58. Cichewicz RH, Kouzi SA, Hamann MT (2000) J Nat Prod 36:29
59. Pezet R, Cuenat P (1996) Am J Enol Vitic 47:287
60. Andlauer W, Kolb J, Siebert K, Furst P (2000) Drugs Exp Clin Res 26:47
61. Aumont V, Krisa S, Battaglia E, Netter P, Richard T, Mérillon JM, Magdalou J, Sabolovic N (2001) Arch Biochem Biophys 393:281
62. Artursson P, Karlsson J (1991) Biochem Biophys Res Commun 175:880
63. Henry C, Vitrac X, Decendit A, Ennamany R, Krisa S, Mérillon JM (2005) J Agric Food Chem 53:798
64. Day AJ, Gee JM, DuPont MS, Johnson IT, Williamson G (2003) Biochem Pharmacol 65:1199
65. Henry-Vitrac C, Desmoulière A, Girard D, Mérillon JM, Krisa S (2006) Eur J Nutr 45:376
66. Soleas GJ, Angelini M, Grass L, Diamandis EP, Golberg DM (2001) Methods Enzymol 335:145
67. Vitrac X, Desmoulière A, Brouillaud B, Krisa S, Deffieux G, Barthe N, Rosenbaum J, Mérillon JM (2003) Life Sci 72:2219
68. Marier JF, Vachon P, Gritsas A, Zhang J, Moreau JP, Ducharme M (2002) J Pharmacol Exp Ther 302:369
69. Yu C, Shin YG, Chow A, Li Y, Kosmeder JW, Lee YS, Hirschelman WH, Pezzuto JM, Mehta RG, van Breemen RB (2002) Pharm Res 19:1907
70. Golberg DM, Yan J, Soleas GJ (2003) Clin Biochem 36:79
71. Walle T, Hsieh F, DeLegge MH, Oatis JE, Walle UK (2004) Drug Metab Dispos 32:1377
72. Wenzel E, Somoza V (2005) Mol Nutr Food Res 49:472
73. Halliwell B, Gutteridge JMC (1994) Nutr Rev 52:253
74. Ross R (1993) Nature 362:801
75. Esterbauer H, Wäg G, Puhl H (1993) Med Bull 49:566
76. Frankel E, Kanner J, German J, Parks E, Kinsella J (1993) Lancet 341:454
77. Frémont L (2000) Life Sci 66:663
78. Delmas D, Jannin B, Latruffe N (2005) Mol Nutr Food Res 49:377
79. Tedesco I, Russo M, Russo P, Iacomino G, Russo GL, Carraturo A, Faruolo C, Moio L, Palumbo R (2000) J Nutr Biochem 11:114
80. Pace-Asciak CR, Hahn S, Diamandis EP, Soleas G, Goldberg DM (1995) Clin Chim Acta 235:207
81. Varache-Lembège M, Waffo-Teguo P, Richard T, Monti JP, Deffieux G, Vercauteren J, Mérillon JM, Nuhrich A (2000) Med Chem Res 10:253
82. Bertelli AAE, Giovannini L, De Caterina R, Bernini W, Migliori M, Fregoni M, Bavaresco L, Bertelli A (1996) Drug Exp Clin Res 22:61
83. Orsini F, Pelizzoni F, Verotta L, Aburjai T (1997) J Nat Prod 60:1082
84. Dobrydneva Y, Williams RL, Blackmore PF (1999) Br J Pharmacol 128:149
85. Olas B, Wachowicz B (2005) Platelets 16: 251
86. Huang YL, Tsai WJ, Shen CC, Chen CC (2005) J Nat Prod 68:217
87. Cox DA, Cohen ML (1996) Pharmacol Rev 48:3
88. Fitzpatrick DF, Hirschfield SL, Coffey RG (1993) Am J Physiol 265:774
89. Andriambeloson E, Kleschyov AL, Muller B, Beretz A, Stoclet JC, Andriantsitohaina R (1997) Br J Pharmacol 120:1053
90. Wallerath T, Deckert G, Ternes T, Anderson H, Li H, Witte K, Forstermann U (2002) Circulation 106:1652
91. Liu JC, Chen JJ, Chan P, Cheng CF, Cheng TH (2003) Hypertension 42:1198

92. Ijima K, Yoshizumi M, Hashimoito M, Kim S, Eto M, Ako J, Liang YQ, Sudoh N, Hosoda K, Nakahara K, Toba K, Ouchi Y (2000) Circulation 101:805
93. Stoclet JC, Chataigneau T, Ndiaye M, Oak MH, Bedoui J E, Chataigneau M, Chini-Kerth VB (2004) Eur J Pharmacol 500:299
94. Klurfeld DM, Kritchevsky D (1981) Exp Mol Pathol 34:62
95. Auger C, Teissèdre PL, Gérain P, Lequeux N, Bornet A, Serisier S, Besançon P, Caporiccio B, Cristol JP, Rouanet JM (2005) J Agric Food Chem 53:2015
96. Demrow HS, Slane PR, Folts JD (1995) Circulation 91:1182
97. Ruf JC, Berger JL, Renaud S (1995) Arterioscler Thromb Vasc Biol 15:140
98. Zou JG, Wang ZR, Huang YZ, Cao KJ, Wu JM (2003) Int J Mol Med 11:317
99. Maxwell S, Cruickshank A, Thorpe G (1994) Lancet 344:193
100. Whitehead T, Robinson D, Allaway S, Syms J, Hale A (1995) Clin Chem 41:32
101. Kondo K, Matsumoto A, Kurata H, Tanahashi H, Koda H, Amachi T, Itakura H (1994) Lancet 344:1152
102. Fuhrman B, Lavy A, Aviram M (1995) Am J Clinic Nutr 61:549
103. Nigdikar SV, Williams NR, Griffin BA, Howard AN (1998) Am J Clin Nutr 68:258
104. Seigneur M, Bonnet J, Dorian B, Benchimol D, Drouillet F, Gouverneur G, Larrue J, Crockett R, Boisseau M, Ribéreau-Gayon P, Bricaus H (1990) J Appl Cardiol 5:215
105. Ruf JC (2004) Biol Res 37:209
106. Sharpe PC, Mcgrath LT, Mcclean E, Young IS, Archbold GPR (1995) Q J Med 88:101
107. Pace-Asciak CR, Rounova O, Hahn SE, Diamandis EP, Elefiherios P, Goldberg DM (1996) Clin Chim Acta 246:163
108. De Rijke J, Demacker PNM, Assen NA, Sloots LM, Katan MB, Stalenhoef AFH (1996) Am J Clin Nutr 63:329
109. Watkins TR, Bierenbaum ML (1997) California wine use leads to improvement of thrombogenic and peroxidation risk factors in hyperlipemic subjects. In: Watkins TR (ed) Wine Nutritional and Therapeutic Benefits. ACS Symp. Series 661, Washington, p 261
110. Shimada K, Watanabe H, Hosoda K, Takeuchi K, Yoshikawa J (1999) Lancet 354:1002
111. Agewall S, Wright S, Douglas RN, Whalley GA, Duxbury M, Sharpe N (2000) Eur Heart J 21:74
112. Lekakis J, Rallidis LS, Andreadou I, Vamvakou G, Kazantzoglou G, Magiatis P, Skaltsounis AL, Kremastinos DT (2005) Eur J Cardiovasc Prev Rehabil 12:596
113. Kinghorn AD, Fong HHS, Farnsworth NR, Mehta RG, Moon RC, Moriarty RM, Pezzuto JM (1998) Curr Org Chem 2:597
114. Bailey GS, Williams DE (1993) Food Technol 47:105
115. Alberts DS, Colvin OM, Conney AH, Ernster VL, Garber JE, Greenwald P (1999) Cancer Res 59:4743
116. Suschetet M, Siess MH, Le Bon AM, Canivenc-Lavier MC (1998) Anticarcinogenic properties of some flavonoids. In: Vercauteren J, Chèze C, Triaud J (eds) Polyphenols Communications 96. Groupe Polyphénols, Bordeaux, France, p 165
117. Delmas D, Lançon A, Colin D, Jannin B, Latruffe N (2006) Curr Drug Targets 7:1
118. Clifford AJ, Ebeler SE, Ebeler JD, Bills ND, Hinrichs SH, Teissèdre PL (1996) Am J Clin Nutr 64:748
119. Dolara P, Luceri C, De Filippo C, Femia AP, Giovannelli L, Caderni G, Cecchini C, Silvi S, Orpianesi C, Cresci A (2005) Mut Res 591:237
120. Jang M, Cai L, Udeani GO, Slowing KV, Thomas CF, Beecher CWW, Fong HHS, Farnsworth NR, Kinghorn AD, Mehta RG, Moon RC, Pezzuto JM (1997) Science 275:218
121. Soleas GJ, Grass L, Josephy PD, Goldberg DM, Diamandis EP (2002) Clin Biochem 35:119
122. Tessitore L, Davit A, Sarotto I, Caderni G (2000) Carcinogenesis 21:1619
123. Mgbonyebi OP, Russo J, Russo IH (1998) Int J Oncol 12:865

124. Elattar TMA, Virji AS (1999) Anti-Cancer Drug Des 10:187
125. Sareen D, Van Ginkel PR, Takach JC, Mohiuddin A, Darjatmoko SR, Albert DM, Polans AS (2006) Invest Ophthalmol Vis Sci 47:3708
126. Fuggetta MP, Lanzilli G, Tricarico M, Cottarelli A, Falchetti R, Ravagnan G, Bonmassar E (2006) J Exp Clin Cancer Res 25:189
127. Lanzilli G, Fuggetta MP, Tricarico M, Cottarelli A, Serafino A, Falchetti R, Ravagnan G, Turriziani M, Adamo R, Franzese O, Bonmassar E (2006) Int J Oncol 28:641
128. Ohyama M, Tanaka T, Ito T, Iinuma M, Bastow KF, Lee KH (1999) Bioorg Med Chem Lett 9:3057
129. Kimura Y, Baba K, Okuda H (2000) Anticancer Res 20:2923
130. Wieder T, Prokop A, Bagci B, Essmann F, Bernicke D, Schulze-Osthoff K, Dörken B, Schmalz HG, Daniel PT, Henze G (2001) Leukemia 15:1735
131. Billard C, Izard JC, Roman V, Kern C, Mathiot C, Mentz F, Kolb JP (2002) Leuk Lymph 43:1991
132. Mishima S, Matsumoto K, Futamura Y, Araki Y, Ito T, Tanaka T, Iinuma M, Nozawa Y, Akao Y (2003) J Exp Ther Oncol 3:283
133. Ito T, Akao Y, Yi H, Ohguchi K, Matsumoto K, Tanaka T, Iinuma M, Nozawa Y (2003) Carcinogenesis 24:1489
134. Liu BL, Inami Y, Tanaka H, Inagaki N, Iinuma M, Nagai H (2004) Chin J Nat Med 2:176
135. Tolomeo M, Grimaudo S, Di Cristina A, Roberti M, Pizzirani D, Meli M, Dusonchet L, Gebbia N, Abbadessa V, Crosta L, Baruchello R, Grisolia G, Invidiata F, Simoni D (2005) Int J Biochem Cell Biol 37:1709
136. Chowdhury SA, Kishino K, Satoh R, Hashimoto K, Kikuchi H, Nishikawa H, Shirataki Y, Sakagami H (2005) Anticancer Res 35:2055
137. Godichaud S, Krisa S, Couronné B, Dubuisson L, Mérillon JM, Desmoulière A, Rosenbaum J (2000) Hepatology 31:922
138. De Ledinghen V, Monvoisin A, Neaud V, Krisa S, Payrastre B, Bedin C, Desmoulière A, Bioulac-Sage P, Rosenbaum J (2001) Int J Oncol 19:83
139. Jang M, Pezzuto JM (1998) Pharm Biol 326:28
140. Jang M, Pezzuto JM (1999) Drug Exp Clin Res 25:65
141. Mehta RG, Hawthorne ME, Steele VE (1997) Methods Cell Sci 19:19
142. Kurumbail RG, Stevens AM, Gierse JK, McDonald JJ, Stegeman RA, Pak JY (1996) Nature 384:644
143. Aggarwal BB, Shishodia S, Sandur SK, Pandey MK, Sethi G (2006) Biochem Pharmacol 72:1605
144. Issa AY, Volate SR, Wargovich MJ (2006) J Food Comp Anal 19:405
145. Waffo-Téguo P, Hawthorne ME, Cuendet M, Mérillon JM, Kinghorn AD, Pezzuto JM, Mehta RG (2001) Nutr Cancer 40:173
146. Bode AM, Dong Z (2004) Mutat Res 555:33
147. Lin HY, Lansing L, Mérillon JM, Davis FB, Tang HY, Shih A, Vitrac X, Krisa S, Keating T, Cao HJ, Bergh J, Quackenbush S, Davis PJ (2006) FASEB J 20:1742
148. Kimura Y, Okuda H (2001) J Nutr 131:1844
149. Kimura Y (2005) In Vivo 19:37
150. Piver B, Fer M, Vitrac X, Mérillon JM, Berthou F, Dreano Y, Lucas D (2004) Biochem Pharmacol 68:773
151. Potter GA, Patterson LH, Wanagho E, Perry PJ, Butler PC, Ijaz T, Ruparelia KC, Lamb JH, Farmer PB, Stanley LA, Burke MD (2002) Br J Cancer 86:774
152. Geahlen RL, McLaughlin JL (1989) Biochem Biophys Res Commun 165:241
153. Burke TR (1994) Stem Cells 12:1
154. Dang O, Navarro L, David M (2004) Shock 21:470
155. Richard N, Porath D, Radspieler A, Schwager J (2005) Mol Nutr Food Res 49:431
156. Ashikawa K, Majumdar S, Banerjee S, Bharti AC, Shishodia S, Aggarwal BB (2002) J Immunol 169:6490

157. Piver B, Berthou F, Dreano Y, Lucas D (2003) Life Sci 73:1199
158. Do QT, Renimel I, André P, Lugnier C, Muller CD, Bernard P (2005) Curr Drug Disc Technol 2:161
159. Huang HS, Lin M, Cheng GF (2001) Phytochemistry 58:357
160. Ramassamy C (2006) Eur J Pharmacol 545:51
161. Gao ZB, Hu GY (2005) Brain Res 1056:68
162. Okawara M, Katsuki H, Kurimoto E, Shibata H, Kume T, Akaike A (2007) Biochem Pharmacol 73:550
163. Parker JA, Arango M, Abderahmane S, Lambert E, Tourette C, Catoire H, Neri C (2005) Nat Genet 37:339
164. Wang Q, Xu J, Rottinghaus GE, Simonyi A, Lubahn D, Sun GY, Sun AY (2002) Brain Res 958:439
165. Ritchie K, Lovestone S (2002) Lancet 360:1759
166. Sisodia SS, Price DL (1995) FASEB J 9:366
167. Selkoe DJ (2001) Physiol Rev 81:741
168. Pereira C, Agostinho P, Moreira PI, Cardoso SM, Oliveira CR (2005) Curr Drug Targets CNS Neurol Disord 4:383
169. Anekonda TS (2006) Brain Res Rev 52:316
170. Jang JH, Surh YJ (2003) Free Radic Biol Med 34:1100
171. Marambaud P, Zhao PH, Davies P (2005) J Biol Chem 280:37377
172. Han YS, Zheng WH, Bastianetto S, Chabot JG, Quirion R (2004) Br J Pharmacol 141:997
173. Jeon SY, Kwon SH, Seong YH, Bae K, Hur JM, Lee YY, Suh DY, Song KS (2007) Phytomedicine 14:403
174. Rivière C, Richard T, Quentin L, Krisa S, Mérillon JM, Monti JP (2007) Bioorg Med Chem 15:1160
175. Ono K, Hasegawa K, Naiki H, Yamada M (2004) J Neurosci Res 75:742

Chapter 3
Research into Isoflavonoid Phyto-oestrogens in Plant Cell Cultures

M.T. Luczkiewicz

Department of Pharmacognosy, Medical University of Gdańsk, al. Gen. J. Hallera 107, 80 – 416 Gdańsk-Wrzeszcz, Poland, e-mail: mlcz@amg.gda.pl

Abstract The purpose of this article is to summarize the present knowledge on biotechnological research into isoflavonoid phyto-oestrogens. The physiological and pharmacological properties of isoflavones are discussed in the Introduction. The accumulation of phyto-oestrogens by plant cell and tissue cultures is reviewed. A special emphasis is put on the influence of basic experimental media and physiological factors on the production of isoflavonoids in in vitro cultures. The potential role of the transformation process as well as the techniques (including infection with bacteria and particle bombardment) and the various technological procedures (elicitation, feeding experiments) in isoflavone biosynthesis is discussed. Moreover, this chapter deals with the in vitro cultures of legume plants oriented for selective accumulation of phyto-oestrogens.

Abbreviations

ABA	Abscisic acid
AOPP	α-Aminooxy-ß-phenylpropionic acid
BAP	6-Benzylaminopurine
CHS	Chalcone synthase
2,4-D	2,4-Dichlorophenoxyacetic acid
GA	Natural gibberellins
IFR	isoflavone reductase
IFS	Isoflavone synthase
PAL	Phenylalanine ammonia-lyase
SH	Schenk-Hildebrandt medium

Ramawat KG, Mérillon JM (eds.), In: *Bioactive Molecules and Medicinal Plants*
Chapter DOI: 10.1007/978-3-540-74603-4_3, © Springer 2008

3.1 Introduction

In the second half of the 20th century, research in cultivating higher plants in vitro led to the development of cultures of single cells and their aggregates, protoplasts, tissues and organs capable of continuous growth on sterile, specially designed growth media [1–3]. Strictly controlled conditions and constant access to stable biological material made it possible to conduct basic research into plant physiology on a much larger scale than before, leading to the explanation of numerous biochemical processes that constitute primary and secondary tissue metabolism [3, 4]. It was also proved that plant biomasses cultivated in vitro are capable of biosynthesis of secondary metabolites typical for intact plants, or that they may serve as sources of entirely new substances not identified in nature [3–10]. In effect, a path was opened for intensive biotechnological research into the potential use of in vitro cultures to produce highly valuable secondary metabolites, including compounds for which medical applications could be found [10–13].

Statistical research indicated that even in developed countries, where the broadly understood chemical synthesis is the basis for the pharmaceutical industry as such, still as much as 25% of all medicines are compounds of natural origin. Steady degradation of the environment, slow growth of plants, common problems with low concentrations of active substances in the cultivated plants are only some of the factors that make it difficult to obtain biologically active substances from natural sources. With the aforementioned obstacles and the need to cultivate tropical species in a temperate climate, in vitro cultures came to be seen as an alternative method for producing secondary metabolites with particularly valuable therapeutic properties [3, 14].

Among the numerous secondary metabolites, it is isoflavonoids, together with alkaloids and terpenoids, that are natural compounds most commonly researched under in vitro conditions [9, 10, 15]. The reason for this may be seen in their multidirectional biological activity [16, 17]. Isoflavones, being recognised phytoalexins and phytoanticipins, play a key role in the defence mechanisms in plants of Fabaceae family. That is why plant biomasses obtained from various species of legume plants became model systems for testing the defence response of plants to broadly defined environmental stresses [18, 19]. Isoflavones are, in addition, secondary metabolites with broad health-promoting activity. In addition to the anti-inflammatory, antifungal and anti-free-radical activities that are typical for the whole group of bioflavonoids, they also have features typical for compounds that inhibit oestrogen β receptors in mammals [16, 17, 20–22]. For this reason, among other applications, they are commonly recommended for the treatment of menopause symptoms [21, 23]. Moreover, some consideration is given to the possible using isoflavones in the prevention and treatment of neoplasms related to the distortion of hormonal balance in the body [16, 17, 22, 24, 25]. Predominantly limited to soy-bean products, the natural sources of isoflavones do not permit mass production of substantial amounts of these compounds, not only for the pharmaceutical industry, but also for the comprehensive pharmacological and toxicological research of the isolated substances

[26]. Therefore, in recent years, only very few reports dealing with biotechnological research indicated the possibility of using in vitro cultures of certain species of the Fabaceae family for large-scale production of phyto-oestrogens [27, 28].

Irrespective of the main goal of a research project, the biosynthesis and metabolism of isoflavonoids is stimulated in in vitro cultures by using both traditional as well as more advanced biotechnological strategies [15, 26]. Such procedures as the modification of the basic composition of the experimental medium, with a focus on growth regulators [29–33] and the regulation of the growth conditions by controlled lighting and temperature, should be mentioned [30, 34, 35]. The natural enzymatic potential of the investigated biomass is also used to induce broadly understood bioconversion with the use of direct and indirect precursors, in order to obtain the final product [36–38].

Based on the fact that isoflavones belong to a class of substances that play an active role in defence processes [18, 19], the biosynthesis and metabolism of these compounds in plant cells is regulated using a wide range of biotic and abiotic elicitors [39–43]. Sample tests of isoflavone accumulation in various types of in vitro cultures (callus, suspension, roots and shoots) provide preliminary insight into the production potential of the particular types of cultures [27, 28, 31–33, 44]. The metabolism of isoflavones was also investigated using transgenic biomasses (suspensions, hairy roots, regenerated organs) obtained through genetic transformation with wild or pre-designed strains of bacteria [31, 45–47], or particle bombardment [48, 49].

The following discussion, based on current literature reports, presents the impact of the particular growth strategies on the biosynthesis and metabolism of isoflavones in in vitro plant cultures.

3.2 The Influence of the Basic Experimental Media on the Biosynthesis of Isoflavones in In Vitro Cultures

Modifications of growth media composition belong to the basic methods of primary secondary metabolism regulation in biomasses cultivated in vitro [50]. These modifications involve both qualitative and quantitative changes in the macro- and microelements, sources of carbon and growth regulators responsible for the particular hormonal balance in the growth system [2, 5, 8, 9, 51].

Most media used to grow legume plants in vitro belong to the so-called rich growth media, such as, for example, Murashige and Skoog, Schenk-Hildebrandt (SH) and Lindsmeyer and Skoog [27, 28, 31–33]. The composition of these media, especially in terms of growth regulators, was selected to support the micropropagation of the investigated species or to ensure maximum growth of the biomasses [44, 52–56].

For instance, in callus and suspension cultures of the most frequently investigated species of the Fabaceae family, such as *Glycine max*, *Cicer arietinum*, *Phaseolus vulgaris*, *Lupinus albus* and *Medicago sativa*, the growth regulators of choice were most often kinetin and 2,4-dichlorophenoxyacetic acid (2,4-D) [27, 28, 37–39, 41, 42]. In the process of shoot formation in papilionaceous plants, also useful were such cytokinins as benzylaminopurine (BAP) and thidiazuron, as well as natural gibberellins (GA_1, GA_3, GA_4, GA_7, GA_9 and GA_{13}). It should be noted here that the effect of the particular growth regulators on shoot initiation and elongation was variable and each time was species specific [44, 55–63].

Only few reported experiments involving in vitro cultures of Fabaceae plants were directly focussed on the influence of the basic composition of the experimental media on the accumulation of isoflavones in the plant material [30–34, 55, 64, 65]. For instance, callus cultures of *Genista* plants grown on modified SH medium accumulated phyto-oestrogens (16 compounds), derivatives of genistein, daidzein and formononetin, in concentrations many times higher than in intact plants [55].

The other experiments showed that a significant increase in the concentration of the discussed compounds in the biomasses can be achieved by reducing the level of mineral salts in the growth media [64–66]. A reduction of ammonia salts and nitrites in the growth media caused *Phaseolus vulgaris* shoot cultures to synthesise several times more phaseolin and kievetone than control biomasses [34]. The process was accompanied by a generally increased expression of genes related to the phenylpropanoid pathway and inhibition of culture growth. Similarly, increased nitrogen content in the medium resulted in the concentration of phytoalexins in bean shoots being reduced by even 9% [34]. An increased concentration of isoflavones in in vitro cultures of *Phaseolus vulgaris* was also achieved when there was a deficit of phosphate, borate and manganese ions in the growth system [65, 66]. The reason for this phenomenon is seen in the so-called "nutrition stress", although not much is known about its mechanism. It is presumed that it inhibits basic growth-related proteins in the culture. In effect, the biochemical pathways are "switched" towards the biosynthesis of particular secondary metabolites [34, 67–69].

Research concerning in vitro cultures of legume plants also showed the effect of certain growth regulators on the biosynthesis of isoflavones in plant biomasses [29–33, 38, 70, 71]. It was proved that 2,4-D stimulated the biosynthesis of free and glycosidated biochanin A and formononetin in a *Cicer arietinum* suspension culture [38, 70]. The aforementioned auxin also induced the production of daidzein in callus cultures of five species of *Psolarea* genus [32] and maackiain and medicarpin in a *Medicago sativa* suspension culture [71]. The beneficial role played by 2,4-D in the process of isoflavone biosynthesis was also noted in callus lines of *Maackia amurensis* [33]. In the presence of auxin, the discussed cultures synthesised as much as four times more daidzein, genistein, formononetin and retuzin than the intact plant. This phenomenon was not observed, however, when the auxin of choice was naphthaleneacetic acid [33]. The described role of 2,4-D in the process of isoflavone biosynthesis seems surprising, because it is quoted in numerous literature reports as a blocker of the key enzyme of the phenylpropanoid pathway – chalcone synthase (CHS) [69, 70].

Moreover, experiments involving in vitro cultures of *Phaseolus vulgaris* [29, 30, 72, 73] showed that growth regulators could selectively regulate the biosynthesis and metabolism of certain isoflavones. For instance, abscisic acid (ABA) and BAP stimulated the biosynthesis of phaseolin in suspension and cotyledon cultures of the common bean. At the same time, unlike BAP, ABA inhibited the synthesis of kievetone in the same biomasses [29, 30, 73]. The authors of these reports concluded that the growth regulators used had a selective effect on the metabolism of the investigated phytoalexins by creating more or less advantageous cytological conditions in the analysed biomasses [29, 30]. Due to the specificity of the experiments, the eventual concentration of kievetone and phaseolin resulted not only from the particular medium modification in terms of growth regulators, but also from variable growth conditions [29, 30, 73].

In all of the aforementioned experiments, the role of particular medium components on the biosynthesis of isoflavones received only fragmentary treatment. Still, these results indicate clearly that media components, and especially the phytohormones, may play an important regulatory role in the biosynthesis of isoflavones in in vitro cultures.

3.3 The Influence of Physical Factors on the Biosynthesis and Accumulation of Isoflavonoids in In Vitro Cultures

The most important physical factors that affect the biosynthesis and metabolism of isoflavones in plant biomasses include ambient temperature and lighting conditions [1, 7–9, 12, 50]. In case of the latter, a crucial role is played by both the length of the electromagnetic wave and the actual exposition of plant material to daylight [34, 50, 74, 75–77].

It is believed that approximately 16 enzymes of the basic phenylpropanoid pathway, including phenylalanine ammonia-lyase (PAL) and CHS, is temporarily induced by the presence of daylight [34, 74, 78, 79]. Therefore, the access to daylight may have a material regulatory impact on the biosynthesis of phenylpropane derivatives, including bioflavonoids, such as isoflavones [30, 78, 79]. This theory was confirmed by research of suspension cultures of *Cicer arietinum*, which synthesised significant amounts of biochanin A and formononetin when grown in daylight [38, 74]. On the other hand, exposure to light does not seem necessary to initiate the biosynthesis of isoflavones, as indicated by their considerable concentration in the roots of intact plants and in in vitro cultures of legume plants kept in complete darkness [35, 46, 47, 49].

Research involving in vitro cotyledon cultures of *Phaseolus vulgaris* showed that the lighting conditions may also selectively regulate the biosynthesis of certain isoflavones [30]. These bean cultures contained clearly heightened accumulations of kievetone (5-hydroxyflavone) only in cotyledons incubated in complete darkness. Conversely, phaseolin (5-deoxyisoflavone) was synthesised in significant quantities mainly in the biomass exposed to daylight. The authors of this paper suggest that while the biosynthesis of phaseolin in bean tissue is directly induced by the presence of daylight, in the case of kievetone, the regulatory mechanisms related to its biosynthetic pathway are more complex. It involves, among others, an inhibitor of kievetone biosynthesis, namely ABA, which in the investigated plant is also synthesised in daylight [30].

The modification of lighting conditions in the callus cultures of several *Genista* plants indicated that the presence or absence of daylight seemed to have a selective effect on the accumulation of ester derivatives of genistein, which are considered to be a storage form of phytoalexins [55].

As research into in vitro cultures of *Phaseolus vulgaris* shows, the exposure of the biomass to ultraviolet (UV) light may also significantly stimulate the biosynthesis of isoflavones [34, 65]. The presented theory assumes that the process involves a UV-activated photochrome [65, 78]. As a result of environmental stress caused by UV radiation, the bean culture increased the biosynthesis of isoflavones, the biological activity of which is related to photoprotective mechanisms [34, 65]. Moreover, in the *Phaseolus vulgaris* biomass, UV radiation stimulated the selective production of highly hydroxylated isoflavones [34], which are characterised by strong antioxidant properties [79–81]. The biosynthesis of isoflavones is also induced in in vitro plant cultures as a result of rapid changes in the temperature of the experimental environment [35, 82, 83].

Non-transformed root cultures of *Glycine max*, chilled to 10°C accumulated even several times more of the basic phyto-oestrogens than the control biomass kept at 25°C [82, 83]. Detailed research showed that in the investigated cultures, the concentration of genistein and daidzein increased approximately by a factor of two and genistein even by a factor of three as a result of thermal stress [35]. It is supposed that the synthesised isoflavones, which have considerable antioxidant properties, function as free-radical scavengers, which are mediators of certain biochemical processes caused by reduced temperature [18, 19, 35, 78, 83].

3.4 The Effect of Technological Procedures on the Biosynthesis and Accumulation of Isoflavonoids in In Vitro Cultures

Apart from the genetic potential of the plant culture and the basic growth conditions (media composition, temperature and lighting), a fundamental effect on the increased production of secondary metabolites in vitro can be produced by so-called technological procedures [1, 3, 5–11, 13, 14, 50]. They involve delivering certain environmental impulses to the biomass, as well as modifying the plant genome. Described below are the most important strategies applied to in vitro cultures of legume plants that may be used to make these biomasses produce more isoflavonoids.

3.4.1 Elicitation

As mentioned above, most literature reports concerning in vitro cultures where isoflavones are the direct or indirect subject describe model experiments that are designed to explain the role of these compounds in the relationship between certain plant species and the environment [15, 18, 19, 26, 48, 78, 84, 85].

Detailed research into plant physiology indicate an important role of the broadly understood isoflavone class in the way that plants belonging to Fabaceae family respond to biological stress such as: poisons, injury, extreme temperatures, pathogen attack or relationship with symbiotic micro-organisms [15, 18, 19]. In order to define the role of isoflavones in the defensive response to these stressful situations, several experiments were carried out in vitro, with the use of cell, tissue and isolated plant organ cultures [38, 39, 43, 67, 78, 86]. The model plants were mostly selected species of the Fabaceae family, such as *Cicer arietinum, Medicago sativa, Lupinus albus, Glycine soje et max* (i.e. plants that are rich in isoflavonoids belonging to either phytoalexins or phytoanticipins) [38, 41–43, 71, 87–98]. Some of the methods of causing biological stress included the use of biotic elicitors in the form of metabolites isolated from symbiotic or pathogenic micro-organisms (exogenous elicitors), or plant isolates (endogenous elicitors) such as methyl jasmonate, salicylic acid, ethylene and glutathione [15, 18, 19, 37–43, 48, 71, 78, 87–102]. Abiotic elicitors in the form of heavy metal salts were also used on numerous occasions [15, 19, 37, 39, 41, 78, 87]. Even though all of the experiments described here were of a basic nature and their goal was not to develop growth systems producing isoflavones of specific pharmacological activity, their highly interesting results can still be used in applied research.

The highly diversified experimental procedures used in in vitro experiments involving Fabaceae plants showed that the biosynthesis of isoflavonoid phytoalexins in the course of the defence response of the plant biomasses depended on the type and quantity of the stress factor used [15, 18, 19, 37–43, 48, 71, 78, 87–102]. This is especially visible in the case of elicitors originating from pathogenic or symbiotic micro-organisms [37–43, 48, 71, 78, 87–102].

When incubating suspension cultures of *Medicago sativa* with an extract from the pathogenic fungus *Phoma medicagensis*, Paiva et al. achieved considerable activation of the phenylpropanoid pathway, leading to the biosynthesis of basic phytoalexins, such as sativan, medicarpin and coumestrol [39]. Also observed was an over 50-fold increase of gene expression in genes related to isoflavonoid metabolism. The increased level of basic phytoalexins was preceded by significantly increased activity of key enzymes related to defence mechanisms in Fabaceae plants, for example PAL, CHS and isoflavone reductase (IFR). It was also unambiguously determined that the free phytoalexins produced in this case were not synthesised de novo, but originated from the hydrolysis of ester phytoanticipins present in the non-elicited plant material. What seems highly relevant from the application perspective is that both the activity of key enzymes and the levels of sativan, coumestrol and medicarpin in *Medicago sativa* suspension culture dropped already after 72 h to the levels identified prior to elicitation.

The incubation of alfalfa suspension culture with the symbiotic fungus *Glomus versiforme* only resulted in a 3–4% increase of PAL, CHS and IFR activity, paired with an equally minor increase in the concentration of the basic phytoalexins. In this case, the described phenomenon was accompanied by a clearly increased concentration of esterified phytoalexins in the tested biomass. Irrespective of the type of elicitor, the determined metabolites were always stored intracellularly [39].

The variable effect of colonising with a symbiotic or pathogenic micro-organism on isoflavone biosynthesis was confirmed by experiments on in vitro cultures of de novo roots of *Glycine soja* [18]. Phillips and Kapulnik [18], like Paiva et al. [39], noted considerable, albeit short-termed increase in activity of the isoflavonoid pathway related to the biosynthesis of considerable amounts of glyceollin as a result of incubating plant material with the pathogenic *Rhizoctonia* fungus. They also found that the actual process of forming a free phytoalexin depended directly on the dose of the elicitor. Small concentrations of the fungal pathogen, like in Dixon's report [39], evoked ester hydrolysis of the phytoanticipin and related release of the phytoalexin, in this case glyceollin. On the other hand, large amounts of the pathogenic fungus evoked immediate de novo biosynthesis of glyceollin [18]. Unlike Dixon's experiments [39], glyceollin was not stored intracellularly, but was always released into the growth media [18].

The metabolism of isoflavones following the incubation of soybean root cultures with such symbiotic strains as *Bradyrhizobium japonicum* or *Sinorhizobium fredii* was also different. At the initial colonisation phase, the roots synthesised considerable amounts of glyceollin, released immediately to the media from the rhizosphere [18]. At the same time, in this case the concentration of glyceollin dropped in much faster than in alfalfa cultures [39] and even reached the level much lower than before elicitation. The hypothesis proposed by the authors of this paper points out the very special relationships between symbiotic micro-organisms and plants. The thesis is that at an early stage of colonisation, it is the cell membrane of a symbiotic fungus that constitutes the source of actual elicitors (i.e. lipo-oligosaccharides) and directly evokes considerable biosynthesis of phytoalexins in the plant biomass. The depolarisation of the plant cell membrane results in phytoalexins being released outside. These compounds activate the biosynthesis of several nodulation factors (*Nod*-genes) in the symbiotic micro-organism, following which their concentration in the medium decreases. Thus, the defence responses of the host are suppressed and the symbiont can actively colonise the plant [18].

The elimination of defence mechanisms while plants are colonised by symbiotic strains was confirmed in an experiment with alfalfa roots elicited at the same time with the pathogenic strain *Rhizoctonia solani* and the symbiotic strain *Glomus intraradices* [90].

During double infection, only trace amounts of medicarpin-3-O-glucoside were identified in the investigated tissues, together with low activity of CHS and IFR, which clearly indicates a suppression of the defence system. This phenomenon occurred in alfalfa roots despite the presence of a pathogen strain that in a single infection evoked a typical defence reaction with a 5- to 10-fold increase in concentration of the basic phytoalexin (i.e. medicarpin 3-O-glucoside) [90].

It was also noted that the suppression of the defence system requires not only that the symbiotic strain in present in the rhyzosphere, but also that it is directly initiated by the colonisation of the plant tissue. Similar observations were made in the course of infecting root cultures of soybean with *Bradyrhizobium japonicum*, *Rhizobium japonicum* and *Azorhizobium japonicum* [103]. The biosynthesis of phytoalexins induced by the presence of a fungal pathogen was also observed to be inhibited during simultaneous infection of soy cotyledons with cyclic glucans from the pathogenic fungus *Phytophtora sojae* and the symbiotic bacteria *Bradyrhizobium japonivum* [95]. Thus, the experiments demonstrated the metabolism of isoflavones when the plant is colonised by a symbiont, as proposed by Phillips and Kapulnik [18].

Many authors who carried out experiments into the biosynthesis of phytoalexins in vitro were concerned not only with the effect of the various types of elicitors on the level of isoflavones synthesised in the biomasses, but mainly investigated the actual mechanism behind the formation and metabolism of phytoalexins [37–43, 71, 87, 88, 91–102]. Depending on the plant material tested and the research procedure used, authors varied in their opinions as to whether stress situations can cause isoflavones to be synthesised de novo or to be produced as a result of storage forms hydrolysis (phytoanticipins) [48, 67, 71, 78, 79, 91–102]. As mentioned earlier, in their work on soybean root cultures, Phillips and Kapulnik concluded that two phytoalexin formation mechanisms were active in a plant, and that one or the other metabolic pathway was activated depending directly on the dose of the elicitor [18].

The authors of the paper on lupin suspension cultures believe otherwise, that as a result of elicitation, the biosynthesis of phytoalexins is always induced de novo [42]. In the presented experiment, as a result of incubation of the investigated cultures with a yeast extract, significant concentrations of medicarpin, the basic phytoalexin, were identified in the biomass, with simultaneous increase of S-adenosyl-L-methionine, which was the source of methyl groups in the biosynthesis of prenylated isoflavones [79]. The addition of selected inhibitors of transmethylation (cycloleucine, sinefungin and tubericidin) selectively inhibited the medicarpin production induced by elicitation. Thus, all of the circumstances indicate that in discussed case the production of medicarpin was the effect of de novo biosynthesis. The authors of this paper also noted that the inhibition of the basic "isoflavonoid" defence mechanism could activate the biosynthesis of other, non-flavonoid protection substances. In the described case it was licodione, which belongs to the family of retrochalcones [42]. A similar process was observed after eliciting *Glycyrhiza echinata* suspension with a fungal naphtochinone [99]. These cultures were characterised by a very rich phenylpropanoid metabolism, therefore the authors propose to use in vitro cultures of *Glycyrhiza echinata* as model systems for testing defence reaction in Fabaceae plants. Biotic elicitation led in the tested cultures to very quick de novo biosynthesis of significant amounts of medicarpin and formononetin, isoflavones produced as part of a typical defence reaction [18, 48, 78, 79, 82, 86]. Relatively late, only after 48 h, was the increased production noted of retrochalcones typical for the tested plant, such as echinatin and licodione. As mentioned earlier, the biosynthesis of the latter two compounds was much more intensive after the basic isoflavonoid pathway was inhibited. Based on the achieved results, the authors concluded that elicitation-induced isoflavone de novo biosynthesis only constitutes a narrow section of a far more complex defence reaction [99].

De novo biosynthesis of phytoalexins was also confirmed in several-day-old seedlings of chickpea (*Cicer arietinum*) subjected to the model elicitor, glutathione [91]. The authors of this paper actually ruled out the formation of medicarpin, maackiain, biochanin A and formononetin from the respective storage forms by effectively blocking the phenylpropanoid pathway at the level of phenylalanine or *trans*-cinnamic acid. After using selected PAL and cinnamic acid inhibitors (α-amino-oxy-β-phenylpropionic acid – AOPP – and 3,4-methylene-dioxycinnamic acid), the biosynthesis of all free phytoalexins was completely inhibited, which proves that in this case, as a result of elicitation, these compounds were synthesised de novo and not by hydrolysis of malonic storage forms [91].

Similar to the reports described above, hairy roots of *Glycine max* incubated with a suspension of a pathogenic bacteria *Fusarium solani* produced de novo significant amounts of the basic phytoalexin, glyceollin. It was confirmed that this compound was always formed from daidzein as a result of direct activation of the basic phenylpropanoid pathway [98].

The hypothesis of de novo formation of phytoalexins as a result of stress was completely negated by the authors of the publication on suspension cultures of *Cicer arietinum* [38]. Apart from adding such elicitors as *Ascochyta rabiei* suspension or yeast polysaccharide fraction to the culture, in this case a PAL inhibitor was used (AOPP). With the basic phenylpropanoid pathway inhibited in this way, de novo formation of phytoalexins was made impossible. Despite this procedure, elicitation led each time to a significant increase in the concentration of free phytoalexins, irrespective of the type of stress factor. It was concluded, therefore, that reaction to stress always leads to the synthesis of phytoalexins as a result of hydrolysis of storage forms [38].

The aforementioned model of isoflavone formation seems to be confirmed by research involving soybean seedlings infected with *Phytophtora megasperma*, which, irrespective of the amount of elicitor used, produced glyceollin through the hydrolysis of daidzein malonate [104].

Also, in soybean cotyledons subjected variably to local influence of lactofen (diphenyl ether), which is the active agent in Cobra herbicide, or glucan from the cell wall of *Phytophtora soje*, genistein malonate and daidzein immediately decomposed and significant amounts of respective aglycones were formed [99]. It was noted that while genistein was the end product of these reactions, daidzein continued to be actively metabolised to the basic phytoalexin of soybean, namely glyceollin. The observed isoflavone metabolism was present both locally, where the stress factor operated, and in cells that were quite distant from the direct elicitation zone. Moreover, the authors of this paper noted that in the defence reaction, the glucan elicitor induced the formation of glyceollin to a much larger extent than genistein. The use of lactofen, on the other hand, resulted in more biosynthesis of genistein than glyceollin. Due to the specific activity of the aforementioned diphenol, related to the formation of activated oxygen species, in this case the hydrolysis of phytoanticipins requires the presence of light [97].

Exhaustive research that largely explains the metabolism of isoflavones in elicited in vitro cultures involved suspension cultures of *Pueraria lobata* [40]. Under elicitation with yeast extract and glycoprotein fraction from the cell wall of *Phytophtora megasperma*, like in the reports quoted earlier [38, 97, 104], a significant increase in the level of free phytoalexins in the culture was observed, together with a 10–15% decrease in the concentration of storage forms (genistin malonate and daidzin malonate). This clearly suggests that the identified free phytoalexins were formed by hydrolysis and were not synthesised de novo.

Moreover, 8 h after elicitation, a decreased concentration of free isoflavones was noted, which is typical for elicited cultures and, interestingly, a significant production of malonic derivatives of genistein and daidzein. The authors of this paper believe that these "secondary" storage compounds were formed fully as a result of de novo biosynthesis, and not by re-esterification of previously hydrolysed phytoalexins. Moreover, the authors suggest that the decreased concentration of free phytoalexins observed after 8 h is related to the incorporation of these compounds into insoluble cell wall fractions, which is a section of the triggered systemic defence response [40]. In order to prove these hypotheses ^{14}C-labelled exogenous daidzein was bioconverted in *Pueraria lobata* suspension, and actively metabolised to ^{14}C-daidzein malonate. After yeast extract elicitation, following a preliminary ester hydrolysis, increased radioactivity was observed in the lignocellulose fraction of the suspension [40].

The complete metabolism of isoflavones forming in the defence response was also described by Edwards et al. for *Medicago sativa* (alfalfa) suspensions elicited with a yeast cell wall extract [105]. In general, the proposed model of phytoalexin metabolism in elicited cultures does not fully confirm any of the previously discussed hypotheses, and even negates some of their elements.

As can be concluded from numerous scientific research projects, elicitation of in vitro cultures of alfalfa, chickpea and lupine leads primarily to the activation of those pathways of phenylpropanoid metabolism that are related to increased methylation and, therefore, the formation of such isoflavonoid derivatives as: medicarpin, maackiain, sativan, variabilin and others [38, 39, 41–43, 87, 90, 91, 96, 100]. Considering the above, in order to achieve full control over isoflavonoid metabolism, Edwards et al. added ^{3}H-methylmethionine to alfalfa suspensions prior to elicitation, and then observed the possible occurrence of radioactivity in the various phytoalexin or phytoanticipin fractions [105]. The authors decided that the elicitation mechanism in various cultures is highly complex, unlike the experiments discussed herein. Phytoalexins formed in the first hours after elicitation were produced by means of hydrolysis of storage forms, like in the case of experiments involving in vitro cultures of *Cicer arietinum* [91] and *Pueraria lobata* [40]. On the other hand, further increase in the concentration of medicarpin, sativan and variabilin was due to their de novo production, as proved by significant radioactivity of these products. This, as we can call it, "second phase" of elicitation is reminiscent of the model of defence response proposed in experiments on in vitro cultures of lupine [42], *Glycyrhiza echinata* [99] and *Cicer arietinum* [91]. Only the phytoalexins biosynthesised de novo, as in the case of *Pueraria lobata* [40] cultures, were released into the medium and "detoxified" there in the process of esterification to their malonic derivatives. The storage of these compounds, on the other hand, was intracellular [105]. Moreover, unlike the *Pueraria lobata* [40] suspension, free phytoanticipins in the alfalfa suspension were not incorporated into the cell-wall fraction [105].

The use of the same elicitor in in vitro cultures of various species of the Fabaceae family proved that irrespective of the stress factor used, the mechanism of observed defence response is in each instance dependent on the particular plant material [94]. Suspension cultures of *Medicago sativa* treated with monoethyl ester of glutathione did not produce the basic phytoalexin (i.e. medicarpin) at all. No increased activity of the enzymes related to the biosynthesis of isoflavones was observed in them either. In contrast, in a suspension culture of bean (*Phaseolus vulgaris*) glutathione activated the entire pathway of phytoalexin synthesis, manifested by a 15–20% more intensive biosynthesis of kievetone, the basic phytoalexin in the bean [94]. An identical effect was obtained after treating the aforementioned cell suspensions with a cell-wall fraction from *Colletotrichum lindemuthianum*. In addition, the authors of this paper observed the metabolism of exogenous glutathione with the endogenous fraction of the above thiol in both suspension cultures. It turned out that the metabolism of the labelled exogenous elicitor in the cultures of bean and alfalfa is different. In beans, exogenous glutathione was evenly incorporated into soluble and insoluble thiol fractions. In alfalfa, on the other hand, most of the exogenous glutathione was incorporated into the insoluble fraction. It is possible that the distortion of internal balance at the level of soluble and insoluble thiols results from the absence of a signal initiating a defence response in alfalfa. Intracellular glutathione did not seem to be involved in a typical defence reaction in either of the suspension cultures. The levels of internal thiols grew too slowly to play their typical part in phytoalexin formation. It is suggested that they may act as antioxidants and scavengers of electrophiles [94].

The type of elicited plant material may determine not only the possible occurrence of a defence response to biological stress, but also the time after which the biosynthesis of phytoalexins occurs [93].

Root cultures of *Glycine max* incubated with zoospores of the pathogen fungus *Phytophthora megasperma* f.sp. *glycinea* synthesised phytoalexins much faster than the suspension of the same plant subjected to an identical stress factor [93]. The authors of this paper believe that in the case of root cultures the defence reaction was quick, closely related to the elicitation area, and local in nature. Contrary to that, it seems that in a suspension, all cells are subject to

Chapter 3 Research into Isoflavonoid Phyto-oestrogens in Plant Cell Cultures

elicitation at the same time, and as a result a delayed systemic defence reaction is induced [93].

The literature data quoted above indicate that the biosynthesis of isoflavonoid phytoalexins induced by applying an elicitor depends not only on the stress factor itself, but also on the type of plant material subject to elicitation (suspension cultures, root cultures, cotyledons, etc.) and the species of the plant itself [37–39, 40–43, 71, 87, 88, 91–102]. Further reports suggest that the process is also affected by the age of the plant biomass [91, 106]. One of the factors that made the proof of this hypothesis possible was the research into hairy roots of *Lotus corniculatus* elicited with glutathione [106]. Unlike suspensions of cell aggregates, hairy roots are heterogenic systems, characterised by the diverse age of the particular roots within a culture [107, 108]. In effect, such a system provides an opportunity to observe the relationship between the age of an elicited de novo root and the level of resulting defence response [106]. Finally, it was found that the PAL activity and the concentrations of vestitol and sativan, the basic phytoalexins identified after eliciting cultures with glutathione increased in proportion to the age of *Lotus corniculatus* roots. Each time, the observed phenomenon was related to the age of the particular roots, and not the age of the culture as a whole. Of the free phytoalexins formed after elicitation, 60–70% were released into experimental medium [106].

The age of the roots was also a significant factor affecting the biosynthesis of phytoalexins in seedling cultures of chickpea elicited with glutathione [91]. Similar to the report described above, older roots synthesised much more phytoalexins than young ones. In addition, the authors observed here that it was not only the amount of synthesised phytoalexins that was related to the age of the culture, but also their type. While the 1- to 2-day elicited roots had increased levels of medicarpin, maackiain, biochanin A and formononetin, the 4- to 6-day seedlings also synthesised cicerin and homoferreirin [91].

The type of elicitor used and its concentration may affect not only the amount of phytoalexins produced in biomasses, but also the type of the synthesised compounds. This is proved by experiments by Phillips and Kapulnik related to seedling cultures of white lupine to which various concentrations of elicitors were added, such as: yeast extract, chitosan, the suspension of the symbiotic bacteria *Rhizobium loti* and copper chloride [41]. The authors identified high concentrations of several isoflavonoid metabolites in the media, namely: prenylated derivatives of genistein and 2'-hydroxygenistein and free aglycones. Such broadly designed experiments unambiguously confirmed that the type of elicitor conditions the level of the defence response. In the case of lupine seedlings, the strongest elicitor in the whole range of concentrations proved to be yeast extract (the concentration of free phytoalexins grew 48 times). The other elicitors induced the biosynthesis of isoflavones far less strongly (3–27 times more).

Moreover, the authors proved that depending on the type, the elicitors may selectively induce the biosynthesis of particular phytoalexins. Significant concentrations of 2'-hydroxygenistein and its prenylated derivatives were observed following incubation of plant material with yeast extract, chitosan or copper chloride. The suspension of *Rhizobium loti*, on the other hand, selectively induced the biosynthesis of genistein and its prenyl derivatives. Based on this observation, the authors concluded that the biosynthesis of genistein derivatives is specific for plants colonised with symbiotic micro-organisms. That is why genistein derivatives were called "symbiotic phytoalexins", unlike the 2'-hydroxygenistein derivatives synthesised during a typical defence response [41].

Similar effects were achieved by Wojtaszek and Stobiecki after using a yeast elicitor and copper chloride on excised cotyledons of white lupine [87]. Also in this case, the content of 2'-hydroxygenistein and its prenylated derivatives increased several times directly after elicitation, with no changes in the concentration of genistein.

Contrary to these reports, Shibuya [109] and Bednarek et al. [43] elicited entire lupine seedlings with yeast extract [43] or copper chloride [109] and observed accumulation of both 2'-hydroxygenistein and its prenylated derivatives, as well as genistein itself. These discrepancies are difficult to account for, due to differences in experimental procedures, including the varying time of elicitation and time intervals at which measurements were taken [43, 109].

All of the aforementioned literature reports indicate a variable way of storing the phytoalexins synthesised as a result of stress [15, 18, 19, 37–43, 48, 71, 78, 87–106, 109]. Given the different experimental procedures used, including the differences in plant material and elicitation type, it is difficult to define the cause of the observed variability.

The authors of the report on eliciting chickpea with yeast extract attempted to partly explain this problem [70, 71]. As the amount of the yeast elicitor in the culture increased, they measured increasing concentrations of phytoalexins in the medium. They link this effect with the possible permeability of cell walls as a result of toxic, high doses of the elicitor, leading in consequence to mechanical leaks of isoflavones into the medium. A similar positive relationship between the rate at which isoflavones were released and the dose of the elicitor was also observed in the cultures of white lupine seedlings [41].

It seems that it is not only the quantity, but also the type of elicitor that affects the storage of phytoalexins synthesised in its presence. Lupine cotyledons elicited with copper chloride [87] synthesised isoflavone phytoalexins of which only 0.09% was released into the medium. Contrary to that, yeast extract evoked a defence response of a more complex nature, in which 70% of the synthesised phytoalexins was released into the medium [109, 110].

The way in which phytoalexins were stored in elicited soybean roots seemed to depend directly on the type of isoflavonoids produced in response to stress [98]. A co-culture of this root tissue with *Fusarium solani*, f sp. *glycines* resulted in significant biosynthesis of genistein, glycitein, coumestrol and glyceollin. The only phytoalexin, 50% of which was released into the medium, was glyceollin. The storage of other compounds was intracellular. A similar relationship was observed in soybean suspension cultures elicited with a pathogenic bacterium of *Fusarium* genus [111]. The authors of this work believe that storage of various phytoalexins is related to their respective functions in a plant's defence system. Considering the above, glyceollin, which takes part in the systemic reaction, is released into the environment, while simple isoflavones responsible for the local defence response are stored inside cells near the infection zone [98, 111].

Also the authors of the publication that dealt with eliciting several-day-old seedlings of chickpea with glutathione broadly discuss the issue of intracellular storage or leakages of the synthesised phytoalexins [91]. In the designed experiment, a complete leak of the synthesised isoflavones from the root biomass into the environment was achieved. It was therefore concluded that the reason lies with the specific nature of the actual plant material used in the experiment, rather than with the quantity of the elicitor. The authors also presumed that in in vitro culture or intact plants, strict compartmentation of the synthesis of secondary metabolites and their storage takes place. Therefore, particular compartments (cells, tissues and organs) may have varying sensitivity to elicitation. For example, phytoalexins synthesised and stored in the elicitation zone, are responsible for a local defence response. The release of phytoalexins from the tissue is then the beginning of a systemic defence mechanism [42, 112]. Although both defence mechanisms are present in each plant biomass, in most cases one dominates over the other. In effect, researchers, who have at their disposal analytical methods of a particular sensitivity, identify phytoalexins either in the tissues or in the media [91].

Incubation of several-day-old seedlings of chickpea with a fungal elicitor (fusicoccin) suggests that the key to the active release of phytoalexins from the plant biomass may be the depolarisation of the plasma membrane and a change in the ion efflux across plasmalemma, cause by elicitation. Fusicoccin is a known inhibitor of the membrane H^+-ATPase responsible for maintaining intracellular pH and electric potential differences across the plasmalemma [113]. In this way, it more or less determines the main ionic permeability properties. That is why the authors of this paper concluded that ATPase may play a role in the signalling pathway leading to the activation of phytoalexins, but in particular actively affects isoflavonoid excretion [89]. That is why fusicoccin in chickpea seedlings induced the release of the synthesised isoflavones into the experimental medium. Identical effects were achieved by changing the acidity of the environment. A pH above 6 always caused a complete release of basic phytoalexins, such as medicarpin, biochanin A, maackiain and formononetin into the medium [89].

The experiments described herein on the biosynthesis of isoflavones in elicited cultures of Fabaceae plants clearly prove that the complete mechanism of the formation and metabolism of phytoalexins has not yet been fully explained. This is partly due to its complex nature and, consequently, the diverse results achieved by different researchers. It is not only related to the actual process in which phytoalexins are biosynthesised, but also the way they are stored. The effects of an external factor (i.e. the external elicitor itself) as well as the internal factors (i.e. the plant material) on these processes can not be neglected either.

It seems that for a biotechnologist working to develop in vitro plant systems that could produce high amounts of isoflavones of particular biological activity, the most important fact is that elicitation can lead to much higher concentrations of these compounds in the biomass. Moreover, it is essential from the technological point of view that it is possible to cause controlled release of the desired isoflavones into the growth medium [37, 41, 42, 92–99].

3.4.2 Supplementation with Biosynthesis Precursors

There are very few reports dealing with in vitro cultures of legume plants that cover supplementation with the so-called distant isoflavonoid precursors [33, 37, 38, 91, 114]. This technological procedure is used predominantly to identify the effect of these precursors on the biosynthesis and metabolism of isoflavonoids. Thus, researchers focus more on clarifying certain stages in isoflavone biosynthesis, rather than improving biomass productivity with regard to the investigated compounds [37, 38, 91, 114]. The precursors most commonly used in experiments concerning in vitro biosynthesis of isoflavones are early metabolites of the phenylpropanoid pathway such as phenylalanine and *trans*-cinnamic acid [33, 37, 38, 91, 114]. Experiments show that the phenomenon of including these compounds in the isoflavonoid pathway depend both on the type of precursor used and, indirectly, the tested growth system [37, 38, 114].

Trans-cinnamic acid added to experimental media was not used by the cultures as a substrate for isoflavone production, irrespective of the initial concentration [37, 38]. This phenomenon seems to be completely independent of the type of plant biomass.

In *Phaseolus vulgaris* suspension elicited with *Colletotrichum lindemutianum*, exogenous cinnamic acid was not only not used in the biogenic pathway leading to the formation of isoflavones, but it also inhibited the metabolism of phytoalexins conditioned by the occurrence of stress factor [37]. The added precursor only induced enzymes that take part in the metabolism of cinnamic acid, such as cinnamic acid 4-hydroxylase and specific glucotransferases. In effect, significant amounts of *trans*-cinnamic, *p*-coumaric and caffeic acids as well as their glucoside derivatives were identified in the bean suspension. Some of the exogenous (^{14}C) *trans*-cinnamic acid was also incorporated into insoluble cellulose and hemicellulose fractions of the biomass [37].

Similar effects of supplementing in vitro cultures with cinnamic acid were observed in the case of *Cicer arietinum* suspension elicited with yeast extract [38]. Also here, cinnamic acid underwent immediate glycosidation, was built into insoluble fractions of cell walls and additionally was fixed with glutathione. The described processes were accompanied by inhibited activity of the basic enzymes that take part in the isoflavone pathway (PAL and isoflavone methyltransferase) [37, 38, 114].

The described experiments refer to biomasses elicited directly before supplementing a precursor. Therefore, the eventual metabolism of cinnamic acid results from the stress response, as proved by the biosynthesis of phenolic acids and their glucosides, of certain antibacterial and antifungal properties [37, 38, 114]. The metabolism of *trans*-cinnamic acid applied to isoflavone-producing cultures could be completely different.

Unlike cinnamic acid, phenylalanine added to root cultures of *Cicer arietinum* was actively incorporated in the phenylpropanoid pathways, leading to the formation of isoflavones. When applied to chickpea cultures, this feeding procedure resulted in increased concentrations of biochanin A, formononetin, homoferreirin and cicerin [91].

A two-fold increase in the total isoflavone content was also achieved in callus cultures of *Maackia amurensis*, supplemented with phenylalanine [33]. In this experiment, the addition of benzoic acid, which does not take part in the biosynthesis of metabolites of phenylpropanoid pathway [78, 79, 88], did not stimulate the accumulation of the investigated compounds [33].

The biotechnological research quoted above proves that the problem of low productivity of in vitro cultures with regard to isoflavonoids could be possibly solved by supplementing the biomasses with selected, early metabolites of the phenylpropanoid pathway, and especially phenylalanine [33, 91].

3.4.3 Biotransformation

Controlled biosynthesis of pharmacologically active isoflavones is carried out in vitro with the participation of selected micro-organisms and legume plant cultures which have enzymatic systems capable of transforming endogenic isoflavones, and also xenobiotics with similar structure, added to experimental media [36, 40, 78, 84, 86, 115–128]. The one-stage or two-stage process of isoflavone biotransformation most often includes isomerisation, dehydrogenation, hydrolysis, hydroxylation, esterification, demethylation and glycosidation of the respective substrates, to finally lead to the formation of compounds with the desired structure [78, 84, 86, 122].

One of the two basic strategies used in these types of experiment is to isolate selected enzymatic fractions or enzymes from plant or bacterial in vitro cultures and then use them in reaction mixtures supplemented with the respective precursors and co-factors in order to obtain the required isoflavone [115, 116, 118, 119, 121, 126–128]. This strategy, which resembles procedures used in strictly chemical technologies, was used to obtain daidzein from liquiritigenin, with the participation of microsomal fraction isolated from *Pueraria lobata* suspension [116]. The two-stage reaction involved transformation of the flavone (liquiritigenin) to 2,7,4'trihydroxyisoflavone with the participation of microsomal mono-oxygenase. In the second stage of the process, the 2,7,4'trihydroxyisoflavone was immediately bioconverted to daidzein, through the respective dehydratase.

The strategy of first isolating microsomal fraction from lupine suspension and then using it in a reaction mixture with genistein and 2'-hydroxygenistein led to the formation, from those two compounds, of products prenylated at positions 6, 8 and 3' [118]. However, flavonols and previously monoprenylated isoflavones, like 6-prenylpolyhydroxyisoflavone, were not biotransformed in the mixture. This indicates the high degree of substrate specificity of prenyltransferases in lupine with respect to particular isoflavonoids. This is confirmed by an analogous experiment using a microsomal fraction of *Phaseolus vulgaris* suspension, which only prenylated isoflavones at positions 2 and 4 [128].

Also, the microsomal fraction of *Cicer arietinum* suspension in the presence of NADPH and molecular oxygen catalysed metoxyisoflavone monohydroxylation at positions 2', 3' and 4'. This bioconversion led to the formation of maackiain and medicarpin from biochanin A and formononetin, respectively [115].

Because of the need to isolate enzymatic fractions, the procedures described above involve multiple stages and are therefore time-consuming [115, 118, 121, 126–128]. Since it is necessary to use expensive and often unstable co-factors (e.g. NADPH), despite good productivity and purity of the products, these processes are not economically viable. That is why they are used predominantly to obtain rare isoflavonoid standards or to explain particular biosynthetic pathways of these compounds [126]. More potential for large-scale bioconversions of biologically active isoflavones lies in technologies using actively growing cultures of legume plants or micro-organisms capable of transforming isoflavonoid substrates added directly to growth media [36, 40, 117, 119, 123, 127, 128].

An example of these types of experiment using cultures of selected microorganisms are *Brevibacterium epidermides* and *Micrococcus luteus* cultures grown in the presence of glycitein, genistein or daidzein [117, 119]. These bacteria, which occur naturally in the ferment used to process soybean pulp into a high-protein product called tempe [120], proved capable of 6-O-demethylation of glycitein and 6-O-hydroxylation of daidzein, glycitein and daidzein. Compared with the substrates, the products, obtained with 60% efficiency, have much higher antioxidant activity [117].

Another interesting biocatalyst seems to be cultures of *Aspergillus niger*, which transforms exogenously supplemented daidzein to 8-hydroxyformononetin and 7,4'-dimethoxyflavone to daidzein [127].

The use of *Fusarium pomhferatum* made it possible to bioconvert exogenous formononetin to daidzein, which has a much stronger phyto-oestrogenic activity [128]. Also, in vitro cultures of anaerobic bacteria dwelling in human intestines, *Eubacterium limosum* (ATCC 8486), were capable of demethylation (with 90% efficiency) of biochanin A, formononetin and glycitein to genistein, daidzein and 6,7,4'-trihydroxyflavone, respectively [121]. All products obtained in a 26-day process had stronger phyto-oestrogenic activity than the original substrates. Unlike processes that naturally occur in the human intestine, the described in vitro cultures were not capable of performing further demethylation of isoflavones, with equol as the end product [121].

A potential for bioconversion in the broadly understood isoflavone class also exists in actively growing in vitro cultures of legume plants [36, 40, 123]. For example, the suspension of *Glycyrrhiza echinata* was capable of transforming exogenous 2,7,4'-trihydroxyisoflavanone to formononetin [123]. The indirect product of this reaction occurring in in vitro cultures was 2,7-dihydroxy-4'-methoxyisoflavone, and not daidzein, as in the intact plant. Thus, in the described growth system, the addition of an exogenous substrate led to the formation of a new, alternative metabolic pathway, leading to the biosynthesis of formononetin [123].

The rare 7-hydroxy-6,4'-dimethoxyisoflavone (afrormosin) was obtained in few-day-old seedlings of *Onobrichis vicifolia*, supplemented with formononetin, daidzin and texasin, and labelled with [Me-^{14}C] [36]. All of the applied substrates were bioconverted in a two- or three-stage process into afrormosin with 90% efficiency [36].

In the case of *Pueraria lobata* suspension, several isoflavonoid aglycones, such as biochanin A, formononetin, genistein and daidzein, were successfully biotransformed with 70% efficiency to respective glucoside and malonate derivatives [40]. Moreover, the methyltransferases and glycosidases present in the culture proved not to be very stable. In addition to isoflavones, they were able to biotransform such flavones as chrysine, acacetin and apigenin and a simple phenol, resacetophenone [40]. As a result of this procedure, in vitro cultures of *Pueraria lobata* were obtained that had much broader secondary metabolism.

It is currently known that enzymes related to the biosynthesis of isoflavones are organised in multi-enzymatic aggregates that not only allow enzyme-to-enzyme reactions, but also condition the appropriate redirection of substrates [48, 128]. The broadly understood compartmentation of isoflavone metabolism, together with feedback inhibition, related here to the accumulation of products, limits the use of growing in vitro cultures to produce particular phyto-oestrogens through bioconversion.

3.4.4 Genetic Modifications

With the development of molecular biology and the genetic engineering strategies that are related to it, research was undertaken into the biosynthesis of isoflavonoid based on transgenic cultures [48, 78, 79, 128].

Part of the discussed research projects deals with hairy root cultures of selected species of the Fabaceae family. In this case, the effect of the actual agroinfection on the biosynthesis and metabolism of isoflavones in previously obtained transgenic roots is identified [26, 31, 45–47, 93, 96, 99, 103, 106, 129]. However, controlled biosynthesis of selected phyto-oestrogens (isoflavones, pterocarpans and coumestans) is the subject of experiments on transgenic plant biomasses in which the donor genome is supplemented with alien genes related to the biosynthesis of the selected enzymes of the phenylpropanoid pathway (e.g. CHS, chalcone isomerase – CHI –, isoflavone reductase) [122, 126, 128–139]. These transgenes are introduced with the use of bacterial vectors [15, 26, 31, 70, 130, 132–135] or particle bombardment [26, 84, 128, 132, 133, 138] of biomasses originating from legume plants [26, 31, 84, 128–130, 132, 134–137, 139] or species that do not normally synthesise isoflavonoids in nature (e.g. *Arabidopsis thaliana* and *Nicotiana tabacum*) [26, 126, 128, 131, 132].

From the technological point of view, the genetic transformation of plants belonging to the Fabaceae family appears to be relatively difficult [43, 129, 134,

136, 138]. Numerous authors see the reason for this fact in the specific defence mechanisms characteristic of legume plants and related to symbiotic processes involving these plants and bacteria of the *Rhizobium* genus [31, 43, 128, 129, 136, 138]. The difficulties lie both in the actual introduction of the transgene into the host cells and in obtaining sterile and yet vital plant cultures at a later stage in the case of bacterial transformations. One important problem is the selection of clones with stable expression of alien structural genes and the possible regeneration of transgenic plants [43, 128, 129, 132–139]. In view of these difficulties, numerous strategies were developed to make it possible to obtain stable transformed plant material [26, 128, 134, 135, 137–139].

The model species in research on genetic transformation of Fabaceae plants include *Glycine max*, *Cicer arietinum*, *Lupinus albus* and *Medicago sativa* (i.e. basic crop plants of this family) [134–139]. Research indicates that the transgenesis process in papilionaceous plants is determined directly by the selection of the transformation technique. It was found that non-vector transformation methods are still not very effective given the nature of the incorporated structural genes (e.g. conditioning frost resistance or the biosynthesis of selected enzymes of the phenylpropanoid pathway) [128, 132, 133, 135–139]. Irrespective of the original biological material (plant species, tissue type), each time the survival rate of the plant biomass was low, so was the percentage of copies of the alien DNA fragment included in the recipient's genome [128, 139]. The authors of the quoted reports concluded that the reason for the numerous failures was acoustic stress suffered by plant cells, and to a smaller degree the subsequent selection of transformants [139]. In view of this, the use of bacterial vectors in the form of wild or pre-constructed strains of the *Agrobacterium* genus is still considered to be the most effective method of genetically transforming Fabaceae plants. In this case the effectiveness of the transgenesis depends both on the type of the original plant material and on the concentration of the bacterial suspension in the co-culture. An important role is also played by the timing of co-culture cultivation and a possible supplementation with transformation "stimulants" [84, 128, 130–138]. In experiments involving *Glycine max* [137] and *Cicer arietinum* [134], the transformation process seemed more effective when young, unorganised plant material was used (embryo or parenchymatic suspension).

The effectiveness of transgenesis in Fabaceae plants was enhanced by a supplement of acetosyringone, which activates *vir* genes in bacteria [134, 135, 137]. However, this phenol selectively inhibited the subsequent formation of somatic embryos from the previously transformed embryo suspension of soybean [137]. The process of soybean embryo transformation was also accelerated by a supplement of such antioxidants as L-cysteine, ascorbic acid, polyvinylpyrolidone and dithiothreitol [138].

The experiments involving soybean cotyledons and embryos demonstrated that the probability of plant material transformation increases significantly after the co-culture is subjected to sonication lasting several seconds [135, 137]. According to various authors, this procedure causes the formation of numerous micro-wounds, both on the surface and within plant tissue. As a result, this

Chapter 3   Research into Isoflavonoid Phyto-oestrogens in Plant Cell Cultures   75

event triggers the synthesis of plant phenols, which increase the accessibility of possible factors that bind bacteria cell walls [135, 137, 140–142].

The transformation process of *Glycine max* and *Cicer arietinum* also depended on the duration of co-culture of plant material and bacteria [134, 135], the pH of the medium [135] and the temperature of the environment [135, 136]. It was found that temperatures below 27°C and co-culture lasting more than 2 days clearly reduced the survival rate of the plant material [135, 136]. A medium pH of above 7, on the other hand, inhibits the activity of the bacterial *vir* gene [135].

Numerous research projects on the transformation of papilionaceous plants showed their poor resistance to the antibiotics used to eliminate bacterial strains from the co-culture [133, 134, 137, 138]. The high survival rate of the transformed plant material can be achieved by removing bacteria form the culture in a gradual way. To do this, post-transformation plant biomasses are cultivated with a growing antibiotic gradient, which builds up in the plants the desired tolerance for the antibacterial agent [134, 136–138].

Using these experimental procedures, several researchers managed to obtain biological material with a different isoflavonoid metabolism than the original Fabaceae plants [26, 31, 84, 128–130, 131, 134–137, 139, 143].

Fusion of the maize transcription factor CRC caused major changes in the levels of isoflavone compounds in a soybean embryo suspension [133]. The activity of the basic enzymes of the isoflavonoid pathway, such as PAL, CHS, CHI and isoflavone synthase (IFS) increased considerably in the transformed suspension. Compared to the original biomass, the transformed cultures produced four times more isoflavones, predominantly daidzein derivatives. Due to the post-transformational activation, flavanone 3-hydroxylase was increased; however, the general concentration of genistein in the suspension decreased. This enzyme effectively competed with flavanone isomerase for a shared precursor (i.e. naringenin). In effect, in the soybean suspension the presence of flavonole derivatives was identified, together with a decreasing accumulation of genistein [133]. A similar experimental procedure used with alfalfa protoplasts [144], which, together with intentional suppression of flavanone 3-hydroxylase, produced transformants synthesising selectively a rich set of isoflavonoids [144]. In both cases, the multigene phenylpropanoid pathway was activated in the plant material, together with the simultaneous redirection of the respective substrates towards phyto-oestrogenic isoflavonoids [133, 144].

Genetic transformation activates particular fragments of the phenylpropanoid pathway not only in papilionaceous plants, which constitutionally produce isoflavonoids, but also in species that do not naturally synthesise these compounds, as they do not have the appropriate enzymatic systems [26, 126, 128, 131, 132].

The introduction of genes that encode the biosynthesis of IFS into embryo suspensions of *Arabidopsis thaliana* [26, 128, 131, 132] led to the accumulation of previously unsynthesised genistein in the recipient's biomass. In a similar experiment, Liu and Dixon [125] found out that in the *Arabidopsis thaliana* suspension the aforementioned phytoalexin was immediately transformed to a respective rhamnoside, glucoside and glucoside-rhamnoside. "Local" enzymes that effectively glycoside the "alien" metabolite are thought to be responsible for this effect [26, 125, 128].

If the host's cells lack the right substrates, the introduction of genes coding certain enzymatic systems is often insufficient to obtain the desired isoflavone. *Nicotiana tabacum* suspension, which has high pro-transformation activity of isoflavone lupine reductase of lupine origin, produced vestitone only after the respective precursor (i.e. 2'-hydroxyformononetin) was added to the culture [126]. At the same time, both the substrate and the product underwent immediate bioconversion to the respective glucoside, with the participation of tobacco glycosidases [126]. Thus, the process of agroinfection has a considerable impact of isoflavonoid fractions in the transformed cultures of legume plants [26, 31, 45–47, 93, 96, 98, 103, 128, 129, 143, 145].

Hairy roots of *Genista tinctoria* established after inoculation of in vitro shoot cultures with *Agrobacterium rhizogenes* strain ATCC 15834 were only capable of selective production of isoliquiritigenin, a daidzein precursor that is absent in the intact plant. This compound was stored entirely within cells and it was not until ABA was added that approximately 80% of it was released into the medium. Consequently, a prototype basket-bubble bioreactor was designed and built to upgrade the scale of *Genista tinctoria* hairy root cultures. With immobilised roots and a new aeration system, large amounts of biomass were obtained that produced a high content of isoliquiritigenin (2.9%) [143].

Hairy roots obtained by infecting lupine seedlings with *Agrobacterium thizogenes* (15834) produced orobol 7,3'-diglucoside, 3'-O-methylorobol 7-O-glucoside, and genistein 7,4'-diglucoside, compounds identical to those synthesised in de novo roots of *Lupinus albus*. However, the concentration of these compounds in transgenic cultures was ten times higher [43]. Transformed roots of seven species of the *Psoralea* genus also synthesised higher amounts of daidzein and coumestrol than intact plants [45]. At the same time, the authors of this experiment found that younger roots selectively synthesised daidzein, while older roots produced higher concentrations of coumestrol. This piece of information may be useful if *Psoralea* hairy roots are used in the large-scale production of both of these phyto-oestrogens.

The differences in the levels of synthesised isoflavones were also noted in non-transformed roots and hairy roots of *Glycine max* [98]. Compared to de novo roots, far more free daidzein was identified in a transgenic culture, with only traces of ester and glycoside derivatives of this compounds, which dominate in the roots of the maternal plant.

Entirely new isoflavonoids, not identified in intact plants, were produced by transgenic roots of *Glycyrrhiza gabra*, *Glycyrrhiza pallidiflora*, *Glycyrrhiza ularensis* and *Glycyrrhiza aspera* [46, 47]. Hairy roots of these species, obtained by infecting young seedlings with *Agrobacterium rhizogenes* pRi 15834 and pGS-Glu1, are characterised by a high content of prenylated isoflavones, such as licoagaroside A, calycosin, and a chalcone (i.e. isoliquiritigenin) [46, 47].

Chapter 3   Research into Isoflavonoid Phyto-oestrogens in Plant Cell Cultures

The examples of biotechnological experiments described above prove that the biosynthesis of isoflavones for which respective precursors, enzymes and coding genes have been identified, can be controlled in a practically unlimited way, and the genetic engineering technology to obtain biomasses selectively producing large quantities of phyto-oestrogens is just a step away.

3.5   In Vitro Cultures of Legume Plants Oriented for Selective Production of Phyto-oestrogens

Of the vast number of research projects dealing with isoflavone biosynthesis and metabolism in in vitro cultures, only few reports are directly oriented at obtaining plant biomasses that could be rich sources of active phyto-oestrogens [27, 28, 31–33, 45, 70, 146]. The key role in these type of experiments is played by the selection of the original plant material. Comparable or larger concentrations of phyto-oestrogens than in intact plants were obtained from in vitro cultures of species that, already in their intact form, were characterised by a rich isoflavonoid metabolism. These included cultures of *Glycine max*, *Maackia amurensis* and *Cicer arietinum*, and *Lupinus*, *Psoralea* and *Genista* species [27, 28, 31–33, 45, 55, 56, 70, 146].

The biosynthesis and distribution of isoflavonoid compounds in a plant, unlike several alkaloids, is not organ related [19, 27, 32, 84]. Therefore, the type of the original explant does not directly affect the content of isoflavones in the initial biomasses. Highly productive biomasses were obtained in in vitro cultures of five *Psoralea* species, in callus originating from cotyledons, hypocotyls and roots [32]. A similar pattern was observed in callus lines of *Maackia amurensis* originating from petioles and terminal meristems of the intact plant [33].

Research indicates that the eventual isoflavone content in in vitro cultures of papilionaceous plants depends directly on the type of culture [32, 33]. It has been proved that callus cultures allow for preparing biomasses in vitro that synthesise isoflavones in higher quantities than the maternal plants [32, 33]. High concentrations of daidzein, genistein, formononetin, retusin, maackiain and medicarpin were achieved in *Maackia amurensis* calli [33]. The average amount of phyto-oestrogens in the investigated cell lines (2%) exceeded the content of isoflavones in the intact plants even 4 times over, and stayed at a constant level over 13 subsequent passages [33].

Callus cultures of six *Genista* species produced more isoflavones than the respective intact plants. The isoflavonoid group produced comprised 14 compounds, with clear domination of genistin. The callus of highest isoflavone content was obtained from *Genista tinctoria*, producing 6.5% of the isoflavones [55].

Liquid root, shoot, embryo and suspension cultures of *Genista tinctoria* cultivated in the modified SH medium, in comparison with the intact plants, had a much higher isoflavone content (six to nine times). Moreover, none of the cultures produced the simple flavones characteristic of the natural plant. It was noted that the ratio of isoflavone accumulation was in this case influenced greatly by tissue differentiation [147].

The highest isoflavone accumulation was observed in *Genista tinctoria* suspension cultures (9.14% of genistin), which is one of the highest reported amounts of a single metabolite produced by an in vitro culture of higher plants [55, 147].

The somaclonal variation, typical for callus biomasses, was not observed in cases of highly productive cell lines of *Glycine max* [28]. Federici et al. selected from 40 callus lines of soybean and obtained biomasses that for 25 years synthesised isoflavones at the stable level of approximately 5% dry weight. In addition, the soybean suspension originating from that callus was characterised by a high (7%) and constant concentration of genistein and daidzein derivatives [28]. *Cicer arietnum* also proved stable in terms of isoflavone production. Unlike the soybean biomass, it synthesised derivatives of biochanin A and formononetin only in similar quantities as produced by the intact plant [70].

On the other hand, considerable somaclonal variation was observed in callus cultures of five species of the *Psolarea* genus, i.e. *Psolarea cinerea, Psolarea macrostachya, Psolarea bituminosa, Psolarea tenex* and *Psolarea obtusifolia* [32]. Irrespective of the original species, all calluses synthesised daidzein in quantities that were five times higher than in intact plants. However, the concentration of daidzein in the calluses changed (0.6–1%) with subsequent passages [32]. The undesirable lack of long-term genetic stability of *Psolarea* calli was overcome by establishing hairy root cultures of these species [45, 146]. Compared to the calli, they not only synthesised more daidzein (ca. 0.8%), but at stable levels over a 2-year period when the culture was maintained. Moreover, the transformed roots produced more plant biomass and, unlike the calli, synthesised daidzein throughout the entire growth cycle, irrespective of the growth phase [45].

The direct effect of the culture type on the isoflavone content in plant biomasses was also noted in in vitro cultures of *Lupinus polyphyllus* and *Lupinus hartwegii* [31]. Berlin et al. [31] compared the production of isoflavones in standard non-transformed suspension cultures and transgenic cultures (roots and suspensions) obtained by transforming lupine tissues with wild strains of *Agrobacterium rhizogenes* and *Agrobacterium tumefaciens*. Irrespective of the culture type, all of the investigated biomasses synthesised a ten-component set of genistein and 2'-hydroxygenistein derivatives in similar quantities as intact plants. It was found that long-term stability in isoflavone production is directly related to the degree of cytodifferentiation of the biomasses. Only the hairy roots of lupine were characterised by a stable content of isoflavones. In both the transgenic and the non-transformed suspensions, the level of isoflavones varied from 0.2 to 0.5% and depended on the degree of biomass aggregation, which varied with time.

The process of transgenesis itself resulted in over four times higher concentration of isoflavones in lupine tissues. Berlin et al. [31] connected this fact with the possible growth of transformed biomasses on media lacking phytohormones, which could adversely affect the biosynthesis of particular phytooestrogens.

In projects on isoflavone biosynthesis, attempts were made to solve the common problem of low phyto-oestrogen content in plant biomasses cultured in vitro. To that end, additional cultivation procedures were undertaken; suspensions were immobilised on beads from calcium alginate [27] and powdered cork tissue [148]. In this way, an attempt was made to create biochemical conditions in the growth system that would be similar to those in intact plants [148]. The addition of cork tissue to the suspension culture of *Sophora flaverscens* caused even a five-fold increase in the biosynthesis of prenylated isoflavones, typical for the investigated species. At the same time, in the described case, the use of cork powder brought an additional effect of biomass permeabilisation, related to the release of 70% of the isoflavonoid metabolites into the cork tissue [148]. Unlike the in vitro culture of *Sophora flaverscens*, immobilising *Glycine max* suspension on calcium alginate beads did not affect the eventual concentration of isoflavones in the biomass. This technological procedure only brought one advantage, that of extending the growth period [27].

These experiments, which resulted in biomasses that produce high concentrations of isoflavones, were only conducted on laboratory scale [31–33, 45, 70, 146]. The only attempts to develop continuous systems for isoflavone production based on plant biomasses pertain to *Glycine max* suspensions and *Genista tinctoria* co-cultures [27, 28, 149].

In order to obtain considerable quantities of soybean phyto-oestrogens on large scale, Ames and Worden [27] constructed a continuous growth system based on a magnetofluidised bed bioreactor. The innovative features of this system include both the construction of the "growth vessel" and the procedure in which the plant biomass is prepared for the technological process [27]. The actual bioreactor chamber was in this case a glass column surrounded with solenoids, producing a magnetic field of specific intensity. Prior to being applied to the bioreactor column, the soybean suspension was immobilised onto the so-called two-phase support, which was made of calcium alginate beads and powdered magnetite [27]. While the role of calcium alginate beads was to create the biochemical conditions to enhance isoflavone production, the addition of magnetite kept the suspension evenly dispersed in the bioreactor's magnetic field. In this way, the suspension was not carried with the stream of medium pumped into the column. The purpose of this type of technological solution was to protect delicate plant suspension cells from the shear stress induced by the growth process. The regulated intensity of the magnetic field made it possible to periodically release the biocatalyst into a receiving vessel and to apply fresh biomass to the column. This growth system made it possible to maintain soybean suspension culture in a continuous way without any detriment to the vitality of the biomass. Unfortunately, the applied growth procedure contributed to a dramatic decrease of isoflavone content in the suspension. In the case of daidzein it was reduced by approximately 50%, and in the case of genistein, by approximately 20%. Such a dramatic inhibition of phyto-oestrogen biosynthesis was not observed in the soybean suspension grown in a standard air-lift bioreactor. The described culture maintained isoflavone production capability at the level of a standard liquid culture [28].

An in vitro growth system based on *Genista tinctoria* shoot and hairy root co-culture was developed in order to produce large amounts of phyto-oestrogens, derivatives of genistein and daidzein. The different tissue inoculation ratios were tested to achieve the best growth of *Genista tinctoria* shoots and roots in the co-culture system. The hairy roots produced large amounts of a single bioflavonoid, isoliquiritigenin (2.5%), which is a daidzein precursor that is absent in the intact plant. Only after the addition of ABA, was isoliquiritigenin almost completely released into the growth medium, from which it was used by the shoots to produce significant amounts of daidzein and daidzin. Moreover, *Genista tinctoria* shoots in the co-culture system, like in a monoculture [147], maintained the ability to produce high amounts of genistin (6.69%) and its derivatives [149]. As a result of the described bioconversion of isoliquiritigenin, the shoots synthesised 38 times more daidzin than the intact plant [149]. The prototype basket-bubble bioreactor was designed to upgrade the scale of the *Genista tinctoria* co-culture. The new device significantly improved the growth parameters and the productivity of both tissues [149].

The few relevant literature reports presented above indicate that legume plants introduced into in vitro cultures are able to maintain the capability to biosynthesise significant amounts of isoflavones. Moreover, it seems that based on modern growth procedures, it is possible to develop a viable growth system that would produce valuable phyto-oestrogens on the basis of "natural bioreactors", that is, plant biomasses.

References

1. Boulter D (1995) Phytochemistry 40:1
2. Bourgaud F, Gravot A, Milesi S, Gontier E (2001) Plant Sci 161:839
3. Maleszy S (2004) Wprowadzenie. In: Maleszy S (ed) Biotechnologia Rośli. Wydawnictwo Naukowe PWN SA, Warszawa, p 15
4. Király Z (1986) Cell wall composition and metabolism. In: Goodman RN, Király Z, Wood KR (eds) The Biochemistry and Physiology of Plant Disease. University of Missouri Press, Columbia, p 105
5. Heinstein PF (1985) J Nat Prod 48:1
6. Alfermann AW, Petersen M (1995) Plant Cell Tiss Org Cult 43:199
7. Stöckigt J, Obitz P, Falkenhagen H, Lutterbach R, Endress S (1995) Plant Cell Tiss Org Cult 43:97
8. Yeoman MM, Yeoman CL (1996) New Phytol 134:553
9. Ramachandra Rao S, Ravishankar GA (2002) Biotechnol Adv 20:101
10. Oksman-Caldentey KM, Inzé D (2004) Trends Plant Sci 9:433
11. Verpoorte R, van der Heijden R, ten Hoopen HJG, Memmelink G (1998) Plant Tiss Cult Biotechnol 4:3
12. Kieran PM, MacLoughlin PF, Malone DM (1997) J Biotechnol 59:39
13. Verpoorte R (2000) Transgenic Res 9:323
14. Verpoorte R (2002) Plant secondary metabolism. In: Verpoorte R, Alfermann AW (eds) Metabolic Engineering of Plant Secondary Metabolism. Kluwer Academic, Dordrecht, Boston, London, p 1
15. Dixon RA, Steele CL (1999) Trends Plant Sci 4:340
16. Qiang Ren M, Kuhn G, Wegner J, Chen J (2001) Eur J Nutr 40:135

17. Grynkiewicz G, Gadzikowska M (2003) Postępy Fitoterapii 10:28
18. Phillips DA, Kapulnik Y (1995) Trends Microbiol 3:58
19. Dixon RA (2001) Nature 411:843
20. Brandi ML (1997) Calcif Tissue Int 61:55*
21. Fritsche S, Steinhart SH (1999) Eur Food Res Technol 209:153
22. Mueller SO, Korach KS (2001) Mechanism of estrogen receptor-mediated agonistic and antagonistic effects. In: Metzler M (ed) The Handbook of Environmental Chemistry, vol. 3. Springer, Berlin, Heidelberg, p 3
23. Wuttke W, Jarry H, Becker T, Schultens A, Christoffel V, Gorkow Ch, Seidlová-Wuttke D (2003) Maturitas 44:9
24. Kuntz S, Wenzel U, Daniel H (1999) Eur J Nutr 38:133
25. Radzikowski Cz, Wietrzyk J, Grynkiewicz G, Opolski A (2004) Postępy Hig Med Dośw 58:128
26. Dixon RA, Ferreira D (2002) Phytochemistry 60:205
27. Ames TT, Worden RM (1997) Biotechnol Prog 13:336
28. Federici E, Touché A, Choquart S, Avanti O, Fay L, Offord E, Courtois D (2003) Phytochemistry 64:717
29. Dixon RA, Fuller KW (1976) Physiol Plant Pathol 9:299
30. Goossens JFV, Vendrig JC (1982) Planta 154:441
31. Berlin J, Fecker L, Rügenhagen C, Sator C, Strack D, Witte L, Wray V (1991) Z Naturforsch 46c:725
32. Bouque V, Bourgaud F, Nguyen C, Guckert A (1998) Plant Cell Tiss Org Cult 53:35
33. Fedoreyev SA, Pokushalova TV, Veselova MV, Glebko LI, Kulesh NI, Muzarok TI, Seletskaya LD, Bulgakov VP, Zhuravlev YN (2000) Fitoterapia 71:365
34. Pinto ME, Casati P, Hsu TP,Ku MSB, Edwards GE (1999) J Photochem Photobiol B: Biol 48:200
35. Janas KM, Cvikrová M, Pałagiewicz A, Szafrańska K, Posmyk MM (2002) Plant Sci 163:369
36. Al-Ani HAM, Dewick PM (1980) Phytochemistry 19:2337
37. Edwards R, Mavandad M, Dixon RA (1990) Phytochemistry 29:1867
38. Barz W, Mackenbrock U (1994) Plant Cell Tiss Org Cult 38:199
39. Paiva NL, Oommen A, Harrison MJ, Dixon RA (1994) Plant Cell Tiss Org Cult 38:213
40. Park HH, Hakamatsuka T, Sankawa U, Ebizuka Y (1995) Phytochemistry 38:373
41. Gagnon H, Ibrahim RK (1997) Phytochemistry 44:1463
42. Daniell T, O'Hagan D, Edwards R (1997) Phytochemistry 44:285
43. Bednarek P, Frański R, Kerhoas L, Einhorn J, Wojtaszek P, Stobiecki M (2001) Phytochemistry 56:77
44. Thiem B (2003) Plant Sci 165:1123
45. Bourgaud E, Bouque V, Guckert A (1999) Plant Cell Tiss Org Cult 56:97
46. Li W, Asada Y, Yoshikawa T (2000) Phytochemistry 55:447
47. Li W, Asada Y, Koike K, Hirotani M, Rui H, Yoshikawa T, Nikaido T (2001) Phytochemistry 58:595
48. Dixon RA, Lamb ChJ, Masoud S, Sewalt VJH, Paiva NL (1996) Gene 179:61
49. Marita JM, Ralph J, Hatfield RD, Guo D, Chen F, Dixon RA (2003) Phytochemistry 62:53
50. Grajek W (2004) Biosynteza metabolitów wtórnych w kulturach in vitro. In: Maleszy S (ed) Biotechnologia Roślin. Wydawnictwo Naukowe PWN SA, Warszawa, p 306
51. Łuczkiewicz M, Cisowski W (2001) Plant Cell Tiss Org Cult 65:57
52. Parker CW, Letham DS, Gollnow BI, Summons RE, Duke CC, MacLeod JK (1978) Planta 142:239
53. Wright MS, Koehler SM, Hinchee MA, Carnes MG (1986) Plant Cell Rep 5:150
54. Song J, Sorensen EL, Liang GH (1990) Plant Cell Rep 9:21
55. Łuczkiewicz M, Głód D (2003) Plant Sci 165:1101
56. Łuczkiewicz M, Piotrowski A (2005) Z Naturforsch 60c:557

57. Gamborg OL, Miller RA, Ojima K (1968) Exp Cell Res 50:151
58. Philips GC, Collins GB (1979) Crop Sci 19:59
59. Dhir SK, Dhir S, Widholm JM (1992) Plant Cell Rep 11:285
60. Komatsuda T, Lee W, Oka S (1992) Plant Cell Tiss Org Cult 28:103
61. Malik KA, Saxena PK (1992) Planta 186:384
62. Botia JM, Ortuño A, Sabater F, Acosta M, Sánchez-Bravo J (1994) Planta 193:224
63. Kaneda Y, Tabei Y, Nishimura S, Harada K, Akihima T, Kitamura K (1997) Plant Cell Rep 17:8
64. Tan SC (1980) Aust J Plant Physiol 7:159
65. Murali NS, Teramura AH (1985) Physiol Plant 63:413
66. Bongue-Bartelsman M, Phillips DA (1995) J Plant Physiol 33:539
67. Ebel J, Schmidt WE, Loyal R (1984) Arch Biochem Biophys 232:240
68. Margna U, Margna E, Vainjarv T (1989) J Plant Physiol 134:697
69. Sakamoto Iida K, Sawamura K, Hajiro K, Asada Y, Yoshikawa T, Furuya T (1993) Phytochemistry 6:357
70. Kessmann H, Barz W (1987) Plant Cell Rep 6:55
71. Kessmann H, Choudhary AD, Dixon RA (1990) Plant Cell Rep 9:38
72. Tillberg E (1974) Physiol Plant 31:106
73. Dixon RA, Bendall DS (1978) Physiol Plant Pathol 13:295
74. Hinderer W, Seitz HU (1988) Flavonoids. In: Constabel F (ed) Cell Culture and Somatic Cell Genetics of Plants, vol 5. Academic Press, San Diego, New York, Berkley, Boston, London, Sydney, Tokyo, p 23
75. Hirata K, Asada M, Yatani E, Miyamoto K, Miura T (1993) Planta Med 59:46
76. Kishima Y, Shimaya A, Adachi T (1995) Plant Cell Tiss Org Cult 43:67
77. Łuczkiewicz M, Zárate R, Dembińska-Migas W, Migas P, Verpoorte R (2002) Plant Sci 163:91
78. Dixon RA (1999) Isoflavonoids Biochemistry, Molecular Biology, and Biological Functions. In: Barton D, Nakanishi K, Meth-Cohn O (eds) Comprehensive Natural Products Chemistry, vol 1. Elsevier, Amsterdam, Lausanne, New York, Oxford, Shannon, Singapore, Tokyo, p 773
79. Winkel-Shirley B (2002) Curr Opin Plant Biol 5:218
80. Beggs CJ, Schneider-Zieberg U, Wellmann E (1985) Plant Physiol 79:630
81. Sallaud C, El-Turk J, Breda C, Buffard D, de Kozak I, Esnault R, Kondorosi A (1995) Plant Sci 109:179
82. Dixon RA, Paiva NL (1995) Plant Cell 7:1085
83. Zhang F Smith DL (1996) J Exp Bot 47:785
84. Dixon RA (1980) Plant tissue culture methods in the study of phytoalexin induction In: Ingram DS, Helgeson JP (eds) Tissue Culture Methods for Plant Pathologists. Blackwell, Oxford, p 185
85. Cen YP, Bornman JF (1990) J Exp Bot 41:1489
86. Wink M (2003) Phytochemistry 64:3
87. Wojtaszek P, Stobiecki M (1997) Plant Physiol Biochem 35:129
88. Gunia W, Hinderer W, Wittkampf U, Barz W (1991) Z Naturforsch 46c:58
89. Armero J, Tena M (2001) Plant Sci 161:791
90. Guenoune D, Galili S, Phillips DA, Volpin H, Chest I, Okon Y, Kapulnik Y (2001) Plant Sci 160:925
91. Armero J, Requejo R, Jorrin J, Lòpez-Valbuena R, Tena M (2001) Plant Physiol Biochem 39:785
92. van Etten HD, Smith DA (1975) Physiol Plant Pathol 5:225
93. Haberder H, Schröder G, Ebel J (1989) Planta 177:59
94. Edwards R, Blount JW, Dixon RA (1991)Planta 184:403
95. Mithöfer A, Bhagwat AA, Feger M, Ebel J (1996) Planta 199:270
96. Minamisawa K, Onodera S, Tinimura Y, Kobayashi N, Yuhashi KI, Kubota M (1997) FEMS Microbiol Ecol 24:49
97. Landini S, Graham MY, Graham TL (2003) Phytochemistry 62:865

98. Lozovaya VV, Lygin AV, Zernova OV, Li S, Hartman GL, Widholm JM (2004) Plant Physiol Biochem 42:671
99. Nakamura K, Akashi T, Aoki T, Kawaguchi K (1999) Biosci Biotechnol Biochem 63:1618
100. Li ZS, Alfenito M, Rea PA, Walbot V, Dixon RA (1997) Phytochemistry 45:689
101. Akashi T, Aoki T, Ayabe S (1998) FEBS Letters 432:287
102. Akashi T, Aoki T, Takahashi T, Kameya N, Nakamura I, Ayabe S (1997) Plant Sci 126:39
103. Stacey G, Sanjuan J, Luka S, Dockendorff T, Carlson RW (1995) Soil Biol Biochem 27:473
104. Graham TL, Kim JE, Graham MY (1990) Plant-Microbe Interact 3:157
105. Edwards R, Daniell TJ, Gregory ACE (1997) Planta 201:359
106. Robbins MP, Hartnoll Morris JP (1991) Plant Cell Rep 10:59
107. Wysokińska H, Chmiel A (1997) Acta Biotechnol 17:131
108. Giri A, Narasu ML (2000) Biotechnol Adv 18:1
109. Shibuya Y, Sugimura Y, Tahara S, Mizutani J (2003) Biosci Biotech Biochem 17:14
110. Gagnon H, Seguin J, Bleichert E, Tahara S, Ibrahim RK (1992) Plant Physiol 100:76
111. Zacharius RM, Kalan EB (1990) J Plant Physiol 135:732
112. Schlieper D, Tiemann K, Barz W (2002) Phytochemistry 29:1519
113. Johansson F, Sommarin M, Larson C (1993) Plant Cell 5:321
114. Daniel S, Tiemann K, Wittkampf U, Bless W, Hinderer W, Barz W (1990) Planta 182:270
115. Hinderer W, Flentje U, Barz W (1987) FEB 04574 214:101
116. Hashim MF, Hakamatsuka T, Ebizuki Y, Sanakawa U (1990) FEBS Lett 271:219
117. Klus K, Börger-Papendorf G, Barz W (1993) Phytochemistry 34:979
118. Laflamme P, Khouri H, Gulick P, Ibrahim R (1993) Phytochemistry 34:147
119. Klus K, Barz W (1995) Arch Microbiol 164:428
120. Tahara S, Tanaka M, Barz W (1997) Phytochemisty 44:1031
121. Hur HG, Rafii F (2000) FEMS Microbiol Lett 192:21
122. Jung W, Yu O, Lau SM, O'Keefe DP, Odell J, Fader G, McGonigle B (2000) Nature Biotechnol 18:208
123. Akashi T, Sawada Y, Aoki T, Ayabe SI (2000) Biosci Biotechnol Biochem 64:2276
124. Kimura Y, Aoki T, Ayabe S (2001) Plant Cell Physiol 42:1169
125. Liu CJ, Dixon RA (2001) Plant Cell 13:2643
126. Cooper JD, Qiu F, Paiva NL (2002) Plant Cell Rep 20:876
127. Miyazawa M, Ando H, Okuno Y, Araki H (2004) J Mol Catal B: Enzym 27:91
128. Yu O, McGonigle B (2005) Biosynthesis Adv Agronom 86:147
129. Stougaard J (2001) Curr Opin Plant Biol 4:328
130. Steele CL, Gijzen M, Qutob D, Dixon RA (1999) Arch Biochem Biophys 367:147
131. Yu O, Jung W, Shi J, Croes RA, Fader GM, McGonigle B, Odell JT (2000) Plant Physiol 124:781
132. Liu CJ, Blount JW, Steele CL, Dixon RA (2002) Proc Natl Acad Sci U S A 99:14578
133. Yu O, Shi J, Hession AO, Maxwell CA, McGonigle B, Odell JT (2003) Phytochemistry 63:753
134. Fontana GS, Santini L, Caretto S, Frugis G, Mariotti D (1993) Plant Cell Rep 12:194
135. Santarém ER, Trick HN, Essig JS, Finer JJ (1998) Plant Cell Rep 17:752
136. Zhang Z, Xing A, Staswick P, Clemente TE (1999) Plant Cell Tiss Org Cult 56:37
137. Yan B, Srinivasa Reddy MS, Collins GB, Dinkins RD (2000) Plant Cell Rep 19:1090
138. Zeng P, Vadnais DA, Zhang Z, Polacco JC (2004) Plant Cell Rep 7:478
139. Pereira LF, Erickson L (1995) Plant Cell Rep 14:290
140. Zhang LJ, Cheng LM, Xu N, Zhao NM, Li CG, Jing Y, Jia SR (1991) Bio Technol 9:996
141. Joersbo M, Brunstedt J (1992) Physiol Plant 85:230
142. Trick HN, Finer JJ (1997) Transgenic Res 6:329
143. Łuczkiewicz M, Kokotkiewicz A (2005) Z Naturforsch 60c:867

144. Loake GJ, Faktor O, Lamb CJ, Dixon RA (1992) Proc Natl Acad Sci U S A 89:9230
145. Berlin J, Rippert M, Mollenschott C, Maywald F, Strack D, Wray V, Sator C (1989) Planta Med 55:685
146. Nguyen C, Bourgaud F, Forlot P, Guckert A (1992) Plant Cell Rep 11:424
147. Łuczkiewicz M, Głód D (2005) Plant Sci 168:967
148. Zhao P, Hamada Ch, Inoue K, Yamamoto H (2003) Phytochemistry 62:1093
149. Łuczkiewicz M, Kokotkiewicz A (2005) Plant Sci 169:862

Chapter 4
Secondary Metabolite Production from Plant Cell Cultures: the Success Stories of Rosmarinic Acid and Taxol

S. Kintzios

Laboratory of Plant Physiology, Faculty of Agricultural Biotechnology, Agricultural University of Athens, Iera Odos 75, 11855, Athens, Greece, e-mail: spiroskintzios@usa.net

Abstract Advances in scale-up approaches and immobilization techniques contribute to a considerable increase in the number of applications of plant cell cultures for the production of compounds with a high added value. The present review handles the cumulative progress in this field and focuses on the most recent developments regarding the in vitro production of plant-derived compounds with cancer chemotherapeutic or antioxidant properties, using rosmarinic acid (RA) and taxol as representative examples. Stimulation of biosynthetic pathways leads to enhanced RA accumulation in vitro. Critical issues are thoroughly discussed, including the dependence of in vitro, compound-specific production on culture growth and differentiation, elicitation strategies, physiological effects of immobilization, and the current status of scale-up production systems.

Keywords Plant tissue culture, Secondary Metabolite, Rosmarinic acid, Taxol, Bioreactor, Cell immobilization

Abbreviations

2,4-D	2,4-Dichlorophenoxyacetic acid
BA	6-Benzyladenine
IAA	3-Indole acetic acid
NAA	Naphthaleneacetic acid
NADPH	Nicotinamide adenine dinucleotide phosphate (reduced)

PAL   Phenylalanine ammonia lyase
RA   Rosmarinic acid

4.1 Introduction: Cell Factories at the Cross Point

Initial efforts that focused on the creation of plants originating from the culture of a plant part, ex vivo, under specific conditions in vitro can be traced back to the dawn of the last century. During the last four decades, tissue culture was further extended for the production of secondary metabolites, where it has demonstrated itself to be an important tool for studying their biosynthesis. Although the production of secondary metabolites from cell cultures has been reported as early as in 1956 [1], real milestones were: (1) the first biotransformation [2], (2) cell immobilization [3], and (3) rosmarinic acid (RA) production in bioreactor-assisted cultures of *Coleus blumei* [4]. In more recent years, immobilization protocols and scale-up techniques have been improved to a quite considerable degree, thus allowing for the development of in vitro metabolite production systems that are functional at a commercial or pre-commercial level. Finally, plant cells are continuously being used as expression systems for manufacturing very important proteins, such as vaccines because (1) higher plants generally synthesize proteins from eukaryotes with correct folding, glycosylation, and activity, and (2) plant cells can direct proteins to environments that reduce degradation and therefore increase stability [5].

Research into the use of plant cell cultures for producing natural products is focused on pharmaceuticals, flavors and fragrances, and fine chemicals [6]. In vitro, the synthesis of secondary metabolites is typically considered as non-growth-associated and takes place when the division of cells in the culture decreases or stops, or when the culture passes from the logarithmic developmental stage to the static phase [7]. This, and the fact that plant cells do not always maintain photoautotrophic growth in vitro, has limited the productivity, and thus the scope of application of cell culture systems in the production of bioactive compounds [8]. As a result, secondary metabolites are still obtained commercially by extraction from whole plants or tissues. Cell factories also face rapidly increasing competition from transgenic higher plants that produce foreign proteins with economic value ("biofactories") [5, 9–12]. The advantages of using higher plants rather than cell cultures include the significantly lower production costs, the already existing infrastructure and expertise for crop management, and the apparent absence of human pathogens (such as viruses). However, in vitro production facilities require far less space than conventional agriculture farms in order to produce the same amount of secondary metabolite on an absolute basis of comparison.

Given, in addition, the possibility of achieving (to a certain degree) a satisfactory standardization of product quality in an environment-independent manner (another considerable advantage over in vivo production), there are promising indications for an emerging large-scale application of plant cell culture for the production of compounds with a high added value. Fueled by advances in both scale-up approaches (essentially overcoming year-long problems related to plant tissue culture in bioreactors) and immobilization techniques, the current trend in pharmaceutical production from in vitro systems is definitely upward. In a commercial sense of application, the most recent developments are focused on plant-derived compounds that are either cancer chemotherapeutic agents or antioxidant supplements (i.e., they belong to either of two groups with significant, though distinct bioactive properties).

The particular progress recorded in the last few years in each field will be analyzed in more detail in the following, using RA and taxol as representative examples.

4.2 Rosmarinic Acid

4.2.1 General Information

RA (α-O-caffeoyl-3,4-dihydroxyphenyllactic acid; Fig. 4.1), a plant secondary metabolite belonging to the class of hydroxycinnamic esters, is a constituent of the Lamiaceae, Boraginaceae, and Apiaceae families [13], although it has been detected in at least 12 other plant families [14]. A promising bioactive substance, it has applications as a food preservative and in medicine due to its functional properties as an antioxidant and antimicrobial substance [15–17].

4.2.2 Historical Development of In Vitro RA Production – a Brief Overview

RA has been reported to accumulate in undifferentiated cells (callus and cell suspensions) of several Lamiaceae species including *C. blumei* [4, 18], *Anchusa*

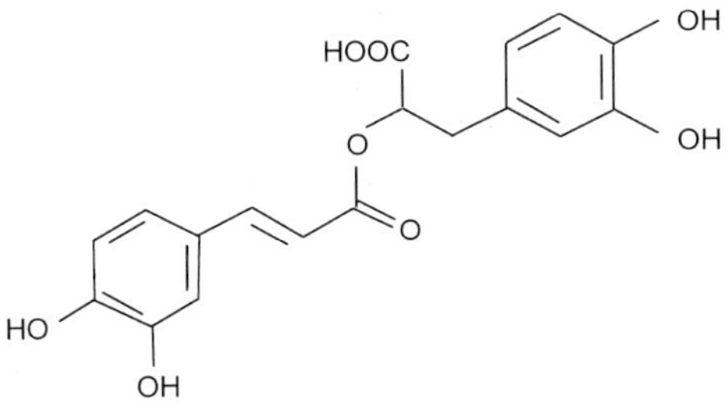

Fig. 4.1 Rosmarinic acid (RA)

officinalis [19], *Lithospermum erythrorhizon* [20], *Orthospiron aristatus* [21], *Ocimum basilicum* [22, 23], *Ocimum americanum* [24], *Ocimum sanctum* [25], *Origanum vulgare* [26], *Salvia miltiorrhiza* [27], *S. officinalis* [28, 29], *S. fruticosa* [29], *Hyssopus officinalis* [30], *Zataria multiflora* [31], *Eritrichium sericeum* [32], and *Anthoceros agrestis* [33]. Accumulation rates vary considerably among different species, even between those belonging to the same genus, whereas cell extracts of some species may also contain other phenolic compounds in addition to RA. For example, cell cultures of *Lavandula vera* produce small quantities of caffeic acid, one of the intermediate substances in the RA biosynthetic chain. Although Banthrope et al. [34] reported that the main phenolic compound synthesized by *L. vera* callus culture was a blue pigment, which they identified as a complex of Fe^{2+} with the isomers of an enol ester of caffeic acid, Kovatcheva et al. [35] reported that the main compound produced was RA. A methanolic extract from fresh callus cells of *L. angustifolia*, cultured in Linsmaier and Skoog (LS) medium, was tested for the presence of RA, caffeic acid, p-coumaric acid, and ferulic acid. The extracts contained a comparatively large amount of RA, smaller amounts of caffeic acids, and only traces of p-coumaric and ferulic acids [35]. In yet another example, cell cultures of *C. blumei* synthesized unidentified phenolic compounds, but in very small quantities [18].

4.2.3 Stimulation of Biosynthetic Pathways Leads to Enhanced RA Accumulation In Vitro

RA is biosynthesized by l-phenylalanine and l-tyrosine [14]; consequently, feeding cell cultures with the amino acid precursors would logically lead to an increase in RA accumulation. Indeed, cell suspensions of *S. officinalis* are capable of producing extremely large quantities of RA under stimulation conditions. Hippolyte et al. [28, 36] found that the growth and production of RA by sage cells was modified by the type of culture medium used. RA production was increased tenfold, attaining 6.4 g·l^{-1} under optimal conditions, representing 36% of the dry weight. Investigation of cell growth kinetics showed that a change in the medium caused shifts in peaks of growth and RA production, and modifications of the cell metabolism. By changing the composition of the culture medium it was possible to manipulate RA production to coincide with cell growth or to begin only when growth had stopped. The addition of phenylalanine to boost RA production gives contradictory results in different species. It reduces RA production by 35% in suspensions of *C. blumei* when added to a standard medium [18], but improves it by 100% when it is added to a medium containing sucrose at insufficient concentrations [4]. Callus cultures of sweet basil (*Ocimum basilicum* L.) have been demonstrated to accumulate RA at levels almost twofold higher than that of the intact plant [37]. Kintzios et al. [22] investigated the accumulation of RA in different types of sweet basil cell cultures (i.e., callus cultures, cell suspensions, and immobilized cell cultures). Leaf-derived suspension cultures grown in a liquid Murashige and Skoog medium were able to accumulate RA at a maximum concentration of 10 mg·g^{-1} dry weight, a value 8.5–11 times higher than for callus cultures and donor plants. This might have been due to feeding culture media with the RA precursor l-phenylalanine (at a concentration of 0.5 g·l^{-1}).

RA accumulation in suspension cultures of *C. blumei* Benth. (Lamiaceae) can be stimulated up to tenfold by raising the sucrose content of the culture medium. Sucrose has a substantial trophic effect and an osmotic effect. Experimenting on suspension cultures of *C. blumei*, Gertlowski and Peterson [38] demonstrated that the level of RA accumulation was dependent on the sucrose content, whereas the growth limitation (e.g., by a limiting supply of phosphate) determined the onset of metabolite accumulation. Cell cultures in medium with 2% sucrose accumulate only about 2–3% of the dry cell weight as RA, whereas up to 19% RA can be obtained in medium with 4% sucrose. In a recent report [39], 16 diversified *Agrobacterium*-transformed callus cultures accumulated RA at concentrations up to 11% of the cell dry weight. Addition of 4 or 5% sucrose stimulated RA synthesis and decreased callus growth. Addition of 0.1 mg·l^{-1} l-phenylalanine stimulated RA production in two lines, but had little effect on the metabolite level in others.

The overall effect of NH_4NO_3, KNO_3, and KH_2PO_4 on the biosynthesis of RA and cell biomass by *L. vera* MM cell suspension was also studied [40]. As a result, modified ingredients of the LS nutrient medium were applied for the cultivation of *L. vera* MM to achieve a maximum RA yield of 1.9 g·l^{-1} (which was 27 times higher compared with the cultivation in the standard LS medium). According to Yang and Shetty [26], proline, proline analogs (such as azetidine-2-carboxylate) and proline precursors (ornithine and arginine) can clearly stimulate RA synthesis in shoot cultures of *Origanum vulgare*. This suggests strongly that perturbing the proline metabolism can help to redirect metabolites from the pentose phosphate pathway toward phenolic acid synthesis.

It is also possible that RA is synthesized by cultured cells as a response to induced stress. Several reports indicate that RA accumulation can be enhanced by the addition of fungal elicitors (e.g., yeast or fungal extracts) or methyl jasmonate [41]. According to Nosov [42], however, if the ecological function of the secondary metabolism predominates in the whole plant, secondary metabolites have no meaningful function in vitro and, therefore, secondary metabolite production should be essentially absent or unstable in this system.

4.2.4 Is RA Biosynthesis Growth Dependent?

The hypothesis of Sakuta and Komamine [43], according to which secondary metabolites may belong to two categories according to their mode of production (i.e., those whose production is growth-related and those whose synthesis takes place after growth and is dependent on cell differentiation), does not seem to apply to RA. Here, the connection between primary metabolism and secondary metabolism seems to be caused by pressure in the culture medium, in which osmotic pressure may play a regulatory role. Other factors may also be involved, such as the type of hormones. Genetic factors may also determine the mode of production [29]. The (rather abundant) reports on the growth-dependence of RA biosynthesis in vitro are frequently contradictory to each other. For example, suspension cultures of *C. blumei* synthesize and accumulate all their RA during only a few days at the end of the growth phase [44]. However, Bauer et al. [39] recently demonstrated that different lines showed different RA accumulation in relation to their growth rate; it was either parallel or inversely related to the tissue growth. In several parallel-run experiments, Kintzios et al. [29] observed that RA accumulation in *S. officinalis* was inversely related to callus growth, whereas, in contrast, RA accumulation in *S. fruticosa* callus, RA accumulation continuously increased in a parallel fashion to callus growth, being only slightly reduced between the 2nd and the 3rd week after callus initiation. The highest values of RA accumulation in the second *Salvia* species were observed both at high callus growth rates (weeks 1–2) and during growth stabilization (weeks 4–5). In another study, Kintzios et al. [45] investigated the effect of antioxidant phenolic compounds produced by sage callus cultures on some physiological parameters of the producing cells. Although cultures demonstrated a continuous growth during an incubation period of 5 weeks, the cell dehydrogenase activity and the cytochrome c oxidase activity of isolated mitochondria declined. During the same period, mitochondrial respiration increased. An analysis of methanolic extracts derived from the callus pieces indicated that the antioxidant activity (Fe^{2+} reduction) in vitro was independent from RA accumulation and was not correlated with the observed pattern of callus growth.

4.2.5 Is RA Accumulation Related to Culture Differentiation?

Although a high level of cellular or tissue differentiation does not seem to be necessary for a satisfactory RA accumulation in vitro, a few studies have provided indications that induction of differentiation in vitro can dramatically enhance culture productivity. In one case, induction of somatic embryogenesis was closely associated with maximum RA accumulation in *S. officinalis* and *S. fruticosa* callus cultures [29] (25.9 and 29.0 g·l^{-1}, respectively – both very high concentrations). At a higher level of morphological differentiation, root cultures of *Ocimum basilicum* induced by *Agrobacterium rhizogenes* produced RA and lithospermic acid (up to 2% of the dry weight) and lithospermic acid B (0.15% of the dry weight) [46]. A similar effect of differentiation has been observed previously by Tanaka et al. [47] and Kelley et al. [48] on RA accumulation in *Hyssopus officinalis* transformed root cultures. On the contrary, highly differentiated shoot cultures of *Ocimum sanctum* demonstrated a higher RA content (>5% dry weight) than in vivo grown plants, but lower than callus cultures [23]. Rady and Nazif [24] observed that the addition of 1 mg·l^{-1} of

6-benzyladenine (BA) to the culture medium caused a considerable increase in RA accumulation in regenerated *Ocimum americanum* shoots than that of the intact plant, although a considerable decline was associated with higher BA concentrations.

4.2.6 Recent Attempts to Scale Up RA Production

Cell suspensions accumulating RA tend to increase their productivity with the volume of the culture vessel. When cultured in 5-l airlift bioreactors for 3 weeks [49], the fresh weight of sweet basil suspension cultures increased by 241%, while RA was accumulated at 29 $\mu g \cdot g^{-1}$ dry weight (Fig. 4.2). During the same period, RA accumulation in bioreactor-micropropagated basil plants reached 178 $\mu g \cdot g^{-1}$ dry weight, which is 44 times higher than in suspension cultures in a 250-ml flask. Since product formation was growth-associated, a minimum inoculum was required for culture initiation, and satisfactory RA accumulation was achieved within a short culture period. In a recent study, Pavlov et al. [40] investigated the effect of dissolved oxygen concentration, agitation speed, and temperature on the RA production by *L. vera* MM cell suspension in a 3-l laboratory bioreactor. A maximal RA yield of 3.5 $g \cdot l^{-1}$ was achieved, which was twofold higher than in shake-flask culture under the same experimental conditions.

Fig. 4.2 Scale-up cell suspension culture of sweet basil (*Ocimum basilicum* L.) in a 5-l airlift bioreactor (Osmotek Lifereactor)

4.2.7 RA Production in Immobilized Cell Cultures

There is only one published report on RA production in immobilized cell cultures, with rather disappointing effects. More analytically, sweet basil cells immobilized in 1.5, 2, or 3% w/v calcium alginate beads accumulated RA at a much-reduced rate (<15 µg·g^{-1}) compared to cell suspensions of tissue cultures [22]. Quite recently, however, a remarkably high RA production (21 mg·g^{-1} dry weight) was achieved by immobilizing sweet basil cells at a high density (approximately 25×10^4 cells·ml^{-1}) in specially designed solid-state bioreactors (personal communication: Georgia Moschopoulou, 2006), a production performance that was 1400 times higher than in basil cells immobilized in beads. Even more significantly, RA was excreted into the culture medium, where it was collected without terminating the culture of immobilized cells. In both cases, RA accumulation in sweet basil did not seem to require cessation of cell growth, as reported for other species. Figure 4.3 demonstrates a comparison of the efficiency of recent scale-up RA production systems from sweet basil.

4.3 Taxol

4.3.1 General Information

Pacific yew (*Taxus baccata* L.) is a relatively small conifer tree. Like mistletoe, it was also favored by the Druids as a sacred tree. Stem segments, needles, and roots contain taxane diterpenes, among them taxol (paclitaxel; Fig. 4.4) and

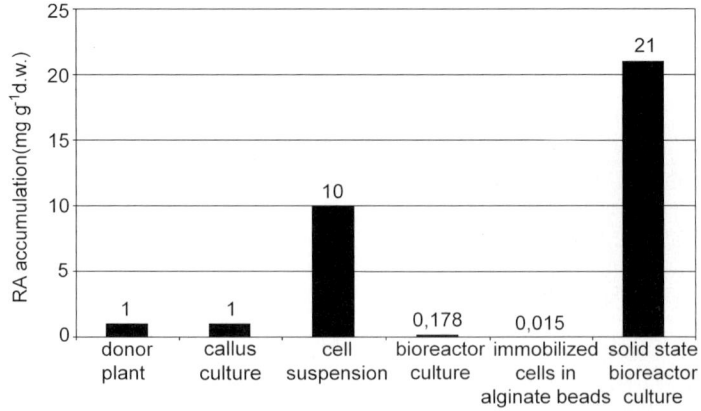

Fig. 4.3 Comparative accumulation of RA in different types of sweet basil cell culture

cephalomannine, which are used against ovarian cancer and have also shown activity against breast, lung, and other cancers [50, 51]. *T. brevifolia* L. is a related species, the needles of which contain the taxol precursor [52]. High taxol and 10-desacetylbaccatin III accumulation has also been detected in other *Taxus* species, such as *T. marei* Hu ex Liu (a giant Taiwan endemic tree) [53], *T. cuspidata* [54], and *T. wallachiana* L. (Himalayan yew) [55].

4.3.2 Historical Development of In Vitro Taxol Production – a Brief Overview

Although several efforts have been reported aiming to the enhanced production of taxol and/or its precursors by tissue culture and other biotechnological approaches [56], the chemical synthesis and hemisynthesis of taxol [57] as well as the availability of other *Taxus* species have reduced the dependence on *T. baccata* as the taxol primary source. The emergence of functional taxol analogs such as taxotere (which is produced by attaching a synthetic side chain to 10-desacetyl baccatin III) has also contributed to a relative alleviation of the taxol supply crisis [58]. Fett-Neto et al. [59] were the first to induce callus cultures of *T. cuspidata* and *T. canadensis* from different tissue explants, such as arils, seeds, young stems, and needles. They were also able to extract taxol at a concentration of 0.02% per dry weight. Jha et al. [60] established cell cultures of *T. wallichiana* Zucc (*T. baccata* ssp *wallichiana* Zucc. Pilg.; Himalayan yew). The cell line NC110, which was derived from the needle leaf of a 40-year-old tree growing in Darjeeling in the Himalayas, produced 0.018% taxol in B5 basal medium supplemented with 2,4-dichlorophenoxyacetic acid (2,4-D), kinetin, and casein hydrolysate. Goleni-

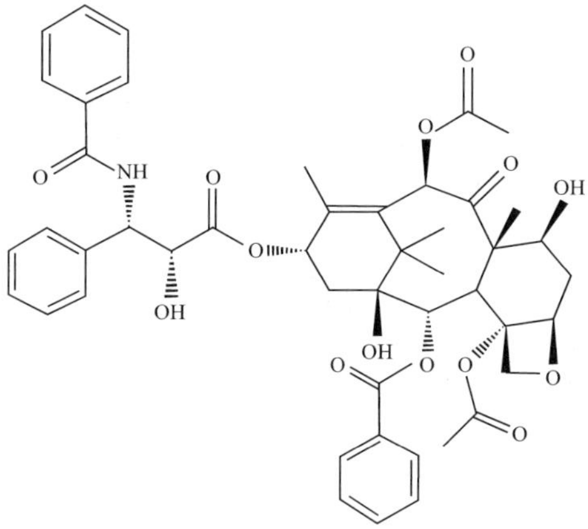

Fig. 4.4 Paclitaxel

owski [61] reported production of taxol ranging from 0.1 to 15 µg·g⁻¹ dry weight of callus cultures from *T. baccata* and Taxus×Media, respectively. The maximum taxol production for suspension cell was 5–6 mg·l⁻¹. Bai et al. [62, 63] isolated 17 known taxoids from callus culturea of *T. cuspidata* cultivated on a modified Gamborg's B5 medium with 0.5 or 1.0 mg·l⁻¹ 1-naphthaleneacetic acid (NAA).

Aoyagi et al. [64] demonstrated that about 30% and 35% of taxol was located in cell walls and/or between the cell wall and cell membrane of *T. cuspidata* suspension cells in the growth phase and in the stationary phase, respectively. About 30% and 43% of paclitaxel in the cells was located in the cell wall of the cells grown in solid culture in the growth phase and in the stationary phase, respectively. In comparison with the cell suspension culture, protoplasts in a static culture and the protoplasts immobilized in agarose gel in shaking culture resulted in a sixfold increase in the extracellular taxol accumulation. Release of taxol and other taxanes into the culture medium can be facilitated by the digestions of the cell wall of cultured cells, as demonstrated for *T. canadiensis* by Roberts et al. [65].

4.3.3 Stimulation of Biosynthetic Pathways Leads to Enhanced Taxol Accumulation In Vitro

The accumulation of taxol and related taxanes in *Taxus* plants is thought to be a biological response to specific external stimuli. Biosynthesis of the N-benzoyl phenylisoserinoyl side chain of the anticancer drug Taxol starts with the conversion of 2S-alpha-phenylalanine to 3R-beta-phenylalanine by phenylalanine aminomutase [66]. A key enzyme in the taxane biosynthetic pathway is taxadiene synthase, which can be elicited by methyl jasmonate. Yukimune et al. [67] observed that significantly increased amounts of paclitaxel and baccatin III were observed in cultured cells of *Taxus* species after exposure to methyl jasmonate, with *T. media* showing the highest paclitaxel content and *T. baccata* showing the highest baccatin III content. Reduction of the keto group at the C-3 position of the structure of methyl jasmonate greatly reduced this activity, whereas cis-jasmone, which does not have a carboxyl group at the C-1 position, had almost no activity.

A time-course analysis by Tabata [68] revealed two regulatory steps in taxane biosynthesis: the taxane-ring formation step and the acylation step at the C-13 position. Methyl jasmonate promoted the formation of the taxane-ring. The same researcher demonstrated that taxol accumulation in *T. baccata* suspension cultures was strongly promoted by methyl jasmonate and silver thiosulfate as an antiethylene compound. The production of paclitaxel reached a maximum level of 295 mg·l⁻¹ in a large-scale culture using a two-stage process. Cell suspensions of *T. cuspidata* and *T. canadensis* are also capable of producing high levels of taxol and related taxanes after elicitation with methyl jasmonate, but successful elicitation leads to loss of cell viability that is apparently related to taxane production itself, rather than to the direct effect of methyl jasmonate [69]. A similar observation has been made previously by Yuan et al. [70], whereas addition of taxol into suspension cultures of *T. cuspidate* resulted in excessive cell apoptosis. Some studies indicate that oxidative stress (in particular the accumulation of intracellular and extracellular H_2O_2) might be one factor promoting taxol biosynthesis. Xu et al. [71] showed that *T. cuspidata* cells responded to oleic acid with oxidative bursts in both intracellular H_2O_2 and extracellular oxygen superoxide (O^{2-}) production. Yin et al. [72] also demonstrated that although the addition of exogenous H_2O_2 barely affected malondialdehyde content and the cell membrane permeability of cultured *T. cuspidata* cells, it led to an increased accumulation of taxol.

Inhibition studies with diphenylene iodonium suggested that the key enzyme responsible for oxidative bursts was primarily NADPH oxidase. Furthermore, Han and Yuan [73] investigated the relationship between active oxidative species and defense responses induced by the shear stress during culture of *T. cuspidata* cells in a bioreactor. They determined that the superoxide burst may account for the change of membrane permeability, whereas H_2O_2 plays an important role in inducing secondary metabolites such as the activation of phenylalanine ammonia lyase (PAL)enzyme and phenolic accumulation. The number of taxoids produced by callus cultures of *T. cuspidata* cultivated on a modified B5 medium in the presence of 0.5 mg·l^{-1} NAA was drastically increased after stimulation with 100 µM methyl jasmonate [63]. This was accompanied by the formation of taxinine NN-11, a new taxane that exhibits significant cytotoxic activity toward 2780 AD tumor cells.

Brincat et al. [74] followed a different approach in order to boost taxol accumulation in vitro: they used specific enzyme inhibitors (cinnamic acid, alpha-aminooxyacetic acid, L-alpha-aminooxy-beta-phenylpropionic acid) against the first enzyme in the phenylpropanoid pathway (i.e., PAL). Cinnamic acid acted quickly in reducing PAL activity by 40–50%, without affecting total protein levels, but it generally inhibited the taxane pathway, reducing taxol by 90% of control levels. Of the taxanes produced, 13-acetyl-9-dihydrobaccatin III and 9-dihydrobaccatin III doubled as a percentage of total taxanes in cells treated with cinnamic acid, when all other taxanes were lowered. The PAL inhibitor alpha-aminooxyacetic acid almost entirely shut down taxol production, whereas L-alpha-aminooxy-beta-phenylpropionic acid had a slightly opposite effect. Thus, it was concluded that the impact of cinnamic acid on taxol is related not to PAL, but rather to a specific effect on the taxane pathway.

Yari Khosroushahi et al. [75] developed a two-stage suspension cell culture of *T. baccata*. During stage I (biomass growth), B5 medium was gradually supplemented with vanadyl sulfate (0.1 mg·l^{-1}), silver nitrate (0.3 mg·l^{-1}), cobalt chloride (0.25 mg·l^{-1}), sucrose (1%), ammonium citrate (50 mg·l^{-1}) and phenylalanine (0.1 mM). At stage II (which started on day 25), methyl jasmonate, salicylic acid, and a fungal elicitor were added to the medium. At stage I, overall taxol amount of biomass growth medium was 13.75 mg·l^{-1}, which was sixfold higher than that of the control medium (B5 medium without supplements). At stage II, elicitor-treated cells produced the highest amount of taxol (39.5 mg·l^{-1}), which was 16-fold higher than that of control medium (2.45 mg·l^{-1}).

4.3.4 Is Taxol Biosynthesis Growth and Differentiation Dependent?

Since the early experimentation with *Taxus* cell suspensions, it became rather apparent that taxol would be produced during late or nongrowth stages of the cultures. Working with *T. cuspidata*, Pestchanker et al. [76] observed that taxol production occurred during the last 7 days of the cultivation period and was not growth-associated. In addition, Jha et al. [60] observed a significant enhancement in the level of taxol obtained from cell cultures of *T. wallichiana* by supplementation of the culture medium with 5 mg·l^{-1} of indole-3-acetic acid (IAA)-phenylalanine instead of 2,4-D without adversely affecting cell growth. IAA-glycine also enhanced taxol levels (0.03%), while IAA alone was ineffective in inducing taxol accumulation. Using three different cell lines with different taxol-producing capacities, it has been demonstrated that 2,4-D and IAA-phenylalanine, when present alone promoted growth and taxol production, but when combined enhanced the biomass to a maximum without simultaneously enhancing taxol accumulation. These early results suggest that a two-stage culture may be beneficial for optimizing taxol accumulation in vitro. While experimenting with suspension cultures of *T. yunnanensis*, Zhang et al. [77] observed that the volumetric yield and productivity of taxol increased with inoculum size, while the specific taxol yield per cell was dependent mainly on inoculum age, with an optimum of 20 days, during the early stationary phase. The highest taxol yield and productivity, 39.8 mg·l^{-1} and 1.9 mg·l^{-1} per day, respectively, were obtained with a 20-day-old inoculum at 200 g formula weight·l^{-1}. Taxol excretion by the cells increased with inoculum age but decreased with inoculum size. More recently, Kim et al. [78] investigated the possibility of cell signaling within the population as a biological trigger for taxol production. They used parental cultures and their subcultures from five different cell lines to test whether a high-taxol-producing culture grows more slowly or dies more rapidly than a low-producing one. These cell lines were of three types: (1) taxol-producing with and without methyl jasmonate, (2) taxol-producing only upon elicitation, and (3) nonproducing. High-producing cultures showed growth inhibition upon subculture, whereas nonproducing cultures and elicited cultures show little growth inhibition. Thus, growth inhibition was due primarily to taxol or taxane accumulation, so that culture components were generated by cells alter culture properties. To assess variability as a function of culture lineage, two groups of replicate cultures were generated either with a mixing of the parental flasks or segregation of parental flasks at each subculture. Although parental culture mixing did not reduce flask-to-flask variation, the production level of taxol in subcultures resulting from mixing inocula was sustained at a higher level relative to segregated subcultures.

4.3.5 Recent Attempts to Scale Up Taxol Production

Ten years ago, Pestchanker et al. [76] demonstrated that taxol was accumulated in suspension cultures of *T. cuspidata* grown in either shake flasks or airlift bioreactors at a volumetric productivity rate of 1.1 mg·l^{-1}·day^{-1}, which was many-fold higher than reported for other *Taxus* sp. suspension cultures by that time. Taxol was released to the extracellular medium as it was produced, with little intracellular retention (≤10%). Although the same taxol titers (22 mg·l^{-1}) could be obtained in both reactor types, nutrient uptake rates were faster in the airlift bioreactor than in shake flasks. However, formation of a growth ring in the bioreactor reduced the yield of cell mass. Luo et al. [79] investigated the effect of dissolved oxygen on taxol production by suspended cell cultures of *T. chinensis* in shake flasks and in a 20-l mechanically agitated bioreactor. The oxygen supply exhibited significant influence on the production of taxol, which increased when the level of dissolved oxygen was increased to 40–60%. An optimum taxol level of 7.2 mg·l^{-1} was obtained by the dissolved oxygen controlled process. Finally, Gong and Yuan [80] found that shear stress during *T. cuspidata* culture in a Couette-type shear reactor induced nitric oxide generation and reduced the activity of glutathione S-transferase, a principal enzyme responsible for antioxidant-mediated detoxification.

4.3.6 Taxol Production in Immobilized Cell Cultures

Fetto-Neto et al. [59] immobilized suspension cultures of *T. cuspidata* onto glass fiber mats, and maintained then as immobilized cultures for 6 months. A maximum taxol production of approximately 0.012% of the extracted dry weight was observed, which was somehow lower than in callus cultures grown under the same experimental conditions.

Yin et al. [72] investigated the dynamic changes in reactive oxygen species (ROS) and taxol production of *T. cuspidata* cells immobilized on polyurethane foam. The taxol content of immobilized cells was fourfold that of suspended cells at day 35. Immobilization shortened the lag period of cell growth and increased H_2O_2 and O^{2-} contents inside the culture microenvironment. More recently [81], the same research group observed distinct spatiotemporal variations of metal ions and taxol production in the immobilized cell culture system. The taxol content in the inner foam layer reached 215 mg·g^{-1}, which was 40-fold higher than that in the outer foam layer and was accompanied by higher intracellular Ca^{2+} and Mg^{2+} contents and a lower intracellular K^+ content.

4.4 Conclusions

The two examples presented in this review are not the only cases of a successful application of plant cell culture for the scale-up production of bioactive compounds. Indeed, several reports provide technically and economically feasible protocols for a plethora of compounds such as catharanthine [82], podophyllotoxin [83] and mistletoe lectins [84]. One might wonder, however, why in vitro plant systems are currently operating in a very limited number of commercial production systems, such as shikonin, ginsenosides, and berberine [8]. It is not difficult to answer such a question: although both bioreactor and immobilization techniques were developed more than 20 years ago, our knowledge on cellular behavior under conditions of scale-up culture are still extremely limited and are mainly restricted to the level of empirically defining growth–productivity relationships. Unfortunately, basic research in this field has rather been neglected, at least compared to advances on the side of applications. As already demonstrated for taxol (and, to a lesser extend, for RA), problems associated with large culture volumes are not only mechanical (i.e., requirements for specific design of culture vessels, reactors, and medium reservoirs, for example) but also relate to the altered state of cellular biochemistry. In other words, scaling up a plant cell culture is a novel source of cellular stress, which can be expressed in several different ways, such as ROS formation and altered cell signaling. Therefore, future challenges lie mainly in understanding the principles of scaled up metabolite biosynthesis. Given the current rapid progress in research of cell biology in vitro, it is not unrealistic to expect that plant cell culture systems will provide a significant source of pharmaceuticals and other chemicals in the next 5–10 years.

References

1. Nickell LG (1958) Science 128:88
2. Alfermann AW, Merz D, Reinhard E (1975) Planta Med Suppl:70
3. Doller G, Alfermann AW, Reinhard E (1976) Planta Med 30:14
4. Zenk MH, Elshagi H, Ulbrich B (1977) Naturwissenshaften 64:585
5. Horn ME, Woodard SL, Howard JA (2004) Plant Cell Rep. 22:711
6. Sajc L, Grubisic D, Novakovic GV (2000) Biochem Eng J 4:89
7. Robins R, Parr A, Richards S, Rhodes M (1986) In: Morris P, Scragg A, Stafford A, Fowler M (eds) Secondary Metabolism in Plant Cell Cultures. Cambridge, Cambridge University Press, p 162
8. Bourgaud F, Gravot A, Milesi S, Gonteir E (2001) Plant Sci 161:839
9. Kusnadi AR, Hood EE, Witcher DR, Howard JA, Nikolov ZL (1998) Biotechnol Prog 14:149
10. Hood EE, Jilka JM (1999) Curr Opin Biotechnol 10:382
11. Daniell H, Streatfield SJ, Wycoff K (2001) Trends Plant Sci 6:219
12. Hood EE, Howard JA (2002) Plants as Factories for Protein Production. Dordrecht, Kluwer
13. De-Eknamkul W, Ellis BE (1987) Phytochemistry 26:1941
14. Petersen M, Simmonds MSJ (2003) Phytochemistry 62:121

15. Deighton N, Glidewell SM, Deans SG, Goodman BA (1993) J Food Sci Agric 63:221
16. Frankel EN, Huang SW, Aeschbach R, Prior E (1996) J Agric Food Chem 44:131
17. Shetty K, Ohshima M, Murakami T, Oosawa K, Ohashi Y (1997) Food Biotechnol 11:11
18. Razzaque A, Ellis BE (1977) Planta 137:287
19. De-Eknamkul W, Ellis BE (1984) Planta Med 50:346
20. Fukui H, Yazaki K, Tabata M (1984) Phytochemistry 23:2398
21. Sumaryono W, Prokasch P, Hartmann T, Nimitz M, Wray V (1991) Phytochemistry 30:3267
22. Kintzios S, Makri O, Panagiotopoulos EM, Scapeti M (2003) Biotechnol Lett 25:405
23. Omoto T, Murakami Y, Shimomura K, Yoshira K, Mori K, Nakashima T (1997) Jpn J Food Chem 4:11
24. Rady MR, Nazif NM (2005) Fitotherapia 76:525
25. Ishimaru K, Murakami Y, Shimomura K (2002) In: Nagata T, Ebizuka Y (eds) Biotechnology in Agriculture and Forestry, Vol. 51. Berlin, Springer-Verlag, p 156
26. Yang R, Shetty K (1998) J Agric Food Chem 46:2888
27. Morimoto S, Goto Y, Shoyama Y (1994) J Nat Prod 57:817
28. Hippolyte I, Mrine B, Baccu JC, Jonard R (1992) Plant Cell Rep 11:109
29. Kintzios S, Nicolaou A, Skoula M (1998) Plant Cell Rep 18:462
30. Murakami Y, Omoto T, Asai I, Shimomura K, Yoshihira K, Ishimaru K (1998) Plant Cell Tiss Organ Cult 53:75
31. Mohagheghzadeh A, Shams-Ardakani M, Ghannadi A, Minaeian M (2004) Fitotherapia 75:315
32. Fedoreyev SA, Veselova MV, Krivoschekova OE, Mischenko NP, Denisenko VA, Dmitrenok PS, Glazunov VP, Bulgakov VP, Tchernoded GK, Zhuravlev YN (2005) Planta Med 71:446
33. Vogelsang K, Schneider B, Petersen M (2006) Planta 223:369
34. Banthrope DV, Bilyard HJ, Watson DG (1985) Phytochemistry 24:2677
35. Kovatcheva E, Pavlov A, Koleva M, Ilieva M, Mihneva M (1996) Phytochem. 43:1243
36. Hippolyte I, Marin B, Baccou JC, Jonard R (1991) Comptes Rendus de l'Academie des Sciences. Series 3, Sciences de la Vie 313:365
37. Makri O, Kintzios S (1999) In: Cassel A (ed) ISHS Working Group Quality Management in Micropropagation "Methods and Markers for Quality Assurance in Micropropagation", University College, Cork/Ireland
38. Gertlowski C, Petersen M (1993) Plant Cell Tiss Org Cult 34:183
39. Bauer N, Leljak-Levanic D, Jelaska S (2004) Z Naturforsch [C] 59:554
40. Pavlov AI, Georgiev MI, Panchev IN, Ilieva MP (2005) Biotechnol Prog 21:394
41. Szabo E, Thelen A, Petersen M (1999) Plant Cell Rep 18:485
42. Nosov AM (1994) Russian Plant Physiol 41:767
43. Sakuta M, Komamine A (1987) In: Vasil I (ed) Cell Culture and Somatic Cell Geneticis of Plants. New York, Academic Press, p 97
44. Petersen M (1992) Planta Med 58:578
45. Kintzios S, Adamopoulou M, Pistola E, Makri O, Delki K, Drossopoulos J (2002) J Herbs Spices Med Plant 9:229
46. Tada H, Murakami Y, Omoto T, Shimomuta K, Ishimaru K (1996) Phytochemistry 42:431
47. Tanaka T, Morimoto S, Nonaka G, Nishioka I, Yokozawa T, Chung HY (1989) Chem Pharm Bull 37:340
48. Kelley CJ, Mahajan JR, Brooks LC, Neubert LA, Breneman WR, Carmack M (1975) J Org Chem 40:1804
49. Kintzios S, Kollias Ch, Straitouris Ev, Makri O (2004) Biotechnol Lett 26:521
50. Rowinsky EK, Cazenave A, Donehower RC (1990) J Natl Cancer Inst 82:1247
51. Kintzios S (2006) Crit Rev Plant Sci 24:1
52. Helfferich C (1993) Alaska Sci Forum 1126
53. Chang SH, Ho CK, Chen ZZ, Tsay JY (2001) Plant Cell Rep 20:496

54. Kim Y, Bang SC, Lee JH, Ahn BZ (2004) Arch Pharm Res 27:915
55. Strobel G, Yang X, Sears J, Kramer R, Sidhu RS, Hess WM (1996) Microbiol 142:435
56. Furmanowa M, Glowniak K, Syklowska-Baranek K, Zgorka G, Jozefczyk A (1997) Plant Cell Tiss Org Cult 49:75
57. Nicolaou KC, Dai WM, Guy RK (1994) Angew Chem Int Ed Engl 33:15
58. Cragg GM (1998) Paclitaxel (Taxol): A Success Story with Valuable Lessons for Natural Product Drug Discovery and Development. John Wiley Sons, New York
59. Fetto-Neto AG, DiCosmo F, Reynolds WF, Sakata K (1992) Biotechnol 10:1572
60. Jha S, Sanyal D, Ghosh B, Jha TB (1998) Planta Med 64:270
61. Goleniowski ME (2000) Biocell 24:139
62. Bai J, Ito N, Sakai J, Kitabatake M, Fujisawa H, Bai L (2005) J Nat Prod 68:497
63. Bai J, Kitabatake M, Toyoizumi K, Fu L, Zhang S, Dai J, Sakai J, Hirose K, Yamori T, Tomida A, Tsuruo T, Ando M (2004) J Nat Prod 67:58
64. Aoyagi H, DiCosmo F, Tanaka H (2002) Planta Med 68:420
65. Roberts SC, Naill M, Gibson DM, Shuler ML (2003) Plant Cell Rep 21:1217
66. Walker KD, Klettke K, Akiyama T, Croteau R (2004) J Biol Chem 279:53947
67. Yukimune Y, Tabata H, Higashi Y, Hara Y (1996) Nat Biotechnol 14:1129
68. Tabata H (2004) Adv Biochem Eng Biotechnol 87:1
69. Kim BJ, Gibson DM, Shuler ML (2005) Biotechnol Prog. 21:700
70. Yuan YJ, Ma ZY, Wu JC (2002) Mol Biotechnol 20:137
71. Xu QM, Cheng JS, Ge ZQ, Yuan YJ (2005) Appl Biochem Biotechnol 125:11
72. Yin DM, Wu JC, Yuan YJ (2005) Appl Biochem Biotechnol 127:173
73. Han RB, Yuan YJ (2004) Biotechnol Prog 20:507
74. Brincat MC, Gibson DM, Shuler ML (2002) Biotechnol Prog 18:1149
75. Yari Khosroushahi A, Valizadeh M, Ghasempour A, Khosrowshahli M, Naghdibadi H, Dadpour MR, Omidi Y (2006) Cell Biol Int 30:262
76. Pestchanker LJ, Roberts SC, Shuler ML (1996) Enzyme Microb Technol 19:256
77. Zhang CH, Wu JY, He GY (2002) Appl Microbiol Biotechnol 60:396
78. Kim BJ, Gibson DM, Shuler ML (2004) 20:1666
79. Luo J, Yu F, Liu L, Wu CD, Mei XG (2001) Sheng Wu Gong Cheng Xue Bao 17:215
80. Gong YW, Yuan YJ (2006) J Biotechnol 123:185
81. Yin DM, Wu JC, Yuan YJ (2006) Biotechnol Lett 28:29
82. St-Pierre B, Vazquez-Flota FA, De Luca V (1999) Plant Cell 11:887
83. Smollny T, Wichers H, Kalenberg S, Shahsavari A, Petersen M, Alfermann AW (1998) Phytochemistry 48:975
84. Kintzios S, Barberaki M, Tourgelis P, Aivalakis G, Volioti A (2002) J Herbs Spices Med Plants 9:217

Chapter 5
Guggulsterone: a Potent Natural Hypolipidemic Agent from *Commiphora wightii* – Problems, Perseverance, and Prospects

K.G. Ramawat (✉), M. Mathur, S. Dass and S. Suthar

Laboratory of Bio-Molecular Technology, Department of Botany,
M. L. Sukhadia University, Udaipur-313001, India, e-mail: kg_ramawat@yahoo.com

Abstract Two isomers of guggulsterone, -E and -Z, have been established as bioactive molecules responsible for the lipid- and cholesterol-lowering activities of oleogum-resin of *Commiphora wightii* (Arnott.)Bhandari (syn. *C. mukul*). Guggulsterone is a safe and effective natural product for hypercholesterolemia that has been used as such for the past 3000 years in Ayurveda. It is obtained from a very slow growing desert tree endemic to the Thar Desert and has become endangered due to its over exploitation. Oleogum-resin is a complex mixture of several classes of compounds including gum, minerals, essential oils, sterols, flavanones, and sterones. Early chemical and pharmacological work was carried out in India, but after approval by the United States Food and Drug Administration as a food supplement, several reports describe a role for guggulsterone in the excretion of cholesterol, involving the farnesoid X receptor, pregnane X receptor, *Cyp-7A1* gene, and the bile salt export pump. Biotechnological approaches have been made to develop micropropagation methods through axillary bud break and somatic embryogenesis, as well as guggulsterone production through cell cultures grown in shake flasks and bioreactors. Field-grown plants show genetic variations, as evident by randomly amplified polymorphic DNA fingerprinting. This review summarizes the research already carried out and that needs to be done to elucidate the biosynthetic pathway, mechanism of action, and biotechnological production of guggulsterone through cell cultures before commercialization of the molecule as a drug.

Keywords *Commiphora wightii*, *C. mukul*, Guggulsterone, Farnesoid X receptor, Micropropagation, Somatic embryogenesis, Production of secondary metabolite

Abbreviations

2iP	2-Isopentenyl adenine
CAR	Constitutive androgen receptor
FXR	Farnesoid X receptor
HPLC	High performance liquid chromatography
LDL	Low-density lipoproteins
LPO	Lipid peroxidation
MS	Murashige and Skoog's
PXR-Ko	Pregnane X receptor
T3	Tri-iodothyronine

5.1 Introduction

The two closely related steroidal ketones, guggulsterone-E (pregna-4,17-diene-3,16-dione) and guggulsterone-Z, are the potent hypolipidemic and hypocholesterolemic bioactive molecules present in the gum resin of *Commiphora wightii* (Arnott.) Bhandari (Syn. *C. mukul, Balsamadendron mukul*). *C. wightii* is a slow-growing, woody, endangered medicinal tree, indigenous to Indian subcontinent, belonging to the family Burseraceae and having the chromosome number 2n=26 [1]. Since it is a very slow growing plant, the returns from the plant are only after several years and thus, not preferred for social forestry. Besides gum-resin, nothing is obtained as forest produce from this plant.

5.2 Distribution

Commiphora species are widely distributed in tropical regions of Africa, Madagascar, and Asia. It is generally distributed in arid regions of Africa and the Indian side of Thar Desert. In the Indian subcontinent, *Commiphora* species occur in Pakistan, Baluchistan, and India. Of the total 185 species, only 3 (*C. wightii, C. stocksii,* and *C. berryi*) have been found in India. *C. wightii* occurs in Rajasthan, Gujrat, and Maharashtra [1].

5.3 Biology

A characteristic feature of the family is the presence of resin ducts in the parenchymatous bark. The plant is a shrub that reaches 3 m in height and has crooked, knotty branches ending in sharp spines. The papery bark peels in flakes from the older parts of the stem, whereas younger parts are pubescent and glandular leaves are trifoliate. Gupta et al. [2, 3] reported apomictic seed development associated with polyembryony in guggul. Female plants set seeds irrespective of the presence or absence of pollen. Hand pollination experiments and embryological studies have confirmed the occurrence of nonpseudogamous apomixis, nucellar polyembryony, and autonomous endosperm formation. It was inferred that apomixis might have a significant role in the speciation of tropical trees. Apomixis may be favored by natural selection if the population densities are low and the distance between individual trees is greater than the permissible cross-pollination range. Another study [3] described the cause of low seed set in *C. wightii* on the basis of pollen–stigma interaction in the nonpseudogamous apomictic plants. They observed that although pollen grains germinated on stigma, the pistil did not support pollen tube growth, perhaps due to an alteration in the orientation of the cells of the transmitting tissue and the absence of proteins in the intercellular matrix. This results in poor seed set. Multiple sapling formation by the germination of polyembryonic seeds of *C. wightii* has also been observed [4], thereby confirming multiple embryo formation in the seeds. The plant population of different localities showed variability on the basis of random amplified polymorphic DNA markers, therefore, selection of desirable characters is important (unpublished data, the authors).

The seed set is about 16% in the plant in the Aravalli Ranges [5] and in drier parts of Western Rajasthan it is even lower. *C. wightii* is an excellent fuel wood and burns even wet due to the presence of resin in the stem. The plant is cut mercilessly by villagers for cooking food and is used with other wet woods to facilitate burning [6]. Due to these biological and social problems, the plant has become an endangered species [7].

5.4 Gum-Resin Production

In *C. wightii*, the balsam or guggul (oleogum-resin) is present in "balsam canals" in the phloem of the larger veins of the leaf and in the soft base of the stem. The development and widening of the gum-resin canal in the young stem occurs schizogenously. Gum is tapped during February to June. Plants over 5 years old with a basal diameter of more than 7.5 cm are suitable. Circular incisions 1.5 cm deep are made on the main branches and stem at a uniform distance of 30 cm apart and at an angle of 60° with the stem. The yellow, fragrant latex oozes out through the incisions and slowly solidifies into vermicular or stalactitic pieces, which are collected manually. Subsequent collections of gum resin are made at intervals of 10–15 days. About 200–500 g dry guggul is usually obtained from a plant in one season. Application of ethephon on the incisions enhances guggul production 22-fold over that obtained in the control condition. Guggul production is highest with the onset of summer (a stress-induced secondary product formation), as supported by observations with brightfield and fluorescence microscopy. But in the long-term, excessive production through ethephon application exhausts the plant and as a result, kills it [8].

5.5 Chemistry

The oleogum-resin exudates from the trunk of *C. wightii* contain a large number of compounds like steroids, diterpenoids, aliphatic esters, gums (carbohydrates), and minerals (Table 5.1). The presence of guggulsterones differentiates *C. wightii* from 184 other *Commiphora* species [9].

The complex oleogum-resin mixture needs stepwise separation [10]. The ethyl acetate- soluble fraction (45%), hereafter in this text referred to as guggul, contains all active constituents, while the insoluble fraction (55%) contains gum, minerals, and other toxic ingredients (Fig. 5.1a). Details of the separation of various guggul components are described in an excellent review [10]. Meselhy [11] used a different approach using column chromatography for separation of guggulsterones and isolated guggulsterone-M. The ketonic part is the most important bioactive fraction (12% of the ethyl acetate-soluble fraction) and contains about two dozen compounds including sterols and guggulsterone-E and guggulsterone-Z (Fig. 5.1b, c). Guggulsterone and guggulsterol, and its derivatives, are known to occur in some other plants and lower animals (Table 5.2). These compounds help during insect molting [12] and are involved in defense of these insects from predators [13].

The two isomers, guggulsterone-E and guggulsterone-Z, are interconvertible, as we recorded in callus and cell cultures of the plant grown under different conditions in our laboratory. Similarly, fungus (*Aspergillus niger*, *Cephalosporium aphidicola*) culture converted guggulsterone-E into guggulsterone-Z and several other derivatives [14].

5.6 Methods of Analysis

The methods reported for the chemical analysis of guggulsterones are not foolproof in the quantification and separation of these molecules because of the presence of a large number of closely related compounds in gum-resin and the plant.

5.6.1 Thin-Layer Chromatography

A method of analysis of guggulsterones both as a bulk drug and in formulations was developed using thin-layer chromatography aluminum plates (silica gel 60 F_{254}) and a solvent system consisting of toluene:acetone (9:1, v/v). This system gave compact spots for guggulsterone-E and guggulsterone-Z (Rf values of 0.38±0.02 and 0.46±0.02, respectively) following double development of chromatoplates with the same mobile phase [15]. Petroleum ether:ethyl acetate (3:1, v/v) or methanol:ethyl acetate (97:3, v/v) can also be used for gum-resin.

Chapter 5 Guggulsterone from *Commiphora wightii* 105

Table 5.1 Chemical constituents of *Commiphora wightii* gum resin

Class	Structure	Reference
Sterols	Guggulsterol-I	[85]
	Guggulsterol-II	[86]
	Guggulsterol-III	[87]
	Guggulsterol-IV	[10]
	Guggulsterol-VI	[11]
	Z-Guggulsterol	
	Guggulsterol-Y	

Table 5.1 *(continued)* Chemical constituents of *Commiphora wightii* gum resin

Class	Structure	Reference
Steroids	Guggulsterone-E; Guggulsterone-Z; Guggulsterone-M; Dehydroguggulsterone-M	[87] [88] [13] [11]
Terpenes and alcohols	Cembrene-a; Mukulol; Myrrhanol A; Myrrhanone A	[89] [35]
Essential Oil	Myrcene; α, ß-pinene; Limonene	[90] [91]

Chapter 5 Guggulsterone from *Commiphora wightii*

Table 5.1 *(continued)* Chemical constituents of *Commiphora wightii* gum resin

Class	Structure	Reference
Flavones	Muscanone; Quercetin	[36] [85] [92]
Acids	Ferulic acid	[85]
Ester mixture	(Z)-5-tricosene-1,2,3,4-tetraol, n=16; (Z)-5-tetracosene-1,2,3,4-tetraol, n=17	[93]
Lignans	Guggullignan-I; Guggullignan-II	[10, 94, 95]

Table 5.2 Guggulsterone and related compounds found in other plant and animal species

Molecule	Source	Reference
Guggulsterol III	Mediterranean Gorgonian *Leptogorgia sarmantosa*	[96]
Guggulsterone	*Ailanthus grandis*	[97]
Z-guggulsterol and derivatives	Defense secretion of *Dytiscus marginalis*	[98]
Z-guggulsterol and derivatives	*Acitus sulcatus*	[98]
Z-guggulsterol and derivatives	Prothoracic defensive gland secretion of *Cybister tripuncatus, Ilybius fenestratus*	[94] [98]
Z-guggulsterol and derivatives	Bark of *Khaya grandifolia*	[99]

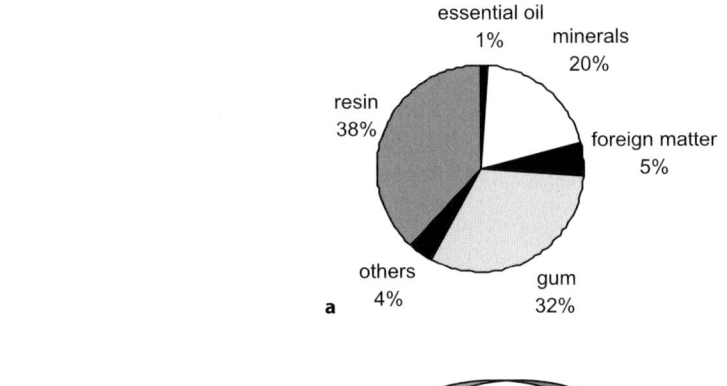

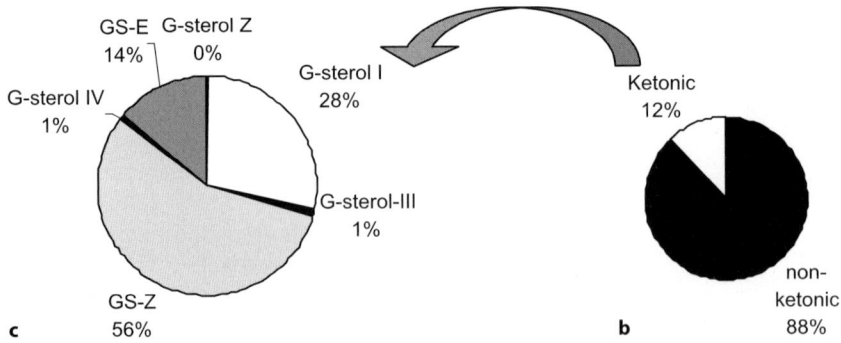

Fig. 5.1a–c Chemical composition of oleogum-resin and ketonic fraction containing guggulsterones. *G* Guggul, *GS-E* guggulsterone-E *GS-Z* guggulsterone-Z

5.6.2 High-Performance Liquid Chromatography

Guggulsterones in the resin of *C. wightii* and market formulations [16] in blood serum [17, 18] and in various nutraceuticals [19] were measured using high-performance liquid chromatography (HPLC). In complex gum-resin, the peak purity is important, as at the 241–245 nm used for quantification by HPLC, guggulsterones absorb ultraviolet light, but the column may not be able to separate guggulsterones very clearly from others components. Therefore, peak purity determination by photo diode array detector or mass spectrometry spectra is desirable.

The HPLC method developed by Mesrob et al. [16] for fingerprinting and quantitative guggulsterone estimation was used in our laboratory with a slightly modified gradient [20]. The separation was accomplished on a C_{18} (5 μm) reverse-phase column using a grasdient solvent system consisting of (A) trifluoroacetic acid in water and (B) 80% (v/v) acetonitrile in solvent "A". The developed HPLC method was found to be sensitive and useful for analysis (Fig. 5.2) of *in vitro* and *in vivo* small samples [21].

5.7 Traditional Therapeutic Uses

Guggul is one of the very ancient Ayurvedic drugs, and was first described in "Atharva Veda" (2000 B.C.). According to Sushrut Samhita, when taken orally, guggul is curative of obesity, liver function, internal tumors, malignant sores and ulcers, urinary complaints, fistula-in-ano, intestinal worms, leucoderma, sinus, edema, and sudden paralytic seizures. It is also considered as a cardiac tonic. Various methods are described to purify guggul for human consumption [22]. Traditionally, guggul is given as is or mixed with other preparations. Microparticles of guggul were formulated by different techniques using chitosan, egg albumin, sodium alginate, ethyl cellulose, cellulose acetate, gelatin, and beeswax. These microparticles were found to be useful for their use in formulations as evident by evaluation on various physicochemical parameters [23].

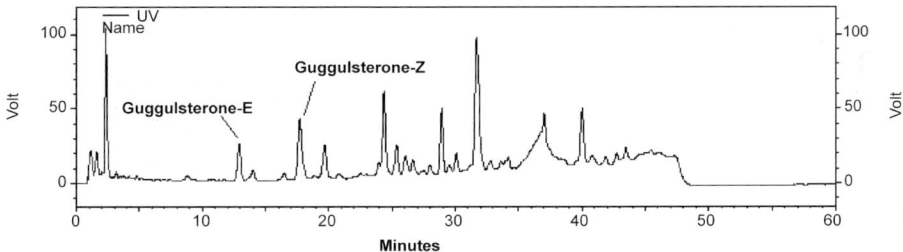

Fig. 5.2 High-performance liquid chromatography profile of gum resin at 245 nm

5.8 Pharmacology

5.8.1 *Animal and Clinical Trials*

Hyperlipidemia refers to an abnormally high concentration of lipids (fats) in the blood, and is correlated with the development of atherosclerosis, the underlying cause of coronary heart diseases and stroke. Hyperlipidemia is caused by abnormal lipid and lipoprotein metabolism.

Satyavati [24–26], who started the initial evaluation work on guggul based on Ayurvedic text, established the hypolipidemic effect using rabbit and human models and emphasized the need to understand the mechanism of action of guggulsterones. There was a surge in publications immediately after the United States Food and Drug Administration approved the drug as food supplement in 1994, including a publication describing guggulsterones as farnesenoid X receptor (FXR) agonists [27]. Several clinical studies conducted in India and elsewhere have demonstrated that administration of guggul significantly lowers low-density lipoprotein (LDL) cholestcrol and triglyceride levels in patients with hyperlipidemia [28–30]. In general, about two dozen studies [31] were carried out using 20–200 patients with high cholesterol (>250 mg/dl) and triglycerides (>200 mg/dl) who were given guggul (1–3 g/day) for 8–12 weeks. Treatment with guggul resulted in a 25–30% reduction in serum cholesterol and triglycerides [29, 31, 32]. Even after so much basic work with animal models and clinical trials, doubts about the efficacy of guggul have been raised [33] and an explanation of its ineffectiveness in some patients is required [34]. Therefore, attention should be focused on the preparation of guggul according to ancient Ayurvedic literature, proper storage conditions and dietary habits, and genotypic reactions. Guggul is a complex mixture of more than 150 compounds and new compounds continue to be reported [35, 36]; thus, the development of fractionation-based bioassays and isolation methods for several compounds are desirable. This is important because the percentage of guggulsterone-E and guggulsterone-Z is very low in ethyl acetate fractions, and the preparations do not usually contain the claimed amount [16, 31, 37].

Guggul is used for many ailments in the Ayurvedic system of medicine, therefore, animal [24, 25, 38, 39] and human models [40, 41] were used to assess its efficacy, considering it safe for human consumption [42]. Hypercholesterolemia was induced in male albino rabbits by feeding them cholesterol (500 mg/kg body weight). The initial works that have led to the establishment of guggul as a modern hypolipidemic agent have been reviewed by Satyavati [26], and subsequently by several other authors [9, 31, 32, 43]. With the exception of a few studies [33, 44], both the earlier [26, 31, 32] and more recent studies describe a significant lowering of the lipid profile in patients with hyperlipidemia. In brief, animals with hypercholesterolemia were given a 1- to 3-g/kg body weight standardized preparation of guggul (containing 2–3% guggulsterone-E and guggulsterone-Z) for 8–12 weeks. This resulted in significant decrease in lipid levels, with the elimination of atherosclerotic symptoms in the aorta. Such results were reported in chicks [45], the domestic pig [46], the presbytis monkey [47], and albino rats [34]. The results using guggul were validated in animal models using isomers of guggulsterone-E and guggulsterone-Z at 25 mg/kg body weight for 10 days, which resulted in a 27% decrease in serum cholesterol levels and a 30% decrease in serum triglyceride levels [48].

The lipid-lowering effect of gum guggul and its ether and ethyl acetate fractions has been examined in several clinical studies in India. These studies have reported a 10–20% decrease in triglyceride levels and a 20–30% decrease in cholesterol levels. Although the guggul or its fraction decreased the lipid profile in each study, variations in lipid reduction in different individuals were recorded, a result that has yet to be explained.

Compared to most cholesterol-reducing drugs currently used in modern medicine, guggulsterone is an extremely safe product, especially when compared to the commonly used statins (cholesterol biosynthesis inhibitors), which inhibits 3-hydroxy-3-methyl-glutaryl-coenzyme A reductase and can lead to liver damage [49]. Experimental evidence indicates that guggulsterone lowers cholesterol, partially by enhancing uptake of excess serum LDL-cholesterol particles by the liver. This is accomplished through receptor-mediated endocytosis at the surface of liver cell membranes. It is evident that guggulsterone increases the catabolism of LDL-cholesterol and reduces the oxidation of LDL-cholesterol via its antioxidant properties, thereby providing protection against cardiovascular diseases [50, 51]. Guggul has also been shown to reduce the stickiness of blood platelets, another biological action associated with a reduced risk of cardiovascular diseases [52].

5.8.2 Mechanism of Action

Cholesterol metabolism is a highly regulated enzymatic pathway. Deregulation of this pathway leads to accumulation of excess cholesterol and results in diseases such as atherosclerosis and gallstone formation. The homeostatic balance between uptake and elimination of cholesterol involves three pathways: *de novo* cholesterol synthesis from acetate, uptake of cholesterol from the intestine, and elimination of cholesterol through the synthesis of bile acids [53].

FXR is a member of the nuclear hormone receptor superfamily and is expressed mainly in the liver, kidney, and the small intestine. Nuclear receptors are ligand-modulated transcription factors. The physiological ligands of FXR are the bile acids, which are also responsible for absorption of dietary lipids and fat-soluble vitamins. FXR regulates target gene activity in response to the ligand. FXR is a bile acid sensor that plays an integral role in bile acid synthesis and transport. The bile salt chenodeoxycholic acid can activate the FXR heterodimer, and FXR is required for the bile-salt-dependent transcriptional control of the human *ABCB11* gene (Bile Salt Export Pump) [54]. In addition, FXR has been shown to inhibit the cholesterol-7-hydrogenase gene (*CYP7A1*) transcription [55]. In addition, activation of FXR lowers plasma triglyceride levels via a mechanism that involves the repression of hepatic *SREBP-1C* expression [56]. FXR regulation of these genes suggests that it plays a key role in maintaining steroid homeostasis. This function has been validated by the development and use of FXR-null mice, which exhibit higher serum hepatic lipid levels; a bile acid diet kills these animals [57]. Guggulsterone treatment decreased hepatic cholesterol in wild-type mice fed a high-cholesterol diet, but was not effective in FXR-null mice. It was concluded that inhibition of FXR activation was the basis for the cholesterol-lowering activity of guggulsterones [27].

Both guggulsterone-E and guggulsterone-Z also display a high affinity for other steroid receptors, including androgen, glucocorticoid, and progesterone receptors with K_i values ranging from 224 to 315 nM [58], thus altering the functions of several steroid receptors. Guggulsterone treatment represses the expression of cytochrome P450 2b10 (*Cyp2b10*) gene expression by inhibiting constitutive androgen receptor (CAR) activity in hepatocytes lacking a functional pregnane X receptor (PXR-Ko). The PXR-CAR plays important role in expression of *Cyp2b10* gene, and the expression of CAR is variable in individuals, which might be the reason for the observed variations in drug metabolism seen in different individuals [59].

Guggulsterone-E and guggulsterone-Z are responsible for lipid-lowering properties in human blood and at least four mechanisms have been proposed to explain their activity. Firstly, guggulsterones might interfere with the formation of lipoproteins by inhibiting the biosynthesis of cholesterol in the liver [60]. Secondly, guggulsterones have been shown to enhance the uptake of LDL by the liver through stimulation of LDL receptor binding activity in the membranes of hepatic cells [48]. Thirdly, guggulsterones increase the fecal excretion of bile acids and cholesterol, resulting in a low rate of absorption of fat and cholesterol in the intestine [60]. Finally, guggulsterones directly stimulate the thyroid gland [61, 62].

Guggul or guggulsterone effectively inhibits the LDL oxidation mediated by either copper ions, free-radicals generated by azo compounds, soybean lipoxygenase enzymatically, or mouse peritoneal macrophages. It was concluded that the combination of antioxidant and lipid-lowering properties of guggul or guggulsterone makes them beneficial against atherogenesis [51].

5.8.3 *Other Potential Activities*

Besides its beneficial hypolipidemic effects, guggul is used to treat obesity, inflammation, arthritis, acne, thyroid gland stimulation, and drug metabolism. Several recent reports establish a significant role for guggulsterones or guggul in antineoplastic activity [63], suppression of activity of antiapoptotic gene products [64, 65], and inhibition of osteoclastogenesis [66], demonstrating its role in cancer cell death. Guggulsterones mediate their action through nuclear receptor signal modulation [66], mitochondrial dysfunction in the absence of caspase activity [63, 65], decreased expression of antiapoptosis gene products [64], and inhibition of mitogen-activated protein kinases [67].

In India, several preparations of guggul are used in the treatment of arthritis and inflammation and have been validated in several animal models. Guggul exhibited significant anti-inflammatory activity in normal and adrenalectomized rats with formaldehyde-induced arthritis [68], in rats and rabbits with Freund's adjuvant arthritis, and in rats with paw edema induced by carrageenan [25]. The methanol extract and the ethyl acetate-soluble fraction were found to demonstrate significant inhibition of nitrous oxide formation in lipopolysaccharide-activated marine macrophages in vitro [11].

A stimulatory effect by guggulsterone on the thyroid gland has been reported, which is also considered as a possible mechanism for its lipid-lowering activity [61, 62]. An increase in tri-iodothyronine (T3) concentration with a concomitant decrease in lipid peroxidation (LPO) was observed in guggulsterone-administered mice [69], suggesting that the guggul-induced increase in T3 concentration is LPO mediated.

5.8.4 Toxicity

Guggul appears to be devoid of acute, subacute, or chronic toxicity in rats, dogs, and monkeys. It is neither a mutagen nor a teratogen [70]. However, side effects like skin rashes and diarrhea have been reported with therapeutic doses, and which differ in individuals. Therefore, formulations should be prepared as per the ancient Ayurvedic literature (e.g., purification in triphala kashaya) to avoid side effects.

5.9 Biotechnological Approaches

5.9.1 Micropropagation

Nothing concrete has come from different programs on micropropagation of *C. wightii* since 1979, demonstrating the difficult nature of the material [5, 6, 71]. Presence of resin on the surface and bacteria in the resin canals make the sterilization of the explants (stem, leaf, or petioles) difficult. Although zygotic embryos give satisfactory results, low seed set, difficulty in dissecting hard fruits (drupe) to obtain the ovule, and the low frequency of embryonic response of the zygotic embryos makes it a difficult material [5]. This makes the establishment of axenic culture in itself a difficult task.

Micropropagation is the most widely used application of plant tissue culture in general and medicinal plants in particular [72, 73]. Clonal propagation as a biotechnological approach is commonly applied for the vegetative propagation of selected materials. No selected germplasm is available in *C. wightii*, since it is a wild plant. However, micropropagation through axillary shoots using nodal explants [74] and seedling explants [75] have been described (Table 5.3). A limited number of shoot formations from these explants and plantlets transferred into pots showed feasibility but not technology development for the plant toward its micropropagation.

5.9.2 Somatic Embryogenesis

Somatic embryogenesis offers several advantages over vegetative propagation by organogenesis and axillary bud break, like the enormous numbers produced in a very short duration, the possibility of scale-up using cell culture in a bioreactor, and the development of artificial seeds through immobilization [76]. Immature ovule explants produced several embryos in cultures very rapidly [50] because of the presence of polyembryony in the fruits [2]. Subsequently, somatic embryogenesis in callus cultures of *C. wightii* was achieved by repetitive reciprocal transfer between Murashige and Skoog's (MS) medium [77] contain-

Table 5.3 *In vitro* studies on *C. wightii*. *MS* Murashige and Skoog's medium, *2,4-D* 2,4-dichlorophenoxyacetic acid, *Kn* kinetin, *BA* benzyl adenosine, *IAA* indole acetic acid, *B5* Gamborg's medium, *2,4,5-T* 2,4,5-trichlorophenoxyacetic acid, *IBA* indole butyric acid, *HF* hormone free

Explant	Medium	Results	Reference
Leaf	MS+2,4-D (0.2)+Kn (0.2)	Production of guggulsterones in immobilized cells	[82]
Nodal region	MS+BA (1.0)+IAA (0.1)	Shoot formation from node	[74]
Zygotic embryo	B5+2,4,5-T (0.5) or IBA (0.5)+Kn (0.5)	Somatic embryogenesis	[50]
Cotyledonary nodal segment of seedling	MS+BA (1.0)+IAA (0.1)+additives	Shoot multiplication from juvenile stem segments	[75]
Zygotic embryo	MS-2+BA (0.25)+ IBA (0.1)	Somatic embryogenesis and plantlet formation	[5]
Callus	MS-2+BA (0.25)+ IBA (0.1)	Formation of resin canal	[81]
Callus	MS-HF+Charcoal MS+Plant growth regulators	Production of guggulsterones in embryogenic cultures and secondary somatic embryogenesis. Morphactin enhances guggulsterones	[78] [21]

ing 2,4,5-trichlorophenoxyacetic acid and kinetin to a medium devoid of any growth regulator. This was the only way to achieve somatic embryogenesis in callus cultures initiated from immature zygotic embryos; treatments with other plant growth regulators did not produce any embryogenesis [5]. Although the frequency of explants producing embryonic culture was low, immature zygotic embryos were the only suitable explants to produce an embryonic callus as mature explants did not produce any type of organogenesis on any of the treatments tested. Maximum growth of embryonic callus was recorded on modified MS medium supplemented with 0.25 mg/l benzyl adenosine and 0.1 mg/l indole butyric acid. Asynchronously growing embryos formed plantlets regularly, which were successfully transferred to field conditions [5]. Somatic embryos produced secondary somatic embryos on hormone-free medium [78]. The development of technology through somatic embryogenesis using callus and cell cultures has been a major focus of our laboratory for last 7 years. Several problems associated with this plant system remain to be resolved before it can be used for mass propagation, namely, asynchronous development of embryos, the low conversion rate, and the survival of plantlets.

5.9.3 Resin Canal Formation

Cytodifferentiation in callus and cell cultures is a prerequisite [79] for the production of secondary metabolites, which are produced in complex tissue systems like laticifers and resin canals. Although laticifer formation and their regulation in callus cultures is known [79, 80], no single report describes resin canal formation in callus cultures. Resin canal formation was not recorded in callus cultures, but its formation was observed during somatic embryogenesis. Resin canals were observed in torpedo-shaped and cotyledonary-stage embryos of *C. wightii*. Early stages were devoid of it. Resin canals formed in somatic embryos were comparable with those formed in the stem [81]. Because of resin canal development, somatic embryos accumulated a threefold higher level of guggulsterones. This provides large quantities of aseptic resin canals of somatic origin, which can be used for guggulsterone production through biotechnological methods, thus relieving pressure on natural resources [78].

5.9.4 Guggulsterone Production

The unavailability of sufficient guggul from natural resources and destruction of the plants initiated the search for alternative methods of guggulsterone production [74, 78, 82]. Callus cultures were initiated on modified MS medium containing 2,4-dichlorophenoxyacetic acid (0.5 mg/l) or 2,4,5-trichlorophenoxyacetic acid (0.25 mg/l) and kinetin (0.1 mg/l) [20]. White, watery, and soft callus developed from leaf explants over 2–3 weeks. They turned off-white to light-brown on separation from the mother explant during the next two to three subcultures. The cultures were grown for three passages on medium containing an antibiotic (tetracycline 250 mg/l) to eliminate contamination of explant origin.

Gum resin is material of commerce and contained about 2–3.3% guggulsterone. Undifferentiated cultures contained the lowest amount of guggulsterones as compared to embryonic cultures and plant parts. Among the various treatments of plant growth regulators, morphactin and 2-isopentenyl adenine (2iP) interacted significantly to enhance callus growth and guggulsterone production by about eightfold in 1-year-old cultures. However, the effect of morphactin on callus growth and guggulsterone production was not uniform over the levels of 2iP tested [21]. Although it was clear that cytodifferentiation might be a prerequisite for high production of guggulsterones, cell cultures were initiated to look into the possibility of guggulsterone production. Optimization of growth and production by varying nitrogen, $CaCl_2$, potassium dihydrogen phosphate, and sugar levels in MS medium resulted in the development of combinations for the growth and the production media [83]. The cell cultures accumulated small amounts of guggulsterones (Fig. 5.3). These cultures were subcultures every 3 weeks with 10% inoculum. Of all the permutations and combinations used for enhancing the guggulsterone contents of the cells, the maximum guggulsterone content and yield per litre were recorded in cells grown in the medium containing a sucrose:glucose combination at the 4% level. This concentration and combination of sugars enhanced the yield of guggulsterones two- to threefold as compared to similar concentrations of sucrose or maltose. It is evident from Fig. 5.4 that guggulsterone content increased, decreased, and then almost stabilized after 1 year of culture. After a few months, dry biomass increased with the age of the cultures. Biomass generation using cells grown in growth medium in vessels of different sizes (250 ml to 2 l) and a stirred tank (2 l) bioreactor has been achieved [84]. Guggulsterone accumulation increased about fourfold during somatic embryogenesis (Fig. 5.5). It ap-

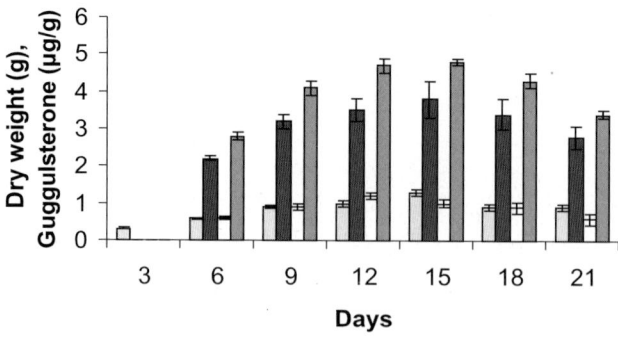

Fig. 5.3 Time course of guggulsterone production and cell dry weight in cell cultures grown in modified Murashige and Skoog's medium

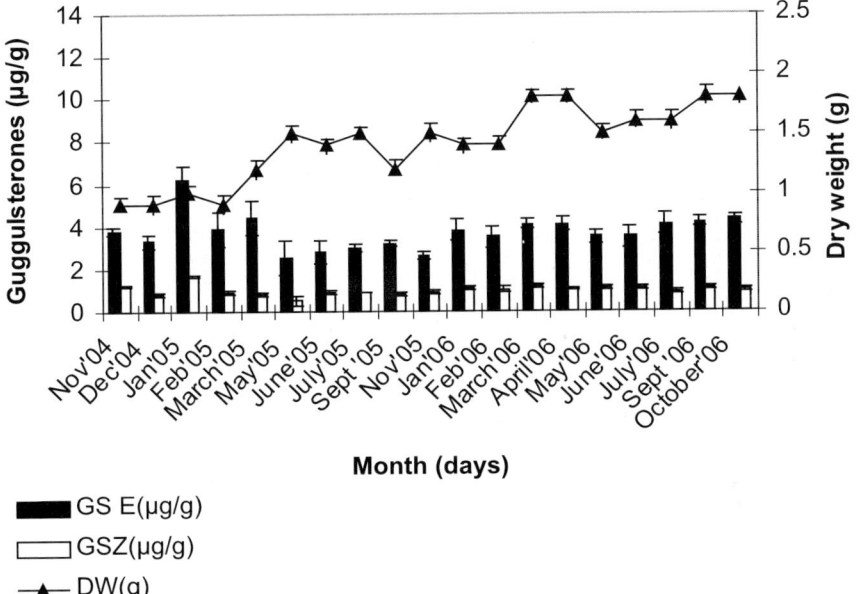

Fig. 5.4 Growth and guggulsterone accumulation in cell cultures grown in passages in CM4 medium during the past 2 years. *GSZ* Guggulsterone-Z, *GSE* guggulsterone-E, *DW* dry weight

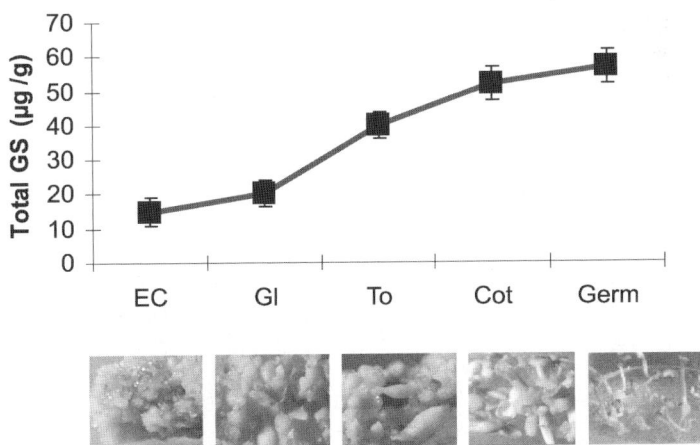

Fig. 5.5 Changes in total guggulsterone accumulation during somatic embryo development. *GS* Guggulsterone, *EC* embryogenic callus, *Gl* globular, *To* torpedo, *Cot* cotyledonary stage, *Germ* geminating embryos

pears that development of the resin canal [81] during somatic embryogenesis increased guggulsterone accumulation and other resin components. The results obtained during past 10 years are summarized in Fig. 5.6, which shows success in some aspects like somatic embryogenesis, cell cultures for guggulsterone production, and selection. However, not all culture types accumulated guggulsterones in high amounts. Currently, conditions for the growth of these cells in production medium using elicitors and immobilization, and histochemical localization of guggulsterones are in process.

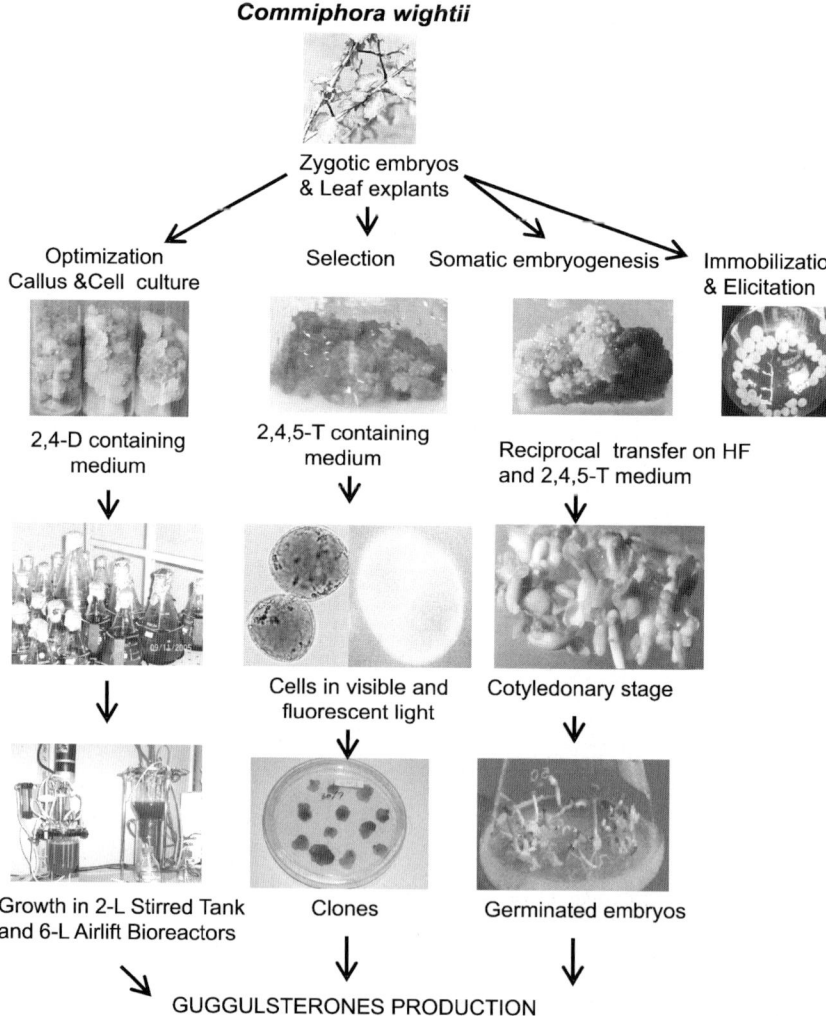

Fig. 5.6 Summary of work on *Commiphora wightii* for guggulsterone production. *2,4-D* 2,4-Dichlorophenoxyacetic acid, *2,4,5-T* 2,4,5-trichlorophenoxyacetic acid, *HF* hormone free

5.10 Future Prospects

There is no other natural product with lipid-lowering properties as effective as guggulsterone, but the plant is a very difficult material to work with using plant tissue culture. The continuous research efforts during the last three decades has resulted in some success in establishment of cell suspension cultures grown up to bioreactor level, developing somatic embryogenesis in callus cultures, micropropagation systems and feasibility of guggulsterones production in cell cultures. However, there are several problems that remain to be resolved before guggulsterones are produced by alternative technology, like somatic embryogenesis in cell suspension cultures, assessment of genetic variability in nature and in the regenerants, regulation of somatic embryogenesis, high yields of guggulsterones through immobilization and elicitation of suitably selected somatic embryos, histochemical localization of producer cells, and identification of stem cells for better embryogenesis. A better understanding of biosynthetic pathways and the use of precursors hold promise toward developing this technology-achieving target of high guggulsterone production.

Acknowledgements

This work was supported by financial assistance (grant No. BT/PR3214/ PBD /17/210/2002) from the Department of Biotechnology, Government of India, New Delhi, and partially by a DST-FIST program for infrastructure development, and UGC-DRS under a special assistance program for medicinal plant research to Prof. K.G. Ramawat.

References

1. Kumar S, Shanker V (1982) J Arid Environ 5:1
2. Gupta P, Shivanna KR, Mohan Ram HY (1996) Ann Bot 78:67
3. Gupta P, Shivanna KR, Mohan Ram HY (1998) Ann Bot 81:589
4. Prakash J, Kasera P, Chawan DD (2000) Curr Sci 78:1185
5. Kumar S, Suri SS, Sonie KC, Ramawat KG (2003) Indian J Exp Biol 41:69
6. Ramawat KG, Bhardwaj L, Tiwari MN (1991) Indian Rev Life Sci 11:3
7. Gupta R, Chadha KL (eds) (1995) Advances in Horticulture – Medicinal and Aromatic plants, vol 11. Malhotra, New Delhi
8. Bhat JR, Nair MNB, Mohan Ram, HY (1989) Curr Sci 59:346
9. Schauss AG, Munson SE (1999) Natural Medicine online (http://www.nat-med.com)
10. Dev S (1987) Sci Age 13
11. Meselhy MR (2003) Phytochemistry 62:213
12. Assad YO, Torto B, Hassanali A, Njagi PGN, Bashir NHH, Mahamat H (1997) Phytochemistry 44:833
13. Dev S (1989) Pure Appl Chem 61:353
14. Rahman A, Choudhary MI, Shaheen F, Asraf M, Jahan S (1998) J Nat Prod 61:428

15. Agrawal H, Kaul N, Paradkar AR, Mahadik KR (2004) J Pharm Biomed Anal 36:33
16. Mesrob B, Nesbitt C, Misra R, Pandey RC (1998) J Chromatogr B Biomed Sci Appl 720:189
17. Singh SK, Verma N, Gupta RC (1995) J Chromatogr B Biomed Sci Appl 670:173
18. Verma N, Singh SK, Gupta RC (1998) J Chromatogr B Biomed Sci Appl 708:243
19. Nagarajan M, Waszkuc TW, Sun J (2001) JAOAC Int 84:24
20. Mathur M, Jain AK, Ramawat KG (2005) Proc Botanical Products in New Millenium Developments Challenges. Dept Chem, Univ Raj, Jaipur India
21. Tanwar YS, Mathur M, Ramawat KG (2007) Plant Growth Reg 51:93
22. Shastry VVS (1976) Bull Indian Inst History Med 6:102
23. Tanwar YS, Gupta GD, Ramawat KG (2006) Pharma Rev 4:124
24. Satyavati GV (1966) MD thesis, BHU, Varanasi
25. Satyavati GV, Dwarkanath C, Tripathi SN (1969) Indian J Med Res 57:1950
26. Satyavati GV (1988) Indian J Med Res 87:327
27. Urizar NZ, Liverman AB, Dodds DT (2002) Science 296:1703
28. Gopal K, Saran RK, Nityanand S, Gupta PP, Hasan M, Das SK, et al (1986) J Assoc Physicians India 34:249
29. Nityanand S, Srivastava JS, Asthana OP (1989) J Assoc Physicians India 37:323
30. Singh R B, Naiz M A, Ghosh S (1994) Cardiovasc Drugs Ther 8:659
31. Ubrich C, Basch E, Szapary, P, Hammerness P, Axentsev S, Boon H, Kroll D, Garraway L, Vora M, Woods J (2005) Comp Ther Med 13:270
32. Urizar NZ, Moore DD (2003) Ann Rev Nutr 23:303
33. Szapary PO, Wolfe ML, Bloedon LT (2003) JAMA 290:765
34. Lata S, Saxena KK, Bhasin V, Saxena RS, Kumar A, Srivastava VK (1991) J Postgrad Med 37:132
35. Kimura I, Yoshikawa M, Kobayashy S, Sukihara Y, Sujuki M, Oominay H, Murakami T, Matsuda H, Doyphote VV (2001) Med Chem Lett 11:985
36. Fatope MO, Al-Burtomani SKS, Ochei JO, Abdulnor AO, Al-Kindy MZ, Takeda Y (2003) Phytochemistry 62:1251
37. Tanwar YS (2006) PhD thesis, MLSUniversity, Udaipur, India
38. Malhotra CL, Agarwal, YK, Mehta VL, Prasad S (1970) Indian J Med Res 58:394
39. Arora RB, Das D, Kapoor SC, Sharma RC (1973) Indian J Exp Biol 11:166
40. Malhotra SC, Ahuja MMS (1971) Indian J Med Res 59:1621
41. Malhotra SC, Ahuja MM, Sundaram KR (1977) Indian J Med Res 65:390
42. Agarwal RC, Singh SP, Saran RK (1986) Indian J Med Res 84:626
43. Shields KM, Moranille MP (2005) Am J Health Syst Pharm 15:1012
44. Dasgupta R (1990) J Assoc Physicians India 38:346
45. Baldwa VS, Bhasin V, Ranka PC, Mathur KM (1981) J Assoc Physicians India 29:13
46. Khanna DS, Agarwal OP, Gupta SK, Arora RB (1969) Indian J Med Res 57:900
47. Dixit VP, Joshi S, Sinha R, Bhargava SK, Verma M (1980) Biochem Exp Biol 16:421
48. Singh V, Kaul S, Chander R, Kapoor NK (1990) Pharmacol Res 22:37
49. Meschino J (2003) Chiropractic 21:10
50. Singh AK, Suri SS, Ramawat KG (1997) Gartenbauwissenschaft 62:44
51. Wang X, Greilberger J, Ledinski G, Kager G, Paigen B, Jurgens G (2004) Atherosclerosis 172:239
52. Mester L, Mester M, Nityanand S (1979) Planta Med 37:367
53. Otte K, Kranz H, Kober I et al (2003) Mol Cell Biol 23:864
54. Cui J, Huang L, Zhao A, Lew JL, Yu J, Sahoo S, Meinke PT, Royo I, Pelaz F, Wright ST (2003) J Biol Chem 278:10214
55. Owsley E, Chiang JYL (2003) Biophys Res Comm 304:191
56. Zhang Y, Lee FY, Barrera G, Lee H, Vales C, Gonzalez FJ, Willson TM, Edwards PA (2006) Proc Natl Acad Sci U S A 103:1006
57. Sinal CJ, Gonzalez FJ (2002) Trends Endocrinol Metabol 13:275
58. Burris TP, Montrose C, Houck KA, Osborne He et al (2005) Mol Pharmacol 67:948
59. Ding X, Staudinger JL (2005) J Pharmacol Exp Ther 314:120

60. Gupta A, Kapoor NK, Nityanand S (1982) Indian J Pharmacol 14:65
61. Tripathi YB, Malhotra OP, Tripathi SN (1984) Planta Med 50:78
62. Tripathi YB, Malhotra OP, Tripathi SN (1988) Planta Med 54:271
63. Singh SV, Zeng Y, Xiao D, Vogel VG, Nelson JB, Dhir R, Tripathi YB (2005) Mol Cancer Ther 4:1747
64. Shishodia S, Aggrawal BB (2004) J Biol Chem 279:47148
65. Samudio I, Konopleva M, Safe S, McQueen T, Andreeff M (2005) Mol Cancer Ther 4:1982
66. Ischikawa H, Aggrawal BB (2006) Clin Cancer Res 12:662
67. Manjula N Gayatri B, Vinaykumar KS, Shankernarayanan NP, Vishwakarma RA, Balkrishnan A (2006) Immunopharmacology 6:122
68. Gujral ML, Sareen K, Reddy GS, Amma MK, Kumari GS (1962) Indian J Med Sci 16:847
69. Panda S, Kar A (1999) Life Sci 65:137
70. Breneton J (1995) Pharmacognosy, Phytochemistry, Medicinal Plants. Lavoisier, Paris.
71. Sharma R, Suri SS, Ramawat KG, Sonie KC (1998) Biotechnological approaches to the medicinal plants of Aravalli Hills with special reference to Commiphora wightii. In: Khan IA, Khanum A (eds) Role of Biotechnology in Medicinal and Aromatic Plants. Ukaaz, Hyderabad, p 140
72. Ramawat KG, Sonie KC, Sharma MC (2004) Therapeutic potential of medicinal plants: an introduction. In: Ramawat KG (ed) Biotechnology of Medicinal Plants, Vitalizer and Therapeutic. Science, USA, p 1
73. Rout GR, Samantray S, Das P (2000) Biotechnol Adv 18:91
74. Barve DM, Mehta AR (1993) Plant Cell Tiss Org Cult 35:237
75. Yusuf A, Rathore TS, Shekhawat NS (1999) Indian J Plant Genet Res 12: 371
76. Jain A, Rout GR, Raina SN (2002) Sci Hort 94:137
77. Murashige T, Skoog F (1962) Physiol Planta 15:473
78. Kumar S, Mathur M, Jain AK, Ramawat KG (2006) Indian J Biotech 5:217
79. Suri SS, Ramawat KG (1995) Ann Bot London 75:477
80. Suri SS, Ramawat KG (1997) Ann Bot London 79:371
81. Kumar S, Sonie KC, Ramawat KG (2004) Indian J Biotech 3:267
82. Phale P, Subramani J, Bhatt PN, Mehta AR (1989) Indian J Exp Biol 27:338
83. Mathur M, Jain AK, Dass S, Ramawat KG (2007) Indian J Biotech 6: (in press)
84. Mathur M, Ramawat KG (2007) Biotech Lett 29:979
85. Bajaj AG, Dev S (1982) Tetrahedron 38:2949
86. Purushothaman KK, Chandrasekharam S (1976) Indian J Chem 14:802
87. Benn WR, Dodson RM (1964) J Org Chem 29:1142
88. Patil VD, Nayak UR, Dev S (1972) Tetrahedron 28:2341
89. Patil VD, Nayak UR, Dev S (1973) Tetrahedron 29:1595
90. Prasad RS, Dev S (1976) Tetrahedron 32:1437
91. Asres K, Tei A, Moges G, Sporer F, Wink M (1998) Planta Med 64:473
92. Kakrani HK (1981) Fitoterapie 52:221
93. Zu N, Kikuzaki H, Sheng S, Sang S, Rafi MM, Wang M, Nakatani N, DiPaola RS, Rosen RT, Ho CT (2001) J Nat Prod 64:1460
94. Chadha MS, Joshi NK, Mamdapur VR, Sipahimalani T (1970) Tetrahedron 26:2061
95. Hanus LO, Rezanka T, Dembitsky VM, Moussaieff A (2005) Biomed Papers 149:3
96. Benvegnu R, Cimino G, De Rosa S, De Stefano S (1982) Cell Mol Life Sci 38:1443
97. Hung T, Stuppener H, Ellmerer-Muller EP, Scholz D, Eigner D, Manandhar MP (1995) Phytochemistry 39:1403
98. Schildknacht H (1970) Angew Chem Int 9:1
99. Adesogan EK, Taylor DAH (1967) Chem Ind (London) 13:65

Chapter 6
Silybum marianum (L.) Gaertn: the Source of Silymarin

P. Corchete

Department of Plant Physiology, Faculty of Pharmacy, University of Salamanca
37007-Salamanca, Spain, e-mail: corchpu@usal.es

Keywords *Silybum marianum*, Silymarin, Fruits, Plant Cell Culture

Abstract The milk thistle *Silybum marianum* (L.) Gaernt, a member of the Asteraceae family, is an herb whose fruits have been used medicinally for over 2000 years. Their properties are due to the presence of silymarin, an isomeric mixture of the flavonolignans silydianin, silychristin, present in two diastereoisomeric forms, A and B, silybin and isosilybin, which also exist as two diastereoisomers: silybin A and B, and isosilybin A and B. The biosynthesis of these compounds is carried out by oxidative coupling catalysed by peroxidase enzymes between the flavonoid taxifolin and the phenylpropanoid coniferyl alcohol. The silymarin content in fruits depends on the milk thistle variety and geographic and climatic conditions in which they grow; however, the relative proportions of individual components is a genetic characteristic associated with specific chemoraces. Extracts of the fruits have traditionally been employed for treating liver disorders. Studies performed in vitro and in vivo have demonstrated the antioxidant activity of silymarin and its ability to stimulate protein synthesis and cell regeneration; thus, silymarin is being used for the treatment of toxic liver damage and for therapy of chronic inflammatory liver diseases and liver cirrhosis. Silymarin also inhibits chemically induced carcinogenesis and shows direct anticarcinogenic activity against several human carcinoma cells; in addition, silymarin shows antidiabetic, hipolipidaemic, anti-inflammatory, cardioprotective, neurotrophic and neuroprotective effects. Further studies with pure components of silymarin will extend its applications. Tissue cultures have been derived from several organs of this species. Silymarin accumulation in cell cultures is lower than in the fruit and can be stimulated by elicitation with yeast extract and/or methyl jasmonate. An extensive metabolic reprogramming occurred upon elicitation: phenylpropanoid, carbohydrate and amino acid metabolism was altered and probably redirected to support the biosynthesis of flavonolignans. Yeast extract promotes the accumulation of choline and α-linolenic acid in cells, suggesting an action on membranes and the involvement of the octadecanoid pathway in the induction of silymarin in *S. marianum* cultures.

Abbreviations

EGTA	Ethylene glycol tetraacetic acid
GSH	Reduced glutathione
HPLC	High-performance liquid chromatography
MeJA	Methyl jasmonate
INOS	Inducible nitric-oxide synthase
NF-κ-β	Nuclear factor-κ-β
NO	Nitric oxide
OA	Okadaic acid
ODC	Ornithine decarboxylase
ROS	Reactive oxygen species
SENCAR	SENsitivity to CAnceR
TGF-β1	Transformation growth factor-beta 1
TLC	Thin-layer chromatography
TNF-α	Tumour necrosis factor-alpha
TPA	12-O-Tetradecanoylphorbol 13-acetate
UV	Ultraviolet
YE	Yeast elicitor

6.1 Introduction

Milk thistle is one of the most ancient of all known herbal medicines. In one form or another, several preparations of the plant, especially the fruits, have been used medicinally for over 2000 years. Its use as a liver-protecting agent dates back to early Greek references. Theophrastus (4th century B.C.) described it under the name of "Pternix" and Dioscorides called it "Sillybom" in his "Materia Medica" (1st century B.C.) and suggested that it should be prepared in a tea "for those that be bitten of serpents". The Roman naturalist Pliny the Elder (23–79 A.D.) mentioned that the juice of the plant mixed with honey was excellent for "carrying off bile". Historical references to the plant are abundant in the herbal literature of the Middle Ages, and by the 18th century, Culpepper recommended it for the plague and for congestive conditions of the liver and spleen [1]. During the 20th century, the use of milk thistle was generalised for the treatment of liver and biliary complaints, including cirrhosis, jaundice, hepatitis and liver poisoning from chemicals or drug and alcohol abuse. Its curative properties have also been applied to other conditions, including stimulating breast-milk production and treating depression [2]. The first attempts to isolate the active components of the fruits began in 1958. In 1968 Wagner et al. [3] successfully isolated a compound designated silymarin, which was later described as a mixture of complex chemicals known as the flavonolignans. The primary components of silymarin, isolated and structurally characterised by Wagner and Seligmann [4], include silybin, silydianin and silychristin, the most active component being silybin [5]. Silymarin is found at several concentrations ranging from 1 to 6% in the ripe fruit. Extensive clinical, histological and laboratory data have confirmed the efficacy of silymarin as a hepatoprotective and antihepatotoxic agent [6, 7]. These well-described properties of silymarin are due to its antioxidative and radical-scavenging activity. Silymarin also inhibits leukotriene production, which explains its anti-inflammatory effect; it stimulates protein synthesis and exerts an antifibrotic action [8–10]. New activities based on specific receptor interactions have recently been reported and there is growing interest in its anticancer and chemopreventive effects, as well as in its hypocholesterolaemic, cardioprotective, neuroactive, and neuroprotective activities [11].

Silybum marianum is currently one of the most popular medicinal plants, and silymarin is perhaps the plant principle best researched in recent years. Biotechnological approaches by means of tissue cultures could be an alternative for the production of flavonolignans. In some studies it has been shown that the biosynthesis of silymarin is severely reduced in cell cultures of this species [12–14]. However, recent advances in the molecular biology, enzymology and fermentation technology of plant cell cultures suggest that these systems could be a viable source of these bioactive secondary metabolites.

6.2 Botany

Silybum derives from a name given to certain edible thistles by the Greek physician Dioscorides. The genus contains two species: *S. marianum* (L.) Gaernt, with variegated leaves, and *S. eburneum* Coss. et Durieux, with totally green leaves [15]. However, the incompatibility barriers that are expected between two different plant species were not found by Hetz et al. [16], and the authors concluded that the genus *Silybum* has only one species: namely, *S. marianum*.

The Latin synonyms for *S. marianum* include *Carduum marianum* L., *Carduus marianum* L. and *Cnicus marianum* L. The most widely known English common name for this species is milk thistle, although other words are also associated with it including bull thistle, heal thistle, Holy thistle, Lady's thistle, Marian thistle, Mary thistle, mild thistle, milk ipecac, Our Lady's Thistle, pig leaves, royal thistle, snake milk, sow thistle, St. Mary's thistle, Venus thistle, variegated thistle or wild artichoke. In Germany it is known as Mariendistel or Frauendistel, Chardon-Marie is the common name in French, and cardo mariano and cardo lechero in Spanish.

S. marianum belongs to the family Asteraceae and is described as an annual, winter annual and biennial herb that grows from 20 to 150 cm high, rarely shorter, glabrous or slightly downy, erect and branched in the upper part. The leaves are alternate, large, white-veined and glabrous with strongly spiny margins. The inflorescences are large and round capitula, solitary at the apex of the stem or its branches, surrounded by thorny bracts. The florets are hermaphrodite, tubular in shape, with a red-purple corolla. The fruits are hard-skinned achenes 6–8 mm long, shiny, generally brownish in colour and with a white silk-like pappus at the apex. The fruits are harvested in July–August after blooming [17]. According to legend, the white pattern on the leaves represents the breast milk of the Virgin Mary, spilt on the plant while she was breastfeeding Jesus during their escape to Egypt. Milk thistle is native to a narrow area of the Mediterranean, but has been grown for centuries throughout Europe. It also grows in India, China, Africa and Australia. The plant was carried to North America by European colonists during the 19th century and is now naturalised in the United States and South America. Formerly cultivated in gardens, it is found in abandoned fields, old pastures and by the roadsides and, in some parts, is considered a problematic invasive weed and a target of classic biological control efforts. Germination occurs in autumn and spring. Seedling establishment is favourable after rainfalls begin, particularly after a dry summer when there is an absence of grass cover, since thistle seedlings require light to thrive [18]. Milk thistle overwinters as a rosette, sometimes reaching 1 m in diameter. Cold temperatures are required for flower production. Each flower head produces approximately 100 seeds, and 10–50 heads are produced per plant, with an average of 3000 seeds per plant, 94% of them viable. The seeds show little to no dormancy requirements, and any dormancy length is affected by temperature and moisture. The seeds remain viable for 9 years or more [19]. Dormancy is induced when the seeds are buried in the soil. The percentage of germination varies from year to year and may be less than 50% [20]. Milk thistle seeds have after-ripening requirements related to the germination temperature, which limits germination to 10–20°C for up to 5 months after harvest. Milk thistle seeds then germinate over a temperature range of 0–30°C [21]. Germination rates are higher in older seeds [18].

6.3 Chemical Composition of *S. marianum* Fruits

The medicinal properties of *S. marianum* are found in the ripened fruits, sometimes mistakenly called seeds, but technically known as achenes. A fairly complete list of the chemical compounds of milk thistle can be found in Dr. Duke's Phytochemical and Ethnobotanical databases and in that author's handbook [22]. Apart from silymarin, whose components will be described in detail below, and other flavonolignans, 20–30% of the fruit is composed of fatty acids, mainly linoleic, linolenic, myristic, oleic, palmitic and stearic acid; 25–30% is protein; 0.038% tocopherol; 0.63% sterols, including cholesterol, campesterol, stigmasterol and sitosterol; together with some mucilage [23].

According to Duke [22], the following flavonoids are also present in the fruit: apigenin, chrysoeriol, eriodictyol, kaempferol, naringenin, quercetin and taxifolin, a synonym for dihydroquercetin.

Silymarin belongs to a group of compounds termed flavonolignans, and their constituents are formed by an oxidative coupling reaction between the flavonoid taxifolin and a phenylpropanoid, usually coniferyl alcohol (Scheme 6.1) [24]. The presence of flavonoid-type compounds was observed in milk thistle fruits by Schindler in 1952 [25]. Later on, the research group of Wagner isolated and identified the hepatoprotective principle of fruits and designated it silymarin or silybin [3, 26]. In 1974, Wagner et al. [27] reported that silymarin was a mixture of three flavonolignans, namely silybin, silychristin and silydianin, and the structures of these compounds were determined by spectroscopic and spectrometric methods. Subsequent work led to the isolation of several other flavonolignans as minor constituents of the same and/or different varieties of *S. marianum* [28–32], including isosilychristin, silymonin, silandrin, silyhermin and neosilyhermins A and B. The structures of all flavonolignans found in fruits (up to 12) can be found in the revision compiled by Kurkin in 2003 [33].

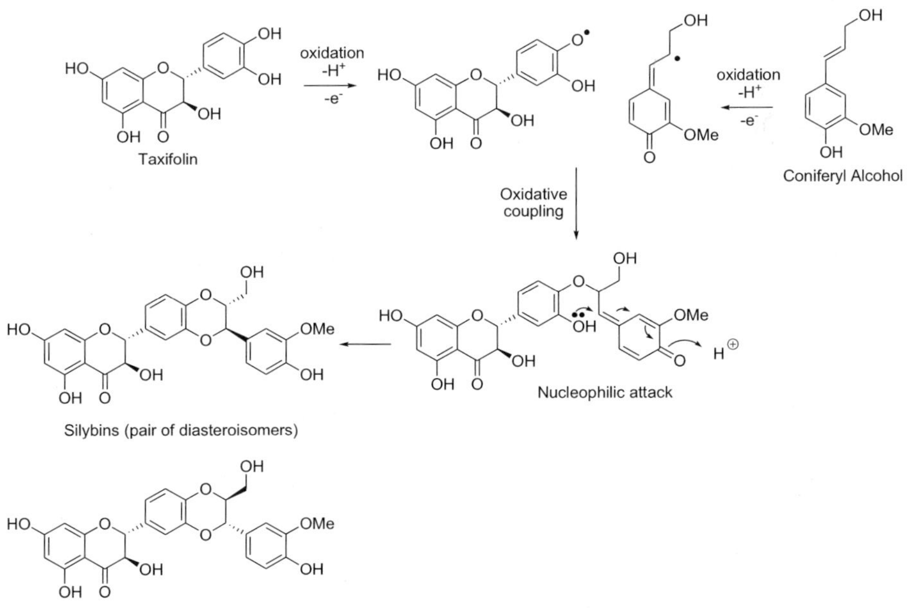

Scheme 6.1 Synthesis of silybins by peroxidative coupling of taxifolin and coniferyl alcohol

From the standpoint of biological activity, silychristin, silydianin and, overall, silybin (silymarin mixture; Structure 6.1) are of the greatest interest and have been subjected to intensive chemical, pharmacological, toxicological and clinical research.

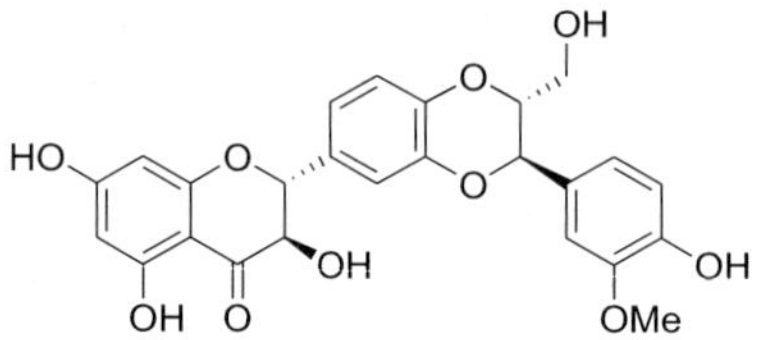

[Structures shown: Silybin A, Isosilybin A, Silybin B, Isosilybin B, Silychristin A, Silydianin, Silychristin B]

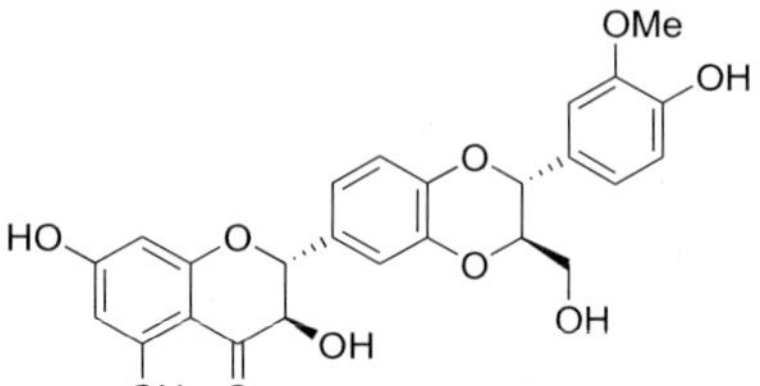

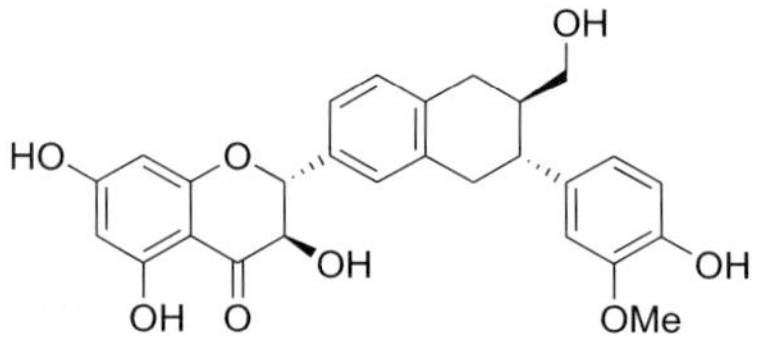

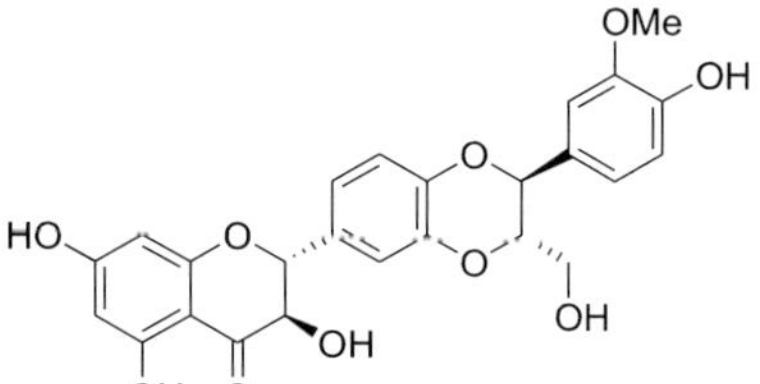

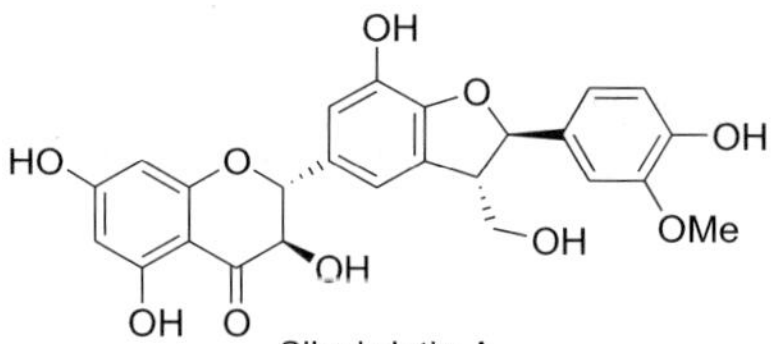

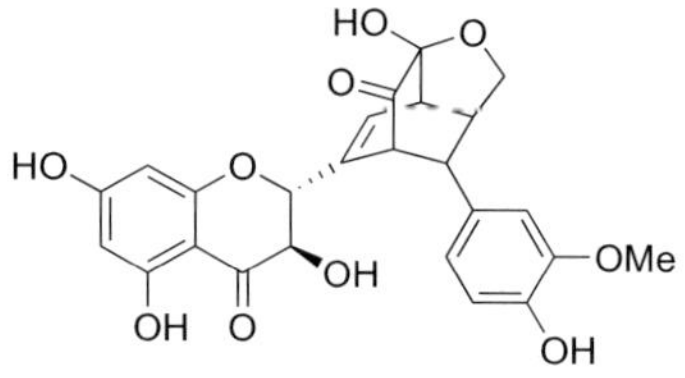

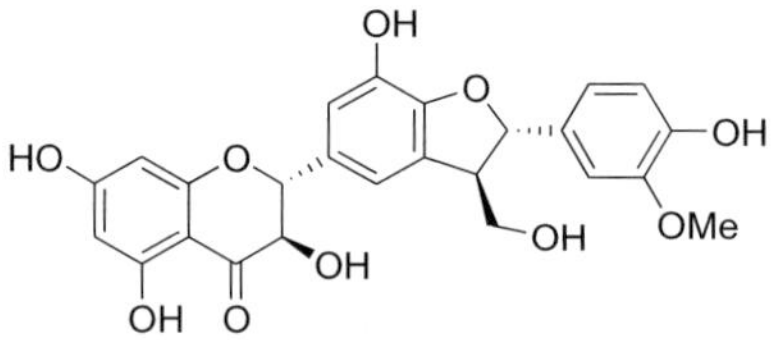

During the isolation and structural determination of silybin, its regiosiomer isosilybin was isolated and purified [27, 34, 35]. Shortly thereafter, it was recognised that silybin and isosilybin exist in two diastereoisomeric forms: silybin A and B and isosilybin A and B [34]. The absolute stereochemistries of these compounds have recently been reported by Lee and Liu [36], who fully isolated pure isomers from a natural source. It has recently been reported that silychristin also exists in two diastereoisomeric forms: silychristin A and silychristin B [37].

According to Freudenberg's hypothesis [38], which is shared by other authors [39–41], the biosynthesis of the flavonolignans that contain a 1,4-benzodioxane moiety, silybin A and B and isosilybin A and B is most achieved by the reaction between taxifolin and coniferyl alcohol in an oxidative process that is catalysed by peroxidase enzymes. The corresponding radicals are formed first, followed by their neither regio- nor enantioselective O-coupling. The final step in the biosynthesis is the thermodynamically controlled nucleophilic attack of the OH group to furnish the 2,3-*trans*-substituted 1,4-benzodioxane skeleton of the silybins [42]. It is not known whether a similar biosynthetic process takes place in the fruits for the synthesis of other flavonolignans of the 1,4-benzodioxane type or, if so, whether the O-coupling step occurs in an enantioselective manner. This would explain the existence of different chemovariants in the species, such as, for example, in white-flower varieties, which contain substantial amounts of silandrin and silymonin [43].

The flavonolignan content in fruits differs sharply depending on the milk thistle variety in question, the site of developmental growth, and on the climatic and physical conditions of the soil since the water and fertilisation regime and yield of silymarin range from as low as 0.5% dry weight in some wild populations to as high as 6% in selected varieties [44, 45]. The individual contribution of each compound to the total of the mixture also varies and seems to depend on climate conditions [46]; however, in two inbred generations Hetz et al. [16] reported genetic stability of the flavonolignan composition of 11 lines and, even the I_4 generation of three lines cultivated under different climatic conditions showed the same relative proportion of flavonolignans. The authors concluded that silymarin composition is a genetically fixed character associated with specific chemoraces. Martin et al. [47] also found that the percentage composition for a given line between years was remarkably consistent and suggested that from a commercial point of view, the selection and optimisation of the yield of a seed line that gave the preferred component profile is the best option for a reliable therapeutic application of milk-thistle fruits.

Extraction of silymarin is usually accomplished by defatting the fruits in a soxhlet device with hexane, followed by extraction with organic solvents, mainly acetonitrile, ethanol, ethylacetate or methanol. It has recently been reported that even water at 85–100°C is effective in extracting flavonolignans from milk thistle without prior defatting, a procedure outlined in the traditional extraction protocol [48]. It should be noted that in all of the procedures described, the flavonoid taxifolin is always present in the extracts at proportions that are never higher than 1.5%. Analysis of the extracted flavonolignans has traditionally been carried out with thin-layer chromatography (TLC) and spectrophotometric determinations [3, 27]. Separation and individual assays of the flavonolignans are now generally performed with high-performance liquid chromatography (HPLC). Several different protocols are described in the literature and most of them make use of reversed-phase systems with acetonitrile or methanol and water acidified with acetic or phosphoric acid as the mobile phase [49, 50]. Other methods are also employed, such as those based on HPLC-ultraviolet (UV) and liquid chromatography-mass spectrometry-mass spectrometry or liquid-liquid extraction combined with negative electrospray tandem mass spectrometry [51, 52].

6.4 Pharmacology of Silymarin

The recommended uses for silymarin (silybin) found in the literature are very varied. In a survey of the available data through different web sites, more than 300,000 entries were retrieved for the effects of silymarin. Among the described actions were: elimination of abscesses, control of allergies (seasonal and food), amelioration of Alzheimer's disease, anticarcinogenic, anticirrhotic and antidepressant, antidote to amanite poisoning; for treatment of constipation, cough, dyspepsia and eczema, as an emetic, encouragement of menstruation, as a galactagogue, for gallbladder and gastrointestinal disorders, hypocholesterolaemic, in immunity, infections, kidney disorders, liver disease, liver disorders, lung ailments, migraine, motion sickness, psoriasis, skin cancer, skin and spleen disorders, sweat-inducing, tonic and diuretic, and more.

As reported in the Introduction, milk thistle fruits have been employed in folk medicine for treating liver disorders. Intensive research into the hepatoprotective effects of milk thistle began about 45 years ago, and Germany was the pioneer country in marketing milk thistle fruits for the treatment of chronic hepatitis of all types [53].

The pharmacological properties of *S. marianum* flavonolignans have been demonstrated in vitro and in animal and human studies, and they have been the subject of several reviews. Perhaps the work that covered the pharmacology of milk thistle in greatest depth is that of Morazonni and Bombardelli [17], which was rooted in the biochemical bases of the pharmacological action(s) of silymarin published by Valenzuela and Garrido in 1994 [54]. Fraschini et al. [55] and Kurkin [33] have also updated the pharmacology and toxicology of the active flavonolignans of *S. marianum*, and Kren and Walterova [11] have recently reviewed the literature concerning new and emerging applications of silymarin.

Before summarising the pharmacological effects of *S. marianum* flavonolignans, it should be pointed out that most in vitro and in vivo studies have been performed with the active component of this plant, silymarin, which is a standardised extract isolated from the fruits that comprises 70–80% silymarin and approximately 20–30% polyphenolic compounds. The silymarin mixture is predominantly composed of silybin A and B (30–50%) together with varying percentages of isosilybin, dehydrosilybin, silychristin, silydianin and, in a smaller proportion, the flavonoid taxifolin [3].

6.4.1 Mechanisms of Action

The pharmacological action of silymarin is based on multifactor mechanisms.

6.4.1.1 Antioxidant Activity

The pharmacological activities of silymarin are assumed to derive mainly from the inhibition of several enzymes involved in the formation of free radicals of oxygen such as xanthine oxidase and cytochrome P-450 enzyme activity [56, 57]. Additionally, silymarin is able to react with hydroxyl radicals [58] and it inhibits O_2^- release in phorbol-myristate-acetate-stimulated human neutrophils through the inhibition of protein kinase C translocation and NADPH oxidase activity [59, 60]. However, silymarin does not scavenge O_2^- radical formation [61]. These properties prevent the peroxidation of membrane lipids and the consequent degeneration of cell membranes. Silymarin also enhances antioxidant systems such as glutathione (GSH) and superoxide dismutase activity [62, 63] and inhibits several peroxidases [64, 65].

6.4.1.2 Effects on Hepatocyte Membranes and Cellular Permeability

In strong connection with the previous point, silymarin stabilises hepatocyte membranes and influences cellular permeability by quantitatively and qualitatively acting on membrane lipids. These effects also vary, depending on the experimental model. Phospholipid synthesis and turnover in the liver of healthy rats is inhibited experimentally by silybin [66], but in animals subjected to acute doses of hepatotoxicants different responses are observed; that is, silymarin neutralises the inhibition of phospholipids caused by ethanol [67] and prevents the observed depletion of phosphatidylethanolamine in acute CCl_4 liver damage [68]. In contrast, silymarin stimulates phosphatidylcholine synthesis and increases the activity of cholinephosphate cytidyl transferase in the liver of normal rats or rats intoxicated with galactosamine [69].

6.4.1.3 Effects on Receptor Binding of Toxins and Drugs

Also at the membrane level, silymarin blocks receptor binding of various toxins and drugs. A unique property of silymarin mixture is the ability to neutralise the poisons phalloidin and α-amanitin from *Amanita phalloides*. The toxins are captured by hepatocytes through the sinusoidal system; phalloidin destroys the external membranes of cells, which leads to a lethal condition within a few hours, and amanitin penetrates inside the cell nuclei and suppresses protein synthesis, causing death 3–5 days after poisoning [70]. In primary cultures of rat hepatocytes, silymarin competitively inhibits the entry of both peptides into cells [71].

6.4.1.4 Stimulation of Protein Synthesis

Several in vivo and in vitro studies have demonstrated that silymarin may stimulate the action of nucleolar polymerase A, resulting in an increase in ribosomal protein synthesis by an unknown mechanism. This effect therefore stimulates the regeneration capacity of the liver and the formation of new hepatocytes [9]. Intriguingly, protein synthesis was only observed in injured livers and not in healthy ones [72].

6.4.1.5 Inhibition of Cell Proliferation in Hepatic Fibrosis

In contrast to the aforementioned stimulatory mechanism, silymarin has been reported to inhibit cell proliferation in hepatic fibrosis and in several lines of cancer cells. The underlying mechanisms for these effects are through the inhibition of mitogenic and cell-survival signalling or modulations of cell-cycle regulators [73–75].

6.4.1.6 Anti-inflammatory Activity

Silymarin interacts with cyclooxygenase and 5-lipoxygenase, enzymes involved in arachidonate metabolism, thus inhibiting leukotriene synthesis. These effects, together with the inhibition of nitric oxide (NO) synthesis, could be responsible for the anti-inflammatory activity of *S. marianum* flavonolignans [76, 77].

6.4.2 Pharmacological Applications

As mentioned above, the main action of the active principles of *S. marianum* is hepatoprotective. Studies of other protective effects suggest future applications of silymarin beyond conditions affecting the liver. A summary of the pharmacological applications is given below.

6.4.2.1 Hepatoprotective Action

Hepatocytes continually cope with a variety of noxious substances that are present not only in ingested food, such as pesticides and artificial additives, but also with a variety of agents including alcohol, cigarette toxins, drugs, heavy metals, pollutants, pharmaceuticals and viral and bacterial antigens. Although this organ is equipped with a sophisticated machinery to get rid of such toxics, if the situation persists, or if the aggression is very acute, the detoxifying systems fails, resulting in hepatocyte damage or destruction. A variety of blood tests can assess the general status of the liver and biliary systems. The most common includes the analysis in serum of alanine aminotransferase and aspartate aminotransferase. These are the enzymes that indicate liver cell damage. The other frequently used liver enzymes are alkaline phosphatase, gammaglutamyltranspeptidase and sorbitol and glutamic dehydrogenases, which indicate obstruction to the biliary system, either within the liver or in the larger bile channels outside the liver [78, 79].

The hepatoprotective action of silymarin has been tested in animal models, mainly rats and mice, in which acute or chronic hepatitis was induced by drugs with well-known mechanisms of action. Silymarin, administered through the intraperitoneal or intravenous route, prevents liver damage in animals treated with a broad range of hepatotoxic drugs, such as, for example, acetaminophen (paracetamol), *Amanita phalloides* toxins, butirophenones, carbon tetrachloride, ethanol, galactosamine, phenotiazines, thallium, the anaesthetic halothane and iron [17]. Some examples are given below.

6.4.2.1.1 Carbon Tetrachloride

This drug induces centrilobular necrosis and steatosis by acting as a strong inducer of peroxidation of unsaturated fatty acids in membrane phospholipids. As a consequence, experimental rats treated with CCl_4 show a two-fold increase in the duration of sleep [80]. In his review, Kurkin [33] reported that the administration of silymarin reduces the extent of sleep, thus indicating that silymarin suppresses CCl_4-induced peroxidation and prevents the inhibition of liver parenchyma enzymes and the increase in serum marker enzymes.

6.4.2.1.2 Acetaminophen

Paracetamol is employed therapeutically as an analgesic and antipyretic; at high doses, however, it causes centrilobular necrosis of the liver, its toxic effect being exerted through lipid peroxidation and a depletion of GSH. Silymarin, given in soluble form as silybin dihemisuccinate, protects paracetamol-treated animals and antagonises in a dose-dependent manner the exhaustion of GSH and lipid peroxidation. Serum enzymes do not increase but remain within control values at all times studied [81, 82].

6.4.2.1.3 Galactosamine

Galactosamine induces acute hepatitis similar to human viral hepatitis in rats. From the biological point of view, silymarin reduces the level of serum enzymes in rats treated with moderate doses of galactosamine; a reduction in histological and ultrastructural alterations at the cellular and subcellular levels was also observed [17].

6.4.2.1.4 Thioacetamide

The administration of this drug to rats causes changes in liver cell structure resembling the pattern of human cirrhosis, with nodule formation and necrosis of parenchyma cells [70]. Silymarin administration increases the survival of thioacetamide-treated animals by 70% and prevents the increase in serum enzymes [33, 17].

6.4.2.1.5 Ethanol

Acute ethanol administration causes an accumulation of reactive oxygen species (ROS), including superoxide and hydroxyl radicals, and hydrogen peroxide [83]; as a consequence, lipid peroxidation of cellular membranes, protein, and DNA oxidation occurs [84–86]. Hepatic steatosis with mild necrosis and an elevation of serum alanine aminotransferase activity are induced, together with a significant decrease in hepatic GSH levels and abnormal cytokine metabolism, especially tumour necrosis factor-alpha (TNF-α) [87]. Accumulated evidence has demonstrated that supplementation with standardised silymarin attenuates these changes in animal models treated with high doses of ethanol [88–90].

Apart from these studies, silymarin also shows hepatoprotective effects against lanthanides, *tert*-butyl hydroperoxide and, as explained before, phalloidin and α-amanitin [55, 33]. From these findings, it may be concluded that silymarin can be used both for the treatment of liver disorders and for the prophylaxis of several diseases caused by the continuous exposure to xenobiotics that cause membrane lipid peroxidation.

6.4.2.2 Hypocholesterolemic Action

There are some data indicating the ability of silymarin to modulate and positively affect lipoprotein metabolism. However, this effect was only seen in hyperlipidaemic rats, while in normal animals parenterally administered silybin did not affect serum cholesterol levels [91]. The hypolipidaemic effects of silymarin and its polyphenolic fraction are manifested in a decrease in cholesterol levels in the liver and plasma in rats fed on a high-cholesterol or high-sucrose diet [92, 93]. Sbolova et al. [94] have recently demonstrated that these changes are due to the fact that silymarin and its polyphenolic fraction inhibit the absorption of dietary cholesterol from the intestine.

6.4.2.3 Chemopreventive and Anticarcinogenic Effects

The antiradical properties of silymarin, and consequently its cytoprotective activity, form the basis for its use in the prevention of the toxic and even carcinogenic effects of many chemicals, not only on the liver, but also on other susceptible organs. In addition, several lines of evidence point to a direct role for silymarin in the treatment of different types of cancer, both as a direct anticarcinogenic agent and as an adjuvant in chemotherapy.

Silybin protection against cisplatin-induced nephrotoxicity has been demonstrated in rats [95]. Joint administration of silybin with the antiarrythmic drug amiodarone decreases some of the side effects of the drug, such as lysosomal phospholipidosis and conjugated diene formation, without attenuating amiodarone activity [96, 97]. Silymarin protects rat cardiomyocytes against the oxidative stress induced by the cardiotoxic anticancer drug doxorubicin [98].

In the pancreas, silymarin has been shown to protect pancreatic cells against alloxane, an agent employed to induce experimental diabetes mellitus. The protective actions seem to be mediated by both the antioxidant properties of silymarin and/or by the elevation in plasma or pancreatic GSH concentrations [99]. The endocrine and exocrine pancreas of rats is also protected from cyclosporine A toxicity by silybin. In addition, glucose-stimulated insulin secretion in this system is significantly reduced after an 8-day treatment period with silybin in vivo, with no increases in blood-glucose concentrations [100]. The inhibitory effect was non-specific, and hence the authors suggested that silymarin might also protect the exocrine pancreas against other insult principles, such as alcohol. Silymarin has the ability to inhibit the production of inflammatory cytokines [101]. Thus, it has recently been demonstrated that silymarin avoids cytokine-mediated toxicity and cytokine-induced impairment of glucose-stimulated insulin secretion by human islets, and these effects can be explained in terms of the ability of the compound to modulate signalling pathways by suppressing certain mitogen-activated protein kinase activities in pancreatic β cells [102]. Taken together, these results indicate that silymarin may be useful as a therapeutic agent for type 1 diabetes.

A strong protective activity of silymarin is seen against 12-O-tetradecanoylphorbol 13-acetate (TPA)- and okadaic acid (OA)-elicited tumour promotion in "sensitivity to cancer" (SENCAR) mouse skin, this effect being caused by the inhibition of the endogenous tumour promoter TNF-α [103]. Silymarin is also known to inhibit chemically induced carcinogenesis in other animal organs, such as the colon [104], tongue [105] and the bladder [106]. In some of these studies, a moderate to statistically significant increase in the activity of the enzymes glutathione S-transferase and quinone reductase was observed, both of these being enzymes that afford protection against the adverse effects of reactive metabolites of procarcinogens [107, 108].

Direct anticarcinogenic activity is exerted by silymarin against human breast carcinoma cells MDA-MB 468 by arresting them at the G1 phase of the cell cycle, possibly through modulation of kinase activities and associated cyclins [109]. The latter authors extended their studies to human prostate carcinoma DU145 cells and their results also suggested that silymarin may exert a strong anticarcinogenic effect against prostate carcinoma by means of impairment of the epidermal growth factor receptor (erbB1) signalling pathway, the induction of cyclin-dependent kinases and a resulting G1 arrest [110]. In the mouse SENCAR model mentioned above, silymarin not only plays a chemoprotective role against TPA tumour promotion, but can also induce the regression of already established tumours [111]. A strong antiangiogenic effect for silymarin has been reported in the colon cancer LoVo cell line by Yang et al. [112], and an antiproliferative and apoptotic effect has also been observed in human colon carcinoma HT-29 cells treated with silybin [113].

The research group led by Agarwal hypothesised that the anticarcinogenic effects of silymarin would be mediated through the suppression of nuclear factor-κ-β activation (NF-κ-β). NF-κ-β is the mediator of many of the effects of TNF-α and regulates the expression of several genes involved in cytoprotection, carcinogenesis and inflammatory processes [114]. This possibility is supported by several lines of evidence. For example, silymarin potently suppresses both NF-κ-β-DNA-binding activity and its dependent gene expression induced by OA in the HepG2 hepatoma cell line. However, in this case TNF-α-induced NF-κ-β activation was not affected by silymarin [115]. The activation of NF-κ-β is regulated by several kinases and the activation of NF-κ-β and kinases is in most cases dependent on the production of ROS [116]. In this context, the role of silymarin would ultimately also lie in its powerful antioxidant properties.

The use of silymarin in adjuvant therapy of cancer has also been reported in several studies. For example, silybin has been shown to enhance G2/M arrest and the induction of apoptosis by doxorubicin, carboxyplatin and cisplatin [117]. The cytotoxic activity of Brostallicin, which has been shown to be enhanced in the presence of high GSH and GSH transferase levels, is enhanced by silybin. The mechanism underlying the interaction involves the apoptotic pathway, since an increase in mitochondrial proteins (Bcl-2 family members) and a decrease in caspase 3 activity are observed with the silybin-brostallicin combination [118].

Although not strictly referring to a chemopreventive effect, silymarin has also been shown to exert a preventive action in UV-induced skin carcinogenesis. For example, silymarin prevented UV-irradiation-induced apoptosis in human malignant melanoma cells [119]; the inhibitory mechanism seemed to be related to caspases and Bcl-2 proteins as well as to extracellular signal-regulated protein kinase [119]. Application of silymarin to mouse skin prior to that of TPA, which induces ornithine decarboxylase activity (ODC), results in a significant inhibition of TPA-induced epidermal ODC activity in a dose- and time-dependent manner [120]. In the review by Katiyar [121] of the anti-inflammatory, antioxidant and immunomodulatory effects of silymarin on skin-cancer prevention, the author suggests that silymarin is a promising chemopreventive and pharmacologically safe agent that could be exploited or tested against skin cancer in humans. Moreover, silymarin may favourably supplement sunscreen protection and provide additional antiphotocarcinogenic protection.

6.4.2.4 Anti-inflammatory Action

The antinflammatory effects of silymarin are based on multiple activities, including mast cell stabilisation, the inhibition of neutrophil migration, Kuppfer cell inhibition and the inhibition of leukotriene and prostaglandin formation through the inhibition of 5-lipoxygenase [122]. In experimental models, silymarin exhibits significant anti-inflammatory and antiarthritic activities in the papaya latex-induced model of inflammation, and mycobacterial adjuvant-induced arthritis in rats through the inhibition of 5-lipoxygenase [76]. Silymarin inhibits NO production and inducible nitric-oxide synthase (iNOS) gene expression in macrophages, and these effects are mediated through the inhibition of NF-κ-β transcription factor. Since NO plays an important role in the pathogenesis of several inflammatory diseases [123], the inhibitory effect of silymarin on iNOS gene expression is suggestive of one of the mechanisms responsible for the anti-inflammatory action of silymarin [77]. Kang et al. [124] provided further insight into the mechanism of the anti-inflammatory effect of silymarin. Silymarin blocks lipopolysaccharide-induced sepsis and the gene expression of inflammatory mediators, such as interleukin-1β and prostaglandin E2, involved in the septic process. In light of their findings, the authors suggested that silymarin can be considered as a possible therapeutic agent for a variety of acute inflammatory diseases. In a recent study, Jeong et al. [125] have shown that silymarin prevents hepatic fibrosis in CCl_4-induced cirrhosis in rats. In this report, a reduction in the number of mast cells (stellate cells), whose proliferation accompanies fibrogenesis, was noted, and the expression of transformation growth factor-beta 1 (TGF-β1), known to be involved in fibroblast proliferation, was modulated in animals treated with silymarin. Previously, silymarin was also shown to downregulate type 1 collagen and TGF-β1 mRNA levels in treated animals [126], suggesting that it would have an effect on collagen-related genes separate from its antioxidant effect. These observations suggest that silymarin would prevent hepatic fibrosis through the suppression of inflammation and hypoxia in hepatic fibrogenesis.

6.4.2.5 Other Actions

The interest aroused in the use of silymarin for the prevention and treatment of neurodegenerative and neurotoxic diseases is increasing considerably as new mechanisms of action are emerging for such compounds. Perhaps the most cited work is that of Wang et al. [127], in which a neuroprotective effect of silymarin against lipopolysaccharide-induced neurotoxicity in mesencephalic mixed neuron-glia cultures was described. As in other reports, the results revealed that silymarin significantly inhibits the production of inflammatory mediators such as TNF-α and NO. At gastrointestinal level, silymarin also shows anti-ulcerogenic activity and immunomodulating activities [128]. In the review by Kren and Walterova [11] it was reported that silymarin interacted with proteins involved in the transport or depletion of drugs and its use is now recommended to avoid multidrug resistance, a serious problem in the treatment of cancer and infections. Recently it has been shown that dehydrosilybin expresses antimalarial activity in vitro and it has been suggested that in the near future, silymarin derivatives associated with already available drugs could be useful for delaying the spread of *Plasmodium falciparum* resistance [129]. Last but not least, silymarin has also found a use in cosmetics in antiageing formulations [130, 131].

6.4.3 Bioavailability

Due to its poor water solubility, the bioavailability of silymarin is low. In rats, a large dose of silybin given orally as plain silymarin remained almost undetectable in the plasma for the 6-h experiment [132]. In a study performed in 1975 by Bülles et al. [133] it was reported that silybin was excreted unmodified in urine, while in bile it was excreted as sulphates or glucuronides. These results were independent of the administration route – oral or intravenous. Research into the excretion kinetics of silymarin, its presence in faeces and intestine, and its long persistence in the liver are indicative of the enterohepatic circulation of silymarin [17].

Pharmacokinetic studies in humans have shown that silymarin absorbed through different routes is distributed into the digestive tract (liver, stomach, intestine and pancreas). The absorption of silymarin from the gastrointestinal tract is about 20–50%. This low bioavailability could be attributed to degradation by gastric fluid [23], poor enteral absorption [134] or its poor water solubility [135]. It has been reported that absorption decreases with age and may only be 10% at the age of 60 years. Peak plasma concentrations are achieved after between 90 min and 4 h, and on average only 10% of total is in the unconjugated form. Approximately 80% of silymarin is excreted in the bile and about 5% in the urine as total silymarin, with a renal clearance of approximately 30 ml/min. Its elimination half-life ranges from 6 to 8 h [136].

In order to improve its bioavailability, silymarin has been incorporated in different dosage forms. Trials have been reported using cyclodextrin [137], salts of polyhydroxyphenylchromanones [135], soluble derivatives [138], complexes with phospholipids [139] and liposomal encapsulation [140]. From all of these formulations, complexes with phospholipids, and specifically a silymarin–phosphatidylcholine complex, have been shown to exhibit much greater lipophilicity and improved penetration across biological membranes [141]. Most studies on the pharmacokinetics of silymarin have been carried out with this formulation, after Malandrino et al. [142], who described greater bioavailability over the silymarin extract alone. Kid and Head [143] reviewed the studies performed on the bioavailability of silybin-phosphatidylcholine complexed as a phytosome.

In animal studies, a large dose of silybin given orally as plain silymarin remained virtually undetectable in the plasma along a 6-h experiment; the same amount of silybin given as silybin-phosphatidylcholine (Siliphos) is detected in the plasma within minutes, and by 1 h its levels have peaked. Its plasma levels remain elevated after the 6-h limit. Silybin from Siliphos remains high in urine at 70 h following oral dosing, while silybin given alone barely rises above detectable levels after 25 h. The amount of silybin reaching the bile from phytosome is six times greater than that coming from non-complexed silybin (13% versus 2%, over 24 h). A certain portion of the phytosomal silybin remains in the liver for at least 24 h [132].

Pharmacokinetic studies conducted with human subjects have revealed a pattern similar to that found in rats. Phytosomal silybin is absorbed four to six times better than the non-phytosome silybin from silymarin and shows a fourfold greater passage through the liver than plain silymarin [144].

6.4.4 Toxicology

Toxicological studies have been carried out in several animal species. No mortality or any signs of adverse effects were observed after the administration of silymarin at oral doses of 20 g/kg in mice and 1 g/kg in dogs. The 50% lethal dose values after intravenous infusion are 400 mg/kg in mice, 385 mg/kg in rats and 140 mg/kg in rabbits and dogs. These data demonstrate that the acute toxicity of silymarin is very low. Similarly, its subacute and chronic toxicities are very low; the compound is also devoid of embryotoxic potential [17]. European manufacturers have developed pharmaceutical-grade milk thistle preparations standardised to a 70–80% silymarin content. From these products the doses of silymarin used in clinical trials range from 280 to 800 mg/day. In the case of the silymarin-phosphatidylcholine complex, 200–400 mg/day is recommended, taken in two or three divided doses daily [145].

In humans, apart from mild gastrointestinal distress and allergic reactions, the side effects of silymarin are uncommon, and serious toxicity has rarely been reported. In an oral form standardised to contain 70–80% silymarin, milk thistle appears to be safe for up to 41 months of use.

6.4.5 Therapeutics

Silymarin is commonly used to treat viral infections and cirrhosis of the liver. However, despite its clear effects in experimental animal models, silymarin has yet to be proved effective in ameliorating human liver disease. Part of the problem is that silymarin has never been adequately evaluated using objective and clinical trials in well-characterised cohorts of patients with well-defined forms of liver disease. In a survey of clinical trials performed over the past few years on acute viral hepatitis, chronic hepatitis, alcoholic liver disease, cirrhosis and toxic-induced liver damage, Rainone [146] noted that the greatest benefit occurs in patients whose cirrhosis is due to alcoholism and in those who have less severe cirrhosis. Randomised trials performed in patients with hepatitis or cirrhosis and trials aimed at determining the use of silymarin as a prophylactic agent to iatrogenic hepatic toxicity resulted in a lowering of serum liver enzymes [147–149]. In one of the largest studies, involving 2637 patients with chronic liver disease, an 8-week treatment with 560 mg/day of silymarin resulted in reductions in serum enzymes and a decrease in the frequency of palpable hepatomegaly [150]. In a systematic review performed by Mayer et al. [151] on studies that addressed silymarin for the treatment of viral hepatitis, it was concluded that the compound does not affect viral load or improve liver histology in hepatitis B or C. The authors recommended that the effect of silymarin should be determined in conjunction with standard antiviral treatments.

In general, the recommended uses for liver diseases are as follows: early treatment for chronic liver problems; rehabilitation from alcohol, solvent or recreational drug abuse; protection of hepatocytes from toxic chemicals, including alcohol, limitation of fatty degeneration and a slowing or reversing of cirrhosis; support treatment for inflammatory liver conditions and cirrhosis.

The efficacy of silymarin as a cytoprotectant in humans has been the subject of only a few and somewhat heterogeneous clinical trials. In one of them, workers exposed to toluene or xylene for years showed reduced levels of serum transaminases upon treatment with Legalon, a standardised flavonolignan mixture containing 60 mg silymarin.

Clinical studies in oncology and infectious disease that are currently under way should help to determine the efficacy and effectiveness of milk thistle. Recent data have reported a third-phase trial in human prostate cancer patients with high levels of prostate-specific antigen.

The ingestion of *Amanita phalloides* causes hypovolaemia and hypoglycaemia, leading to severe hepatic damage or death. Silymarin inhibits the binding of the toxins to hepatocytes and interrupts the enterohepatic circulation of the toxins. The hepatoprotective action of silymarin in animals (dogs, rabbits, rats, mice) intoxicated with phalloidin is evident, after both protective and curative treatment. However, if the time interval between the administration of the toxic substance and start of treatment increases, the efficacy of silymarin decreases; after 30 min its curative effect is negligible [152]. In one controlled study carried out on humans, silymarin given before the ingestion of this mushroom was effective in 100% of cases. As an antidote given within 24–36 h after poisoning, silymarin was found to reduce liver damage and prevent death [153]. However, further experimental studies aiming at determining the amount of ingested mushroom and the time elapsed before the administration of treatment are needed to clarify its role.

6.5 Biotechnology

As reported in many earlier reviews [154, 155], studies on the production of plant secondary metabolites by callus and cell suspension cultures have been carried out on increasing scales since the end of the 1950s. The 1970s and 1980s of the past century were fruitful in the development of this technology, and interesting approaches to industrial applications were reported during those years [156]. Despite the historical importance of *S. marianum*, efforts to culture this plant in vitro began in the late 1970s with the pioneering study of Becker and Schrall [157], in which a protocol for the initiation and maintenance of tissue and suspension cultures from cotyledons and sterile plantlets was described. A Murashige and Skoog medium with different hormones was employed to support growth, optimum results being obtained with the use of naphthalene acetic acid and kinetin. TLC was used to assay flavonolignans in cultures and with this technique no typical milk thistle compounds were detected. In addition, silymarin, present in primary explants, was not identified in long-term subcultures.

In the same year (1977), Schrall and Becker [40] reported that by feeding taxifolin and coniferyl alcohol, silybin could be isolated from cultures. Horseradish peroxidase and a cell-free extract of suspension cultures were also able to synthesise silybin from the precursors in the presence of hydrogen peroxide.

After this initial work, the literature concerning milk thistle tissue culture has been scarce. Fevereiro et al. [158] analysed the protein content in milk thistle cultures without referring to flavonolignan production. In 1991, the same authors reported the presence of silybin in cultures grown in Gamborg medium supplemented with glutamic acid instead of potassium nitrate. This work was somewhat reduced in scope, and no further references were given. Guinea and Pizarro [159] published two papers in which the effect of the hormones 2,4-dichlorophenoxyacetic acid and kinetin on peroxidase and polyphenoloxidase activities were studied in an attempt to find a parameter that might serve as a marker for silymarin production. These studies did not continue and no further references have been found.

Some publications concerning several aspects of the in vitro culture of milk thistle have appeared recently. Liu and Cai [160] and Hetz et al. [161] referred to the isolation and culture of protoplasts. A protocol for the successful regeneration of plants from mesophyll and suspension culture protoplasts was offered in this latter work. Radice and Caso [162] studied organogenesis and somatic embryogenesis in cotyledon cultures, and Iqbal and Srivastava [163] reported the conditions for the micropropagation of this plant from different explant sources. Alikaridis et al. [14] employed seed explants to produce transformed (hairy) and untransformed root cultures. Both by TLC and HPLC, silybin, isosilybin, silychristin and silydianin were found in untransformed cultures; in contrast, isosilybin and traces of silychristin and silydianin were identified in hairy root cultures. Compared to the whole fruit, less than 0.004% of flavonolignans was quantified per dry weight of cultures. Hasanloo et al. [164, 165] have also studied silymarin production in cultures, but these references were published in local journals with limited international access.

In 1999, our research team reported the influence of the composition of the growth medium in the accumulation of silymarin in cell suspensions of *S. marianum*. The conditions for the initiation and maintenance of cultures were described, as well as the HPLC protocol for the separation of silymarin components. The flavonolignan content in suspension was lower than the values obtained for fruits (0.2–0.4% dry weight versus 3% in fruits). Although production decreased progressively over repeated subcultures, the silymarin content was higher in the stationary phase [13]. In that work, the presence of flavonolignans in culture medium was not reported. However, it was later seen that varying amounts of these compounds can indeed be detected during the 1st week after subculture. The contribution of individual components of the silymarin group to the total flavonolignan content was different in cells and in culture medium. Silychristin was more abundant (95%) in cells than silybins, while in the culture medium a more balanced ratio (50%) was observed [166].

One of the strategies most commonly employed to improve secondary metabolite production in in vitro cultures is manipulation of the components of the culture medium. Silymarin accumulation was not enhanced either upon using different concentrations of KNO_3, KH_2PO_4 or iron or by changes in the levels of sugars in the nutrient medium. However, application of nutrient stress through the elimination of calcium ions did reduce growth and promoted flavonolignan production [13]. The addition of ethylene glycol tetraacetic acid (EGTA), a specific calcium chelator able to reduce the availability of extracellular calcium, increased the accumulation of silymarin in a concentration-dependent manner. The presence of verapamil, a Ca^{2+} channel blocker, and La^{3+}, an ion known to block the entrance of calcium into cells, in calcium-containing medium also enhanced silymarin production, although to a lesser extent than EGTA. Treatment of cultures with 5 or 10 µM of the calcium ionophore A23187 did not alter silymarin production. Prolonged incubation times or higher concentrations reduced the accumulation of silymarin, probably due to a toxic effect since cells exhibited necrosis 24 h after treatment with 20 µM of the ionophore.

Different concentrations of ruthenium red, thapsigargin and TMB-8, inhibitors of intracellular calcium movement, induced the release of silymarin into the culture medium, and a sustained accumulation effect was also observed in cells [165]. It thus seems that in milk thistle cultures, calcium, both external and internal, negatively controls silymarin production.

Feeding *S. marianum* cultures with the precursors phenylalanine, ferulic acid, naringenin, taxifolin or coniferyl alcohol at 100 µM, a concentration much lower than that normally added to other plant cultures, was toxic to cells; for this reason, only concentrations below 10 µM could be employed. At these non-toxic concentrations, the addition of precursors did not improve silymarin production. The lack of effectiveness could hence be partly due to the low concentration employed, which probably reduced the usefulness of such precursors, and also to the fact that they were not taken up by the cells and remained unchanged in the culture medium. The use of this strategy in milk thistle cultures is therefore limited.

Elicitation is another of the strategies employed to increase the accumulation of secondary metabolites. Treatment of *S. marianum* suspensions with a crude extract of yeast elicitor (YE) improved the production of silymarin and caused the release of silymarin into the culture medium to a level about threefold higher than that of the control. Jasmonic acid potentiated the yeast extract effect and one of the jasmonic acid derivatives, methyl jasmonate (MeJA), strongly promoted the accumulation of silymarin, thus indicating that the octadecanoid pathway is presumably involved in elicitation responses [167]. Recently, Hansaloo et al. [165] also reported the effects of jasmonic acid on silymarin production in milk thistle cultures. However, the content of the article is not available through common public data bases.

MeJA seemed to act in several steps of the metabolic pathway of flavonolignans and its stimulating effect was totally dependent of de novo protein synthesis. Chalcone synthase, the entry point enzyme on flavonoid biosynthesis, was slightly enhanced by MeJA; however, the enzyme did not appear to be crucial in the silymarin elicitation process, since its increase was not dependent on de novo protein synthesis [167].

S. marianum cell cultures do not employ conserved signalling components in the transduction of the elicitor signal to downstream responses such as silymarin production. The elicitation of cells is not associated with an oxidative burst, and activation of phosphorylation/dephosphorylation cascades do not mediate in the elicitation mechanism (unpublished observations). Silymarin production in elicited cultures can be improved by pretreatment with the calcium agonists employed in the aforementioned study [165], but the enhancing effect is neither additive nor synergistic. Unexpectedly, the ionophore, although lacking any effect when administered alone, exerts the same effect as the agonists (unpublished results). From these observations, it appears that neither an external source of calcium nor any internal movement of the cation is necessary in the mechanism of silymarin elicitation in S. marianum cultures. Clearly, the signalling pathway(s) involved in elicitor-induced flavonolignan production should be investigated further.

In vivo, exogenous silymarin is rapidly degraded apoplastically. *In vitro* tests have shown that cell extracts and, to a greater degree, the spent medium could degrade silymarin compounds in the presence of H_2O_2. However, the oxidation efficiency is not modified by elicitation. S. marianum peroxidases are also able to perform the oxidative coupling of taxifolin and coniferyl alcohol to silybins. The synthetic activity is mainly associated with the extracellular compartment and elicitation does not modify oxidative coupling yield. Peroxidases also catalyse the dimerisation of taxifolin. Therefore, peroxidases may contribute to the maintenance of the constitutive levels of flavonolignans in cultures, although these enzymes do not seem to participate in the accumulation process in elicited cell cultures [41].

A comprehensive metabolomic profiling of S. marianum cell cultures elicited with YE or MeJA for the production of silymarin has been carried out using one- and two-dimensional nuclear magnetic resonance spectroscopy [168]. With these techniques, both the temporal quantitative variations in the metabolite pool in yeast-extract-elicited cultures and the qualitative differences in cultures treated with both types of elicitors are observed. YE and MeJA cause a metabolic reprogramming that mainly affects carbohydrate metabolism; sucrose levels decrease dramatically and those of glucose levels increase, these changes being dependent on de novo protein synthesis. YE acts differentially on amino acid metabolism and promotes the accumulation of choline and α-linolenic acid in cells, suggesting an action on membranes and the involvement of the octadecanoid pathway in the induction of silymarin in S. marianum cultures. Phenylpropanoid metabolism is altered by elicitation but, depending on the elicitor, a different phenylpropanoid profile is produced.

Elicitation induces supply pathways from primary metabolism and secondary metabolism, thus providing substrates for the rapid and increased production of flavonolignans. This approach offers the possibility of identifying candidate components of the signalling route, which is presumably involved in stimulation of the constitutive pathway of silymarin. Further studies should address this crucial point in order to bypass the low productivity yield of metabolites in milk thistle cell cultures.

Acknowledgements

This work was financed in part by Ministerio de Ciencia y Tecnología, Spain (BFI2000-1362). The author is grateful to Dr. Luis San Roman and Dr. Jorge Fernández Tarrago for critical reading of the chapter, and also wishes to express her gratitude to Dr. Angeles Sánchez Sampedro and Dr. Rafael Pelaez for their personal and scientific support.

References

1. Grieve M (1931) A Modern Herbal. Jonathan Cape, London
2. Foster S (1991) Milk Thistle: Silybum marianum. Botanical series No. 305. Austin Texas
3. Wagner H, Hörhammer L, Munster R (1968) Arzneim Forsch 18:688
4. Wagner H, Seligmann O (1985) Liver therapeutic drugs from Silybum marianum. In: Chang HM, Yeung HW, Tso WW, Koo A (eds) Advances in Chinese Medicinal Materials Research. World Scientific, Singapore
5. Hikino H, Kiso Y, Wagner H, Fiebig M (1984) Planta Med 50:248
6. Valenzuela A, Guerra R, Videla LA (1986) Planta Med 52:438
7. Flora K, Hahn M, Benner K (1998) Am J Gastroenterol 93:139
8. Hikino, H. Kiso Y.(1988) Natural products for liver disease. In: Wagner H, Hikino H, Farnsworth NR (eds) Economic and Medicinal Plant Research. Academic Press, New York, p 39
9. Sonnenbichler J, Zetl I (1986) Biochemical effects of the flavanolignane silibinin on RNA, protein and DNA synthesis in rat livers. In: Cody V, Middleton E, Harbourne JB (eds) Plant Flavonoids in Biology and Medicine: Biochemical, Pharmacological, and Structure-Activity Relationships. Alan R. Liss, New York, p 319
10. Dehmlow C, Erhard J, de Groot H (1996) Hepatology 23:749
11. Kren V, Walterova D (2005) Biomed Pap Med Fac Univ Palacky Olomouc Czech Repub 149:29
12. Fevereiro P, Pais MSS, Cabral JMS (1991) Planta Med 57:2
13. Cacho M, Moran M, Corchete P, Fernández-Tárrago J (1999) Plant Sci 144:63
14. Alikaridis FD, Papadakis D, Pantelia K, Kephalas T (2000) Fitoterapia 71:379
15. Tutin TG (1976) Flora Europaea 4:249
16. Hetz E, Liersch R, Schieder O (1995) Planta Med 61:54
17. Morazzoni P, Bombardelli E (1995) Fitoterapia 46:3
18. Groves RH, Kaye PE (1989) Aust J Bot 37:351
19. Sindel BM (1991) Weed Res 31:189
20. Parsons WT (1973) Noxious Weeds of Victoria. Inkata, Melbourne
21. Young JA, Evans RA, Hawkes RB (1978) Weed Sci 26:395
22. Duke JA (1992) Handbook of Phytochemical Constituents of GRAS Herbs and Other Economic Plants. CRC, Boca Raton, Florida
23. Blumenthal M, Goldberg A, Brinckmann J (2000) Herbal Medicines. Expanded Comission E Monographs. American Botanical Council, Austin, Texas

24. Pelter A, Hänsel R (1968) Tetrahedron Lett 25:2911
25. Schindler H (1952) Arzneim Forsch 2:295
26. Wagner H, Hörhammer L, Münster R (1965) Naturwissenschaften 52:305
27. Wagner H, Diesel P, Seitz M (1974) Arzneim Forsch 24:466
28. Hänsel R, Kaloga M, Pelter A (1976) Tetrahedron Lett 17:2241
29. Kaloga M (1981) Z Naturforsch 36:262
30. Szilagi I, Tétényi P, Antus S, Seligmann O, Chari VM, Seitz M, Wagner H (1981) Planta Med 43:121
31. Fiebig M, Wagner H (1984) Planta Med 50:310
32. Mericli AH (1988) Planta Med 54:44
33. Kurkin VA (2003) Pharm Chem J 37:189
34. Pelter A, Hansel R (1975) Chem Ber 108:790
35. Quercia V, Pierini N, Valcavi U, Caponi R, Innocenti S, Tedeschi S (1983) Chromatography in biochemistry, medicine and environmental research, 1. In: Frigerio A (ed) Proceedings of the First International Symposium on Chromatography in Biochemistry, Medicine and Environmental Research. Elsevier Scientific, Amsterdam, p 1
36. Lee DYW, Liu YZ (2003) J Nat Prod 66:1171
37. Smith AW, Lauren DR, Burgess EJ, Perry NB, Martin RJ (2005) Planta Med 71:877
38. Freudenberg K, Neish AC (1968) Constitution and Biosynthesis of Lignins.Springer Verlag, New York
39. Hänse R, Rimpler H (1968) Dtsch Apoth Ztg 108:1985
40. Schrall R, Becker H (1977) Planta Med 32:27
41. Sánchez-Sampedro MA, Fernández-Tárrago J, Corchete P (2007) J Plant Physiol 164:669
42. Samu S, Nyiredy S, Baitz-Gacs E, Varga Z, Kurtan T, Dinya Z, Antus S (2004) Chem Biodivers 1:1668
43. Sharma DK, Hall IH (1991) J Nat Prod 54:1298
44. Hammouda FM, Ismail SI, Hassan NM, Zaki AK, Kamel A (1993) Phytother Res 7:90
45. Schulz V, Hansel R, Tyler VE (1997) Rational Phytotherapy: A Physicians' Guide to Herbal Medicine. Springer, Berlin, p 306
46. Vömel A, Hölzl J, Ceylan A, Marquard R (1977) Z f Acker Pflanzenbau 144:s90
47. Martin RJ, Lauren DR, Smith WA, Jensen DJ, Deo B, Douglas JA (2006) N Z J Crop Hort Sci 34:239
48. Duan L, Carrier DJ, Clausen EC (2004) Appl Biochem Biotech 114:559
49. Tittle G, Wagner H (1978) J Chromatogr 153:227
50. Kvasnicka F, Biba B, Sevcik R, Voldrich M, Kratka J (2003) J Chromatogr 990:239
51. Wallace SN, Carrier DJ, Clausen EC (2005) Phytochem Anal 16:7
52. Khan NA, Wu NF (2004) Rapid Commun Mass Spectrom 18:2960
53. Weiss RF (1988) Herbal Medicine. Beaconsfield, UK
54. Valenzuela A, Garrido A (1994) Biol Res 27:105
55. Fraschini F, Dermartini G, Esposti D (2002) Clin Drug Invest 22:51
56. Beckmann-Knopp S, Rietbrock S, Weyhenmeyer R, Bocker RH, Beckurts KT, Lang, W, Hunz M, Fuhr U (2000) Pharmacol Toxicol 68:250
57. Sheu SY, Lai CH, Chiang HC (1998) Anticancer Res 18:263
58. György I, Blazovics A, Feher J, Földiak G (1990) Radic Phys Chem 36165
59. Varga Zs, Czompa A, Kakuk G, Antus S (2001) Phytother Res 15:608
60. Varga Zs, Újhelyi L, Kiss A, Balla J, Czompa A, Antus S (2004) Phytomedicine 11:206
61. Cos P, Ying L, Calomme M, Hu JP, Cimanga K, Van Poel B, Pieters L, Vlietinck AJ, Van den Berghe D (1998) J Nat Prod 61:71
62. Valenzuela A, Aspillaga M, Vial S, Guerra R (1989) Planta Med 55:420
63. Muzes G, Deak G, Lang I, Nekam K, Gergely P, FeherJ (1991) Acta Physiol Hung 78:3
64. Greimel A, Koch H (1977) Experientia 33:1417

65. Fiebrich F, Koch H (1979) Experientia 35:1548
66. Montanini I, Castigli E, Arienti UG, Porcellati G (1977) Ed Sci 32:141
67. Porcellati G, Montanini I, Roberti I, Castigli E (1977) Epatologia 23:215
68. Muriel P, Mourelle M (1990) J Appl Toxicol 10:275
69. Schriewer H, Weinhold F (1979) Arzneim Forsch 29:791
70. Vogel G (1977) New Natural Products and Plant Drugs with Pharmacological, Biological or Therapeutical Activity. Springer, Berlin, Heidelberg, New York
71. Münter K, Mayer D, Faulstich H (1986) Biochim Biophys Acta 860:91
72. Sonnenbichler J, Scalera F, Sonnenbichler I, Weyhenmeyer R (1999) J Pharm Exp Ther 290:1375
73. Singh RP, Agarwal R (2002) Antioxid Redox Signal 4:655
74. Singh RP, Tyangi AK, Zhao J, Agarwal R (2002) Carcinogenesis 23:499
75. Zhu W, Zhang JS, Young YF (2001) Carcinogenesis 22:1399
76. Gupta OP, Sing S, Bani S, Sharma N, Malhotra S, Gupta BD, Banerjee SK, Handa SS (2000) Phytomedicine 7:21
77. Kang JS, Jeon YJ, Kim HM, Han SH, Yang KH (2002) J Pharmacol Exp Ther 302:138
78. Vogel G, Trost W, Braatz R, Odenthal KP, Brusewitz G, Antweiler H, Seeger R (1975) Arzneim Forsch 25:82
79. Schriewer H, Badde R, Roth G, Rauen HM (1973) Arzneim Forsch 23:160
80. Hahn G, Lehamnn HD, Kürten M, Uebel H, Vogel G (1968) Arzneim Forsch 18:689.
81. Campos R, Garrido A, Guerra A (1989) Planta Med 55:417
82. Muriel P, Garciapina T, Perez-Alvarez V, Mourelle M (1992) J Appl Toxicol 12:439
83. Nordmann R, Ribière C, Rouach H (1992) Free Radical Biol Med 12:219
84. Kurose I, Higuchi H, Kato S, Miura S, Watanabe N, Kamegaya Y, Tomita K, Takashi M, Horie Y, Fukuda M, Mizukami K, Ishii H (1997) Gastroenterology 112:1331
85. Navasumrit P, Ward TH, Dodd NJ, O'Connor PJ (2000) Carcinogenesis 21:93
86. Rouach H, Fataccioli V, Gentil M, French SW,Mirimoto M, Nordmann R (1997) Hepatology 25:351
87. McClain CJ, Cohen DA (1989) Hepatology 9:349
88. Valenzuela A, Lagos C. Schmidt K, Videla LA (1985) Biochem Pharmacol 34:2209
89. Lieber CS, Leo MA, Cao Q, Ren C, DeCarli LM (2003) J Clin Gastroenterol 37:336
90. Song Z, Deaciuc I, Song M, Lee DYW, Liu Y, Ji X, McClain C (2006) Clinical and Experimental Research 30:407
91. Nassuato G, Iemmolo RM, Lirussi F, Orlando R, Giacon L, Venuti M, Strazzabosco M, Csomos G, Okolicsanyi L (1983) Pharmacol Res Commun 15:337
92. Skottova N, Vecera R, Urbanek K, Vana P, Walterova D, Cvak C (2003) Pharmacol Res 47:17
93. Skottova N, Kazdova L, Oliyarnyk O, Vecera R, Sobolova L, Ulrichova J (2004) Pharmacol Res 50:123
94. Sobolova L, Skottova N, Vecera R, Urbanek K (2006) Pharmacol Res 53:104
95. Bokemeyer C, Felss LM, Dunn T, Voigt W, Gaedeke J, Shmoll HJ, Stolte H, Lentzen H (1996) Br J Cancer 74:2036
96. Ágoston M, Örsi F, Feher E, Hagymási K, Orosz Z, Blázovics A, Fehér J, Vereckei A (2003) Toxicology 190:231
97. Chlopčíková A, Psotová J, Miketová P, Šimánek V (2004) Phytother Res 18:107
98. Psotová J, Chlopčíková S, Grambal F, Šimánek V, Ulrichová J (2002) Phytother Res 16:S63
99. Soto C, Mena R, Luna J, Cerbon M, Larrieta E, Vital P, Uria E, Sanchez M, Recoba R, Barron H, Favari L, Larag A (2004) Life Sci 75:2167
100. Schönfeld J, Weisbrod B, Müller MK (1997) Cell Mol Life Sci 53:917
101. Darville MI, Eizirik DL (1998) Diabetologia 41:1101
102. Matsuda T, Ferreri K, Todorov I, Kuroda Y, Smith CV, Kandeel F, Mullen Y (2005) Endocrinology 146:175
103. Zi X, Mukhtar H, Agarwal R (1997) Biochem Biophys Res Comm 239:334

104. Gershbein LL (1994) Anticancer Res 14:1113
105. Yanaida Y, Kohno Y, Yoshida K, Hirose Y, Yamada Y, Mori H, Tanaka T (2002) Carcinogenesis 23:787
106. Vinh PQ, Sugie S, Tanaka T, Hara A, Yamada Y, Katayama M, Deguchi T, Mori H (2002) Jpn J Cancer Res 93:42
107. Agarwal R, Mukhtar H (1992) Chemical carcinogenesis in skin: causation, mechanism and role of oncogenes. In: Wang RGM, Knaak JB, Maibach HI (eds) Health Risk Assessment: Dermal and Inhalation Exposure and Absorption of Toxicants. CRC, Boca Raton, Florida, p 291
108. Zhao J, Agarwal R (1999) Carcinogenesis 20:2101
109. Xiaolin Z, Feyes D, Agarwal R (1998) Clin Cancer Res 4:1055
110. Xiaolin Z, Grasso AW, Hsing-Jien K, Agarwal R (1999) Cancer Res 59:622
111. Singh, RP, Tyagi AK, Zhao J, Agarwal R (2002) Carcinogenesis 23:499
112. Yang SH, Lin JK, Chen WS, Chiu JH (2003) J Surgical Res 113:133
113. Agarwal C, Singh RP, Dhanalakshmi S, Tyagi AK, Tecklenburg M, Sclafani RA, Agarwal R (2003) Oncogene 22:8271
114. Manna SK, Mukhopadhyay A, Van NT, Aggarwal B (1999) J Immunol 15:6800
115. Saliou C, Rihn B, Cillard J, Okamoto T, Packer L (1998) FEBS Lett 440:8
116. Stancovski I, Baltimore D (1997) Cell 91:299
117. Dhanalakshmi S, Agarwal P, Glode LM, Agarwal R (2003) Int J Cancer 106:699
118. Pook SH, Toh CK, Mahendran R (2006) Cancer Lett 238:146
119. Li LH, Wu LJ, Zhou B, Wu Z, Tashiro S, Onodera S, Uchiumi F, Ikejima T (2004) Biol Pharm Bull 27:1031
120. Agarwal R, Katiyar SK, Lundgren DW Mukhtar H (1994) Carcinogenesis 15:1099
121. Katiyar S (2005) Int J Oncol 26:169
122. De La Puerta R, Martinez E, Bravo L, Ahumada MC (1996) J Pharm Pharmacol 48:968
123. Kleemann R, Rothe H, Kolb-Bachofen V, Xie QW, Nathan C, Martin S, Kolb H (1993) FEBS Lett 328:9
124. Kang JS, Jeon YJ, Park SK, Yang KH, Kim HM (2004) Biochem Pharmacol 67:175
125. Jeong DH, Lee GP, Jeong WI, Do SH, Yang HJ, Yuan DW, Park HY, Kim KJ, Jeong KS (2005) World J Gastroenterol 11:1141
126. Jia JD, Bauer M, Cho JJ, Ruehl M, Milani S, Boigk G, Riecken EO, Schuppan D (2001) J Hepatol 35:392
127. Wang MJ, Lin WW, Chen HL, Chang YH, Ou HC, Kuo JS, Hong JS Jeng KG (2001) Eur J Neurosci 16:2103
128. Alarcon de Lastra C, Martin MJ, Marhuenda E (1992) J Pharm Pharmacol 44:929
129. de Monbrison F, Maitrejean M, Latour C, Bugnaset F, Peyron F, Barron D, Picon S (2006) Acta Trop 97:102
130. Bombardelli E, Spelta M (1991) Fitoterapia 62:115
131. Baumann L (2005) J Invest Dermatol 125:xii
132. Morazzoni P, Montalbetti A, Malandrino S, Pifferi G (1993) Eur J Drug Metab Pharmacokinet 18:289
133. Bülles H, Bulles J, Krumbiegel G, Mennicke WH, Nitz D (1975) Arznei Forsch 25:902
134. Comoglio A, Tomasi A, Malandrino S, Poli G, Albano E (1995) Biochem Pharmacol 50:1313
135. Madaus R, Halbach G, Trost W (1976) US Patent 3 994 925
136. Weynhenmeyer R, Mascher H, Birkmayer J (1992) Int J Pharmacol Ther Toxicol 30:134
137. Valcavi U, Monterosso V, Caponi R, Bosone E, Wachter W, Szejtli J (1993). US Patent 5 198 430
138. Giorgi R, Conti M, Pifferi G (1989) US Patent 4 886 791
139. Gabetta B, Bombardelli E, Pifferi G (1988) US Patent 4 764 508
140. El-Samaligy MS, Afifi NN, Mahmoud EA (2006) J Pharm 308:140

141. Livio S, Seghizzi R, Pifferi G (1990) 4th European Congress of Biopharmaceutics and Pharmacokinetics, April 17–19, Geneva
142. Malandrino S, Pifferi G (1990) Drugs Fut 15:226
143. Kid P, Head K (2005) Alt Med Rev 10:193
144. Barzaghi N, Crema F, Gatti G, Pifferi G, Perucca E (1990) Eur J Drug Metab Pharmacokinet 15:333
145. Pepping J (1999) Am J Health-Syst Pharm 56:1195
146. Rainone F (2005) Am Fam Physician 72:1285
147. Vailati A, Aristia L, Sozzé E, Milani F, Inglese V, Galenda P (1993) Fitoterapia 64:219
148. Velussi M, Cernigoi AM, De Monte A, Dapas F, Caffau C, Zilli M (1997) J Hepatol 26:871
149. Palasciano G, Portincasa P, Palmieri V, Ciani D, Vendemiale G, Altomare E (1994) Curr Ther Res 55:537
150. Albrecht M, Frerick H, Kuhn U,Strenge-Hesse A (1992) Z Klin Med 47:87
151. Mayer KE, Myers RP, Lee SS (2005) J Vir Hepatt 12:559
152. Desplaces A, Choppin J, Vogel G, Trost W (1975) Arznei Forsch 25:89
153. Hruby K, Csomos G, Fuhrmann M, Thaler H (1983) Hum Toxicol 2:183
154. Zenk MH (1978) The impact of plant cell culture in industry. In: Thorpe TA (ed) Frontiers of Plant Tissue Culture. The International Association of Plant Tissue Culture, Calgary, Canada, p1
155. Fowler MW (1988) Plant cell culture: natural products and industrial application. In: Russel GE (ed) Biotechnology of Higher Plants. Intercept, Reinu Unido, p 107
156. Zenk MH, El-Sahgi H, Arens H, Stöckgt J, Weiler EW, Deus B (1977) Formation of the indole alkaloids serpentine and ajmalicine in cell suspension cultures of Catharanthus roseus. In: Barz W, Reinhard E, Zenk MH (eds) Plant Tissue Culture and Its Bio-technological Application. Springer-Verlag, Berlin, Heidelberg, New York, p 27
157. Becker H, Schrall R (1977) Planta Med 31:185
158. Fevereiro P, Cabral JMS, Fonseca MMR, Novais JM, Pais MSS (1986) Biotechnol Lett 8:19
159. Guinea LMC, Pizarro SA (1987) An Real Acad Farm 53:609
160. Liu S, Cai QG (1990) Acta Bot Sin 32:19
161. Hetz E, Huancaruna Perales PE, Liersch R, Schieder O (1995) Planta Med 61:554
162. Radice S, Caso OH (1997) Biocell 21: 59
163. Iqbal SM, Srivastava PS (2000) J Plant Biochem Biotech 9:81
164. Hasanloo T, Khavari-Nejad RA, Majidi E, Shams-Ardekani MR (2005) Thirteenth Iranian Biology Conference and First International Biology Conference, 23–25 August, Guilan University, Rasht
165. Hasanloo T, Khavari-Nejad R.A, Majidi E, Shams Ardakani MR (2006) Iran J Pharm Sci 2:206
166. Sánchez-Sampedro MA, Fernández-Tárrago J, Corchete P (2005) J Plant Physiol 162:1177
167. Sánchez-Sampedro MA, Fernández-Tárrago J, Corchete P (2005) J Biotech 119:60
168. Sánchez-Sampedro MA, Kim HK, Choi YH, Verpoorte R, Corchete P (2007) J Biotech 130:133

Chapter 7
The Production of Dianthrones and Phloroglucinol Derivatives in St. John's Wort

A. Kirakosyan[1] (✉), D.M.Gibson[2], and P.B. Kaufman[1]

[1]Department of Cardiac Surgery, University of Michigan. 1150 West Medical Center Dr., Ann Arbor, Michigan 48109-0686, USA, e-mail: akirakos@umich.edu
[2]USDA Agricultural Research Service, Plant Protection Research Unit, U.S. Plant, Soil, and Nutrition Laboratory, Tower Road, Ithaca, NY 14583, USA

Abstract The effectiveness of the phytochemical arsenal of St John's wort (*Hypericum perforatum*) may be due to the plant's use of interacting phytochemicals to accomplish many complementary tasks. *H. perforatum* produces several types of biologically active compounds, including the hypericins, a family of dianthrones, and the hyperforins, a family of prenylated acylphloroglucinols. These compounds are known for their multitarget activities. While the pharmaceutical benefits of these compounds are obvious, the physiological functions of these compounds in the plant itself have yet to be elucidated. In this chapter, we address several important topics relevant to the production of dianthrones and phloroglucinol derivatives in St. John's wort. We discuss up-to-date information concerning the biosynthesis of hypericin and hyperforin in the mature plant, and highlight biotechnology-driven initiatives concerned with the production of hypericin and hyperforin. This treatise concludes with a perspective on new directions for hypericin and hyperforin production in St. John's wort. The primary conclusion that arises from this treatise is that although we still do not understand fully how hypericins and hyperforin are synthesized in St. John's wort, the biotechnological aspects involved in regulating the production of these compounds in intact plants and *in vitro* cultures have been elucidated over the past several years.

7.1 Introduction

In recent decades, natural product research has developed in a particular and precise manner in order to identify compounds for the treatment of several diseases such as cancer, depression, heart failure, inflammation, and even discovering natural products with strong antiviral actions. There are many medicinal plants that produce chemically more complicated secondary metabolites in nature than is possible to synthesize chemically, or is even feasible. St. John's wort, *Hypericum perforatum* has been used as an herbal remedy for many disorders from ancient times (as early as B.C.). It has particular interest for phytochemistry because of the large range of compounds that are active phytopharmaceuticals.

This plant produces several types of biologically active compound, including the hypericins, a family of dianthrones localized within specialized glands found predominantly on flowers and leaves, and the hyperforins, a family of prenylated acylphloroglucinols localized in the reproductive structures of the plant. St. John's wort also produces other secondary metabolites that include flavonoids, procyanidins, tannins, essential oils, amino acids, phenylpropanoids, xanthones, and other water-soluble components. Hypericin and pseudohypericin have been demonstrated to have antiviral and anticancer activity. Moreover, photodynamic hypericin activities displayed under the influence of light are used for therapy in various diseases. Hyperforin and adhyperforin, the major phloroglucinol derivatives in this plant, display potent antimicrobial and antidepressive activity. Hyperforin exhibits antidepressant activity by a novel mechanism of action, antibiotic activity against Gram-positive bacteria, and antitumoral activity *in vivo*.

While the physiological functions of these compounds in the plant itself have still not been revealed, the pharmaceutical benefits of these compounds are obvious and have been introduced progressively for the treatment of mild or moderate depression, or for various kinds of viral infection.

The seasonal harvesting of this plant, loss of biodiversity, variability in quality, and contamination issues currently trigger the search for alternative methods for the production of hypericins and hyperforins. Thus, we will discuss in this chapter the production of these important compounds by means of plant biotechnology.

7.2 Dianthrone and Phloroglucinol Derivatives Family of Compounds in *Hypericum perforatum*

Hypericin is considered to be a family of light-activated anthraquinones. However, in comparison to anthraquinone, hypericin has a much more extensive aromatic system, where three rings have been replaced by eight rings (Fig. 7.1a). Its molecular formula is $C_{30}H_{16}O_8$. Because the hypericin ring system is fully aromatic, hypericin can be expected to be planar. However, the native structure is much more complex because of interactions with other components. Hypericin is a red dye that forms salts with sodium and potassium, known as hypercinates. It is soluble in acetone, methanol, dimethylsulfoxide, ethanol, ethyl

Chapter 7 The Production of Hypericins and Hyperforins

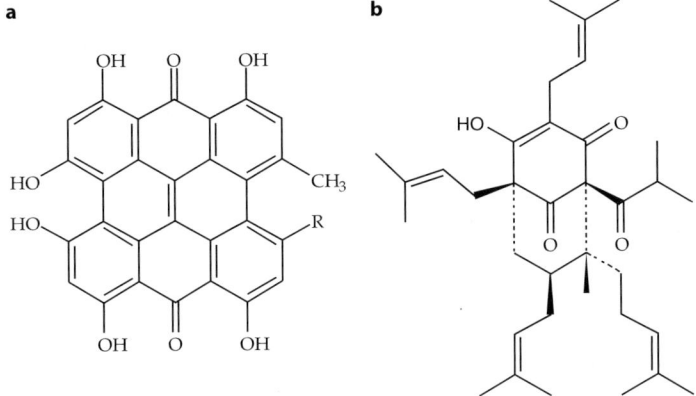

Fig. 7.1 a Chemical structures for hypericin (R=CH$_3$), pseudohypericin (R=CH$_2$OH) and **b** hyperforin

acetate, or aqueous alkali solutions, but partly insoluble in water and methylene chloride. The resulting solution is bright red and exhibits red fluorescence. Hypericin produces singlet oxygen and other excited state intermediates, which indicates that it should be a very efficient phototoxic agent.

Pseudohypericin differs from hypericin at one carbon where a hydroxyl group is substituted for hydrogen, making the compound slightly more polar. Plant extracts typically contain small amounts of the immediate precursors, protohypericin and protopseudohypericin, which are converted to hypericin and pseudohypericin within 2 h in the presence of light [1, 2].

Other derivatives, such as cyclopseudohypericin and isohypericin, can also be recovered from plant extracts of *H. perforatum* in trace amounts [3, 4].

Hyperforin (a polyprenylated acylphloroglucinol derivative; Fig. 1b) is one of the major components (2–4%) of the dried herb. The structure of hyperforin bears no relation to that of hypericin. Instead, it is bicyclic, oxygenated, and unsaturated, but surprisingly not aromatic. The molecular formula of hyperforin is C$_{35}$H$_{52}$O$_4$. It is classified as a derivative of phloroglucinol. Hyperforin is quite sensitive to air oxidation and its content in the herbal drug may vary. Other hyperforin derivatives, including adhyperforin and furohyperforin – an oxygenated analog, also known as orthoforin and furanoforin, respectively, are also present. They are found in the lipophilic fraction of *Hypericum* extracts, but it is still unclear whether they are natural products or artifacts of extraction and isolation [5–9].

Hyperforins are very unstable in purified form in lipophilic solutions, and they rapidly degrade within 25 min [10]. Two new acylphloroglucinol type compounds, secohyperforin and secoadhyperforin, were identified as minor constituents of St. John's wort [10, 11]. No chemical synthesis has been described yet so far for hyperforin and its derivatives.

7.2.1 Botany of Hypericum

St. John's wort is one of about 450 species in the genus *Hypericum*, a diverse group that includes shrubs and even sizeable trees as well as herbs. St. John's wort is an attractive herb with yellow flowers, opposite leaves bearing black glands, and stems bearing two vertical lines. It is an autotetraploid (having four sets of chromosomes, due to doubling of the chromosomes of a single parent species rather than to hybridization) and occasionally hybridizes with its diploid sister species, *H. maculatum*. It is native to Eurasia and is now abundant on several continents. Its hardiness and ability to reproduce asexually allows it to spread quickly. St. Johns wort is a tap-rooted perennial weed that reproduces sexually by seeds and vegetatively by short runners that originate from the crown of the plant. Plants can grow from 25 to 125 cm tall with numerous, rust-colored branches that are woody (suffrutescent) at the base. The taproot may reach depths of 100–125 cm. Lateral roots grow 5–7.5 cm beneath the soil surface, but they may reach depths of 75 cm. Leaves are opposite, sessile (without petioles), entire, elliptic to oblong and generally not more than 2.5 cm long. Flowers occur in open, flat-topped corymb-type inflorescences. Flowers are perfect (having both male stamens and female pistils as reproductive parts). They are bright yellow with five linear, lanceolate, acute, or acuminate sepals, and five petals. Petals are 8–12 mm long, typically twice as long as the sepals, and bear black glands along the margins. Stamens are numerous and arranged in three groups. An egg-shaped, three-valved, capsule-type fruit bursts open at maturity and releases many seeds. Peak flowering occurs during mid to late June, but flowering begins in May and continues through September. Developing seed capsules become green, moist, and sticky At maturity, the capsules are rusty brown in color. Each capsule can contain 400–500 seeds. Seeds may remain viable in the soil for up to 10 years. A mature plant may produce up to 30 flowering stems annually. New crowns may be produced from lateral root buds. Crowns are usually well-spaced, ranging from 12 to 37 per square meter.

7.2.2 Medicinal Uses of Hypericin and Hyperforin

The most common modern medical use of St. John's wort is for the treatment of mild to moderate depression. Clinical trials based on several studies, however, still have produced controversial results regarding both the active principles and mode of action at target sites. There are multiple trials showing positive results, in which St. John's wort was compared directly to prescription drugs (tricyclic antidepressants and selective serotonin reuptake inhibitors) [12]. These trials showed equal or nearly equal activity with fewer side effects for St. John's wort. On the other hand, some recent trials showed no benefit or frequently negative results [13]. The mechanism of its antidepressant activity has yet to be established.

Hypericins are the active, antiviral components of the extracts [14–17]. Hypericin, however, seems to be more effective than pseudohypericin as an antiviral agent [18]. Hypericin and pseudohypericin have been shown to be effective against hepatitis C virus *in vitro* [19] and for possible HIV treatment [20, 21]. Hypericin was found to be a potent and irreversible inhibitor of specific tyrosine kinases, including protein kinase C [22].

The biological activities of the hypericins in *H. perforatum* are thought to be a result of their photodynamic properties [23]. The photodynamic therapy (PDT) and the anticancer actions of hypericin in St. John's wort have recently been successfully introduced [24, 25]. The most important idea in this regard is that hypericin plays a photosensitizer role in photodynamic therapy. PDT refers to the use of low-energy, visible, and near-infrared light to treat various pathological conditions including wound healing, nerve regeneration [26], and several types of cancer [27]. It was demonstrated experimentally that irradiation of hypericin-treated mice led to tumor growth inhibition [28]. Similar results have been obtained with human tumor cell lines, where hypericin was taken up by the tumor cells and made them more vulnerable to the killing effects of light [24].

In recent laboratory tests (reported by "Sunwin International Neutraceuticals Company" http://www.findarticles.com/p/articles/mi_pwwi/is_200604/ai_n16117822), hypericin has proven to be effective as a treatment for poultry infected with strands of avian flu. As reported, hypericin cured all poultry infected by the deadly H5N1 avian flu. The *in vitro* experiments indicated that when H5N1 and H9N2 avian flu viruses were treated with different concentrations of hypericin, the positive results reached as high as 99.99%. However, any specifics for *in vivo* studies have not yet been mentioned. It is noteworthy that *in vitro* concentrations of hypericin are many orders of magnitude above blood levels that are compatible with life. If this result can be substantiated in specific trials for *in vivo* studies, it will apparently raise the bar for hypericin/or pseudohypericin demand and stimulate research for a search for new ways to produce such compounds biotechnologically.

Hyperforin and related phloroglucinol derivatives have been identified as the probable antidepressive components of therapeutically used alcoholic *Hypericum* extracts [12, 29]. Hyperforin has been reported to be a main antibiotic constituent of crude *H. perforatum* extracts [30, 31]. For example, hyperforin is reported to be more effective than sulfonilamide in treatment against infection by *Staphylococcus aureus* when tested *in vivo* and *in vitro* [32]. A novel activity of hyperforin, namely its ability to inhibit the growth of tumor cells by induction of apoptosis, has also been reported [33].

Several observations confirm the view that the antidepressant action of *Hypericum* extracts depends mainly on hyperforin or its derivatives [34]. The spectrum of its primary activity, however, may be due to the other components or a mixture of chemistries that act synergistically in the body.

7.3 Biotechnology for the Production of Hypericin and Hyperforin

In order to better understand the current status of hypericin and hyperforin production, we will discuss (1) biosynthesis, (2) site-specific localization and site of biosynthesis of these compounds, (3) environmental factors regulating hypericin and hyperforin production, (4) hypericin production in response to biotic and abiotic stresses, and finally, (4) our future views for biotechnology in connection with polyketide production in St John's wort.

The chemical synthesis of hypericin or its derivatives is possible and has been reported [35]; however, there are economic considerations and other issues that have to be resolved, such as possible synergism within the complex native mixture, prior to its industrial application. In contrast, there are no data in the literature to date that provide information on the total synthesis of hyperforin. Furthermore, the production of hypericin and hyperforin has not been established by means of biotechnology. These two obstacles constitute significant barriers for the economically feasible production of these important pharmaceuticals in industry.

7.3.1 Biosynthesis of Hypericin and Hyperforin in Mature Plants

Two major phytochemicals of *H. perforatum* – hypericin and hyperforin – are thought to be products of polyketide biosynthesis [36, 37]. Hypericin and pseudohypericin are produced from dimerized emodinanthrone, presumably via phenol oxidation that undergoes further oxidation to hypericins, which are considered to be the end-products of complex polyketide pathways (Fig. 7.2). As an immediate precursor, emodinanthrone is the first cyclization product of the chain. It is formed by the condensation of one acetyl coenzyme A (CoA) molecule and seven malonyl CoA molecules using a polyketide pathway [38]. Polyketides are chemically diverse, but all plant-derived polyketides are produced in the cytosol via the acetate pathway using enzymes called polyketide synthases (PKSs). They catalyze the initial steps in polyketide formation via the condensation of a starter (usually acetyl CoA) and extender molecules (usually malonyl CoA), resulting in a chain with carbonyl groups present. Acetate, propionate, and sometimes butyrate units are used as the building blocks, which are subsequently linked to a specific starter substrate [39–41]. PKSs share striking gene homologies, structural similarities, and chemical strategies. They are classified into several main groups that are based on type I, II, and III synthases, found in bacteria, fungi, and plants. Recently, benzophenone synthase (BPS) was characterized from *H. androsaemum* cell cultures. This result indicates that BPS is a novel member of the super family of plant PKSs, also termed type III PKSs [42].

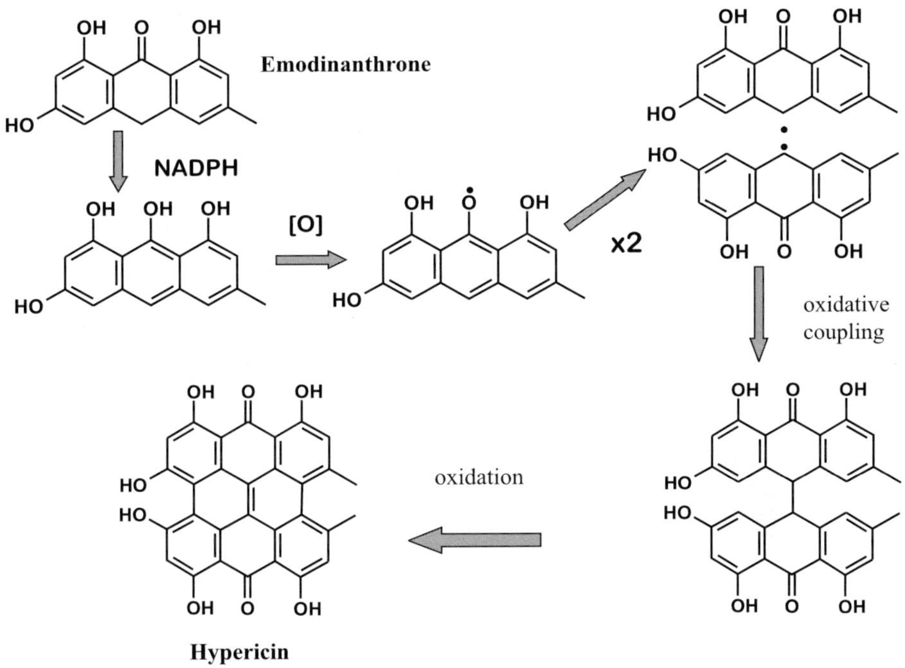

Fig. 7.2 The hypericin formation through oxidative coupling

The specific enzymes and intermediates involved in hypericin and hyperforin biosynthesis in *H. perforatum* have not been identified; nor is there any information concerning the regulation and molecular biology of the pathways involved. Moreover, the rate-limiting step for precursor regulation at the enzymatic level is still unknown.

Although hyperforin is not related structurally to hypericin, it bears similarity to a simpler bacterial ketide-derived compound, monoacetylphloroglucinol. It has been demonstrated recently that the biosynthesis of hyperforin involves five isoprenoid moieties, which are derived entirely, or predominantly (>98%), via the deoxyxylulose phosphate pathway, while the phloroglucinol moiety is generated via a polyketide-type mechanism with isobutyryl-CoA as starter molecule [37]. Formation of the hyperforin nucleus was detected in cell-free extracts from *H. calycinum* cell cultures [43].

In nature, the skeleton of hyperforin is formed by isobutyrophenone synthase (BUS) from isobutyryl-CoA; three molecules of malonyl-CoA are also involved in this biosynthesis [44]. The first prenylation step is catalyzed by a soluble and ion-dependent dimethylallyltransferase. BUS catalyzes the condensation of isobutyryl-CoA with three molecules of malonyl-CoA to give a linear tetraketide intermediate that is cyclized by intramolecular Claisen condensation to yield phlorisobutyrophenone. The first prenylation step was detected in *H. calycinum* cell cultures [45]. During cell culture growth, the formation of hyperforin was preceded by increases in BUS and prenyltransferase activities.

The other important issue is unraveling the original site in the plant where biosynthesis may occur. To date, site-specific biosynthesis of hypericin and hyperforin has not been fully investigated. There are several reports that hypericins are localized within specialized glands found predominantly on flowers and leaves, and that hyperforins are localized in the reproductive structures of the plant. Still, it is not clear where the biosynthesis of these compounds is initiated. An ultrastructural comparison of several *Hypericum* chemotypes and shoot culture lines was employed in our or other studies in order to reveal if there are any anatomical or morphological differences that occur. Such studies could explain the differences in production of hypericin and hyperforin both in different *Hypericum* chemotypes and within the same variety, as well as in different shoot cultures derived from the seeds of genetically distinct populations of *H. perforatum*. In our study, anatomical and ultrastructural differences were examined using light microscopy, scanning electron microscopy, and transmission electron microscopy [46].

The above-cited studies show that hypericin-containing black glands consist of a peripheral sheath of flattened cells surrounding a core of interior cells that are typically dead at maturity. The peripheral cells of glands of a high polyketide-producer *Hypericum* shoot culture line (designated as HP-3) appear less flattened than those of the glands of other low polyketide-producer lines. This observation suggests that the peripheral cells are involved in hypericin production. Moreover, an observed developmental phenomenon shows that the peripheral cells undergo a transformation into interior cells. This has not been reported previously by others. The fact that the size of the peripheral cells may correlate with polyketide metabolite production adds a new hypothesis for the delineation of the actual site of hypericin and hyperforin biosynthesis. This observation confirms other results that apply to intact *Hypericum* plants [47–49]. Our study of the ultrastructure of black glands in *H. perforatum* shoot cultures has revealed interesting differences between different lines and between our *Hypericum* lines and other, previously reported chemotypes of *H. perforatum*.

7.3.2 *Plant Cell Biotechnology*

In this section, we shall consider several possible strategies for the production of hypericin and hyperforin. Generally, the major biotechnology strategies for plant secondary metabolite production include the following:
1. Plant screening for natural product accumulation.
2. Use of high producer plants (elite germplasm) for initiation of callus cultures.
3. Establishment of cell suspension cultures.
4. Analysis of metabolite levels in cell suspension cultures.
5. Selection of cell lines based on single cells.
6. Analysis of culture stability.

7. Further improvement of product yields: development of processes for scaling-up and bioreactor design.
8. Possible biotransformation: the biotransformation of low-value compounds into high-value end products.

Research to date employing these or other strategies has been only marginally successful for many plant-derived natural products. There are several reasons for drawbacks in the use of plant cell biotechnology for the secondary metabolite production. To date, only a few compounds have been produced by application of plant cell cultures in bioreactors, such as the production of Taxol, a chemotherapeutic drug, via plant cell culture [50, 51]. While several kinds of bioreactors have been introduced and recommended by bioengineers for plant cell cultures and secondary metabolite production, only a few industrial applications have been initiated to date. The extensive application of such production strategies is restricted because of economic and technological issues. It is well known that processes using large-scale plant cell cultures could be economically feasible, provided the cells have a sufficiently high growth rates coupled with significant metabolite production rates. While this statement promises future prospects for application, some common problems still need to be adequately resolved.

Such restrictions in large-scale processing are based on the fact that not all desired natural products can be produced by plant cell biotechnology. Biochemical and molecular studies on several model systems (presumably different types of cell cultures) involved may not only answer fundamental questions on how and why these compounds are synthesized, but also, how it will be possible to upregulate their production. Generally, an explanation for the different biosynthetic capacities of either cell suspension or callus cultures is that cells do not produce some compounds until full or partial cell differentiation has occurred. On the other hand, it is possible that the production of some compounds could be triggered in a critical situation where the biosynthetic ability of the cells can be turned on under the influence of biotic or abiotic factors. Critical to cell suspension culture biosynthetic potential is how, and from what part of the intact plant, the cell cultures are derived. If the cells are derived from a reproductive part of the plant that synthesizes a particular metabolite(s), this kind of cell culture may have a higher potential for the production of the desired metabolites. However, this scenario is not a general rule. This is because the type of culture can be reversed to the initial stage, or be changed to a "nondifferentiated or nonproducing" stage, due to the influence of many physical or chemical parameters. For example, media formulation, optimal physical parameters, or even elicitation, could trigger the biosynthesis and production of some metabolite(s). Selection of a high-producer, a genetically and epigenetically stable model system will have more advantages in future trials.

In the following, we introduce a model system, where *H. perforatum* shoot cultures are cultivated in a liquid medium. This constitutes an ideal system for the production of hypericin and hyperforin. Shoot cultures are very easily grown in liquid Murashige and Skoog medium; even after short cultivation periods, they form a substantial biomass, approximately 30 g fresh weight in 100 ml of medium. In addition, such a system allows for extensive manipulation by elicitors, or other stimulatory agents. Our recent results concerning elicitation of *H. perforatum* shoot cultures shows three- to five-fold increases in the production of total hypericins, including pseudohypericin and hyperforin. There are several reports from our group that show that the yeast cell wall glycoprotein, mannan, as well as cork pieces stimulate hypericin and pseudohypericin production in liquid cultivated shoot cultures. In this study, hyperforin production was also considered. In addition to mannan and cork, we also used inert agar cubes (approximately 0.5 cm^3) based on the assumption that agar can stimulate the production of metabolites by a simple mechanical influence on cells, also known as the "massage effect". Figure 7.3 presents results from this study. It was observed that agar cubes cause significant stimulation of the production of all compounds compared to other treatments. There was a five-fold increase in the levels of hyperforin, about a three-fold increase in those of pseudohypericin, and a two-fold increases in those of hypericin compared to control levels. Almost a similar level of stimulation is shown with cork pieces. Based on these results, we have concluded that the stimulatory mechanism of these two agents may act in a similar way, presumably due to a mechanical interaction between the agar cubes or cork pieces and the shoot cultures. These agents float in the medium and thus have immediate contact with shoot culture

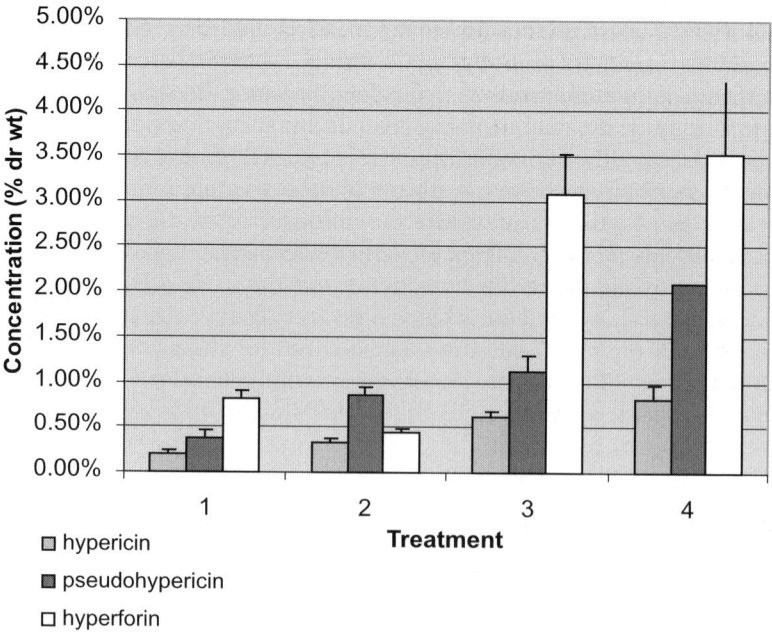

Fig 7.3 The production of hypericin, pseudohypericin and hyperforin in control (1), and elicitor treated with mannan (2), cork (3), and agar cubes (4) in liquid cultivated shoot cultures. *dr wt* Dry weight

surfaces. This is an important finding because mechanical interactions do arise in bioreactors and may greatly influence the productivity. Mannan, in contrast, shows the smallest effect of stimulation of the production of these metabolites, as compared to cork pieces or agar cubes. The stimulatory mechanism for mannan probably has a different origin, as shown by our previous results [52].

The elicitation of production of hypericin and hyperforin in liquid cultivated shoot cultures with bacterial polysaccharides from plant growth promoting rhizobacteria (PGPR) was recently performed in collaboration with our colleague, Beatriz Ramos, from the University San Pablo, Spain. These rhizobacteria had previously been shown to increase growth and secondary metabolite levels in *H. perforatum* seedlings (Ramos, personal communication). We evaluated three different bacterial polysaccharide fractions from plant growth-promoting rhizobacteria to test their ability to increase hypericins levels in shoot cultures of *H. perforatum*. Three bacterial polysaccharide fractions were extracted from the growth culture media of the bacterial strain. Shoot cultures were cultivated in the same liquid culture media used for elicitation with three concentrations of each fraction from bacteria and nutrient broth. After 15 days of incubation, shoots were harvested, extracted, and analyzed for metabolite content by high-performance liquid chromatography. All three PGPR fractions significantly increased pseudohypericin levels. The effect was dose-dependent, being more marked with lower concentrations, and significant differences were found in different fractions. No increases in hypericin levels were observed with fractions derived from culture media that were free of bacteria.

There is one interesting example of *H. perforatum* cell suspension cultivation involving globular structures that may have more practical applications in biotechnology. This is a different system for the cultivation of plant cells than has been introduced heretofore, and it is now being studied extensively. The enhancement of hypericin and pseudohypericin production in liquid-cultivated cell aggregates is possible and differs from that in shoots or the callus [53]. Suspension cultures of *H. perforatum* with compact globular structures have a higher total content of these secondary metabolites than unorganized cell suspension cultures. Moreover, long-term cultivation of globular cell cultures shows further accumulation of the desired compounds. Similar reports about such globular structures, and their biosynthetic abilities, have appeared for two other plant systems; namely, *Catharanthus roseus* (Madagascar pink) and *Rhodiola sachalinensis* – in which compact globular structures constitute a very good system for the synthesis of other kinds of secondary metabolites [54, 55].

7.3.3 *Influences on Hypericin and Hyperforin Productivity by Other Factors*

Biotic factors are among the environmental factors that affect to a great extent the production of phytochemicals. Therefore, it is highly probable that there is a relationship here with the defensive responses that are manifested either with phytoalexin production or with compounds produced along the signal transduction pathway. An approach by which to characterize the biotic parameters that may elicit the plant's defensive mechanisms may be revealed by an analysis of the expression of certain genes involved in the process and by a correlation between gene induction and expression with particular metabolite levels, whenever such genes are identified and characterized.

Applied environmental stress factors, in addition to biotic factors, can cause the upregulation of the biosynthesis of secondary metabolites in both intact plants and in cell cultures. The biosynthesis of hypericin and hyperforin may be influenced by genetic, metabolic, and environmental parameters. Several studies have reported variations in hypericin levels worldwide [34] that highlight and estimate the genetically diverse varieties of this plant. Several other factors can also influence the production of hypericins and hyperforin. These include light intensity, light quality, and temperature [34]. For example, the effect of light intensity on the levels of leaf hypericins was examined for *H. perforatum* grown in a sand culture system with artificial lighting [56]. This study clearly demonstrates that increasing the light intensity results in a continuous increase in the levels of leaf hypericins.

In addition, other important factors are thought to affect or modulate the production or yield of hypericin and hyperforin; these include climate, stage of plant development, method of processing and storage of plant material, methods by which the plant material is harvested and processed, and compound extraction procedures.

In our case-study experiment involving the influence of light, dark, red (650 nm peak transmittance), and far-red (750 nm peak transmittance) wavelengths on hypericin production by shoot cultures given under controlled environments, we determined whether or not production of hypericin and pseudohypericin are affected by the respective light treatments. Following 10 days of the respective light treatments, we found that: (1) hypericin levels were not significantly changed after light treatment, and (2) in general, a combination of light/dark treatment enhanced total hypericin levels in shoots as compared to other treatments. Finally, neither light/dark treatments nor phytochrome-mediated (light/dark, red or light/dark, far-red treatments) appear to play a significant role in regulation of hypericin production in shoot cultures (Table 7.1).

Table 7.1 Levels of hypericins (mg·g^{-1} dry weight biomass) present after 10 days in light-treated shoot cultures (light, dark, red, and far-red wavelengths) of *Hypericum perforatum* (mean ± SD, $n=3$)

Treatment	Pseudohypercin	Hypericin	Total hypericins
Light	4.073 ± 0.925	0.318 ± 0.072	4.391 ± 0.997
Dark	2.940 ± 0.268	0.229 ± 0.021	3.169 ± 0.288
Light/dark	5.285 ± 1.508	0.412 ± 0.118	5.697 ± 1.626
Light/dark, far-red 10 min	2.162 ± 0.409	0.169 ± 0.032	2.331 ± 0.441
Light/dark, red 10 min	4.031 ± 0.701	0.314 ± 0.055	4.345 ± 0.756

Similarly, results reported in [57], showed that hypericin and pseudohypericin levels were not significantly different from each other when plants were grown either in direct light (185 µE·m^{-2}·s^{-1}) or under reduced light intensity (88 µE·m^{-2}·s^{-1}), except for the fact that lower light levels may cause an increase in hypericin and pseudohypericin biosynthesis, especially in the case of pseudohypericin in plantlets grown at 25°C. Generally, the effect of light intensity may be closely linked to the effect of temperature, but the differences cited in these reports may also be due to the use of different cell lines, types of cultures, or differences in extraction and harvest methodologies.

7.3.4 New Directions for Hypericin and Hyperforin Production

Metabolic and genetic engineering could play a crucial role in plant cell biotechnology for the production of hypericin or hyperforin. This approach can be utilized, however, when metabolic pathways of hypericin or hyperforin biosynthesis are fully elucidated, rate-limiting enzymes are characterized, and finally, when genes encoding such enzymes are cloned (see illustration in Fig. 7.4).

While such restrictions still persist, there may be several alternative ways in biotechnology to produce these compounds. One approach could be based

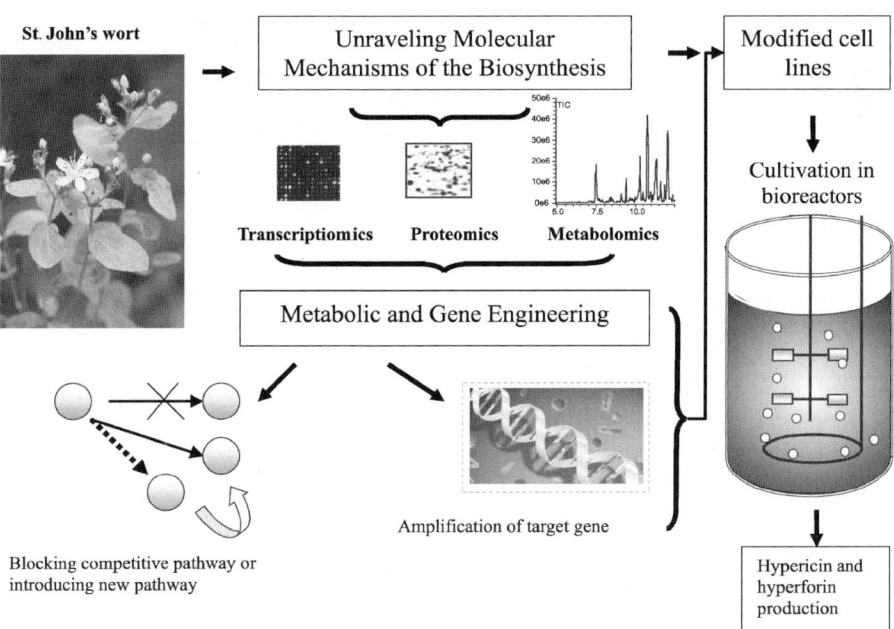

Fig. 7.4 Illustration of possible ways for the production of hypericin and hyperforin by means of biotechnology

on the micropropagation of genetically stable and reverse-tested high producer lines (*in vitro* cultivation followed by greenhouse cultivation and the reverse) of *H. perforatum* elite germplasm. This may be achieved by means of field or greenhouse cultivation and subsequent harvesting and extraction of the desired compounds year-around. This approach may be combined with the use of special photobioreactor systems (culture chambers with controlled physical and chemical parameters), greenhouse cultivation, or field cultivation, if appropriate. We have recently described a new method involving shoot cultures growing in photobioreactors. Here, it has been shown that shoots (with newly developed root parts) can be easily cultivated in several platforms, extensively grown (until critical biomass occurs) and frequently harvested. This estimation is based on the fact that shoot cultures of *H. perforatum* are excellent sources of hypericin and hyperforin. These experiments are still in progress in our current investigations. However, at present, photobioreactor technology, as described here, may not be economically feasible because of its high costs for operation relative to other means of production.

The failure to produce high levels of the desired products by plant cell cultures is also due to our insufficient knowledge as to how plants regulate natural product biosynthesis. Therefore, a requirement to elucidate the molecular basis for the regulation mechanisms is essential for the production of secondary metabolites by means of biotechnology.

7.4 Conclusions

In this chapter, we have addressed several important topics relevant to the production of dianthrones and phloroglucinol derivatives in St. John's wort (*H. perforatum*). These topics include the occurrence of families of dianthrone and phloroglucinol derivatives compounds in *H. perforatum*, the botany of St. John's wort, the medicinal uses of hypericin and hyperforin, the biosynthesis of hypericin and hyperforin in the mature plant, and biotechnology-driven initiatives concerned with the production of hypericin and hyperforin. This treatise concludes with a perspective on new directions for hypericin and hyperforin production from St. John's wort.

The primary conclusion that emanates from this treatise is that although we still do not understand fully how hypericins and hyperforin are synthesized in St. John's wort, nor how these compounds act at target sites in alleviating specific diseases, the biotechnology aspects involved in regulating the production of these compounds in the intact plants and *in vitro* shoot and cell cultures have been elucidated over the past several years. With additional inputs from molecular biology, proteomics, and metabolomics, we should see even more successful breakthroughs concerned with the biotechnology of hypericin and hyperforin production in St. John's wort in the near future.

References

1. Kurth H, Spreemann R (1998) Adv Ther 15:117
2. Sirvent T, Gibson D (2000) J Liq Chromatogr 23:251
3. Poutaraud A, Gregorio F, Tin V, Girardin P (2001) Plant Med 67:254
4. Schütt H (1996) Morphologische, Phytochemische und Botanische Untersuchungen Zur Selektion Hypericin, Pseudohypericin und Flavonoidreicher *Hypericum Perforatum* L. taemme, Ph. D. thesis, University of Berlin
5. Maisenbacher P, Kovar KA (1992) Plant Med 58:351
6. Repcak M, Martonfi P (1997) Biol Bratisl 62:91
7. Verotta L, Appendino G, Belloro E, Jakupovic J, Bombardelli E (1999) J Nat Prod 62:770
8. Verotta L, Appendino G, Jakupovic J, Bombardelli E (2000) J Nat Prod 63:412
9. Fuzzati N, Gabetta B, Peterlongo F, Strepponi I (1999) HPLC-TSP-MS Characterization of Hyperforin-like Phloroglucinols from *Hypericum perforatum*. In: Luijendijk TJC, Verpoorte R (eds) 2000 Years of Natural Products Research – Past, Present and Future. Vrije Univeristeit, Amsterdam, The Netherlands, p 432
10. Erdelmeier CAJ, Klessing K, Renzi S, Hauer H (1999) New Hyperforin Analogues from *Hypericum perforatum* and A Stable Dicyclohexylammonium Salt of Hyperforin. In: Luijendijk TJC, Verpoorte R (eds) 2000 Years of Natural Products Research- Past, Present and Future. Vrije Univeristeit, Amsterdam, The Netherlands, p 423
11. Erdelmeier CAJ (1998) Pharmacopsychiatry 31:2
12. Singer A, Wonnemann M, Müller WE (1999) J Pharmacol Exp Ther 290:1363
13. Linde K, Mulrow CD, Berner M, Egger M (2005) Cochrane Database Syst Rev 2: CD000448
14. Upton R, Graff A, Williamson E, Bunting D, Gatherum DM, Walker EB, Butterweck V, Liefländer U, Nahrstedt A, Winterhoff H, Cott J (1997) St. John's Wort, *Hypericum Perforatum*: Quality Control, Analytical and Therapeutic Monograph, American Herbal Pharmacopoeia, Santa Cruz, CA
15. Bombardelli E, Morazzoni P (1995) Fitoterapia 66:43
16. Cohen P, Hudson J, Towers G (1996) Experientia 52:180
17. Lopez-Bazzocchi I, Hudson JB, Towers GHN (1991) Photochem Photobiol 54:95
18. Vandenbogaerde AL, Delaey EM, Vantieghem AM, Himpens BE, Merlevede WJ, De Witte PA (1998) Photochem Photobiol 67:119
19. Prince AM, Pascual D, Meruelo D, Liebes L, Mazur Y, Dubovi E, Mandel M, Lavie G (2000) Photochem Photobiol 71:188
20. Butterweck V, Petereit F, Winterhoff H, Nahrstedt A (1998) Plant Med 64:291
21. Bork PM, Bacher S, Schmitz ML, Kaspers U, Heinrich M (1999) Plant Med 65:297
22. Agostinis PA, Vandenbogaerde A, Donella D, Pinna L, Lee K, Goris J, Merlevede W, Vandenheede JR, de Witte P (1995) Biochem Pharmacol 49:1615
23. Diwu Z (1995) Photochem Photobiol 61:529
24. Vanderwerf WM, Saxton RE (1996) Laryngoscope 106:479
25. Lavie G, Toren A, Meruelo D, Stackievicz R, Hazen S, Mandel M (1998) Seventh Biennial Congress of the International Photodynamic Association, July 7–9, Nantes, France. Abstract RC23
26. Lubart R, Eichler M, Lovi R, Friedman H, Shainberg A (2005) Photomed Las Surg 23:3
27. Dolmans DE, Fukumura D, Jain RK (2003) Nat Rev Cancer 3:380
28. Vandebogaerde AL (1996) Anticancer Res 16:1619
29. Chatterjee SS, Bhattacharya SK, Wonnemann M, Singer A, Mueller WE (1998) Life Sci 63:499
30. Ebrey RJ (1999) Lancet 354:777
31. Schempp CM, Pelz K, Wittmar A, Schoepf E, Simon JC (1999) Lancet 353:2129

32. Aizeman BE, Smirnov VV, Bondarenko AS (1984) (eds) Fitoncides and Antibiotics of Higher Plants. Naukova Dumka, Kiev
33. Schempp CM, Kirkin V, Simon-Haarhaus B, Kersten A, Kiss J, Termeer CC, Gilb B, Kaufmann T, Borner C, Sleeman JP, Simon JC (2002) Oncogene 21:1242
34. Kirakosyan A, Sirvent TM, Gibson DM, Kaufman PB (2004) Biotech Appl Biochem 39:71
35. Vollmer JJ, Rosenson J (2004) J Chem Educ 81:1450
36. Eckerman S, Schröder G, Schmidt J, Strack D, Erada RA, Helariutta Y, Elomas P, Kotilainen M, Kilpeläinen I, Proksch P, Teeri TH, Schröder J (1998) Nature 396:387
37. Adam P, Arigoni D, Bacher A, Eisenreich W (2002) J Med Chem 45:4786
38. Chen ZG, Fujii I, Ebizuka Y, Sankawa U (1995) Phytochemistry 38:299
39. Hopwood DA (1997) Chem Rev 97:2465
40. Khosla C, Gokhale RS, Jacobsen JR, Cane DE (1999) Ann Rev Biochem 68:219
41. Shen B (2000) Top Curr Chem 209:1
42. Liu B, Falkenstein PH, Schmidt W, Beerhues L (2003) Plant J 34:847
43. Klingauf P, Beuerle T, Mellenthin A, El-Moghazy SAM, Boubakir Z, Beerhues L (2004) Phytochemistry 66:139
44. Beerhues L (2006) Phytochemistry 67:2201
45. Boubakir Z, Beuerle T, Liu B, Beerhues L (2004) Phytochemistry 66:51
46. Kornfeld A, Kaufman PB, Lu CR, Gibson DM, Bolling SF, Warber SL, Chang S, Kirakosyan A (2007) Plant Physiol Biochem 45:24
47. Curtis JD, Lersten NR (1990) New Phytologist 114:571
48. Onelli E, Rivetta A, Giorgi A, Bignami M, Cocucci M, Patrignani G (2002) Flora 197:92
49. Liu W, Lu H, Hu ZH (2002) Acta Bot Sin 44:649
50. Brincat MC, Gibson DM, Shuler ML (2002) Biotechnol Prog 18:1149
51. Zhong JJ (2002) J Biosci Bioeng 94:591
52. Kirakosyan A, Hayashi H, Inoue K, Charchoglyan AG, Vardapetyan RR (2000) Phytochemistry 53:345
53. Vardapetyan HR, Kirakosyan AB, Charchoglyan AG (2000) Biotechnologia 4:53
54. Verpoorte R (1996) In: DiCosmo F, Misawa M, (eds) Plant Cell Culture Secondary Metabolism. CRC Press, Bocca Raton, FL, p 203
55. Xu JF, Ying PQ, Han AM, Su ZG (1999) Plant Cell Tiss Org Cult 55:53
56. Briskin D, Gawienowski M (2001) Plant Physiol Biochem 39:1075
57. Sirvent T (2001) Hypericins: A Family of Light-activated Anthraquinones in St. John's Wort and Their Importance in Plant/pathogen/herbivory Interactions. PhD thesis, Cornell University

Chapter 8
Production of Alkaloids in Plant Cell and Tissue Cultures

D. Laurain-Mattar

Groupe S.U.C.R.E.S., U.M.R. 7565 C.N.R.S., Nancy-Université,
BP 239, 54506 Nancy-Vandoeuvre, France,
e-mail: dominique.laurain-mattar@pharma.uhp-nancy.fr

Keywords *Papaver somniferum*, *Atropa belladonna*, *Leucojum aestivum*, Morphinan alkaloids, Tropane alkaloids, Galanthamine, *Agrobacterium*, Hairy roots, Organogenesis, HPLC

Abstract A low or no productivity of alkaloids in plant cell cultures can be explained by an insufficient level of cell differentiation. The first strategy described in this chapter for improving isoquinoline alkaloid accumulation is organogenesis and somatic embryogenesis induced by the addition of exogenous growth regulators in *Papaver somniferum* and *Leucojum aestivum* cell cultures. The second strategy described is the transformation of medicinal plants (*Atropa belladonna*, *Papaver somniferum*, and *Leucojum aestivum*) using *Agrobacterium rhizogenes* to form hairy root cultures, which carry with them with the benefits of fast growth and rates of alkaloid production equal to or greater than that found for the intact plant.

8.1 Introduction

Plants produce a broad variety of chemical compounds that have huge economical importance. Each plant species has its own specific set of secondary metabolites that are not involved in the basic metabolic processes of the living cells, but are involved in the interaction of the producing organism with its environment [1]. There are about 100000 known compounds that have been extracted from plants, with about 4000 new ones being discovered every year [2]. The largest group consist of the terpenoids and the second largest group is formed by the alkaloids, comprising many drugs and poisons.

For the pure compounds isolated from plants, pharmaceutical application is one of the most important uses. About 25% of modern drugs are of natural plant origin [3]. Compounds are either used directly following isolation from the plant, or are chemically converted after isolation. For some of the natural compounds, the availability of the plants is limited and this has prompted the search for an alternative way to obtain the valuable secondary metabolites using biotechnological processes. Production in plant tissue culture is economically feasible for certain compounds [2], particularly for compounds with a high value like paclitaxel, a major anticancer drug. Alkaloids are also compounds with a high value. Indeed, they are found in low concentrations in plants and are difficult to purify from plant extracts.

Several possibilities for a biotechnological production can be considered:
1. Plant cell and tissue cultures.
2. Transgenic microorganisms.
3. Transgenic plants or plant cell cultures.
4. Isolated enzymes.

Some compounds, like shikonine [4] and paclitaxel [5], can be produced with the technology of large-scale plant cell culture. However, other secondary metabolites, particularly alkaloids, are produced at low concentrations in plant cell cultures. The low or lack of productivity of these desired compounds can be explained by an insufficient level of cell differentiation to allow a production of secondary metabolites. In plants, there is a clear correlation between cellular differentiation and secondary metabolism [2]. *In vitro* shoot and root cultures established in culture media containing various combinations of growth regulators (auxins and/or cytokinins) are able to produce the same secondary metabolites as the intact plant, and by medium optimization even higher levels have been obtained in certain cases.

In this chapter, two strategies leading to cell differentiation are presented with regard to improving alkaloid production. The first one is to induce organogenesis and somatic embryogenesis by the addition of exogenous growth regulators in the culture medium. This strategy was applied in *Papaver somniferum* and *Leucojum aestivum* in order to improve the productivity of cell and tissue cultures of these medicinal plants. The second strategy is the transformation of plants with *Agrobacterium rhizogenes*. These soil bacteria are capable of infecting plant cells and are responsible for the induction of cell differentiation and the proliferation of root growth, the so-called "hairy roots". These transformed roots offer an interesting approach to produce similar or higher yields of alkaloids as compared with untransformed roots.

The works reported here focus on *Atropa belladonna* (Solanaceae), *Papaver somniferum* (Papaveraceae) and *Leucojum aestivum* (Amaryllidaceae) for the main reason that these plants produce several alkaloids of pharmacological and biotechnological importance.

The tropane alkaloids, hyoscyamine and scopolamine, which accumulate in *Atropa belladonna*, are of great interest for the pharmaceutical industry. Scopolamine is medicinally the most important, mainly because it is used as the starting material for the semi-synthesis of several important drugs. Both scopolamine and hyoscyamine possess strong parasympatholytic activity, blocking parasympathic action by binding to the muscarinic acetylcholine receptors in synapses, without exerting any intrinsic activity. It is worthy of note that Solanaceae tissue cultures often grow vigorously and regenerate more easily than do those of many other medicinal plants. Almost all tissue culture systems known for plants have been realized with Solanaceae; these include root cultures, shoot cultures and de-differentiated cells as callus or cell suspension cultures (for a review see [6]).

Papaver somniferum var. *album* L., an annual plant belonging to the family Papaveraceae, is cultivated on a commercial scale for the extraction of morphinan alkaloids (Fig. 8.1) of pharmaceutical interest, in particular morphine, which has strong analgesic properties, and codeine, which also has analgesic effects and can be used to alleviate cough. Many studies have been performed on *in vitro* cultures of this plant with the aim of providing an alternative source for alkaloid production [7–9], and the results showed the major influence of the cell differentiation level upon the biosynthesis of benzylisoquinoline alkaloids [10].

Leucojum aestivum L. (summer snowflake; Amaryllidaceae) alkaloids are known to exhibit a wide range of biological activities, which include analgesic, antiviral [11], antimalarial, antineoplastic [12] and central nervous system effects. Recently, galanthamine, an isoquinoline alkaloid (Fig. 8.2) common to this family has been shown to possess cholinesterase inhibitory activity (acetylcholinesterase, AChE) and has been introduced on the market as an important

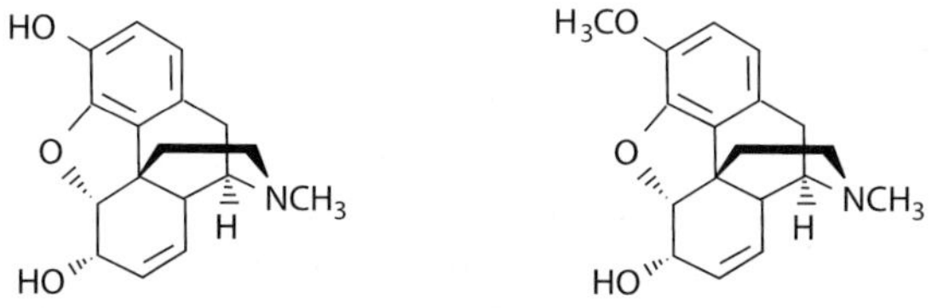

Fig. 8.1 Morphinan alkaloid structures

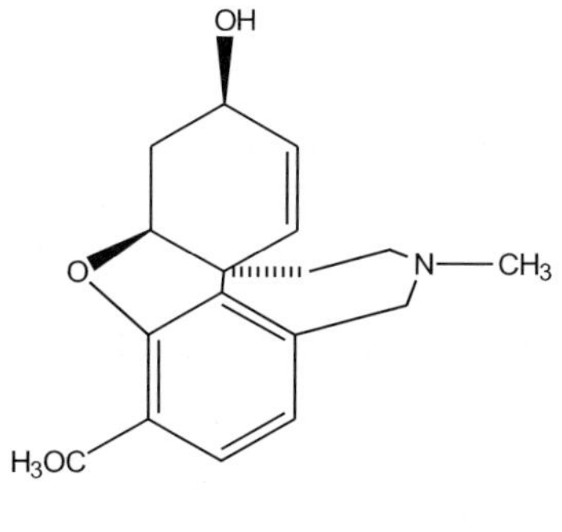

Morphine   Codeine

Fig. 8.2 Galanthamine structure

anti-Alzheimer drug. Alzheimer's disease is one of the most common causes of the loss of mental function broadly known as dementia. This type of dementia is characterized, among others, by degeneration of cholinergic neurons. It was found that levels of the neurotransmitter acetylcholine in the brain are significantly lower in people suffering from Alzheimer's disease. One therapeutic approach to enhance cholinergic neurotransmission is to increase the availability of acetylcholine by inhibiting AChE, the enzyme that degrades acetylcholine in the synaptic cleft. Galanthamine hydrobromide is the most recently approved AChE inhibitor – in Europe by the European registration bureau and in the USA by the Food and Drug Administration – for the symptomatic treatment of Alzheimer's disease. For pharmaceutical needs, galanthamine is obtained mainly by chemical synthesis. Several total synthesis methods have been reported to produce this drug [13] (for a review see [14]); however, the synthetic route is complicated and thus expensive, as galanthamine has three asymmetric carbons, thus requiring a stereochemically controlled synthesis. Galanthamine is also extracted from the bulbs of some Amaryllidaceae plants such as *Leucojum* (the main commercial source in Bulgaria), *Narcissus* and *Galanthus* (snow drops), in levels of 0.01–2% dry weight (D.W.). In the face of growing demand, the supply of this isoquinoline alkaloid is a major problem. Production of galanthamine by *Leucojum* tissue cultures could be an alternative way to obtain this valuable metabolite, as has been shown with *Narcissus* [15].

8.2 Correlation Between Organogenesis, Somatic Embryogenesis and Isoquinoline Alkaloid Accumulation

Tissue cultures of different explants of the poppy plant have been reported in the literature (i.e. seedling hypocotyls [7, 16, 17], seedling roots, stalk and capsule [18]). Callus tissues have been obtained and the presence of alkaloids has been detected [19–21]. However, other investigations have demonstrated the absence of alkaloids in *Papaver somniferum* tissue cultures [18, 22, 23]. This discrepancy has several different explanations: the use of different analytical methods with varied sensibilities, the use of different *Papaver somniferum* cultivars [7] and the analysis of somatic tissue cultures at various stages of differentiation. It is known that organogenesis or embryogenesis are accompanied by morphinan alkaloid accumulation [24, 25]. We have also established *Papaver somniferum* tissue cultures showing different degrees of differentiation controlled by various hormonal conditions. Two genotypes of *Papaver somniferum* were tested [10]. Hypocotyls and roots from seedlings were found to be interesting explants from which to obtain cellular developments. Many roots developed on calli growing on a medium containing α-naphthalene acetic acid (1 mg/l) + kinetin (0.1 mg/l) for the PS genotype (seeds of *Papaver somniferum* var. *album* from Botanical Garden, Nancy, France), while somatic proembryos redifferentiated on calli issued from PS 1639 genotype (dihaploid seeds of *Papaver somniferum* 1639, gift of Sanofi Society). Three-month-old tissue cultures issued from explants of the two genotypes were examined by high-performance liquid chromatography (HPLC) for their alkaloid content. Dramatic variations of the alkaloid contents were observed. Different factors influenced these variations: growth regulator concentration and association, the type of explants and the genotype. As Yoshikawa and Furuya [8] found, no morphinan alkaloids were detected in the unorganized tissues issued from cotyledons, hypocotyls or roots of the two genotypes PS and PS 1639. However, a lot of roots differentiated from calli, initiated from PS cotyledons, produced the highest level of alkaloids ($10^{-2}\%$), including $0.35\times10^{-3}\%$ morphine, $8.5\times10^{-3}\%$ codeine, $0.7\times10^{-3}\%$ thebaine and $0.35\times10^{-3}\%$ papaverine. Even through the results were different with the two genotypes used, differentiated tissues (roots or somatic embryos) were required for morphinan alkaloid biosynthesis. It has been suggested that the regulation of specific alkaloid biosynthesis genes could be controlled by specific developmental programs [26].

In vitro cultures at different stages of morphogenesis were established from leaves of *Leucojum aestivum* and assayed to determine their galanthamine content [27, 28]. A suitable HPLC method for qualitative and quantitative determination of galanthamine in both *in vitro* and *in vivo* extracts has been developed. A correlation was also observed between the state of differentiation and the galanthamine content of the tissue cultures. No galanthamine was detected in the roots grown *in vitro*, while all bulblets grown *in vitro* showed the presence of this alkaloid with dramatic variations in concentration levels, depending

Table 8.1 Galanthamine content of *in vivo* grown bulbs, embryogenic calli, *in vitro* bulblets and roots of *Leucojum aestivum* L. after 3 months of culture on Murashige and Skoog medium containing various growth regulators. *2,4-D* 2,4-Diclorophenoxyacetic acid, *BAP* benzyladenine, *ANA* α-naphthalene acetic acid, *D.W.* dry weight

Extracts		Galanthamine ($10^{-3}\%$ D.W.)
In vivo bulbs		0
Embryogenic calli	2,4-D (5 µM) + BAP (5 µM)	0
	2,4-D (25 µM) + BAP (0.5 µM)	44.8
	2,4-D (10 µM) + BAP (10 µM)	73
In vitro bulblets	Without growth regulators	1.1
	ANA (10 µM) + BAP (0.5 µM)	6.8
	ANA (0.5 µM) + BAP (0.5 µM)	4.74
In vitro bulblets developed on hairy roots	Clone 1	22.1
	Clone 2	10.3
	Clone 3	22.2
	Clone 4	34.6
	Clone 5	51.3
In vitro roots	ANA (10 µM) + BAP (0.5 µM)	0

upon the growth substance balance (Table 8.1). The best galanthamine content (0.073% D.W.) was obtained with embryogenic calli cultivated with 2,4-dichlorophenoxyacetic acid (10 µM) combined with benzyladenine (10 µM). It is worth noticing that all bulblets grown *in vitro*, initiated with or without growth regulators, contained galanthamine ($1.1–51.3 \times 10^{-3}$% D.W.), but no galanthamine was detected at the beginning in the *in vivo* grown bulbs used for the establishment of the *in vitro* cultures. In contrast to the differentiation of bulblets, the differentiation of roots did not allow galanthamine synthesis.

8.3 Hairy Roots and Tropane and Morphinan Alkaloid Accumulation

The second strategy used to improve alkaloid accumulation in tissue cultures is to induce a cell differentiation by modulating the action of plant cell endogenous growth regulators with *rol ABC* genes of *Agrobacterium rhizogenes* [29, 30]. In particular, *rol ABC* genes induce a higher rhizogenesis and could act indirectly on cell growth and on alkaloid production [31–33]. The growth of *Agrobacterium-rhizogenes*-transformed root cultures was independent of exogenous phytoregulators addition. The hairy root phenotype was the result of inserting the T-DNA region of the Ri-plasmid of *Agrobacterium rhizogenes* into the plant genome. Tropane alkaloid biosynthesis was correlated with root differentiation [34]. For this reason, hairy root formation offers an interesting approach to the production of these secondary metabolites. Hairy root cultures have demonstrated their ability to rapidly produce biomass as well as high contents of tropane alkaloids [35–37]. When root cultures were induced by different *Agrobacterium* strains, substantial variation in tropane alkaloid formation and growth characteristics as well as somaclonal variation occurred repeatedly [31, 38–40]. In *Atropa belladonna* hairy roots, high tropane alkaloid production has been obtained after infection with wild strains of *Agrobacterium rhizogenes* [39, 41–44]. The *rol ABC* genes were sufficient to sustain strong growth and high alkaloid production (8 mg·g^{-1} D.W.), with scopolamine concentrations rising to 2.5-fold those of hyoscyamine [31]. The *rol C* gene alone played a significant role in the hairy root growth rate (17-fold increase) [32]; however, this effect was much lower than that induced by the *rol ABC* genes together (75-fold increase). In contrast, the *rol C* gene alone was as sufficient as the rol *ABC* genes together (12-fold times more than in untransformed roots) to stimulate the biosynthesis of tropane alkaloids in *Atropa belladonna* hairy root cultures (Table 8.2).

Hyoscyamine was predominant with a 0.1:0.6 scopolamine:hyoscyamine ratio excepted for the C2 and the C4 root lines, in which the scopolamine:hyoscyamine ratio was respectively 2.5:1. The scopolamine content of the C2 root line is quite similar to those observed in hairy root cultures of *Datura candida* hybrid [45] and much higher than others obtained with *Atropa belladonna* hairy root cultures [41, 44, 46, 47].

Table 8.2 Higher hyoscyamine and scopolamine contents of hairy root lines of *Atropa belladonna* after transformation with *Agrobacterium rhizogenes* 15834 (C1, C2), *Agrobacterium tumefaciens* rol ABC (C3, C4) and *Agrobacterium tumefaciens* rol C (C5)

Hairy root lines	Hyoscyamine (% D.W.)	Scopolamine (% D.W.)	Reference
C1	0.6	0.10	Bonhomme et al. (2000) [31]
C2	0.24	0.60	Bonhomme et al. (2000) [31]
C3	0.40	0.15	Bonhomme et al. (2000) [31]
C4	0.25	0.25	Bonhomme et al. (2000) [31]
C5	0.55	0.33	Bonhomme et al. (2000) [32]

Once established, hairy root cultures have proven to be more stable in metabolism during repeated subcultures than comparable cell suspension cultures [48]. Hairy root cultures of many other medicinal plants obtained by transformation with *Agrobacterium rhizogenes* were examined as potential sources of high-value pharmaceuticals (for a summary, see [49]).

Contrary to *Atropa belladonna*, *Papaver somniferum* is a plant that is difficult to transform. The previous results reported above [10] show the major influence of the cell differentiation level upon the biosynthesis of benzylisoquinoline alkaloids. For this reason, hairy roots offer an interesting approach to the production of similar or higher yields of alkaloids as compared with untransformed roots. For the first time, *Papaver somniferum* hairy root cultures have been established after transformation of hypocotyls with the hypervirulent *Agrobacterium rhizogenes* strain, LBA 9402 [33]. The total alkaloid content (morphine, codeine and sanguinarine) was higher in hairy roots (0.46±0.06% D.W.) than in untransformed roots (0.32±0.05% D.W.) and some of the alkaloids were excreted into the liquid culture medium. The accumulation of sanguinarine only in hairy root cultures could be related to a stress-induced response due to the transformation process. Indeed, sanguinarine is thought to take part in the chemical defence system of *Papaver somniferum* [50]. Sanguinarine also accumulated in *Papaver somniferum* cell suspension cultures treated with fungal elicitors [50, 51].

8.4 Conclusion and Perspective

While it is generally possible to introduce most plants into tissue culture, the production of adequate levels of particular secondary metabolites, like alkaloids, may be problematic. The induction of cell differentiation by the addition of exogenous growth regulators in the culture medium improves alkaloid production. However, this process is time-consuming and therefore it can be used only for the production of compounds with a high value. The transformation of medicinal plants using *Agrobacterium rhizogenes* to form hairy root cultures has the potential benefits of fast growth and rates of alkaloid production equal to or greater than that found for the intact plant. Moreover, hairy root cultures can be scaled-up for bioreactor production to allow for the large-scale recovery of alkaloids or other compounds with pharmacological activities. In the future, advances in molecular methods and in knowledge relating to a secondary metabolite pathway can lead to the use of metabolic engineering as a means of directly modifying pathways for increased alkaloid biosynthesis. All enzymatic steps of alkaloid biosynthesis must be characterized. The genes encoding the enzymes and the corresponding regulatory gene sequences also await characterization. Most metabolic steps in tropane alkaloid formation have been elucidated using radioactive precursors and subsequent metabolite analysis [6]; however, only two enzymes specific to the biosynthesis of hyoscyamine have been isolated and characterized. Today, others alkaloid biosynthetic pathways are better known, with more enzymes and genes having been isolated, sequenced and characterized. An example is the recent metabolic engineering of benzoquinoline alkaloid biosynthesis based on the particular knowledge of pathway enzymes [52, 53]. The up-regulation of biosynthetic pathways using regulatory genes, and the development of short bioconversion pathways in microbes are areas likely to be exploited for the production of compounds of high value.

References

1. Harborne JB (1999) Classes and functions of secondary products from plants. In: Walton NJ, Brown DE (eds) Chemicals from Plants, Perspectives on Secondary Plant Products. Imperial College Press, London, pp 1–25
2. Verpoorte R, van der Heijden R, ten Hoopen HJG, Memelink J (1999) Biotechnol Lett 21:467–479
3. Payne GF, Bringi V, Prince C, Shuler ML (1991) The quest for commercial production of chemicals from plant cell culture. In: Payne GF, Bringi V, Prince C, Shuler ML (eds) Plant Cell and Tissue Culture in Liquid Systems. Hanser, Munich, pp 1–10
4. Fujita Y, Tabata M (1987) Secondary metabolites from plant cells-pharmaceutical applications and progress in commercial production. In: Green CE, Somers DA, Hacket WP, Biesboer DD (eds) Plant Tissue and Cell Culture. Alan R. Liss, New York, USA, pp 169–185
5. Tabata H (2004) Adv Biochem Eng Biotechnol 87:1–23
6. Dräger B (2007) Biotechnology of solanaceae alkaloids: a model or an industrial perspective? In: Kayser O, Quax W (eds) Medicinal Plant Biotechnology, From Basic Research to Industrial Applications, Vol 1. Wiley-VCH, pp 237–265
7. Schuchmann R, Wellmann E (1983) Plant Cell Rep 2:88–91
8. Yoshikawa T, Furuya T (1985) Planta Med 2:110–113
9. Yoshimatsu K, Shimomura K (1992) Plant Cell Rep 11:132–136
10. Laurain-Mattar D, Gillet-Manceau F, Buchon L, Nabha S, Fliniaux MA, Jacquin-Dubreuil A (1999) Planta Med 65:167–170

11. Gabrielsen B, Monath TP, Huggins JW, Kefauver DF, Pettit GR, Groszek G, Hollingshead M, Kirsi JJ, Shannon WM, Schubert EM, Dare J, Ugarkar B, Ussery MA Phelan MJ (1992) J Nat Prod 55:1569–1581
12. Weniger B, Italiano L, Beck JP, Bastida J, Bergonon S, Codina C, Lobstein A, Anton R (1995) Planta Med 61:77–79
13. Guillou C, Beunard J.L, Gras E, Thal C (2001) Angew Chem Int Edit 40:4745–4746
14. Marco-Contelles J, do Carmo Carreiras M, Rodriguez C, Villarroya M, Garcia AG (2006) Chem Rev 106:116–133
15. Sellés M, Bergonon S, Viladomat F, Bastida J, Codina C (1999) Plant Cell Rep 18:646–651
16. Ilahi I (1983) Pakistan J Bot 15:13–18
17. Galewsky S, Nessler C.L (1986) Plant Sci 45:215–222
18. Furuya T, Ikuta A, Syono K (1972) Phytochemistry 11:3041–3044
19. Ranganathan B, Mascarenhas AF, Sayagaver BM, Jagannathan V (1963) Growth of Papaver somniferum in vitro. In: Maheshwari P (ed) Plant Tissue and Organ Culture. International Society of Plant Morphology, University of Delhi, India, pp 108–110
20. Carew DP, Staba EJ (1965) Lloydia 35:1–26
21. Staba EJ (1969) Recent Adv Phytochem 2:75–106
22. Ikuta A, Syono K, Furuya T (1974) Phytochemistry 13:2175–2179
23. Morris P, Fowler MS (1980) Planta Med 39:284–285
24. Staba EJ, Zito S, Amin M (1982) J Nat Prod 45:256–262
25. Kamo KK, Kimoto W, Hsu AF, Mahlberg PG, Bills DD (1982) Phytochemistry 21:219–222
26. Facchini PJ, Bird DA (1998) In Vitro Cell Dev Biol Plant 34:69–79
27. Diop MF, Ptak A, Chrétien F, Henry M, Chapleur Y, Laurain-Mattar D (2006) Nat Prod Commun 1:475–479
28. Diop MF, Hehn A, Ptak A, Chrétien F, Doerper S, Gontier E, Bourgaud F, Henry M, Chapleur Y, Laurain-Mattar D (2007) Phytochem Rev 6:137–141
29. Piñol MT, Palazón J, Cusido R, Serrano M (1996) Bot Acta 109:133–138
30. Palazón J, Cusido RM, Gonzalio J, Bonfill M, Morales C. Piñol T (1998) J Plant Physiol 153:712–718
31. Bonhomme V, Laurain-Mattar D, Lacoux J, Fliniaux MA, Jacquin A (2000) J Biotechnol 81:151–158
32. Bonhomme V, Laurain-Mattar D, Fliniaux MA (2000) J Nat Prod 63:1249–1252
33. Le Flem-Bonhomme V, Laurain-Mattar D, Fliniaux MA (2004) Planta 218:890–893
34. Endo T, Yamada Y (1985) Phytochemistry 24:1233–1236
35. Jouhikainen J, Lindgren L, Jokelainen T, Hiltunen R, Teeri TH, Oksman-Caldentey K-M (1999) Planta 208:545–551
36. Payne J, Hamill JD, Robins RJ, Rhodes M-JC (1987) Planta Med 53:474–478
37. Deno H, Yamagata H, Emoto T, Yoshioka T, Yamada Y, Fujita Y (1987) J Plant Physiol 131:315–324
38. Sevon N, Hiltunen R, Oksman-Caldentey KM (1998) Planta Med 64:37–41
39. Aoki T, Matsumoto H, Asako Y, Matsunagra Y, Shimomura K (1997) Plant Cell Rep 16:282–286
40. Giulietti AM, Parr AJ, Rhodes M-JC (1993) Planta Med 59:428–431
41. Kamada H, Okamura N, Satake M, Harada H, Shimomura K (1986) Plant Cell Rep 5:239–242
42. Hashimoto T, Yun DJ, Yamada Y (1993) Phytochemistry 32:713–718
43. Sharp JM, Doran PM (1990) J Biotechnol 16:171–186
44. Jung G, Tepfer D (1987) Plant Sci 50:145–151
45. Christen P, Roberts MF, Phillipson JD, Evans WC (1989) Plant Cell Rep 8:75–77
46. Hartmann T, Witte L, Oprach F, Toppel G (1986) Planta Med 52:390–395
47. Rothe G, Garske U, Dräger B (2001) Plant Sci 160:1043–1053
48. Toivonen L (1993) Biotechnol Progress 9:12–20
49. Sevon N, Oksman-Caldentey KM (2002) Planta Med 68:859–868
50. Cline SD, Coscia CJ (1988) Plant Physiol 86:161–165

51. Eilert U, Kurz WGW, Constabel F (1985) J Plant Physiol 119:65–76
52. Millgate AG, Pogson BJ, Wilson IW, Kutchan TM, Zenk MH, Gerlach WL, Fist AJ, Larkin PJ (2004) Nature 431:413–414
53. Page J. (2005) Trends Biotechnol 23:331–333

Chapter 9
Bacopa monnieri, a Nootropic Drug

M. Rajani

B.V. Patel Pharmaceutical Education and Research Development (PERD) Centre, Thaltej, Ahmedabad - 380 054, Gujarat, India, e-mail: rajanivenkat@hotmail.com

Abstract *Bacopa monnieri* L. Pennell (family: Scrophulariaceae) is a reputed drug of Ayurveda. It is used in traditional medicine to treat various nervous disorders; it is also used as a stomachic, a digestive, rejuvenate, for promoting memory and intellect, for skin disorders, and as an antiepileptic, antipyretic, and analgesic. In a sector study by the Export–Import Bank of India, *B. monnieri* was placed second in a priority list of the most important medicinal plants, evaluated on the basis of their medicinal importance, commercial value, and potential for further research and development. Based on the traditional claims on *B. monnieri* as a memory enhancer, many classical and proprietary preparations are now available on the market. In the last two decades, *B. monnieri* has been studied extensively for its chemical constituents, its efficacy has been established in several *in vivo* and *in vitro* models, and randomized clinical trials have also been carried out. This article reviews the work carried out on the chemical, pharmacological, clinical, and biotechnological aspects of this plant.

9.1 Introduction

Bacopa monnieri L. Pennell (Fig. 9.1; synonyms: *Bramia monnieri* Pennell, *Moniera cuneifolia* Michx., *Herpestis monneira* (Linn.) H.B. & K. *Herpestis spathulata* Blume., *Gratiola monniera* Linn., *Lysimachia monniera* Linn.), family Scrophulariaceae, commonly known as Water Hyssop, brahmi, jal brahmi, and nir-brahmi, is a reputed drug of Ayurveda. It is used in traditional medicine for various nervous disorders [1, 2]. It is used as a stomachic, digestive, rejuvenate, for promoting memory and intellect, for skin disorders, as an antiepileptic, antipyretic, and analgesic [2, 3].

In a priority list of the most important medicinal plants, evaluated on the basis of their medicinal importance, commercial value and potential for further research and development, *B. monnieri* was placed second according to a sector

Fig. 9.1 *Bacopa monnieri* L. Pennell

study by the Export-Import Bank of India [4]. According to an estimate, the annual requirement of the plant was projected to be about 12,700 tonnes of dry material, valued at approximately Rs 15 billion [5].

In the last two decades, *B. monnieri* has been studied extensively for its chemical constituents, and its activity has been established in several *in vivo* and *in vitro* models; randomized clinical trials have also been carried out. Based on the traditional claims of *B. monnieri* as a memory enhancer, many classical and proprietary preparations are now available on the market. This article reviews the work on the chemical constituents, pharmacological studies, clinical studies, and biotechnological studies on this plant. This article also includes a part of an earlier review by us [6], which covered work on this plant up to 2003.

9.2 Chemical Constituents

The plant contains a complex mixture of dammarane type of triterpenoidal saponins (Fig. 9.2), with jujubogenin or pseudojujubogenin moiety as aglycones (Fig. 9.3). The saponins differ in the sugar moieties. Important saponins include bacoside A_1, bacoside A_2, bacoside A_3, [7–9], bacopasaponins A–D [10, 11], bacopasaponins E and F [12], bacopasaponin G [13], bacopasides I and II [14], bacopasides III–V [13, 15], bacopasides VI–VIII [16], bacopaside N1, bacopaside N2, and bacopaside X [17]. Of these saponins, bacopasaponins A, E, and F and bacopaside VIII are jujubogenin bisdesmosides [10–12, 16].

Hou and coworkers [13] named 3-O-[{6-O-sulfonyl-β-D-glucopyranosyl-(1→3)}-α-L-arabinopyranosyl] pseudojujubogenin as bacopaside III, whereas Chakravarty and coworkers [15] named 3-O-[α-L-arabinofuranosyl-(1→2)-{β-D-glucopyranosyl jujubogenin as bacopaside III. Until this matter of nomenclature is resolved, it is better to identify these two compounds by their IUPAC nomenclatures.

Bacosides A and B were reported earlier from this plant, to which the physiological activity has been attributed (see review [6]). Bacoside A was said to be levorotatory, while bacoside B is dextrorotatory [6]. Subsequently, it was estab-

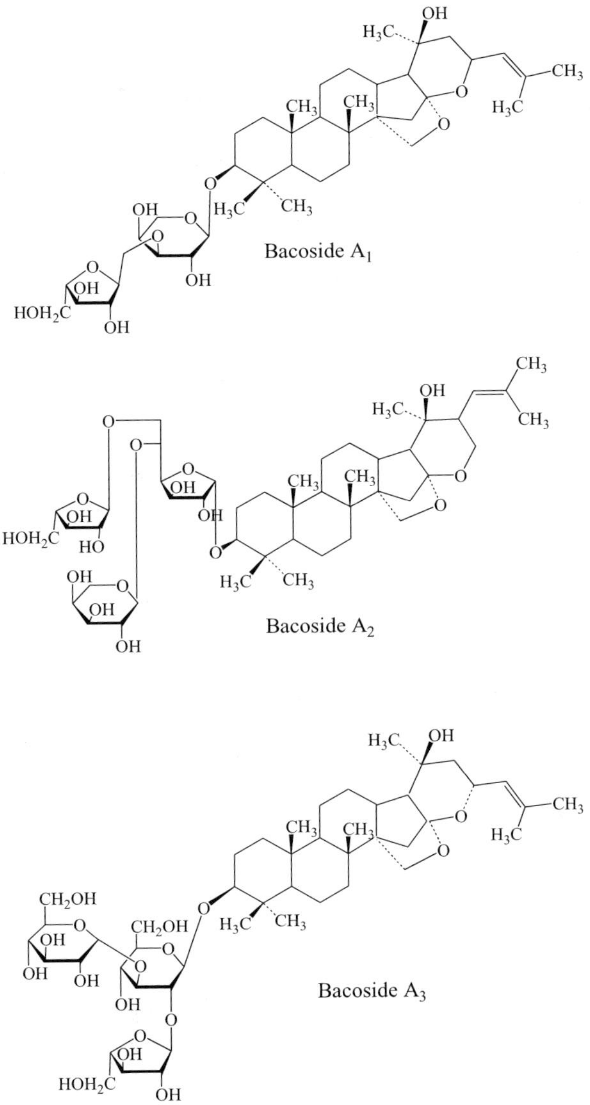

Chapter 9 *Bacopa monnieri*, a Nootropic Drug

Fig. 9.2 Some chemical constituents of *Bacopa monnieri* – saponins

lished that bacoside A is a mixture of four triglycosidic saponins: bacoside A_3, bacopaside II, and bacopasaponin C [17–19], while, according to the earlier report of Rastogi and Kulshrestha [20, 21] bacoside A is a mixture of bacoside A_2 and A_3. Bacoside B is composed of four minor saponins: bacopasides N_1, N_2, IV, and V. It was felt that there was a need to establish the identities of bacosides A and B as putative bioactive saponins of the *B. monnieri* plant [22], since these two mixtures and extracts standardized to these two are commercially available and have been used in many pharmacological and clinical studies (see section 9.3.1 and 9.3.2). The different saponins and their IUPAC names are given in Table 9.1.

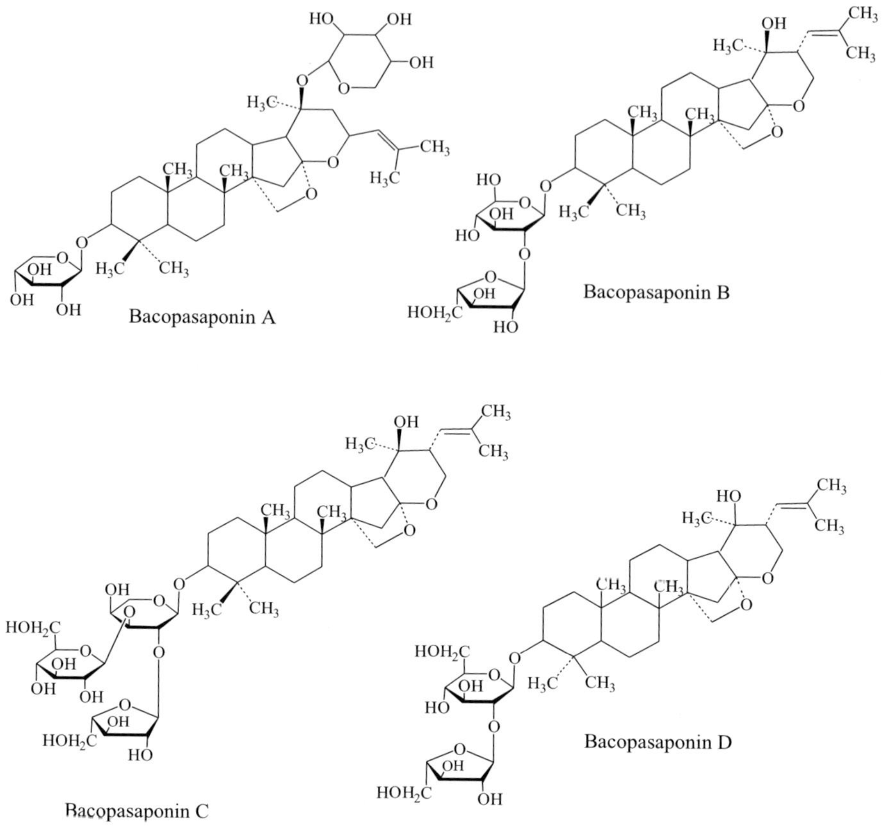

Fig. 9.2 *(continued)* Some chemical constituents of *Bacopa monnieri* – saponins

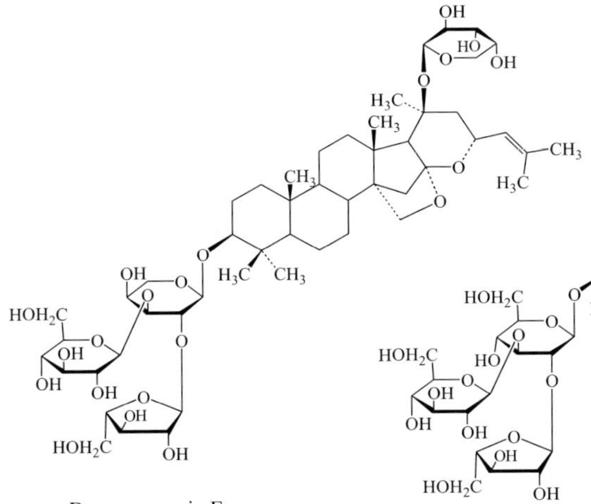

Chapter 9 *Bacopa monnieri*, a Nootropic Drug

Bacopasaponin E

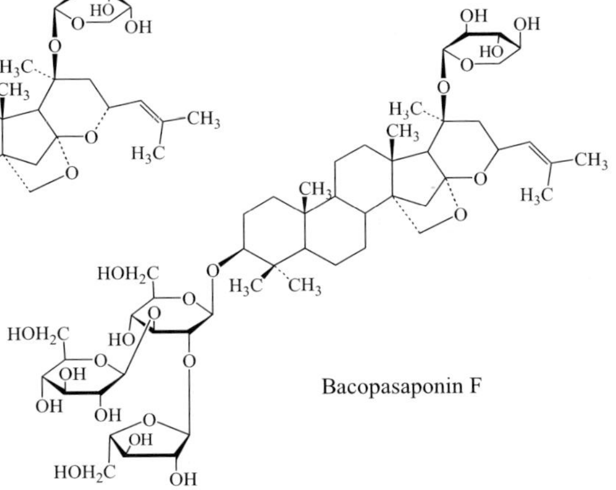

Bacopasaponin F

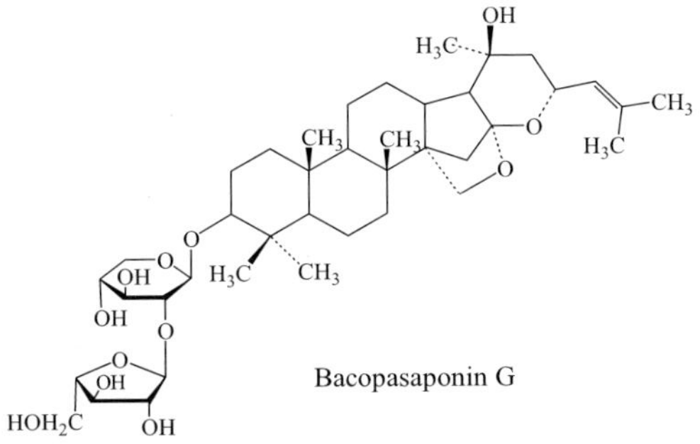

Bacopasaponin G

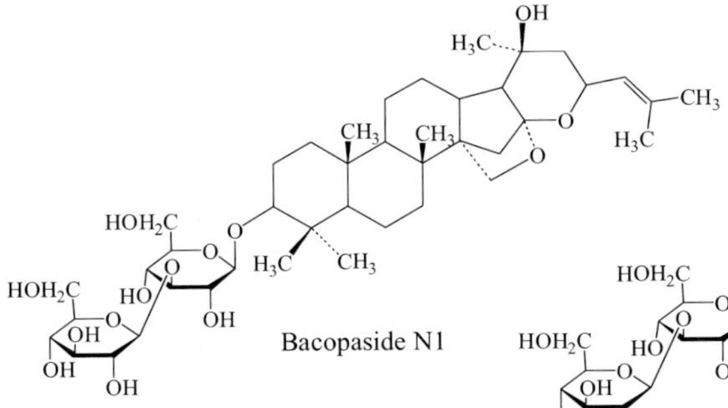

Bacopaside N1

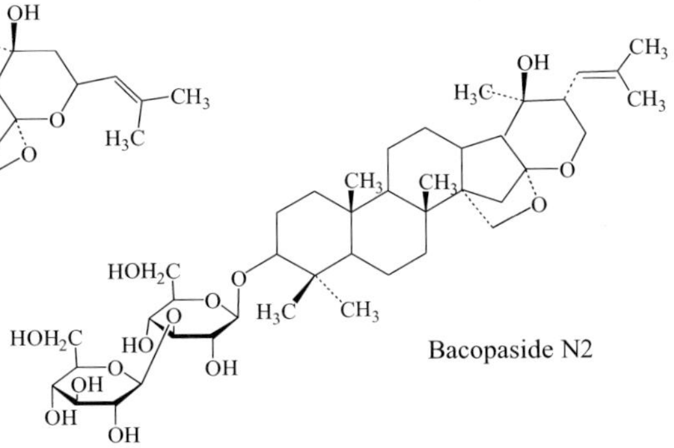

Bacopaside N2

Fig. 9.2 *(continued)* Some chemical constituents of *Bacopa monnieri* – saponins

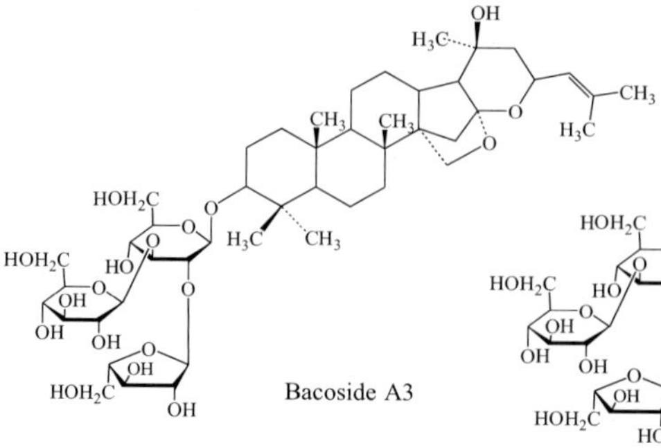

Bacoside A3

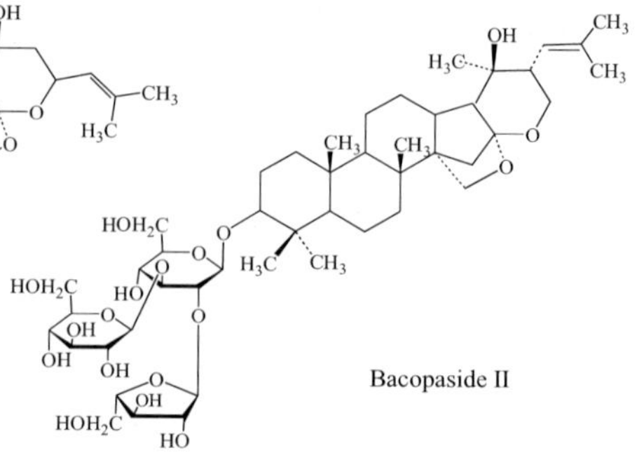

Bacopaside II

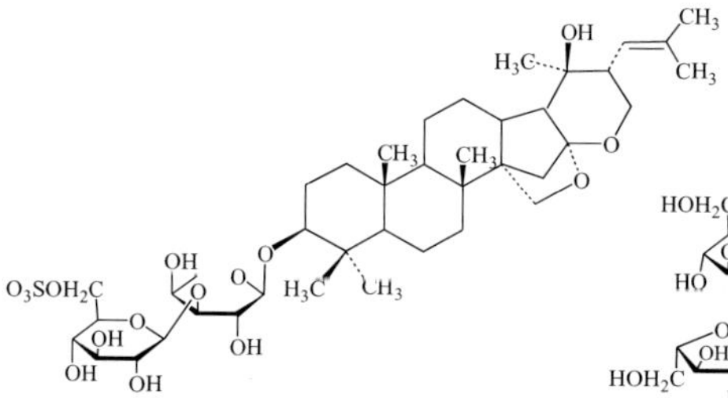

Bacopaside III
[13]

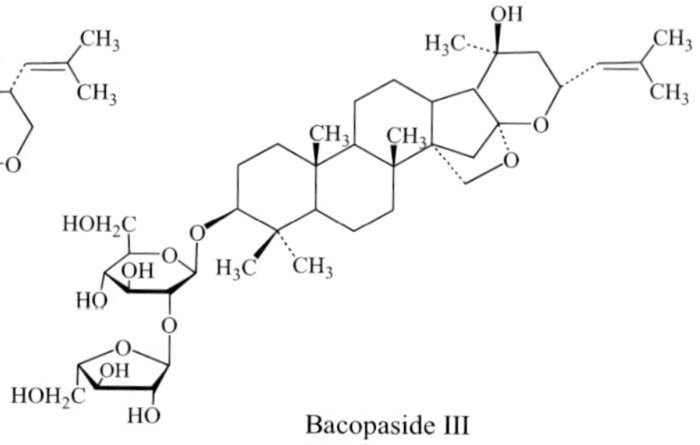

Bacopaside III
[15]

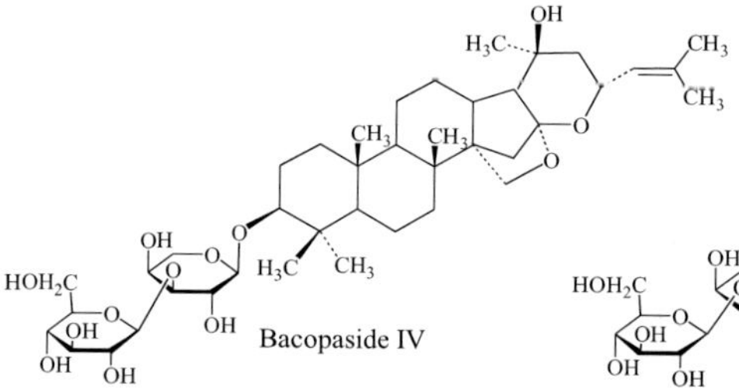

Bacopaside IV

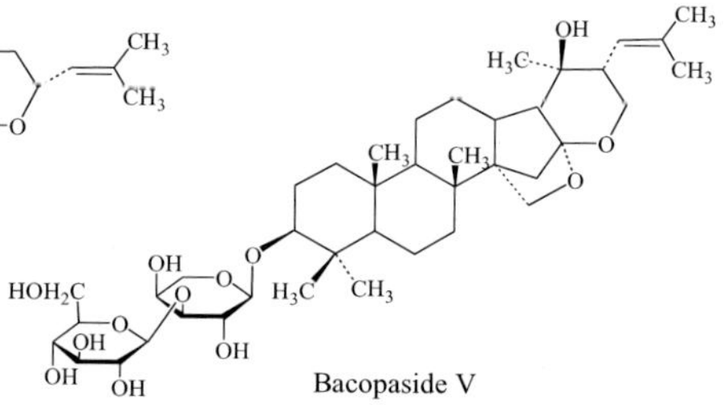

Bacopaside V

Fig. 9.2 *(continued)* Some chemical constituents of *Bacopa monnieri* – saponins

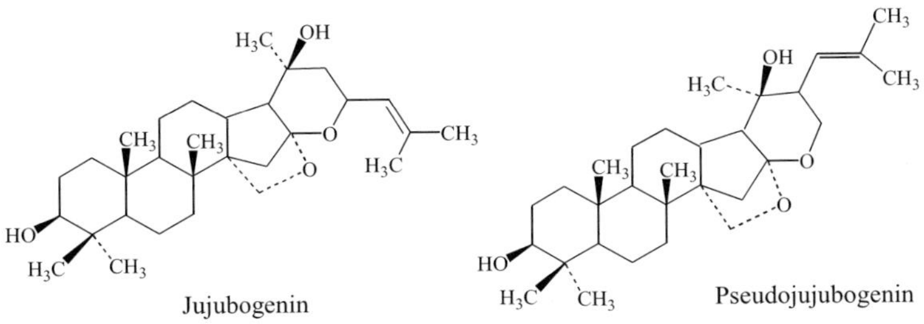

Chapter 9 Bacopa monnieri, a Nootropic Drug

Jujubogenin

Pseudojujubogenin

Fig. 9.3 Aglycones of triterpenoidal saponins

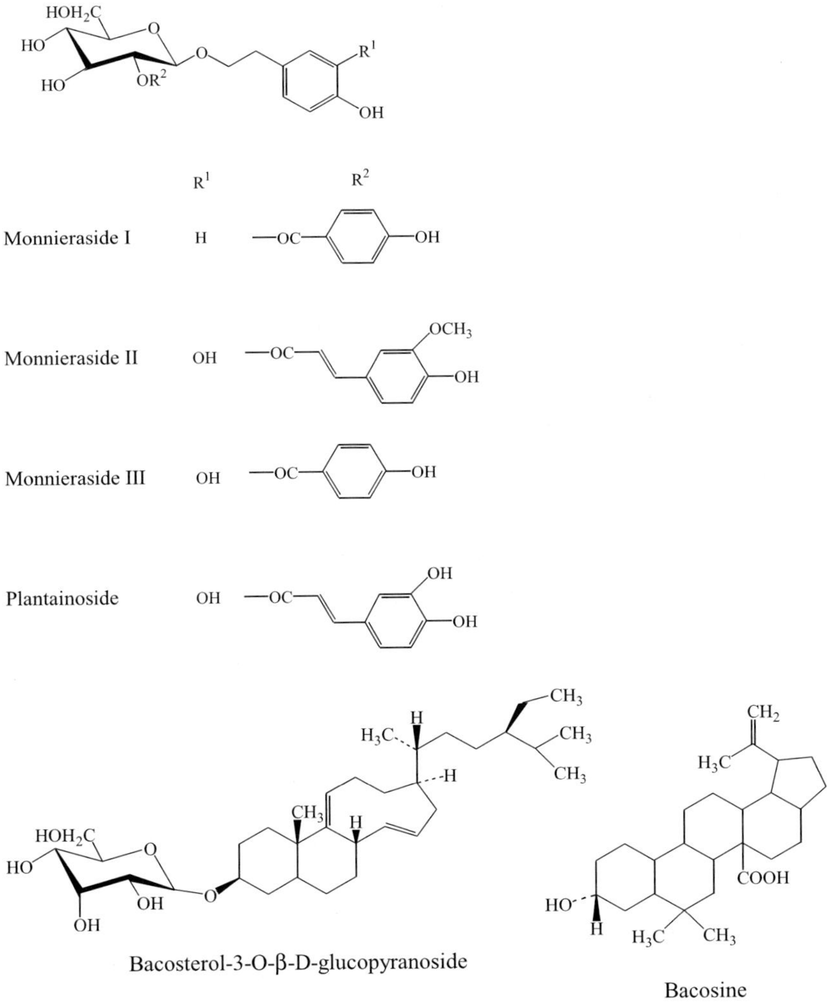

	R^1	R^2
Monnieraside I	H	—OC—⟨C6H4⟩—OH
Monnieraside II	OH	—OC—CH=CH—⟨C6H3(OCH3)⟩—OH
Monnieraside III	OH	—OC—⟨C6H4⟩—OH
Plantainoside	OH	—OC—CH=CH—⟨C6H3(OH)⟩—OH

Bacosterol-3-O-β-D-glucopyranoside

Bacosine

Fig. 9.4 Some other chemical constituents of *Bacopa monnieri*

Table 9.1 Details of some chemical constituents

Glycoside	Description (IUPAC name)	Reference
Bacoside A_1	3-O-[α-L-arabinofuranosyl (1→3)-α-L-arabinopyranosyl]-jujubogenin	[7]
Bacoside A_2	3β-O-[α-L-arabinofuranosyl(1→6)-O-[α-L-arabinopyranosyl-(1→5)-O-α-D-glucofuranosyl)oxy]	[9]
Bacoside A_3	3β-[O-β-D-glucopyranosyl-(1→3)-O-[α-L-arabinofuranosyl-(1→2)]-O-β-D-glucopyranosyl) oxy] jujubogenin	[8,17]
Bacopasaponin A	3-O-α-L-arabinopyranosyl-20-O-α-L-arabinopyranosyl-jujubogenin	[10]
Bacopasaponin B	3-O-[α-L-arabinofuranosyl (1→2)-α-L-arabinopyranosyl] pseudojujubogenin	[10]
Bacopasaponin C	3-O-[β-D-glucopyranosyl-(1→3)-{α-L-arabinofuranosyl-(1→2)}-α-L-arabinopyranosyl] pseudojujubogenin	[10, 16, 17, 33]
Bacopasaponin D	3-O-[α-L-arabinofuranosyl (1→2)-β-D-glucopyranosyl] pseudojujubogenin	[11]
Bacopasaponin E	3-O-[β-D-glucopyranosyl-(1→3)]-{α-L-arabinofuranosyl-(1→2)-α-L-arabinopyranosyl]-20-O-α-L-arabinopyranosyl) jujubogenin	[12, 17]
Bacopasaponin F	3-O-[β-D-glucopyranosyl-(1→3)-{α-L-arabinofuranosyl-(1→2)}-β-D-glucopyranosyl]-20-O-(α-L-arabinopyranosyl) jujubogenin	[12, 17]
Bacopasaponin G	3-O-[α-L-arabinofuranosyl-(1→2)-α-L-arabinopyranosyl] jujubogenin	[13]
Bacopaside I	3-O-[α-L-arabinofuranosyl-(1→2)-{6-O-sulfonyl-β-D-glucopyranosyl-(1→3)}-α-L-arabinopyranosyl pseudojujubogenin	[14, 16, 17, 33]
Bacopaside II	3-O-[α-L-arabinofuranosyl-(1→2)-{β-D-glucopyranosyl-(I→3)}-β-D-glucopyranosyl] pseudojujubogenin	[14, 16, 17, 33]
Bacopaside III[a]	3-O-[{6-O-sulfonyl-β-D-glucopyranosyl-(1→3)}-α-L-arabinopyranosyl] pseudojujubogenin	[13, 17]
Bacopaside III[a]	3-O-[α-L-arabinofuranosyl-(1→2)-{β-D-glucopyranosyl jujubogenin	[15]
Bacopaside IV	3-O-[β-D-glucopyranosyl-(1→3)-α-L-arabinopyranosyl] jujubogenin	[13, 15, 17]
Bacopaside V	3-O-[β-D-glucopyranosyl-(1→3)-α-L-arabinopyranosyl] pseudojujubogenin	[13, 15, 17]

[a]Bacopaside III has been identified as two different compounds by two different groups of researchers.

Table 9.1 *(continued)* Details of some chemical constituents

Glycoside	Description (IUPAC name)	Reference
Bacopaside VI	3-O-[6-O-sulfonyl-β-D-glucopyranosyl (1→3)]-α-L-arabinopyranosyl] pseudojujubogenin	[16]
Bacopaside VII	3-O-{β-D-glucopyranosyl-(1→3)-[α-L-arabinofuranosyl-(1→2)]-α-L-arabinopyranosyl}jujubogenin	[16]
Bacopaside VIII	3-O-{β-D-glucopyranosyl-(1→3)-[α-L-arabinofuranosyl-(1→2)]-β-D-glucopyranosyl}-20-α-L-arabinopyranosyl jujubogenin	[16]
Bacopaside X	3-O-α-L-arabinofuranosyl-(1→2)-{β-D-glucopyranosyl-(1→3)}-α-L-arabinopyranosyl] jujubogenin	[17]
Bacopaside N1	3-O-[β-D-glucopyranosyl-(1→3)-β-D-glucopyranosyl] jujubogenin	[17]
Bacopaside N2	3-O-[β-D-glucopyranosyl-(1→3)-β-D-glucopyranosyl] pseudojujubogenin	[17]
Monnieraside I	α-O-[2-O-(4-hdroxybenzoyl)-β-D-glucopyranosyl]-4-hydroxyphenylethanol	[34]
Monnieraside II	α-O-[2-O-(3-methoxy-4-hdroxycinnamoyl)-β-D-glucopyranosyl]-3,4-dihydroxyphenylethanol	[34]
Monnieraside III	α-O-[2-O-(4-hdroxybenzoyl)-β-D-glucopyranosyl]-3,4-dihydroxyphenylethanol	[34]
Bacosterol glycoside	Bacosterol-3-O-β-D-glucopyranoside	[33]

[a]Bacopaside III has been identified as two different compounds by two different groups of researchers.

In earlier reports, several aglycones (bacogenins A1–A4 and ebelin lactone) have been mentioned as the products of hydrolysis of the saponins from *B. monnieri*. However, these were subsequently found to be artifacts formed during hydrolysis (described in detail in the review on the plant by Rajani et al. [6]). It has now been established that jujubogenin and pseudojujubogenin are the genuine sapogenins of the triterpenoid glycosides of *B. monnieri*.

Other chemical constituents of the plant are d-mannitol [23], hersaponin, betulic acid [24–27], alkaloids brahmine and herpestine [28] (cited in [23] and [25]) [29–30], flavonoids luteolin-7-glucoside, glucuronyl-7-apigenin and glucuronyl-7-luteolin [31, 32], luteolin-7-O-β-glucopyranoside [33]; stigmasterol, β-sitosterol, stigmastanol [24], bacosterol glycoside, bacosterol [33], bacopasides A–C [13], phenylethanoid glycosides monnierasides I–III, and plantainoside B [34], a triterpene bacosine that is lup-20(29)-ene-3α-ol-27-oic acid [33, 35, 36] (Fig. 9.4).

9.3 Analysis of Saponins of *B. monnieri*

For the quantification of different saponins from the plant, an ultraviolet spectrophotometric method [37], high-performance liquid chromatography (HPLC) coupled to nuclear magnetic resonance and mass spectrometry [38], a thin-layer chromatography (TLC) densitometric method for the determination of bacoside A using HP-TLC [39] and several HPLC methods have been reported [17, 20]. Bhandari and coworkers [40] described an HPLC method for the quantification of bacoside A, which separated its three components bacoside A_3, bacopaside II (an isomer of bacopasaponin C) and bacopasaponin C. Recently, Murthy and coworkers [41] determined 12 saponins of the plant using HPLC. The method enables simultaneous determination of all 12 saponins. Of these 12 saponins, bacoside A_3, bacopasides II and I, bacopaside X, bacopasaponin C, and bacopaside N_2 were found to be major saponins, while bacopasaponins F and E, bacopaside N_1, and bacopasides III–V were minor, with the total saponin content of the plant ranging from 5 to 6%.

Srinivasa and co-workers [42] demonstrated the presence of luteolin in the plant by a TLC densitometric method using HP-TLC.

9.3.1 Pharmacological Studies

In Ayurveda, *B. monnieri* has been used to promote memory and intellect, to treat psychoneurological disorders, and as a rejuvenator. In the last four decades the plant and the compounds isolated from it, especially the saponins, have been studied extensively regarding their memory-enhancing activities and their ability to improve cognitive function, including some clinical trials conducted to establish the activity. The pharmacological activities of the plant have been reviewed and published [3, 6, 43].

B. monnieri was shown to improve motor learning capabilities in rats [44]. An alcoholic extract of the plant was shown to improve acquisition, consolidation, and retention of memory when tested on learning response in rats [45]. It augmented cognitive function and mental retention capacity in a foot-shock-motivated brightness discrimination response, active conditioned avoidance response, and Sidman continuous avoidance response in rat [44–49], and bacosides A and B have been shown to be responsible for this activity [49,50]. *B. monnieri* elevated the cerebral glutamic acid levels and caused a transient increase in levels of γ-aminobutyric acid [51, 52].

B. monnieri was shown to have anticholinesterase activity, cognitive-enhancing activity, and antidementic properties [53]. An alcoholic extract of the plant and a commercial sample of a mixture of bacosides were shown to facilitate acquisition or learning in mice, using the water-maze test. Bacosides were found to reverse the anteretrograde experimental amnesia induced by scopolamine, sodium nitrite, and BN52021 (a platelet-activating factor, PAF, receptor antagonist). Suggested possible mechanisms of action were increase in acetylcholine levels, hypoxic conditions, and PAF synthesis in each of the cases of induced experimental amnesia, respectively [54].

The plant exhibited a tranquilizing effect in rat and dog [55, 56]. It caused smooth-muscle relaxation and showed antispasmodic activity [55].

Ethanolic extract of *B. monnieri* afforded a neuoprotective role against aluminum-induced toxicity and prevented oxidative stress induced by aluminum in the hippocampus of rats [57], and the activity was comparable to that of α-deprenyl (a monoamine oxidase-B inhibitor and neuroprotectant used in Parkinson's disease). The extract used in that study comprised 55–60% bacosides [58]. The extract was shown to inhibit lipid peroxidation, protein oxidation, and lipofuscin accumulation. Coadministration of the extract with aluminum was shown to reverse the aluminum-induced oxidative stress and ultrastructural changes in the hippocampus [57, 59] and prevent the accumulation of lipid and protein damage. Furthermore, the reduced activity of the endogenous antioxidant enzymes due to treatment was restored to normal levels [59].

B. monnieri was shown to have antioxidant activity and has the potential to enhance the activity of endogenous antioxidant enzymes. An ethanolic extract of the plant showed potent antioxidant properties [60]. A standardized extract of the plant (containing 50% bacoside A) increased the levels of superoxide dismutase (SOD), catalase, and glutathione peroxidase activities in the frontal cortical and striatal areas and in the hippocampus of rat brain (5–10 mg/kg, taken orally), and the activity was found to be comparable to deprenyl [61]. Bacoside A and B were found to increase protein kinase and protein and serotonin (5-HT) levels and lower the norepinephrine level in the hippocampus [62, 63]. *B. monniera* showed thyroid-stimulating activity, lowered the levels of lipid peroxidation, and increased the activity of superoxide dismutase and catalase [64].

Since it has been claimed that *B. monnieri* is efficacious in the treatment of memory loss, its potential benefit in the treatment of Alzheimer's disease was studied regarding amyloid plaque pathology in the brain of PSAPP transgenic mice [65]. Short- and long-term treatment with *B. monnieri* extract reduced brain Aβ (amyloid beta) 1-40 and 1-42 levels in the cortex and reversed the behavioral deficits in PSAPP mice; the study indicates the potential of *B. monnieri* as a therapeutic agent for Alzheimer's disease [65].

Aqueous and ethanolic extracts of the plant did not protect from the chemical- or electric-shock-induced seizures, but reduced their severity [66]. They did not produce any sedation or antidepressant effect, but potentiated the barbiturate-induced hypnosis. They also increased the pentylenetetrazole-induced seizure latency, suggesting that a longer duration of treatment with the extracts of the plant may be beneficial in petit mal epilepsy [66]. In an earlier study, hersaponin (isolated from *B. monnieri*) was shown to have sedative and hypnotic potentiating activities in mice [67]. It reduced the norepinephrine and 5-HT contents in rat brain [68].

The alcoholic extract exhibited cardiotonic, vasoconstrictor, sedative, and neuromuscular blocking action (LD_{50} 0.33 g/kg, administered intraperitoneally, i.p. in rats), while the saponin fraction exhibited cardiotonic action in normal and hypodynamic frog hearts, sedative action in rats and guinea pigs, and spasmodic action in rabbit and guinea pig ileum and rat uterus (LD_{50} 25 mg/kg, i.p. in albino rats). In these experiments the mortality observed in higher doses was attributed to a possible respiratory failure due to neuromuscular blocking [69]. The alcoholic extract showed a positive inotropic effect in frog heart. In rabbit heart it increased the coronary outflow, heart rate, and amplitude of contraction. It decreased the motility of rabbit jejunum [70].

The plant extract and alkaloid fraction were found to inhibit respiration in rat brain homogenate in a dose-dependent manner and it was shown that 5-HT and lysergic acid diethylamide had no influence on this effect [71]. The alkaloid fraction produced skeletal muscle spasm, initial stimulation and then eventual depression of respiration, initial stimulation of autonomic ganglia followed by blockade, rigidity, and convulsions in mice [72]. Aqueous extract of the leaf exhibited dose-dependant depressive activity [73].

B. monniera exhibited a protective role on morphine-induced brain mitochondrial enzyme activity by maintaining normal enzyme levels [74]. An alcoholic extract of the plant was shown to alleviate withdrawal symptoms from morphine in vitro in guinea pig ileum. The extract was administered 15 min prior to exposure to morphine; the withdrawal symptoms were precipitated by the addition of naloxone to induce contraction and were reduced by the addition of B. monnieri extract (1 mg/ml) [75].

Standardized extract of B. monnieri (containing 55–60% bacosides) was shown to have adaptogenic activity. It normalized the stress-induced elevation in the levels of plasma glucose and creatine kinase, and in adrenal gland weight [76]. In acute stress and chronic unpredictable stress models, this extract of B. monnieri was found to normalize the stress-mediated transient deregulation of plasma corticosterone and monoamine changes in the brain. However, the extract did not regulate the levels of dopamine in the chronic model [77]. A mixture of bacosides A and B was shown to reduce the stress by modulating the activities of the 70-kDa heatshock protein (Hsp70), cytochrome P450 and SOD in different regions of the brain in the stress model of cold-hypoxic restraint in rats [78]. Treatment with bacosides brought about an increase in the activity of SOD in the brain regions, and no change was seen in the expression of Hsp70, while there was increase in the expression of cytochrome P450, although the magnitude of increase in expression of cytochrome P450 was less with bacoside treatment. Pretreatment for 7 days with bacosides prior to stress resulted in: (1) a decrease in the expression of Hsp70 expression in all brain regions, especially the hippocampus, (2) an increase in SOD activity in the cerebral cortex and the rest of the brain, and (3) a reduction in SOD activity in the cerebellum and hippocampus; the expression of cytochrome P450 was maintained at control levels [78].

These effects have implications with regard to the memory-enhancing activity of the plant, which is mainly attributed to its anxiolytic effects [79].

Singh and coworkers [62] reported that bacosides A and B showed antidepressant properties. Recently, Zhou et al. [16] reported that bacopasides I and

II and bacopasaponin C (all three having pseudojujubogenin as aglycone) exhibited antidepressant activity, while bacopaside VII (having jujubogenin as aglycone) did not have any antidepressant activity when tested on forced swimming and tail-suspension models in mice.

Anbarasi and coworkers [80–84] studied the protective effects of bacoside A on cigarette-smoking-induced changes in the brain. Cigarette smoking is known to cause free-radical-mediated damage in the heart and brain. Bacoside A was shown to inhibit lipid peroxidation, improve the activity of ATPases [80], and prevent the structural and functional impairment of mitochondrial membranes in the brains [81] of rats that had been subject to chronic exposure to cigarette smoke. Furthermore, bacoside A prevented cigarette-smoking-induced Hsp70 expression and neuronal apoptosis [82]. Bacoside A has also been found to prevent the cigarette-smoke-induced activity of creatine kinase (CK) and its isoforms (CK-MM, CK-MB, CK-BB), sensitive markers used in the assessment of cardiac and cerebral damage. Membrane damage due to cigarette-smoke-induced lipid peroxidation in the heart and brain is the cause of leakage of CK, and the protective effect of bacoside A on the functional and structural integrity of the membrane is attributed to its antioxidant activity [83]. The antioxidant activity of the plant has been shown to be one of the mechanisms of action of *B. monnieri*, as free radicals and lipid peroxidation are implicated in many disease conditions including Alzheimer's and in the damage caused by cigarette smoking. In a chronic cigarette smoke model, administration of bacoside A improved the antioxidant status, as evidenced by increased levels of reduced glutathione, vitamins C, E, and A, and activities of SOD, catalase, glutathione, glutathione peroxidase [84]. Nicotine, the active compound of cigarette smoke, is known to cause damage including genomic instability and the generation of free radicals. *B. monnieri* afforded protection against these parameters by enhancing the antioxidant status [85].

One of the underlying mechanisms of action of *B. monnieri* is its antioxidant activity, evident from several different studies, as discussed above. Superoxide radicals were inhibited in polymorphonuclear cells in vitro by 70% methanolic extract and its butanolic fraction (IC_{50} 69.33 μg/ml and 82.66 μg/ml, respectively). Bioassay-guided fractionation of the butanolic fraction led to the isolation of bacoside A_3 with an IC_{50} of 10.22 μg/ml and a maximum inhibition of 91.66% at 50 μg/ml. Bacopasaponin C, a structural analog of bacoside A_3, also inhibited the superoxide in polymorphonuclear cells, although it was found to be less potent, with an IC_{50} of 54.16 μg/ml [86]. The methanolic extract of the plant had free-radical-scavenging activity and provided protection against DNA damage in human nonimmortalized fibroblasts [87].

Bacosides A and B have thus been shown to be responsible for different activities. In section 9.2 (Chemical Constituents), it was noted that bacosides A and B are mixtures. In light of that knowledge, it is essential to exercise caution in attributing activity to these mixtures. Further experiments with pure compounds are needed to establish the active principles.

The methanolic extract of the plant (containing 38% bacoside A) showed dose-dependent antiulcerogenic activity when tested in various gastric ulcer models. It exhibited both prophylactic and curative effects. The activity was shown to be due to the augmentation of the defensive mucosal factors, like increasing mucin secretion, increasing the life span of mucosal cells (it decreased cell shedding), and antioxidant activity [88]. Furthermore, it showed anti-*Helicobacter pylori* activity (dose: 1 mg/ml) *in vitro* and increased the prostanoid (prostaglandins – PGE and PGI2) levels in human colonic mucosal incubates *in vitro*, which may contribute to the antiulcerogenic activity of the extract [89].

The bronchovasodilatory activity of *B. monnieri* was studied in detail by Dar and Channa [90–92]. Ethanolic extract of *B. monnieri* showed bronchovasodilatory activity with a concurrent involvement of calcium channels, β-adrenoceptors, and prostaglandins [90–92]. Several fractions and subfractions from the ethanolic extract showed bronchovasodilatory activity, which appears to be mediated by interference with calcium ion movement. Activity-guided fractionation led to the isolation of betulinic acid, which exhibited inhibition of tracheal pressure and heart rate in rats [93].

The ethanolic extract of *B. monnieri* (100 mg, i.p.) exhibited a very good anti-inflammatory activity against carrageenan-induced paw edema in mice and rats, and it selectively inhibited PGE_2-induced inflammation [94]. Bacosine, isolated from the plant, exhibited a moderate analgesic activity and was found to be opioidergic in nature [95].

The *n*-butanol extract of the plant was shown to have good antibacterial activity against a battery of human pathogens and cattle pathogens tested *in vitro* [96]. Betulinic acid isolated from *B. monnieri* showed good antifungal activity against *Alternaria alternata* and *Fusarium fusiformis*. Betulinic acid also showed phytotoxic activity in inhibiting the root growth of wheat seedlings [27].

Different extracts of the plant, bacosides A and B, bacopasides I, II, and X, and bacopasaponin C, showed potent activity in a brine shrimp lethality assay (an assay that is predictive of potential anticancer) activity [97, 98]. In addition, the bacoside A fraction and its individual components were found to be more active than the bacoside B fraction [99, 17]. The ethanolic extract of the plant exhibited anticancer activity against Walker carcinosarcoma 256 in rat [100] and sarcoma-180 cell culture [101]. The methanolic extract was shown to have potent mast cell stabilization activity; this activity was comparable to that of disodium cromoglycate [102]. Bacopasaponin C was tested for its antileishmanial properties in different delivery modes (i.e., niosomes, microspheres, and nanoparticles), and it was found to be highly active in all forms. However, bacopasaponin C showed nephrotoxicity, while in vesicular form it was found to be safe. The activity was found to be parasite specific [103].

9.3.2 Clinical Studies

Several clinical studies have been conducted to evaluate the traditional claims and study the potential of *B. monnieri* in enhancing the cognitive functions. In one clinical study on 35 cases of anxiety neurosis, *B. monnieri* was shown to provide significant relief from symptoms and a quantitative reduction in anxiety level; it also improved mental functions. The treatment reduced the levels of urinary vinyl mandelic acid and corticoids. The antianxiety effect was attributed to a possible adaptogenic property of the plant [104]. The plant was also found to be effective in revitalizing intellectual functions in children [105].

In a double-blind, randomized, placebo-controlled clinical study (76 adults, 40–65 years old), the plant showed a significant effect on retention of new information, while attention, verbal and visual short-term memory, the retrieval of pre-experimental knowledge, everyday memory function, and anxiety levels were unaffected [106].

The effect of acute administration of low standard dose of *B. monniera* extract was tested in healthy subjects. In a double-blind, placebo-controlled, independent-group clinical trial, *B. monnieri* extract administered acutely (300 mg; $n=18$) did not show any effect on cognitive functioning in normal healthy subjects in tests examining attention, working, short-term memory, verbal learning, memory consolidation, executive process, planning and problem solving, information processing speed, motor responsiveness, and decision-making [107]. Acute administration of a combination of 50% ethanolic extract of *B. monnieri* (300 mg; the extract equivalent to 3 g of dry herb containing not less than 50% combined bacosides A and B) and *Ginkgo biloba* extract (120 mg) did not show any effect on cognitive function in normal subjects [108].

In a double-blind, placebo-controlled, independent-group design, chronic administration of the extract of *B. monniera* (300 mg) for 3 months caused a significant improvement in information processing, learning, and memory consolidation in normal healthy subjects (46 healthy subjects, 11 males and 35 females). The treatment significantly improved the speed of visual information processing, as measured by the inspection time task, learning rate, and memory consolidation, with maximal effect evident after 12 weeks. From the results, it was suggested that *B. monnieri* improves higher-order cognitive processes that are critically dependent on the input of information from the environment such as learning and memory [109].

In another clinical study (double-blind, placebo-controlled, independent group), even a chronic 4-week treatment with tablet containing a combination of 300 mg of *B. monnieri* extract (equivalent to 3 g dry leaf standardized to contain 67.5 mg bacosides) and 120 mg of *Ginkgo biloba* extract did not show any cognitive-enhancing effects in healthy subjects [110].

In a double-blind placebo-controlled and phase-1 clinical trial, bacosides were well tolerated by normal healthy male human volunteers in single dose and in multiple doses administered for 4 weeks [48].

9.3.3 Concluding Remarks

The different activities of *B. monnieri* have been studied extensively in suitable animal models and one group of underlying mechanisms of action that emerged from these studies includes free-radical-scavenging activity, antioxidant activity, and augmentation of SOD, catalase and glutathione peroxidase activities. Although the outcome of clinical trials has not been conclusive, the results are promising. Since it is intended for chronic administration, it is necessary to conduct detailed safety studies of the plant.

9.4 Biotechnology and Tissue Culture Studies on *B. monniera*

In an effort to study the genetic diversity in the germplasm of *B. monnieri*, 24 accessions from different geographical regions of India and one introduction from Malaysia maintained at the Central Institute of Medicinal and Aromatic Plants, Lucknow, India, were analyzed for random amplified polymorphic DNA variation. The similarity between accessions was found to be in the range of 0.8–1.0, which is indicative of a narrow genetic base and a low to medium level of polymorphism. The individual accessions could be differentiated, showing differences in morphological and growth properties at the DNA level. The low level of genetic variation was attributed to the interplay between sexual and vegetative reproduction and similarity of local environments in the habitats of the plant [111].

The plant has very high morphogenic potential, because of which explants from it respond very readily to treatment with plant growth regulators. Several protocols have been reported for the rapid multiplication of shoots and shoot cultures, and for the micropropagation of the plant [112–117] (reviewed by Rajani et al. 2004 [6]). In most of the tissue culture studies, a very good response was obtained by 6-benzyl amino purine in producing adventitious shoots from different explants [45–47, 49] and somatic embryogenesis [114], which led to the development of successful protocols for the mass propagation of the plant [114, 115].

As *B. monnieri* has a high morphogenic potential, the explants readily regenerated shoot buds in medium containing small amounts of cytokinins like 6-benzyl amino purine or kinetin [113]. With a high regenerative potential, the stem and leaf explants showed a tendency to regenerate shoots and/or roots even in media that are known to support callus initiation and growth (e.g., media supplemented with 2,4-dichlorophenoxyacetic acid), while the callusing response of the explants was very poor. The best response for shoot regeneration from both stem and leaf explants was obtained with 2 µM 6-benzyl amino purine in Murashige and Skoog (MS) medium gelled with 0.2% Gelrite, where profuse induction of multiple shoot buds was obtained in 9 days of incubation. Furthermore, after 3 weeks of shoot culture growth, when the shoots were harvested and the explant base was transferred into fresh medium, it continued to expand exponentially and regenerate new shoot buds. In addition, the leaf explants from shoot cultures proved to be superior explant material compared to the leaf explants from field-grown plants. The shoot cultures performed better when grown on medium gelled with Gelrite (0.2%) rather than with agar

(0.7%). The phytochemical profile of the regenerated shoots was found to be similar to that of field-grown plants, as revealed by TLC analysis [113].

In an interesting study, *B. monnieri* was shown to respond to the antibiotic trimethoprim and the fungicide bavistin, with bavistin inducing a good number of shoots from internodal explants [116]. However, the growth of the regenerated shoots was stunted and for further optimum shoot growth it required the supplementation of plant growth regulators. This calls for further exploration of the potential of bavistin in shoot regeneration, especially since the induction response itself was not gradual and was a concentration-dependant response, causing a sudden surge of induction at 300 mg/l in internode and leaf explants [116].

Cell suspension cultures were established from leaf explants of *in vitro* plants, and two cell lines were identified that showed a 5- to 6-fold increase in fresh and dry weight in 40 days of culture in MS medium supplemented with 1 mg/l α-naphthalene acetic acid and 0.5 mg/l kinetin. In these two cell lines, bacosides A and B were detected from the 10th day of culture, and they progressively accumulated up to the 40th day. Bacoside A content was found to be higher than bacoside B all through the culture period, with a maximum of 1% dry weight basis (dwb) in 40-day cultures. Bacoside B accumulated up to 0.25–0.37% dwb in 40-day cultures [118].

Ali and coworkers [119–121] used *B. monnieri* grown *in vitro* as a model system for studying the effects of cadmium on plants, toward an effort to establish metal-tolerant plants, since cadmium is a widespread pollutant. It was found that the cultures could become acclimatized to cadmium through gradual exposure to increased concentrations of cadmium [120]. Cadmium caused a reduction in photosynthetic rate, stomatal conductance, and internal carbon dioxide concentration, and supplementation with zinc improved these parameters under cadmium stress [120, 121]. There was a gradual increase in the accumulation of cadmium with an increase in concentration and duration of treatment, with maximum accumulation in root [122]. Cadmium was shown to accumulate in the cell wall. Cadmium induced an increase in total protein content in the cultures (up to 50 μM cadmium), and caused the accumulation of proline [119, 120]; it was suggested that these two parameters can serve as stress indicators [121]. At higher concentrations (200 μM) cadmium caused oxidative damage with increased lipid peroxidation, decreased chlorophyll and protein content, and a concomitant induction in the activities of SOD, ascorbate peroxidase, and guicol peroxidase, perhaps to combat the oxidative stress caused by cadmium. However, there was a significant reduction in catalase activity [122, 123]. Cadmium was found to induce phytochelatins in the root and leaf [123]. Cadmium was shown to inhibit the growth of regenerants *in vitro*; copper partially alleviated this negative effect [119].

In a study of the adaptation of *B. monnieri* regenerants to salinity stress *in vitro*, the proline content of the plants was found to increase sixfold compared to the control, while photosynthetic rate, fresh mass, and root length of the regenerants decreased [124].

Tetraploid plants were generated by treating nodal explants from *in vitro*-grown plants with colchicine followed by culture in MS medium supplemented with 0.25 mg/l 6-benzyl amino purine. The tetraploid obtained had larger leaves and flowers, but its growth was slow. The tetraploid obtained was named as a new variety, Ali INTA-JICA [125].

Acknowledgments

The author is grateful to Professor Harish Padh, Director of the B.V. Patel PERD Centre, for the use of the facilities.

References

1. Chopra IC, Handa KL, Sobti SN (1956) Indian J Pharm 18:364
2. The Ayurvedic Pharmacopoeia of India, Part I, vol. II, (1999) Government of India, Ministry of Health Family Welfare, Department of Indian Systems of Medicine, New Delhi, India, pp 25–26
3. Gupta AK, Sharma M, Tandon N (eds) (2004) Reviews on Indian Medicinal Plants Vol 4. Indian Council of Medical Research, New Delhi, pp 4–23
4. Export-Import Bank of India (1997) Indian Medicinal Plants: A Sector Study. Occasional paper No. 54, 2. Quest Publications, Bombay, India
5. Ahmad RU (1993) Medicinal plants used in ISM – Their procurement, cultivation, regeneration and import/export aspects: A report. In: Govil JN, Singh VK, Hashmi S (eds) Medicinal Plants: New Vistas of Research (Part 1). Today Tomorrow Printers and Publisher, New Delhi, pp 221–225
6. Rajani M, Shrivastava N, Ravishankara MN (2004) Brahmi (Bacopa monnieri (L.) Pennell.) – a Medhya Rasayana drug. In: New Concepts in Medicinal Plants. Science Publishers, Enfield (NH), USA, p 89–110
7. Jain P, Kulshreshtha DK (1993) Phytochemistry 33:449
8. Rastogi S, Pal R, Kulshreshtha DK (1994) Phytochemistry 36:133
9. Rastogi S, Kulshreshtha DK (1998) Indian J Chem 38B:353
10. Garai S, Mahato SB, Ohtani K, Yamasaki K (1996) Phytochemistry 42:815
11. Garai S, Mahato SB, Ohtani K, Yamasaki K (1996) Phytochemistry 43:447
12. Mahato SB, Garai S, Chakravarty AK (2000) Phytochemistry 53:711
13. Hou CC, Lin SJ, Hsu FL (2002) J Nat Prod 65:1759
14. Chakravarty AK, Sarkar T, Masuda K, Shiojima K, Nakane T, Kawahara N (2001) Phytochemistry 58:553
15. Chakravarty AK, Garai S, Masuda K, Nakane T, Kawahara N (2003) Chem Pharm Bull 51: 215
16. Zhou Y, Shen YH, Zhang C, Su J, Liu RH, Zhang WD (2007) J Nat Prod 70:652
17. Sivaramakrishna C, Rao VC, Trimurtulu G, Vanisree M, Subbaraju GV (2005) Phytochemistry 66:2719
18. Basu, NK, Pabrai PR (1947) Q J Pharm Pharmacol 20:137
19. Deepak M, Sangli GK, Arun PC, Amit A (2005) Phytochem Anal 16:24
20. Rastogi S, Pal R, Kulshreshtha DK (1994) Phytochemistry 36:133
21. Rastogi S, Kulshreshtha DK (1998) Indian J Chem 38B:353
22. Deepak M, Amit A (2004) Phytomedicine 11:264
23. Sastry MS, Dhalla NS, Malhotra CL (1959) Indian J Pharm 21:303

24. Chatterji N, Rastogi RP, Dhar ML (1963) Indian J Chem 1:212
25. Rastogi RP, Dhar ML (1960) J Sci Ind Res 19B:455
26. Jain P, Kulshreshtha DK (1993) Phytochemistry 33:449
27. Chaudhuri PK, Srivastava R, Kumar Sunil, Kumar Sushil (2004) Phytother Res 18:114
28. Bose KC, Bose NK (1931) J Indian Med Assoc 1:60
29. Basu NK, Walia JS (1944) Indian J Pharm 6:84:91
30. Basu NK, Pabrai PR (1947) Q J Pharm Pharmacol 20:137
31. Schulte KE, Ruecker G, El-Kersch M (1972) Phytochemistry 11:2649
32. Proliac A, Chabaud A, Raynaud J (1991) Pharm Acta Helv 66:153 – cited in Indian Herbal Pharmacopoeia, Vol. I. 1998, IDMA, Mumbai, India
33. Bhandari P, Kumar N, Singh B, Kaul VK (2006) Chem Pharm Bull 54:240
34. Chakravarty AK, Sarkar T, Nakane T, Kawahara N, Masuda K (2001) Chem Pharm Bull 50:1616
35. Vohora SB, Khanna T, Athar M, Ahmad B (1997) Fitoterapia 68:361
36. Ahmed B, Rahman A (2000) Indian J Chem 39B:620
37. Pal R, Sarin JPS (1992) Indian J Pharm Sci 54:17
38. Renukappa T, Roos G, Klaiber B, Kraus WV (1999) J Chromatogr A 847:109
39. Gupta AP, Mathur S, Gupta MM, Kumar S (1998) J Med Arom Plant Sci 20:1052
40. Bhandari P, Kumar N, Gupta AP, Singh B, Kaul VK (2006) Chromatographia 64:599
41. Murthy PBS, Raju VR, Ramakrisana T, Chakravarthy MS, Kumar VK, Kannababu S, Subbaraju V (2006) Chem Pharm Bull 54:907
42. Srinivasa H, Bagul MS, Padh H, Rajani M (2004) Chromatographia 60:131
43. Vikas Kumar (2006) Phytother Res 20:1023
44. Prakash JC, Sirsi M (1962) J Sci Industr Res 21C:93
45. Singh HK, Dhawan BN (1982) J Ethnopharmacol 5:205
46. Singh HK, Dhawan BN (1978) Indian J Pharmacol 10:72
47. Singh HK, Dhawan BN (1994) Pre-clinical Neuro-psychopharmacological Investigations on Bacosides: A Nootropic Memory Enhancer. Update Ayurveda-94, Bombay, India, 24–26th February 1994, pp 74
48. Singh HK, Dhawan BN (1997) Indian J Pharmacol 29:S359
49. Singh HK, Rastogi RP, Srimal RC, Dhawan BN (1988) Phytotherapy Res 2:70
50. Russo A, Borrelli F (2005) Phytomedicine 12:305
51. Shukla B, Khanna NK, Godhwani JL (1987) J Ethnopharmacol 21:65
52. Handa SS (1996) Rasaayana drugs. In: Handa SS, Kaul MK (eds) Supplement to Cultivation and Utilization of Medicinal Plants. RRL, Jammu-Tawi, pp 510–524
53. Das A, Shankar G, Nath C, Pal R, Singh S, Singh H (2002) Pharmacol Biochem Behav 73:893
54. Kishore K, Singh M (2005) Indian J Exp Biol 43:640
55. Aithal HN, Sirsi M (1961) Indian J Pharm 23:2
56. Ganguly DK, Malhotra CL (1967) Indian J Med Res 55:473
57. Jyoti A, Sharma D (2006) NeuroToxicology 27:451
58. Pal R, Sarin JPS (1992) Indian J Pharm Sci 54:17
59. Jyoti A, Sethi P, Sharma D (2007) J Ethnopharmcol 20:56
60. Tripathi YB, Chaurasia S, Tripathi E, Upadhyay A, Dubey GP (1996) Indian J Exp Biol 34:523
61. Bhattacharya SK, Bhattacharya A, Kumar A, Ghosal S (2000) Phytother Res 14:174
62. Singh HK, Srimal RC, Srivastava AK, Garg NK, Dhawan BN (1990) Proceedings of the Fourth Conference on the Neurobiology of Learning and Memory, California, 17–20 October, 1990. pp 79
63. Dhawan BN, Singh HK (1996) International Convention of Biological Psychiatry, 8–10 January, 1996, Bombay, India: pp 21, Abstr NR 59
64. Kar A, Panda S, Bharti S (2002) J Ethnopharmacol 81:281
65. Holcomb LA, Dhanasekaran M, Hitt AR, Yough AK, Riggs M, Manyam BV (2006) J Alzheimers Dis 9:243

66. Rao GMA, Karanth KS (1992) Fitoterapia 63: 399
67. Malhotra CL, Das PK, Dhalla NS (1960) Arch Int Pharmacodyn Ther 129:290
68. Malhotra CL, Prasad K, Dhalla NS, Das PK (1961) Indian J Pharm Pharmacol 13:447
69. Malhotra CL, Das PK (1959) Indian J Med Res 47:294
70. Khanna T, Ahmad B (1992) Proceedings of a Conference on Trends in Molecular Cellular Cardiology, Lucknow, 4–5 May, 1992. pp 20
71. Dhalla NS, Sastry MS, Malhotra CL (1961) Indian J Med Res 49:781
72. Das PK, Malhotra CL, Dhalla NS (1961) Indian J Physiol Pharmacol 5:136
73. Indurwade NH, Biyani KR (2000) Indian J Med Sci 54:339
74. Sumathi T, Govindasamy S, Balakrishna K, Veluchamy G (2002) Fitoterapia 73:381
75. Sumathi T, Nayeem M, Balakrishna K, Veluchamy G, Devaraj SN (2002) J Ethnopharmacol 82:75
76. Rai D, Bhatia G, Palit G, Pal R Singh, Singh HK (2003) Pharmacol Biochem Behav 75:823
77. Sheikh N, Ahmad A, Siripurapu KB, Kuchibhotla VK, Singh S, Palit G (2007) J Ethnopharmcol 111:671
78. Chowdhuri KD, Parmar D, Kakkar P, Shukla R, Seth PK, Srimal RC (2002) Phytother Res 16:639
79. Das A, Shankar G, Nath C, Pal R, Singh S, Singh H (2002) Pharmacol Biochem Behav 73:893
80. Anbarasi K, Vani G, Balakrishna K, Shyamala Devi CS (2005) J Biochem Mol Toxicol 19:59
81. Anbarasi K, Vani G, Shyamala Devi SC (2005) J Environ Pathol Toxicol Oncol 24:225
82. Anbarasi K, Kathirvel G, Vani G, Jayaraman G, Shyamala Devi S (2006) Neuroscience 138:1127
83. Anbarasi K, Vani G, Balakrishna K, Shyamala Devi CS (2005) Vascular Pharmacology 42:57
84. Anbarasi K, Vani G, Balakrishna K, Shyamala Devi CS (2006) Life Sciences 78:1378
85. Vijayan V, Helen A (2006) Phytother Res 21:378
86. Pawar R, Gopalakrishnan C, Bhutani KK (2001) Plant Med. 67:752
87. Russo A, Izzo AA, Borrelli F, Renis M, Vanell A (2003) Phytother Res 17:870
88. Sairam K, Rao CV, Babu MD, Goel RK (2001) Phytomedicine 8:423
89. Goel RK, Sairam K, Babu MD, Tavares IA, Raman A (2003) Phytomedicine 10:523
90. Dar A, Channa S (1997) Phytother Res 11:323
91. Dar A, Channa S (1997) Phytomedicine 4:319
92. Dar A, Channa S (1999) J Ethnopharmacol 66:167
93. Channa S, Dar A, Yaqoob, Anjum S, Sultani Z, Atta-ur-Rahman (2004) J Ethnopharmacol 86:27
94. Channa S, Dar A, Anjum S, Yaqoob M, Atta-ur-Rahman (2006) J Ethnopharmacol 104:286
95. Vohora SB, Khanna T, Athar M, Ahmad B (1997) Fitoterapia 68:361
96. Ravikumar S, Nazar S, Nuralshiefa A, Abideen S (2005) J Environ Biol 26:383
97. McLaughlin JL, Chang CJ, Smith DL (1992) Am Chem Soc Sympos Ser 534:114
98. McLaughlin JL, Rogers LL, Anderson JE (1998) Drug Inform J 32:513
99. D'Souza P, Deepak M, Rani P, Kadamboor S, Mathew A, Chandrashekar AP, Agarwal A (2002) Phytother Res 16:197
100. Bhakuni DS, Dhar ML, Dhar MM, Dhawan BN, Mehrotra BN (1969) Indian J Exp Biol 7:250
101. Elangovan V, Govindasamy S, Ramamoorthy N, Balasubramanian K (1995) Fitoterapia 66:211
102. Samiulla DS, Prashanth D, Amit A (2001) Fitoterapia 72:284
103. Sinha J, Raay B, Das N, Medda S, Garai S, Mahato SB, Basu MK (2002) Drug Delivery 9:55

104. Singh RH, Singh L (1980) J Res Ayur Siddha (Q) 1:133
105. Sharma R, Chaturvedi C, Tewari PV (1987) J Res Edu Indian Med 6:1
106. Roodenrys S, Booth D, Bulzomi S, Phipps A, Micallef C, Smoker J (2002) Neuropsychopharmacology 27:279
107. Nathan PJ, Clarke J, Lloyd J, Hutchison CW, Downey L, Stough C (2001) Hum Psychopharmacol Clin Exp 16:348
108. Maher BFG, Stough C, Shelmerdine A, Wesnes K, Nathan PJ (2002) Hum Psychopharmacol Clin Exp 17:163
109. Stough C, Lloyd J, Clarke J, Downey LA, Hutchison CW, Rodgers T, Nathan PJ (2001) Psychopharmacology 156:481
110. Nathan PJ, Tanner S, Lloyd J, Harrison B, Curran L, Oliver C, Stough C (2004) Hum Psychopharmacol Clin Exp 19:91
111. Darokar PM, Khanuja PS, Shasany AK, Kumar S (2001) Genet Resour Crop Ev 48:555
112. Ali G, Purohit M, Mughal MH, Iqbal M, Srivastava PS (1996) Plant Tiss Cult Biotech 2:208
113. Shrivastava N, Rajani M (1999) Plant Cell Rep 18:919
114. Tiwari V, Singh BD, Tiwari KN (1998) Plant Cell Rep 17:538
115. Tiwari V, Tiwari KN, Singh BD (2001) Plant Cell Tiss Org Cult 66:9
116. Tiwari V, Tiwari KN, Singh BD (2006) Plant Cell Rep 25:629
117. Mohapatra HP, Rath PS (2005) Indian J Exp Biol 43:373
118. Rahman LU, Verma PC, Singh D, Gupta MM, Banerjee S (2002) Biotech Lett 24:1427
119. Ali G, Srivastava PS, Iqbal M (1998) Biol Plantarum 41:35
120. Ali G, Srivastava PS, Iqbal M (2000) Biol Plantarum 43:599
121. Ali G, Srivastava PS, Iqbal M (2001) Bull Environ Contam Toxicol 66:342
122. Singh S, Eapen S, D'Souza SF (2006) Chemosphere 62:233
123. Mishra S, Srivastava S, Tripathi RD, Govindarajan R, Kuriakose SV, Prasad MNV (2006) Plant Phyisol Biochem 44:25
124. Ali G, Srivastava PS, Iqbal M (1999) Biol Plantarum 42:89 Escandon AS, Hagiwara JC, Alderete LM (2006) Electronic J Biotechnol 9(3 special issue):181

Chapter 10
Chemical Profiling of *Nothapodytes nimmoniana* for Camptothecin, an Important Anticancer Alkaloid: Towards the Development of a Sustainable Production System

R. Uma Shaanker[1,2,3,6] (✉), B.T. Ramesha[1,2], G. Ravikanth[2,3], R.P. Gunaga[4], R. Vasudeva[3,4] and K.N. Ganeshaiah[2,3,5,6]

[1]Department of Crop Physiology, University of Agricultural Sciences, GKVK Campus, Bangalore 560065, India, e-mail: rus@vsnl.com

[2]School of Ecology and Conservation, University of Agricultural Sciences, GKVK Campus, Bangalore 560065, India

[3]Ashoka Trust for Research in Ecology and the Environment, #659, 5th A Main, Hebbal, Bangalore 560024, India

[4]Department of Forest Biology, College of Forestry, Sirsi 581401, India

[5]Department of Genetics and Plant Breeding, University of Agricultural Sciences, GKVK Campus, Bangalore 560065, India

[6]Jawaharlal Nehru Centre for Advanced Scientific Research, Jakkur, Bangalore 560 065, India

Abstract Camptothecin (CPT), a pyrrolo quinoline alkaloid, is one of the most promising anticancer drugs of the 21st century. The compound was first isolated from the Chinese deciduous tree, *Camptotheca acuminata*. CPT exhibits a broad spectrum of antitumor activity both under *in vitro* and *in vivo* conditions. Irinotecan (CPT11) and Topotecan (TPT), two water-soluble derivatives of CPT, have been approved by the United States Food and Drug Administration for treating colorectal and ovarian cancers as well as against several types of brain tumor in children. Although CPT has been reported to exist in several species, the highest concentration (about 0.3%) to date has been realized from *Nothapodytes nimmoniana*. The tree commonly referred to as "stinking tree" is native to warmer regions of South India. In the last few decades, driven by the enormous demand for CPT, there has been a decline of at least 20% in the population, leading to red listing of the species. In recent years, efforts have been initiated in India to identify high-yielding individuals and populations of *N. nimmoniana* in its natural distribution range with the ultimate aim of using these lines to develop clonal orchards, as well as in developing *in vitro* production systems. In this chapter, we briefly review the overall status of *N. nimmoniana* as a source of CPT. Drawing upon existing literature as well as ongoing work at our laboratory, we discuss the basic patterns of accumulation of CPT in *N. nimmoniana*. We review the population variability for CPT accumulation along the distributional range of the species in the Western Ghats, India. Using a relatively new tool, namely the ecological niche model, we predict the chemical hot-spots of the species in the Western Ghats and offer a test of this prediction. Finally, we discuss strategies for a sustainable model of extraction of CPT from *N. nimmoniana*.

Keywords Camptothecin, Chemical profiling, HPLC, DIVA-GIS, *Nothapodytes nimmoniana*, Western Ghats

Abbreviations

CPT	Camptothecin
FDA	United States Food and Drug Administration
GIS	Geographic Information System

10.1 Introduction

Nature has been recognized as a rich source of medicinal compounds for hundreds of years. Today, a vast range of drugs that represent the cornerstones of modern pharmaceutical care are either natural products or have been derived from them [1]. It is estimated that over 50% of all drugs (and their derivatives and analogs) in clinical use are higher-plant-derived, natural products [2]. According to the World Health Organization, about 80% of the people in developing countries still rely on traditional medicine for their primary health care, and about 85% of such medicines involve the use of plant extracts. In other words, an incredibly large number of people (about 3.5–4 billion) in the world rely on plants as source of drugs [1, 3].

In recent years, with the advent of newer tools, including high-throughput screening for bioactive molecules, there is a resurgence of interest in mining higher plants for a variety of metabolites. In fact, nowhere has the effort been more pronounced than in the National Cancer Institute (USA), which has screened over 435,000 plants for antineoplastic effects [4]. Plant-based natural products have played a significant role in the development of contemporary cancer chemotherapy. Several novel antitumor compounds, including taxols, camptothecin (CPT) and its derivatives, maytansine, tripdiolide, homoharringtonine, vinblastine, vincristine, indicine-N-Oxide, baccharin, podophyllotoxin derivatives, and etoposide, are being extracted from plant sources [5,

6]. Considering the enormous significance that these compounds hold, several laboratories worldwide have been striving to intensively mine such compounds and standardize methodologies for their large-scale production.

Among the plant-derived compounds, CPT, a pyrrolo quinoline alkaloid, has been used extensively as a novel antitumor compound. CPT is lauded as one of the most promising anticancer drugs of the 21st century [7]. CPT exhibits a broad spectrum of antitumor activities under both *in vitro* and *in vivo* conditions [8]. CPT and its analogs in the presence of topoisomerase-I produce DNA damage by binding to and stabilizing a covalent DNA-topoisomerase-I complex in which one strand of DNA is broken [5–7, 9–11].

Irinotecan (CPT11) and Topotecan (TPT), two water-soluble derivatives of CPT, have been approved by the United States Food and Drug Administration (FDA) for treating small-cell lung cancer, colorectal cancer, and ovarian cancer [11–17]. They have also been approved by the FDA for the treatment of acquired immune deficiency syndrome [18].

CPT was first discovered in the Chinese deciduous tree, *Camptotheca acuminata* (Nyssaceae) [19]. The other plant species from which CPT is isolated are *Merriliodendron megacarpum* [20] and *Nothapodytes nimmoniana* Graham [21], both belonging to the family Icacinaceae, *Ophirrohiza mugos* [22] and *O. pumila* [23] from the family Rubiaceae, *Eravatamia heyneana* [24], belonging to Apocynaceae, and *Mostuea brunonis* [25], belonging to the family Loganiaceae (Table 10.1). However to date, the highest content of CPT has been realized from *N. nimmoniana* (about 0.3% on a dry weight basis) [21].

The market demand for Irinotecan and Topotecan has been ever increasing and has currently reached approximately US$ 1000,000,000, which represents approximately 1 tonne of CPT in terms of natural material [1, 26]. Most of this demand is currently met from plantations of *C. acuminata* that have been established extensively in China. In India, however, *N. nimmoniana* remains the main source of CPT. Based on current market price, it is estimated that the *N. nimmoniana* available along the northern part of the Western Ghats *per se* is worth over US$ 350,000,000 (Ganeshaiah and Uma Shaanker, unpublished data). While official records are not available, it is reliably learnt that the tree is extensively harvested from the Western Ghats and the billets exported for commercial extraction of CPT. In fact, it is estimated that in the last decade alone, there has been at least a 20% decline in the population, leading to the red listing of the species [27, 28]. Indiscriminate felling of trees for short-term gains could perhaps lead to the loss of elite individuals and populations that could otherwise potentially serve as sources of high CPT.

With no synthetic source of this alkaloid and with an increasing global demand, it has become imperative that the demand for CPT is met from a sustainable supply rather than the current destructive harvesting. Among the various approaches, prospecting for populations and/or individuals of the species for higher yields of the alkaloid could potentially help in establishing high-yielding clonal orchards and in developing *in vitro* production systems, thereby relieving the pressure on natural populations. Toward this end, in recent years,

Table 10.1 Camptothecin (*CPT*; % dry weight) content in different plant species and tissues (Adapted from Ramesha et al., unpublished data). *HPLC* High-performance liquid chromatography, *DAD* diode array detection, *ESI* electrospray ionization, *UV* ultraviolet, *IR* infrared, *NMR* nuclear magnetic resonance, *MS* mass spectrometry

Plant species	Tissue analyzed	CPT (% dry weight)	References	Chromatographic analysis
Camptotheca acuminata	Young leaves	0.4–0.5	[51]	HPLC
	Seeds	0.30	[51]	HPLC
	Bark	0.18–0.2	[51]	HPLC
	Young leaves	0.24–0.30	[52]	HPLC
	Hairy roots	0.1	[11]	HPLC
	Callus	0.20–0.23	[53]	HPLC
Camptotheca lowreyana	Young leaves	0.39–0.55	[52]	HPLC
	Old leaves	0.09–0.11	[52]	HPLC
Camptotheca yunnanensis	Young leaves	0.25–0.44	[52]	HPLC
	Old leaves	0.059	[52]	
Ervatamia heyneana	Wood and stem bark	0.13	[24]	HPLC
Merriliodendron megacarpum	Leaves and stem	0.053	[20]	HPLC
Ophiorrhiza pumila	Young roots	0.1	[54]	HPLC
	Hairy roots	0.1	[54]	
Ophiorrhiza mungos	Whole plant	0.0012	[22]	HPLC
Ophiorrhiza rugosa	Albino plants	0.1	[55]	HPLC
	Normal plant grown *in vitro*	0.03	[55]	HPLC
Mostuea brunonis	Whole plant	0.01	[25]	HPLC
Pyrenacantha klaineana	Stems	0.0048	[56]	HPLC
Nothapodytes foetida	Stem wood	0.14–0.24	[57]	HPLC
	Shoot	0.075	[34]	HPLC
	Plant	0.048	[58]	HPLC-DAD-ESI
Nothapodytes nimmoniana	Stem bark	0.3	[21]	UV, IR, NMR and MS
	Leaves	0.081	[33]	HPLC
	Stem bark	0.236	[33]	HPLC
	Root bark	0.333–0.775	[33]	HPLC
	Stem wood	0.14	[33]	HPLC
	Root wood	0.18	[33]	HPLC

attempts have been made to chemically characterize populations of *N. nimmoniana* along the distributional range of the species in the Western Ghats, India, with the ultimate aim of identifying populations/individuals with high CPT yields. High-yielding sources can be used to produce material for clonal multiplication and to develop cell lines with a high CPT yield. Attempts have also been made to identify the ecological correlates of CPT accumulation and identify the ecological niche of *N. nimmoniana* in Western Ghats and identify the "hot-spots" of CPT accumulation. The latter can guide collection of accessions for conservation as well as for use in sustainable models of extraction of CPT.

In this chapter, we briefly review the overall status of *N. nimmoniana* as a source of CPT. We draw upon existing literature as well as ongoing work in our laboratory to review the population variability for CPT along the distributional range of the species in the Western Ghats, India. We examine whether the population variability reflects intrinsic genetic variations or whether they are related to the ecological correlate of the populations. Using a relatively new Geographic Information System (GIS) technique, namely the ecological niche modeling tool, we predict the chemical hot spots of the species in the Western Ghats and offer a test for this prediction. Finally, we discuss a possible sustainable model of extraction of CPT from *N. nimmoniana*.

10.2 *N. nimmoniana:* Ecology and Distribution

N. nimmoniana, Graham, formerly known as *N. foetida* Sleumer and *Mappia foetida* Meirs, is a small tree belonging to the family Icacinaceae (Fig. 10.1). The genus *Nothapodytes* includes *N. obtusifolia*, found in China, *N. montana*, distributed in Thailand, north-eastern Sumatra, western Java, and western Sumbava, and *N. pittosporoides*, distributed in China and Indonesia [29]. It

Fig. 10.1 *Nothapodytes nimmoniana* (Photograph courtesy of Dr. G. Ravikanth)

has also been reported in Taiwan [30]. *N. nimmoniana*, commonly referred to as "Stinking Tree," is native to warmer regions of South India. It is reported in the western parts of the Deccan peninsula, North Bengal, and Assam [31] (Fig. 10.2). The tree is distributed in the shola forests in Nilgiris and present in both the Western Ghats and the Eastern plateau. It is also distributed in Sri Lanka, Myanmar, Indonesia, and Thailand [27].

The species exhibits a wide array of breeding systems including male, female, hermaphrodite, monoecious, andromonoecious, gynomonoecious, and trimonoecious individuals [27]. The trees flower during July–August, and most of the early flowering trees are dioecious, whereas late flowering trees are monoecious, hermaphrodite, and a mixture of other breeding types [27]. The fruits ripen during November–December and germinate during May–June after the onset of monsoon rainfall.

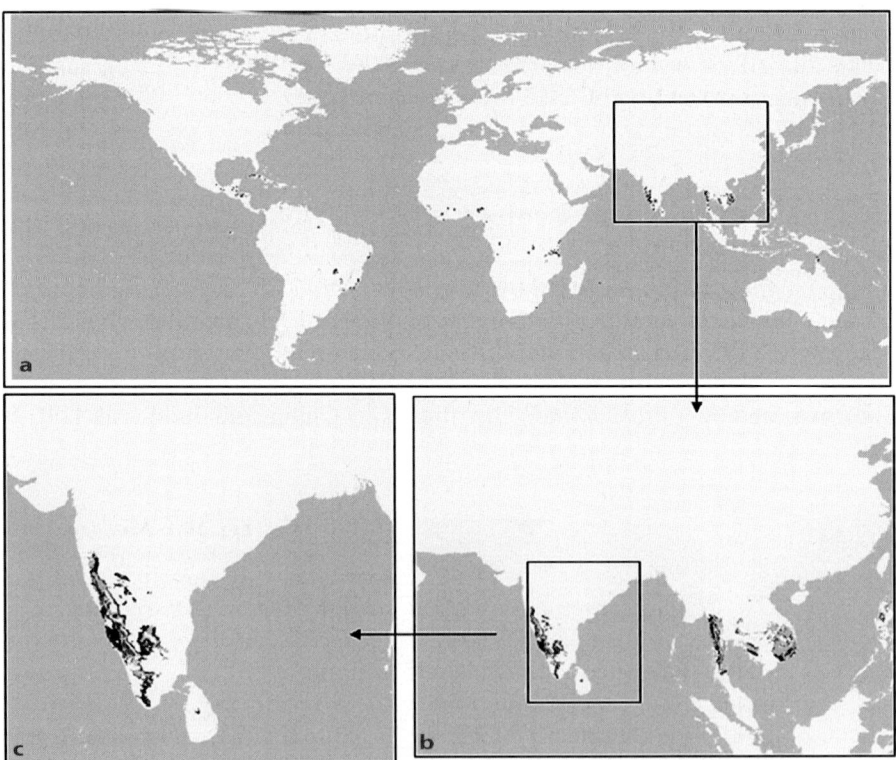

Fig. 10.2 Predicted distribtution of *N. Nimmoniana* in (**a**) the world (**b**) south-east Asia and (**c**) in the Western Ghat region of South India

10.3 Basic Patterns of Accumulation of CPT in *N. nimmoniana*

Although CPT has been reported in over nine species, the basic patterns of accumulation of CPT have been well documented only in *C. acuminata* and, to a lesser extent, in *N. nimmoniana*. For example, Yan et al. [6] reported the highest levels of CPT in leaves of *Camptotheca. acuminata*. CPT content was at least tenfold higher in younger than older leaves [6]. In fact, the high concentration of CPT in leaves has reportedly led to the poisoning of goats that browse on the leaves and even the honey bees foraging on the floral rewards [6]. Although the precise mechanism of transport and storage of CPT is not yet fully understood, it is conjectured that CPT is synthesized in the leaves and sequestered in old and dead tissues [6]. At the cellular level, CPT is localized in mesophyll and subpalisade layers of young leaves [6]. It has also been reported to be localized in vacuoles of young and older leaves [32].

The basic patterns of accumulation of CPT in *N. nimmoniana* have been characterized with respect to the age and sex of the plant and plant parts [33]. Among the various plant parts, the inner root bark is reported to yield the highest CPT content, followed by the inner stem bark. The average CPT content in the inner root bark is about $0.33\pm0.21\%$, compared to $0.23\pm0.15\%$ in inner stem bark (Fig. 10.3). The CPT content in the root and stem wood is significantly lower than that of the respective inner bark tissue. While the root wood contained $0.18\pm0.09\%$, the stem wood contained only $0.14\pm0.12\%$ CPT. Seeds on an average contained only about 0.17% CPT [33]. The CPT content in 2-year-old seedlings was highest in the root tips (0.4%), followed by leaves and stem (0.2%). The CPT content did not differ between the old and the young leaves. There was no difference in the CPT content between the sexes [33]. These studies reaffirm the earlier findings of Govindachari and Vishwanathan [21], who reported the highest yields of CPT from roots of *N. nimmoniana*. In

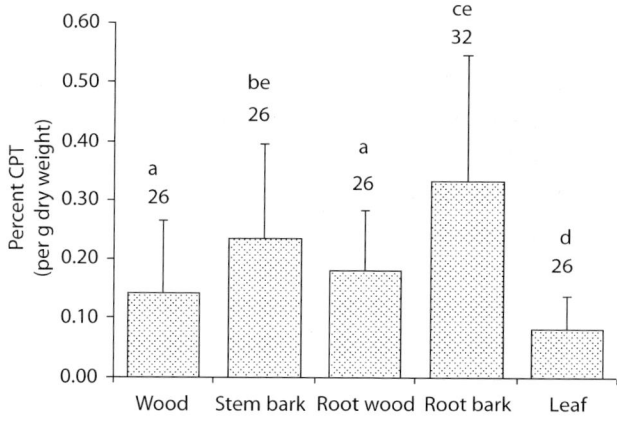

Fig. 10.3 Mean percent camptothecin (CPT) per gram dry weight in different tissues of *N.nimmoniana*. The numbers on the histograms indicate number of trees used in the analysis. Respective histograms with dissimilar letters indicate a significant difference in CPT content (t-test $p<0.05$; redrawn from Padmanabha et al. [3])

fact traditionally, CPT has been extracted from root, root bark, and fruits [34]. Fairly good amounts (0.10%) of the alkaloid have also been reported from seeds [34]. Quite obviously, because of the relatively low levels of CPT in leaves, extraction of CPT from these trees has been mostly destructive, involving the felling of the trees.

10.4 Chemical Profiling of Populations of *N. nimmoniana* for CPT

While *N. nimmoniana* forms one of the richest sources of CPT, commercial production of the alkaloid is still limited for want of high-yielding lines. Prospecting for high-yielding individuals or populations across the distributional range of the species could help in using the identified high-yielding lines for clonal multiplication and commercial production of CPT. Toward this end, recently, Suhas et al. [35] chemically profiled populations of *N. nimmoniana* along the Western Ghats, a mountain chain running parallel to the west coast of south India, and considered as one of the 34 mega-biodiversity hot spots of the world [36]. Based on primary and secondary data sources, the occurrence of the species in the Western Ghats was digitized on a GIS platform. While the species occurs along the length of the Western Ghats, it is clear that the distribution is not uniform; certain parts, namely the southern and central Western Ghats have a greater density of records of distribution. Based on the relative distribution, Suhas et al. [35] analyzed 11 populations from 8° to 15° N latitude (Fig. 10.4). For each of the 11 populations, 10–15 trees were sampled randomly and the CPT estimated in the inner stem and root bark tissues, respectively.

Significant variation exists among populations with respect to their mean CPT content, both in stem bark (one-way ANOVA, $P<0.004$) and root bark ($P<0.001$). The levels of CPT in stem bark ranged from as low as 0.03% to as high as 2.7%, with an overall mean of 0.7%. The mean CPT content in the root bark ranged from 0.003 to 1.41%, with an overall mean of 0.48%. The northern Kerala populations had the highest CPT content both in their stem bark ($1.10\pm0.462\%$) and root bark ($0.93\pm0.359\%$). The CPT content of stem bark was significantly positively correlated with that of the respective root bark ($n=126$; $r=0.320$, $P<0.05$). Finally, the frequency distribution of CPT content over all populations was highly positively skewed (Fig. 10.5).

Suhas et al. [35] found no clear relationship between CPT content and the girth size of trees. In 7 of the 11 populations, there was no relationship; however, in 3 of the remaining 4 there was a significant positive relationship ($r=0.678$, $r=0.762$, $r=0.728$; all $P<0.05$), and in 1 it was negatively related ($r=-0.728$; $P<0.05$). Thus, the differences in CPT content among populations and individuals could not be attributed to age or size class differences. Suhas et al. [35] also showed that even after normalizing for girth differences among the trees, if any, the CPT content expressed as CPT/girth was significantly different among the populations ($P=0.0016$). The mean CPT content of populations was not correlated with latitude, longitude or altitude of their occurrence and collection.

Chapter 10   Chemical Profiling for camptothecin   205

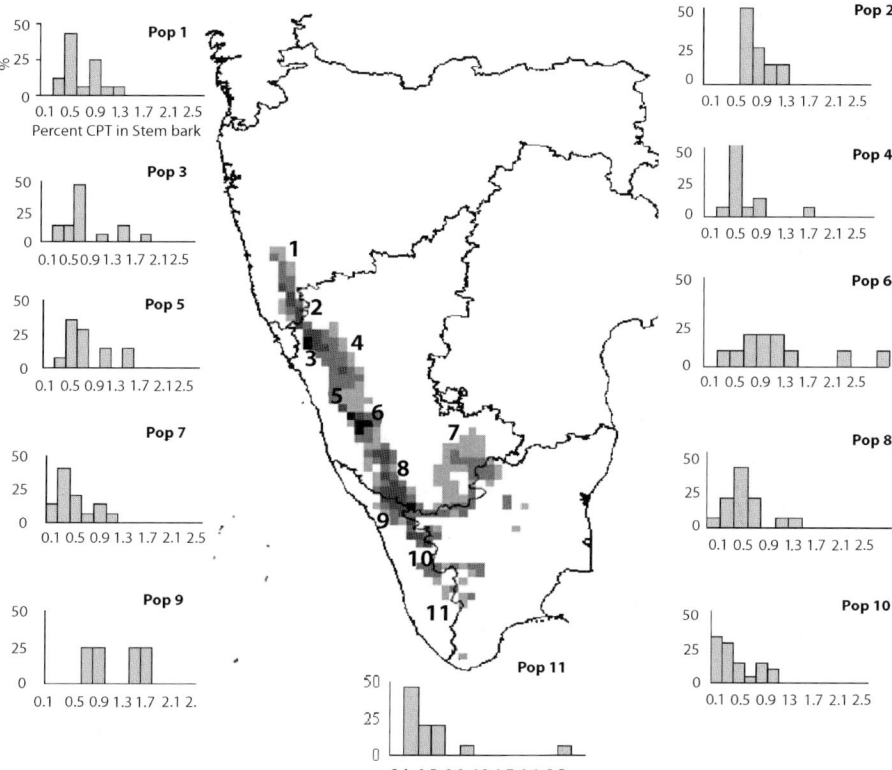

Fig. 10.4 Density distribution map and frequency distribution of percent CPT in the stem bark of *N. nimmoniana* in 11 different populations in the Western Ghats, India. The density distribution map was developed based on 64 points of occurrence of the species using a Geographic Information System (GIS) platform. The different shades of gray indicate the relative concentration of records of the species in the Western Ghats (light to dark indicating increasing concentration). The classification of the latitudinal gradient of the species into the different zones is purely for the purpose of discussion in the text (*x*-axis: percent CPT; *y*-axis: frequency of individuals; adapted from Suhas et al. [35])

CPT content in *C. acuminata* was found to vary significantly across latitude [37, 38]. In a more recent analysis, Ramesha et al. (unpublished data) conducted a forward, step-wise regression for CPT content using 19 climatic variables averaged over 30 years for each of the collection sites. Only two variables, namely mean temperature of the driest and wettest quarters of the year, significantly explained the differences in stem bark CPT among the populations; for root bark CPT, only one variable, namely the mean monthly temperature, was significant. Similar studies conducted in *C. acuminata* showed that CPT content varied significantly with several environmental variables such as temperature, evaporation capacity, and precipitation. Low temperature and precipitation was found to increase the CPT content [39].

The studies of Suhas et al. [35] provide one of the most exhaustive chemical screening of *N. nimmoniana* for CPT. The study assumes significance in

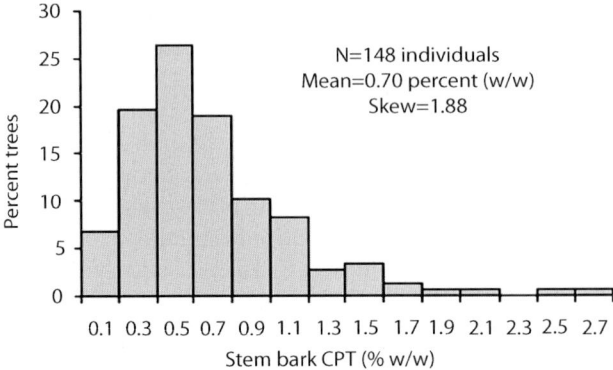
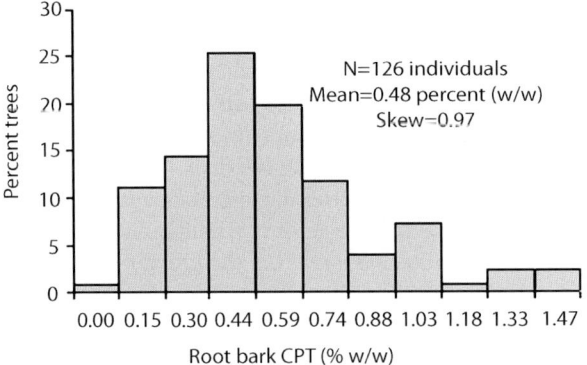

Fig. 10.5 Frequency distribution of CPT (%, w/w) in stem bark (3a) and root bark (3b) of *Nothapodytes nimmoniana* (Adapted from Suhas et al. [35])

that it is perhaps the first to report at least five- to eightfold more CPT in *N. nimmoniana* than has hitherto been reported. Of the 148 individuals assayed, 23 yielded more than 1% CPT. These estimates are nearly three- to eightfold more than what has been reported hitherto in the literature [21]. The study has demonstrated a significant population level variation in CPT content – a tool kit that can be exploited for developing clonally multiplied material from the identified high-yielding populations. While it will be important to examine whether these differences reflect the intrinsic genetic predisposition of populations to synthesize and accumulate CPT, preliminary analyses do indicate a genetic basis. Clearly, more studies will be required to examine this issue critically. Populations in the northern Western Ghats had the highest mean CPT and least intrapopulation variation, based on the analysis of both the stem and root bark. These populations could be important source material for developing high-yielding clonal materials. Furthermore, it will be important to study the heritability of the accumulation patterns across generations by analyzing the parent–offspring regression in the accumulation of CPT. It would be interesting to investigate the proximate/ultimate reasons for the enormously high levels of CPT produced by these trees, as a first step towards domesticating the species for obtaining high CPT yields.

10.5 Modeling Habitat Suitability for CPT Production

One of the key challenges in prospecting for high-yielding sources of specific plant metabolites is to develop algorithms or approaches that can help predict hot spots of distribution of the metabolite. Prediction of hot spots and its subsequent validation not only helps to focus efforts in collecting material from such sites, but also serves to prioritize sites at which plants can be domesticated or conserved. Unfortunately, few studies have seriously modeled the conditions that might help predict the spatial distribution of metabolites.

Recently, a GIS-based approach called the ecological niche model has been used to model the spatial distribution of a given species and offer predictions on the habitat suitability of the species. Using specific algorithms, the model iteratively identifies habitats over a landscape that match best the climatic variables corresponding to sites of known occurrence of the species. Accordingly, habitats are classified from those that are highly suitable (highest match) to those that are not suitable (least match) for the potential occurrence or invasion of the species [40, 41]. The ecological niche models have been used successfully in a variety of scenarios, including in locating rare and threatened species, and in rationalizing the choice of habitats for species reintroduction [42, 43].

At our laboratory, attempts have been made to extend the use of ecological niche modeling tools to offer predictions on the spatial distribution of plant metabolites. An underlying assumption of this application is that, plants would be selected to accumulate secondary metabolites at sites predicted to be highly suitable for the given species compared to sites that are predicted to be unsuitable. Thus, one would expect that a phytochemical such as santalols is best produced in sites suitable for the growth of sandal trees and not in those that are predicted to be unsuitable. Recently, Prakash Kumar [44] modeled the distribution of *Withania somnifera* in south India and showed that individuals in sites predicted to be highly suitable accumulated higher levels of withaferin-A and withanolide-A compared to individuals in sites that were predicted to be unsuitable or poorly suitable.

Figure 10.6 shows the predicted habitat suitability for *N. nimmoniana* in the Western Ghats. It is evident that not all regions in the Western Ghats are uniformly suitable for the species. In fact, within the Western Ghats, certain areas (Fig. 10.6, dark areas) are highly suitable and others (Fig. 10.6, grey areas) are unsuitable. In fact, two distinct sites in the central and northern Western Ghats are predicted to be excellent in their match to the habitat requirements of the species. Analysis of the CPT content of individuals occurring in the different habitat suitability areas indicated that individuals in highly suitable areas accumulated significantly higher levels of CPT compared to those that occurred in unsuitable or poorly suitable areas (Fig. 10.7). Furthermore, over 60% of the trees that accumulated greater that 1% CPT were all from regions predicted to be highly suitable (Fig. 10.8). In summary, these results have demonstrated for the first time the utility of the ecological niche models in predicting the spatial richness of plant metabolites, and hold several important implications.

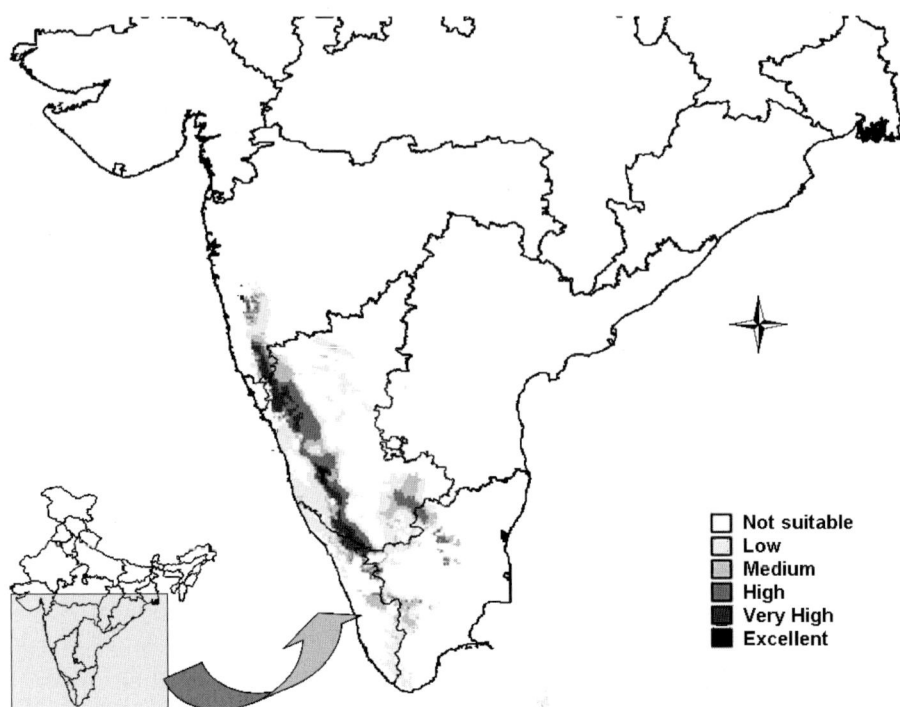

Fig. 10.6 Predicted habitat suitability map of *Nothapodytes nimmoniana* in Western Ghats, India. The different shades of grey indicate different habitat suitability categories as given in the legend

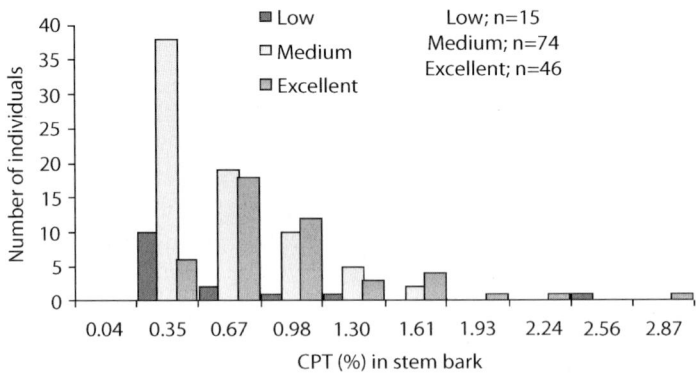

Fig. 10.7 Frequency distribution of CPT (%, w/w) in the stem bark of *Nothapodytes nimmoniana* individuals from different habitat suitability categories (Kolmogorov-Smirnov test low vs. excellent, $D_{max}=0.53$, $p=0.001$)

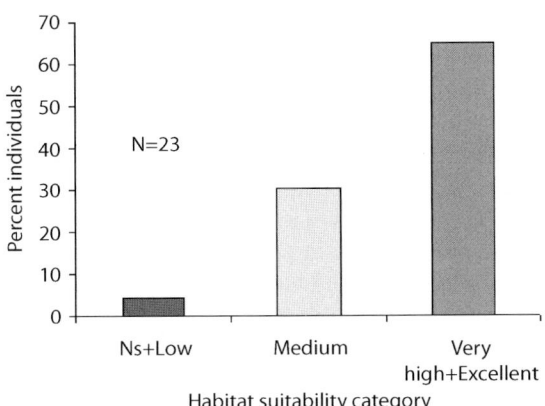

Fig. 10.8 Graph showing percentage of individuals of *Nothapodytes nimmoniana* (with CPT more than 1% in the stem bark) grouped into different habitat suitability categories (NS = not suitable)

For instance, the results raise interesting prospects for further research on how the habitat suitability or otherwise can influence the accumulation of a secondary metabolite. Do ecologically good habitats serve as areas in which the species are genetically predisposed to synthesizing secondary metabolites and other defense compounds that can lead to a potentially higher fitness of the populations? The outputs of the ecological niche model provide a powerful handle and direction to further explore newer populations of *N. nimmoniana* in areas/regions that have not yet been sampled from in search for higher CPT yields. The results have important implications for intelligent prospecting for economically important secondary metabolites such as CPT.

10.6 Development of a Sustainable Extraction Approach

Realizing the ever increasing demand for CPT, plantations of *C. acuminata* have been established in China since 1993 to supply material for CPT extraction [37]. In India, the major source of CPT continues to be *N. nimmoniana*. However, since there are no commercial plantations, all of the demand is sourced from the trees extracted destructively from the natural populations of *N. nimmoniana*. This of course is not sustainable in the long run because of loss of standing populations of the trees in the distributional range of the species. While no detailed inventory of the distribution and abundance of the tree is available in the Western Ghats, it is conjectured that the estimated demand may not be met solely by sourcing trees from their natural populations. Clearly, strategies need to be developed that can ensure a sustained supply of CPT from *N. nimmoniana*.

Several approaches could be deployed to ensure the sustainable extraction of *N. nimmoniana*. For example, establishment of captive plantations using clonally multiplied material from high-yielding lines could greatly contribute to the rising demand for the compound without jeopardizing the naturally occurring populations. In fact, towards this end, as mentioned elsewhere in this chapter, efforts are being made to identify high-yielding lines and populations from the distributional range of the species in India. The recent discovery of population variability for CPT accumulation in *N. nimmoniana* holds immense promise in developing high-yielding clonal orchards and other captive plantations [35].

Extraction of renewable plant parts such as leaves and fruits instead of bark could be one of the possible approaches to sustain the extraction. However, because of the extremely low levels of CPT in the leaves and fruits of trees, this is not economically attractive. However, more recently, Santosh et al. (unpublished data) demonstrated that the CPT content of leaves could be strongly related to the age of the plant, just as was shown for *C. acuminata* [37]. Thus, the leaves of seedlings could accumulate relatively higher levels of CPT than those of juveniles and adults. It is estimated that 1-year-old seedlings producing about 15–20 g of leaf biomass can easily yield about 50 mg of CPT (Santosh et al., unpublished data). Although the economics of extraction need to be analyzed further, it appears that by simple ratooning of the seedling crop once a year, CPT could be extracted on a sustainable basis.

Sustainable extraction of CPT could also be arrived by exploring several *in vitro* production systems. For example, stabilization of cell cultures from high-yielding individuals could help develop *in vitro* production systems. However the economic viability of this approach will depend upon optimizing several protocols, including the sustained growth of the cell culture, *in vitro* elicitation of CPT, and CPT yields. A more recent and exciting possibility has emerged from the discovery that CPT is also produced by an endophytic fungus, *Enterophospora infrequens*, which is associated with *N. nimmoniana* [45]. Cultivation of the fungus and optimizing production of CPT in a liquid culture system could potentially lead to an economically viable and ecologically sustainable model of supply of CPT.

10.7 Conclusions

Plants have been a major source of pharmaceutically important compounds worldwide. It is estimated that even today, 11 of the top 20 best-selling drugs are being derived from plants. In most cases, and particularly in the biodiversity rich regions of the world, the plants are nearly entirely sourced from the wild. For example, in India, of about 880 medicinal plants that are traded for various uses, 538 (61%) are sourced from the wild only [46]. Indiscriminate harvesting of these species has already led to a serious threat to these species. It is estimated that about 100 species (58 of which are globally threatened) of medicinal plants in the Western Ghats, a mega-biodiversity center in south India, might already be highly threatened due to excessive harvesting [28]. Thus, unless alternative approaches are developed, the extraction of most of the medicinal plants to meet global demands will be unsustainable. The extraction of CPT from natural populations of *N. nimmoniana* is a case in point. As mentioned elsewhere, due to increasing harvesting pressures from the wild, this species is already listed as endangered [28]. If global demands for CPT are to be met, it is essential to develop sustainable models of extraction, including developing clonal orchards for high-yielding elite lines, using alternate but renewable biomass resources such as leaves and fruits, and finally in developing *in vitro* production systems. Toward this end, as described in this chapter, several initiatives could be taken that explore all possibilities to meet the global demand sustainably. With the identification of high-yielding populations of *N. nimmoniana* in the Western Ghats, it is possible to develop high-yielding clonal orchards and establish captive plantations to meet the demand outside of the natural populations. These "elite" trees could also be focused toward deriving tissue material for *in vitro* production systems, as was done for several other systems, for example taxane from *Taxus wallichiana* [47, 48] and for podophyllotoxin from *Podophyllum peltatum* [49, 50].

Acknowledgments

The authors wish to thank the Department of Biotechnology, Government of India for supporting the program and the Forest Departments of the states of Karnataka, Kerala and the Maharastra for facilitating the fieldwork in the Western Ghats.

References

1. Raskin I, Ribnicky DM, Momarnytsky S, Ilic N, Poulev A, Borisjuk N, Brinker A, Moreno DA, Ripoll C, Yakoby N, O'Neal JM, Cornwell T, Pastor I, Fridlender B, (2002) Trends in Biotechnology 20:522
2. Farnsworth NR, Morris RW (1976) Am J Pharm Educ 148:46
3. Fransworth NR (1988) Biodiversity Wilson 24:83
4. Daniel SF, Fransworth NR (2001) Environmental Health Perspective 109:69
5. Nalawade SM, Abhay PS, Chen YL, Chao LK, Hsin ST (2003) Bot Bull Acad Sin 44:79
6. Yan XF, Yang W, Yu T, Zhang Y, Dai (2003) Bot Bull Acad Sin 44:99
7. Li SY, Adair KT (1994) *Camptotheca acuminata* Decaisne Xi Shu (Chinese Happy Tree) – A Promising Anti-tumour and Antiviral Tree for the 21st Century (ed. Stephen, FA, Henry M) Rockwell Monograph, Austin State University, Texas, 1994
8. Vladu B, Woynarowski J, Manikumar G, Mansukhlal C, Wani MEW, Daniel D, Von Hoff, Wadkins RM (2000) Molecular Pharmacology 57:243
9. Wall ME, Wani MC, Cook CE, Palmer KH, McPhail AT, Sim GA (1966) J Am Chem Soc 88:3888
10. Cragg GM, Boyd MR, Khanna RR, Kneller R, Mays TD, Mazan KD, Newman DJ, Sausville EA (1999) Pure Appl Chem 71:1619
11. Lorence A, Craig LN (2004) Phytochemistry 65:2731
12. Masuda N, Fukuoka M, Kusunoki Y (1992) J Clin Oncol 10:1225

13. Abigers D, Catimel G, Kozimor R (1995) Ann Oncol 6:141
14. Lilenbaum RC, Ratain MJ, Miller AA, Hargis JB, Hollis DR, Rosner GL, O'Brien SM, Brewster L, Green MR, Schilsk RL (1995) J Clin Oncol 13:2230
15. Romanelli SP, Perego G, Pratesi N, Carenini M, Tortoreto ZF (1998) Cancer Chemother Pharmacol 41:385
16. Clements MK, Jones CB, Cumming M, Doud SS (1999) Cancer Chemother Pharmacol 44:411
17. Giovanella BC, Steblin JS, Wall ME, Wani MC, Nicholas AW, Liu LF, Siber R, Potmesil M (1989) Science 246:1046
18. Priel E, Showalter SD, Blair DG (1991) AIDS Res. Hum. Retroviruses 7:65
19. Wall ME, Wani MC (1968) J Org Chem 34:1364
20. Arisawa M, Gunasekera SP, Cordell GA, Farnsworth NR (1981) Planta Med 43:404
21. Govindachari TR, Viwanathan N (1972) Phytochemistry 11:3529
22. Tafur S, Nelson JD, DeLong DC, Svoboda GH (1976) Lloydia 39:261
23. Aimi, Hoshino NH, Nishimura M, Sakai S, Haginiwa J (1990) Tetrahedron Lett 31:5169
24. Gunashekera SP, Badawi MM, Cordell GA, Farnsworth NR, Chitnis M (1979) J Nat Prod 42:475
25. Dai JR, Cardellina JH, Boyd MR (1999) J Nat Prod 62:1427
26. Watase I, Hiroshi S, Mami Y, Kazuki S (2004) Plant Biotechnology 21:337
27. Hombegowda HC, Vasudeva R, Georgi PM, Uma Shaanker R, Ganeshaiah KN (2002) Curr Sci 83:1077
28. Ravi Kumar, Ved DK (2000) (eds) 100 Red listed Medicinal Plants of Conservation Concern in Southern India, Foundation for revitalization of local Health Traditions, Bangalore
29. Ku KY, Tang TC (1980) Zhing Cao Yao 11:476
30. Wu TS, Leu YL, Hsu HC, Ou LF, Chen CC, Chen C, Ou JC, Wu YC (1995) Phytochemistry 39:383
31. Biswanath D, Padma SR, Kashinatham A, Srinivas KVN (1997) Indian Journal of Chemistry 36:208
32. Nolte BA (1999) M.S. Thesis, Texas A&M University, College Station, TX
33. Padmanabha BV, Chandrasekar M, Ramesha BT, Hombegowda HC, Gunaga R, Suhas S, Vasudeva R, Ganeshaiah KN, Uma Shaanker R (2006) Curr Sci 90:95
34. Roja G, Heble MR (1994) Phytochemistry 36:65
35. Suhas S, Ramesha BT, Ravikanth G, Rajesh PG, Vasudeva R, Ganeshaiah KN, Uma Shaanker R (2007) Curr Sci 92:1142
36. Myers N, Russell AM, Cristina GM, Gustavo ABF, Jennifer K (2000) Nature 403:853
37. Liu Z, John CA, Viator HP, Roysell J, Constantin J, Stanley BC (1999) Physiol Plant 105:402
38. Liu Z, Stanley BC, Wayne JB, Yu Y, Raysell JC, Mathew JF, John CA(1998) Bot Bull Acad Sin 44:265
39. Yan XF, Wang Y, Yu T, Zhang YH, Yin LJ (2002) J Instu Anal 21:15
40. Ganeshaiah KN, Narayani B, Nilima N, Chandrasekara K, Swamy M, Uma Shaanker, R (2003) Curr Sci 85:1526
41. Hijmans RJ, Guarino L, Mathur P, Jarvis A (2004), DIVA-GIS. Free GIS for biodiversity research. http://www.diva-gis.org/
42. Hijmans RJ, Cruz E, Rojas L, Guarino L (2001) Manual. International Potato Center and International Plant Genetic Resources Institute, Lima, Peru
43. Hijmans RJ, David MS (2001) American Journal of Botany. 88:2101
44. Prakash Kumar UM (2007) M.Sc Thesis University of Agricultural Sciences, Bangalore
45. Puri SC, Handa G, Gupta RK, Srivastava TN, Somal P, Sharma SN (1999) J. Indian Chem. Soc 76:370

46. Ved DK (2003) Medicinal plants trade in India- An overview. In: Ankila JH, Gladwin CJ, Uma Shaanker R (eds) Proceedings of the conference on Policies, management, Utilization and conservation of non-timber forest products (NTFP's) in the South Asia region. Ashoka Trust for Research in Ecology and Environment, Bangalore and Forestry Research support Program for Asia and the Pacific, Food and Agriculture Organisation, Bangkok
47. Swapna M, Biswajit G, Timir BJ, Sumita J (2002) Planta Med 68:757
48. Poupat C, Ingrid H, Francoise G, Alain A, Daniel G, Pierre P (2000) Planta Med 66:580
49. Rita MM, Bedir E, Barrett H, Burandt JC, Canel C, Khan IA (2002) Planta Med 68:341
50. Rita MM, Hemanth L, Ebru B, Muhammed M, Kent C (2001) Trends New Crops New Uses 36:527
51. Lopez-Meyer M, Nessler CL, McKnight TD (1994) Planta Med 60:558
52. Li S, Yi Y, Wang Y, Zhang Z, Beasley RS (2002) Planta Med 68:1010
53. Wiedenfeld H, Furmanowa M, Roeder E, Guzewska J, Gustowski W (1997) Plant Cell Tiss Org Cult 49:213
54. Saito K, Sudo H, Yamazaki M, Koseki-Nakamura M, Kitajima M, Yamazaki M, Takayama H, Aimi N (2001) Plant Cell Rep 20:267
55. Vineesh VR, Fijesh PV, Jelly Louis, Jaimsha VK, Jose Padikkala (2007) Curr Sci 92:1216
56. Zhou BN, Hoch JM, Johnson RK, Mattern MR, Eng WK, Ma J, Hecht SM, Newman DJ, Kingston DGI (2000) J Nat Prod 63:1273
57. Aiyama R, Nagai H, Nokata K, Shinohara C, Sawada S (1988) Phytochemistry 27:3663
58. Yamazaki Y, Urano A, Sudo H, Kitajima M, Takayama H, Yamazaki M, Aimi N, Saito K (2003) Phytochemistry 62:461

Chapter 11
Colchicine – an Overview for Plant Biotechnologists

S. Ghosh and S. Jha (✉)

Centre of Advanced Study in Cell and Chromosome Research, Department of Botany, University of Calcutta, 35 Ballygunge Circular Road, Calcutta, 700019, India, e-mail: sjbot@caluniv.ac.in

Abstract Secondary metabolites contribute to plant fitness by interacting with the ecosystems, and thus play a major role in the survival of the plant in its environment. Besides the importance for the plant itself, several secondary metabolites are commercially available as fine chemicals, such as drugs, dyes, flavors, fragrances, and insecticides. Due to their extensive biological activities, plant secondary metabolites have been used for centuries in traditional medicine. Alkaloids are an important group of secondary metabolites, of which colchicine is a useful agent in the treatment of acute attacks of gout. Apart from inhibiting the assembly of microtubules, the major biological effects of colchicine include leukocyte diapedesis, lysosomal degranulation, and inhibition of proliferation of fibroblasts as well as collagen transport to the extracellular space. In this way it relieves the pain associated with acute gout, decreases interleukin-l production in patients with primary biliary cirrhosis, and is used in the prevention or treatment of amyloidosis, scleroderma, and chronic cutaneous leukocytoclastic vasculitis. Colchicine prevents recurrences of acute pericarditis in adults and children. Colchicine is sold under different brand/trade names in different countries as a medicine for gout. The increased demand for colchicine has stimulated great interest in known plant sources of colchicine, new sources, and the route by which the alkaloid could be synthesized. Colchicine and related compounds have been found in several genera such as *Colchicum*, *Merendera*, *Androcymbium*, *Gloriosa*, and *Littonia*. Plant secondary products are vulnerable to fluctuations in value, depending on the effects of climate, pests, and diseases on the producing crops, or political changes in the producing countries. This has raised interest in developing biotechnological processes for the industrial production of such fine chemicals. However, very few reports are available on biotechnological approaches toward the biological production of such an important alkaloid. The evolving commercial importance of colchicine in recent years is sure to pave ways to extensive study by plant biotechnologists leading to exciting opportunities to engineer colchicine metabolism in plants.

Keywords Alkaloid, Biotechnology, Colchicine, *Colchicum autumnale*, *Gloriosa superba*, In vitro culture, *Sandersonia*, Secondary metabolite

Abbreviations

2,4-D	2,4-Dichlorophenoxyacetic acid
FDA	United States Food and Drug Administration
fr. wt.	Fresh weight
IAA	Indole acetic acid
IBA	Indole butyric acid

11.1 Introduction

Plants produce an amazing diversity of low-molecular-weight compounds [1]. Although the structures of close to 50,000 have already been elucidated [2], there are probably hundreds of thousands of such compounds. Only a few of these are part of "primary" metabolic pathways (those common to all organisms), the rest are termed "secondary" metabolites [1]. In biology, the concept of "secondary metabolite" can be attributed to Kossel [3]; he was the first to define these metabolites as opposed to primary ones. Thirty years later, an important step forward was made by Czapek [4], who dedicated an entire volume of his "plant biochemistry" series to what he named "endproduckt". According to him, these products could well be derived from nitrogen metabolism by what he called "secondary modifications" such as deamination. Compared to the main molecules found in plants, these secondary metabolites were soon defined by their low abundance, often less than 1% of the total carbon, or a storage usually occurring in dedicated cells or organs. In the middle of the 20th century, improvements in analytical techniques such as chromatography allowed the recovery of more and more of these molecules, and this was the basis for the establishment of the discipline of phytochemistry [5]. For a long time these compounds were regarded as waste products that had interesting structures and, in many cases, exploitable biological properties. However, a rapidly increasing body of experimental and circumstantial evidences indicates that most secondary metabolites are important for the overall fitness of the plant that produces them. These compounds largely contribute to plant fitness by interacting with ecosystems, and thus play a major role in the survival of the plant in its environment [5]. Secondary metabolites are involved in resistance against pests and diseases, attraction of pollinators, interactions with symbiotic microorganisms, among many others [6]. They have been described as being antibiotic, antifungal, and antiviral, and are therefore able to protect plants from pathogens (phytoalexins), and are also antigerminative or toxic to other plants (allelopathy). In addition, they constitute important ultraviolet-absorbing compounds, thus protecting leaves from light-induced damage [7]. They also act on animals, such as insects (antifeeding properties) or even cattle for which forage grasses such as clover or alfalfa can express estrogenic properties and interact with fertility [8, 9]. In addition to the importance for the plant itself, secondary metabolites also are of interest because they determine the quality of food (color, taste, and aroma) and ornamental plants (flower color and smell) [10]. Various health-improving effects and disease-preventing activities of secondary metabolites have been reported, including antioxidative and cholesterol-lowering properties [10]. Several secondary metabolites are commercially available as fine chemicals, for example drugs, dyes, flavors, fragrances, and insecticides. Some of these phytochemicals are quite expensive because of their low abundance in the plant [10]. Due to their many biological activities, plant secondary metabolites have been used for centuries in traditional medicine. Nowadays, they correspond to valuable compounds such as pharmaceutics, cosmetics, fine chemicals, or more recently nutraceutics [5].

Although about 100,000 plant secondary metabolites are already known, only a small percentage of plants species have been studied for the presence of secondary metabolites. In most cases, such studies are also limited to one or only a few classes of secondary metabolites. Based on the database NAPRALERT, it is estimated that about 15% of the approximately 250,000 known plant species have been subjected to phytochemical study, whereas less than 5% have been screened for 1 or more biological activities [11–13].

Plant secondary compounds are usually classified according to their biosynthetic pathways [9]. Three large molecule families are generally considered: the phenolics, terpenes and steroids, and alkaloids. A good example of a widespread metabolite family is given by phenolics: because these molecules are involved in lignin synthesis, they are common to all higher plants. However, other compounds such as alkaloids are sparsely distributed in the plant kingdom and are much more specific to defined plant genus and species. This narrower distribution of secondary compounds constitutes the basis for chemotaxonomy and chemical ecology [5]. The largest group of secondary metabolites discovered so far consists of the terpenoids, comprising more than one-third of all known compounds. The second largest group is formed by the alkaloids, comprising many drugs and poisons [13].

Alkaloids, an important group of secondary metabolites, are a structurally diverse class of low-molecular-weight nitrogenous compounds that are found in many plants [14] and often exhibit physiological activity. Plants that produce alkaloids and their extracts have been exploited for their medicinal and toxic properties for a long time. Modern examples of widely used plant-derived alkaloids include analgesics (morphine and codeine), stimulants (caffeine and nicotine), anticancer agents (vincristine, vinblastine, and camptothecin derivatives), gout suppressant (colchicine), muscle relaxant (C-tubocurarine, antiarrhythmic ajmaline), antibiotic (sanguinarine), and sedative (scopolamine). Other important alkaloids of plant origin include cocaine and the synthetic O,O-acetylated morphine derivative heroin [14]. While most alkaloids are formed from amino acids such as phenylalanine, tyrosine, tryptophan, ornithine, and arginine, they can be derived from several substrates (e.g., purines for caffeine). In plants, over 12,000 alkaloid structures have already been elucidated [15], providing drug companies with a diverse set of structures that are valuable for pharmacological screening [10, 11].

11.2 The Alkaloid Colchicine

Colchicine, the main alkaloid of *Colchicum autumnale* (Liliaceae), is a useful agent in the treatment of acute attacks of gout [16]. Colchicine is a tricyclic alkaloid, the main features of which include a trimethoxyphenyl ring (A ring), a seven-membered ring (B ring) with an acetamide at the seventh position, and a tropolonic ring (C ring). Chemically, colchicine is (S)-N-(5,6,7,9-tetrahydro-1,2,3,10-tetramethoxy-9-oxobenzo(a)heptalen-7-yl) acetamide [17, 18] with the molecular formula $C_{22}H_{25}NO_6$ (Fig. 11.1; molecular weight 399.43; melting point 151–152°C). Colchicine consists of pale yellow scales or powder; it darkens on exposure to light due to photoisomerization and formation of α-, β-, and γ-lumicolchicines [19, 20].

Colchicine is soluble in water, freely soluble in alcohol and in chloroform, and slightly soluble in ether [21]. The optimum storage temperature for colchicine is –15° to –25°C, in dark-colored bottles.

11.3 Toxicity of Colchicine

Colchicine is a toxic alkaloid that requires careful use and disposal. It is highly toxic and is suspected to be carcinogenic [22]. Overdose of colchicine leads to the delayed onset of multiorgan failure [23] and is frequently fatal [24, 25]. There is no specific treatment and the chances of survival can only be influenced by early and aggressive gastrointestinal decontamination [22, 26]. Doses of 7–60 mg are generally fatal [27]; symptoms are visible in about 4 h, death occurring in about 4 days. Colchicine has been known to kill a human adult after ingestion of a single dose of 3 mg [28]. Colchicine poisoning resembles arsenic poisoning; the symptoms include burning in the mouth and throat, diarrhea, stomach pain, vomiting, and kidney failure [29].

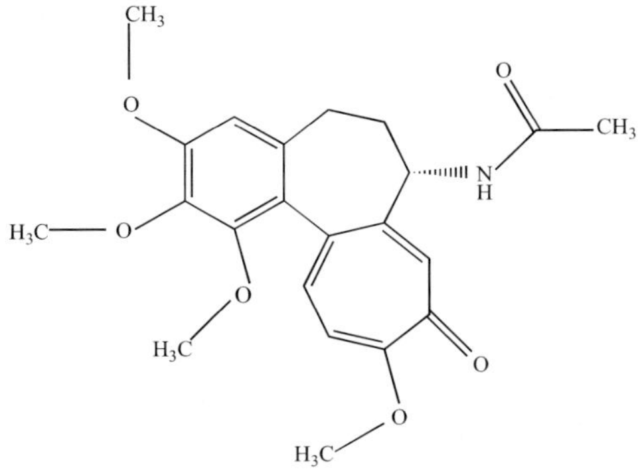

Fig. 11.1 Colchicine [N-{5,6,7,9-tetrahydro-1,2,3,10-tetramethoxy-9-oxobenzo(.alpha.) heptalen-7-yl}-acetamide]

11.4 Biological Effects of Colchicine

An early experimenter with plants and colchicine was Charles Darwin [30], who applied the drug to "insectivorous" and "sensitive" plants. The reactions in leaf movements were tested, but no conclusive results were obtained for colchicine [31]. Later, Pernice in 1889 [32] described the action of colchicine on mitosis. According to many references, Malden was the first scientist to observe the effects of colchicine on mitosis because he reported that the drug appeared to "excite karyokinesis" [33] in white blood cells. Dixon and Malden prepared an excellent report on the effects of colchicine on blood [34].

The major biological effects of colchicine are:
1. Interferes with most leukocyte functions including diapedesis (ameboid movement) [35], mobilization, lysosomal degranulation [36], and most importantly leukocyte chemotaxis [37]. These effects may be mediated at least in part by a decrease in the expression of adhesion molecules on neutrophil membranes.
2. Inhibits lysosomal degranulation and increases the level of cyclic AMP, which decreases both the chemotactic and the phagocytic activity of neutrophils [38].
3. Binds to and inhibits the assembly of microtubules [39], thereby interfering with secretory mechanisms such as the release of histamine from mast cells.
4. Limits leukocyte activity by binding to tubulin, a cellular microtubular protein, thus inhibiting protein polymerization [40].
5. Colchicine reduces the inflammation and relieves the pain associated with acute gout [36], and it is used primarily for this purpose. It is most effective when used at the onset of symptoms.
6. Inhibits proliferation of fibroblasts [41] and decreases interleukin-1 production in patients with primary biliary cirrhosis [42, 43].

7. Inhibits collagen transport to the extracellular space, hence its use in the prevention or treatment of amyloidosis [44, 45] and scleroderma.
8. Colchicine is effective in controlling chronic cutaneous leukocytoclastic vasculitis [46].
9. Prevents recurrences of acute pericarditis in adults and children [47–49], thereby replacing prolonged administration of corticosteroids [50].

It has long been known that gout and urate metabolism are linked and that gout is caused by the deposition of micocrystals of sodium urate in the joints [51, 52], causing inflammation and pain. In humans and other primates, uric acid is the final metabolite in the breakdown of purines [53]. When this metabolic pathway becomes overwhelmed, from either an enzymatic deficiency or an increase in dietary purines, uric acid cannot be efficiently eliminated from the body. The poorly soluble uric acid crystallizes, initiating a response from macrophages and leukocytes. The phagocytosis of urate crystals by the macrophages and leukocytes, mainly neutrophils, stimulates the release of cytokines and interleukins, leading to inflammation and the distinctive symptoms. Chemotactic factors are released and attract more neutrophils. Colchicine therapy diminishes the metabolic activity of leukocytes, resulting in reduced phagocytosis of urate microcrystals, therefore interrupting the cycle of new crystal deposition [54]. It also inhibits neutrophil motility and activity, leading to a net anti-inflammatory effect. The precise mechanism by which colchicine relieves the intense pain of gout is not known [52]. However, it is believed that the major relief of pain involves colchicine's major pharmacological action: inhibiting microtubule assembly. The microtubules are vital for the formation of spindle fibers during mitosis and meiosis, the intracellular transport of vesicles and proteins, flagella reassembly, ameboid motility, and other cellular processes. Inhibition of ameboid motility prevents macrophage and leukocyte migration and phagocytosis, thereby presumably preventing the inflammation and pain of gout. *Colchicum* extract was first described as a treatment for gout in De Materia Medica of Padanius Dioscorides [55]. Colchicine has the United States Food and Drug Administration (FDA)-approved use to treat gout. It is one of the active ingredients of ColBenemid, antigout tablets marketed by Merck and Co.

Colchicine is approved by the FDA for the treatment of familial Mediterranean fever [18, 56], amyloidosis [44, 45] sarcoidosis [57], Behcet's syndrome [58], and scleroderma [59]. Side effects include gastrointestinal upset [60] and neutropenia [61]. Starting the drug early during an attack of gout can exacerbate the symptoms. There have been published reports that show the benefit of using colchicine in treating major recurrent aphthous stomatitis and preventing further recurrences of ulcers [62, 63]. It is an antimitotic agent that suppresses cell division by inhibiting karyokinesis. It inhibits spindle formation by arresting the polymerization of tubulin proteins and thereby checks karyokinesis [64]. Tubulin (molecular mass approximately 10,000 Da) is a protein consisting of two forms, α and β, which form dimers, and these dimers polymerize to form long filaments of microtubules. When colchicine binds to the tubulin dimers, the dimers are unable to form the microtubules. The structure of the high-affinity binding site on one of the nonidentical subunits of tubulin, probably the β-subunit, is not known, but it is believed that binding induces a conformational change in the protein, thereby inhibiting polymerization [65]. Colchicine has been used widely in plant breeding to induce polyploidy for a long time [31]. It is used extensively to produce double-haploid plants from another culture [66–70].

Because cancer cells divide much more rapidly than normal cells, cancers are more susceptible to be poisoned by mitotic inhibitors such as colchicine, paclitaxel, and podophyllotoxin. However, colchicine has proven to have a fairly narrow range of effectiveness as a chemotherapy agent, although it is occasionally used in veterinary medicine to treat cancers in some animals. It is reported to markedly increase the susceptibility of cancer cells to x-rays, presumably due to its action on mitosis [17].

Colchicine has no direct action upon the heart [71, 72], but is among other things a capillary poison. Large doses cause an ascending paralysis of the central nervous system accompanied by vasomotor and respiratory paralysis [22, 24, 73–75]. Peripheral neuritis, neuromyopathy, and myopathy have been reported due to colchicine intoxication [73, 76–80]. The fact that the base induces regression of tumors in mice and effectively treats spontaneous tumors in dogs has led to an investigation of its effect on cell division in normal and malignant cells. Unfortunately, the inhibition of cell division is not specific for tumor cells and the dosage necessary to arrest the growth of a transplanted tumor approaches the lethal dose for the host. High doses can also damage bone marrow [81] and lead to anemia [82]. All colchicine derivatives have proved to be less potent than the parent alkaloid. The alkaloid increases the erythrocyte sedimentation rate of human blood [83] and causes changes in the acid-soluble ribonucleotides in rats [84].

Table 11.1 Brand/trade names of colchicine in different countries

Australia	Colgout (Protea); Colchicine (Medical Research); Colcin (Knoll); Coluric (Nelson)
Canada	Colchicine (Abbott)
France	Colchicine (Houde ISH); Colchineos (Houde ISH)
Germany	Colchicum-Dispert (Kali-Chemie)
South Africa	Colchicine Houdse (Roussel)
USA	Colchicine (Abbott, Barr Lab., Danbury, Lilly, Rugby, Towne, United Research Laboratories)

11.5 Colchicine as a Medicine

Colchicine is available as tablets and, in some countries, as injectable solutions. Tablets contain mostly 0.5–0.65 mg colchicine. Formulations with 1 mg per tablet are also available. Sterile solutions containing 0.5 mg/ml are also available for intravenous injection. Colchicine is sold under the different brand/trade names in different countries as medicine for gout (Table 11.1).

However, use of colchicine as a medicine brings forth the following drug interactions [85, 86]:
1. Colchicine may inhibit the gastrointestinal absorption of some drugs such as isoniazid, quinidine, and sulfisoxazole.
2. Colchicine induces vitamin B12 malabsorption (possibly by reducing the number of intrinsic factor receptors in the intestinal mucosa) [87].
3. Inhibitors of cytochrome P450 (cimetidine, ketoconazole, erythromycin, diltiazem, grape fruit, etc) ⇒⇓ Colchicine metabolism ⇒⇑ Colchicine blood levels.
4. Inducers of cytochrome P450 (rifampin, phenobarbital, phenytoin, etc) ⇒⇑ Colchicine metabolism and clearance ⇒⇓ Colchicine blood levels.
5. Substrates of cytochrome P450 3A4, such as cyclosporin and tacrolimus, may be affected by the coadministration of colchicine.

11.6 Botanical Use of Colchicine

Since chromosome segregation is driven by microtubules, colchicine is also used for inducing polyploidy in plant cells during cellular division by inhibiting chromosome segregation during meiosis; half the resulting gametes therefore contain no chromosomes, while the other half contain double the usual number of chromosomes (i.e., diploid instead of haploid as gametes usually are), and lead to embryos with double the usual number of chromosomes (i.e. tetraploid instead of diploid). While this would be fatal in animal cells, in plant cells it is not only usually well tolerated, but in fact frequently results in plants that are larger, hardier, faster growing, and in general more desirable than the normally diploid parents; for this reason, this type of genetic manipulation is frequent in breeding plants commercially. In addition, when such a tetraploid plant is crossed with a diploid plant, the triploid offspring will be sterile (which may be commercially useful in itself by requiring growers to buy seed from the supplier), but can often be induced to create a "seedless" fruit if pollinated (usually the triploid will also not produce pollen, therefore a diploid parent is needed to provide the pollen). This is the method used to create seedless watermelons, for instance. On the other hand, colchicine's ability to induce polyploidy can be exploited to render infertile hybrids fertile, as is done when breeding triticale from wheat and rye. Wheat is typically tetraploid and rye diploid, with the triploid hybrid being infertile. Treatment with colchicine of triploid triticale gives fertile hexaploid triticale.

11.7 Chemistry of Colchicine

The useful biological properties and novel structure of this compound have resulted in considerable effort being directed toward its synthesis. Extracts of *Colchicum* species were first referred to in 1550 B.C. by the Egyptians [31]. Colchicine alkaloid was first isolated in 1820 by the two French chemists P.S. Pelletier and J. Caventou [88] and they regarded it as veratrine. Thorough descriptions characterizing crystalline colchicine were prepared by Zeisel in 1883 [89] and by Houde in 1884 [90]. The formula $C_{22}H_{25}NO_6$ was proposed by Séris in 1947 [91]. The absolute configuration was determined by Corrodi and Hardegger in 1955 [92].

Colchicine was isolated by Oberlin in 1857 [93] from seeds of *C. autumnale* and has subsequently been obtained from other *Colchicum* species and numerous other members of the family Liliaceae. Since 1930, the increased demand for colchicine has stimulated great interest in known plant sources of colchicine, for new sources and the route by which the alkaloid could be synthesized. During this period, much work was done by Šantavý and his colleagues on the plant sources of colchicine. Advancement was made in colchicine chemistry when Adolph Windaus, after a long series of investigations, set forth the concept of a three-ring structure [91]. By degradation studies, Windaus [94] assigned to colchicine a 9-methylphenethrene structure, and his formula for ring A has not been disputed. Ring A is aromatic, six-carbon with three associated methoxyl groups. The rest of the formula did not agree with the chemical behavior of colchicine. It was then proposed that ring B was seven-membered, and further evidence was produced to support this [95, 96]. Final proof of the cycloheptane structure of ring B was obtained by the synthesis of dl-colchinol methyl ether [97], N-acetylcolchinol methyl ether [98], and deaminocolchicine acid anhydride [99], the degradation products of colchicine. Dewar [100] postulated a tropolone structure for ring C and was responsible for coining the term "tropolone" for cycloheptatrienolone. Convincing evidence for this was given by comparison of tropolone and colchicine [101] and it was proved that ring C was 7-membered by the synthesis of octahydrodemethoxydesoxyes-acetamidocolchicine, a degradation product of colchicine in which ring C remains intact [102, 103].

To determine the biosynthetic pathway of colchicine, feeding experiments have been employed using isotopically labeled compounds [104]. Reports on *C. byzantium* and *C. autumnale* indicate that acetate is the precursor of the N-acetyl group [105, 106] and that phenylalanine forms ring A and carbons 5, 6, and 7 of ring B [107–112]; further confirmation was added by Hill and Unrau [113]. Peripheral methyl groups were produced from methionine in *C. byzantium* [105, 106] and from methanol in *C. autumnale* [113]. The tropolone ring was demonstrated to be derived from tyrosine by ring-expansion of a C6–C1 unit [107–110, 114].

(–)-Colchicine has only one stereogenic center: carbon-7. The designation of this carbon is S, according to the common Cahn-Ingold-Prelog rules. However, colchicine is also asymmetric due to axial chirality. The single bond between the A and C rings is rotationally restricted; this restriction adds a degree of asymmetry to the molecule. In 1933, Kuhn [115] designated this type of stereoisomerism as atropisomerism (from Greek "a" meaning not and "tropos" meaning turn). The designation of this asymmetry is "aS" or "aR," according to the rules of molecular asymmetry, in which the "a" stands for axial chirality [116]. In colchicine, the C–C bond between the A and C rings is the chiral

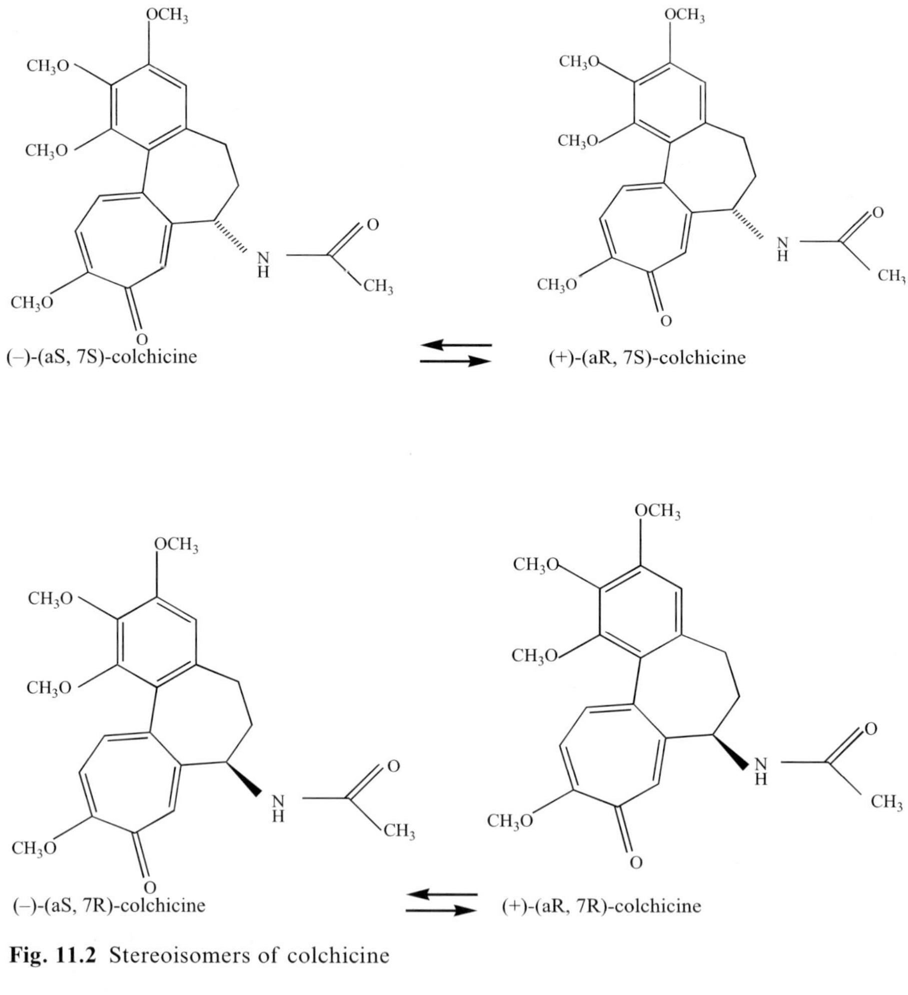

Fig. 11.2 Stereoisomers of colchicine

axis. In light of this molecular asymmetry, colchicine has four stereoisomers, as shown in Fig. 11.2. Each pair has either the R or S configuration at C-7. (–)-(aS, 7S)-Colchicine, the natural isomer, can interconvert between the two conformational isomers aR and aS, given enough energy.

11.8 Occurrence

Many early, nonspecific alkaloid tests and toxicity studies of several genera belonging to the family Liliaceae have been conducted as incontrovertible evidence for the presence of colchicine [117]. Most of these errors have been corrected by the work of Šantavý [118–120] who, by advanced techniques, reinvestigated many of the plants reported to contain colchicine. From his studies, colchicine and related compounds have been found only in the genera *Colchicum, Merendera, Androcymbium, Gloriosa* and *Littonia*. Table 11.2 gives a summary of the occurrence of colchicine reported since 1946.

11.9 Biotechnological Approaches for the Production of Colchicine

C. autumnale is the first plant species exploited for the production of colchicine in vitro. Callus culture of *C. autumnale* was induced on Linsmaier-Skoog medium [121] containing 2,4-dichlorophenoxyacetic acid (2,4-D) and kinetin, as reported by Hunault [122]. However, no data were presented on the accumulation of colchicine. Hayashi et al. [123] published the first report on the production of colchicine by plant tissue culture. Undifferentiated callus tissues were induced from flowering shoots of *C. autumnale* in the presence of 2,4-D and later cultured on liquid Murashige-Skoog medium [124] containing indole butyric acid (IBA) and kinetin. On replacement of IBA with indole acetic acid, the callus generated mostly rootlets. Hayashi et al. [123] also studied the effects of growth substances and nutritional factors on the formation of colchicine. The only effective carbon source for growth and colchicine formation (~5 μg/g fresh weight, fr. wt.) in *C. autumnale* was found to be 3% sucrose. Although nitrate or ammonium as the sole nitrogen source inhibited the formation of colchicine, growth and colchicine accumulation in vitro were better (~6 μg/g fr. wt.) with 20 mM ammonium plus 40 mM nitrate. Addition of SO_4^{2-} markedly increased the accumulation of colchicine to ~40 μg/g fr. wt. Yoshida et al. [125] added various precursors of colchicine to cell suspension cultures of *C. autumnale* and reported that phenylalanine, tyrosine, and methionine had no effect on colchicine production, but p-coumaric acid, tyramine, and demecolcine increased colchicine content.

Table 11.2 Plants containing colchicine

	Plant species	Occurrence	Organ	References
1	*Androcymbium gramineum* McBr.	South Africa	Corms Seeds	[131]
2	*Androcymbium melanthioides* var. *stricta* Bak.	Africa	Corms Seeds	[132]
3	*Colchicum vernum* Ker-Gawl	Eastern Europe	Corms Leaves Flowers	[133]
4	*Colchicum aggripinum* Bak.	Central Europe	Corms	[134]
5	*Colchicum alpinum* Lam *et* D.C.	European Alps	Corms	[133]
6	*Colchicum aeranarium* W.K.	Hungary	Corms Seeds	[134]
7	*Colchicum autumnale* L.	Western Europe	Corms, Seeds Flowers	[134] [135] [134]
8	*Colchicum autumnale* var. *album* Hort.	–	Corms	[134]
9	*Colchicum autumnale flore pleno* Hort.	–	Corms	[134]
10	*Colchicum autumnale major* Hort.	–	Corms	[134]
11	*Colchicum autumnale minor* Hort.	–	Corms	[134]
12	*Colchicum* hybrids var. Disraeli var. Lilac wonder var. The Giant var. Violet Queen	–	Corms Flowers Corms Corms Corms	[134]
13	*Colchicum bornmulleri* Freyn	Asia Minor	Corms	[134]
14	*Colchicum cilicicum* Hayek	Asia Minor	Corms	[134]
15	*Colchicum cornigerum* Täckh, *et* Drar	U.A.R.	Corms Seeds	[136]
16	*Colchicum crocifolium* Schott *et* Kotschy	Southeast Persia	Corms	[134]
17	*Colchicum hierosolymitanum* Feinbr.	Israel	Corms	[137]
18	*Colchicum kesselringii* Rgl	U.S.S.R.	Aerial parts	[138]
19	*Colchicum lactum* Stev.	Caucasus	Corms	[139]
20	*Colchicum lusitanicum* Brot.	North Africa and Iberia	Corms, Flowers	[134]
21	*Colchicum luteum* Bak.	N. India	Corms	[134]
22	*Colchicum macedonicum* Kos	–	Seeds	[140]
23	*Colchicum montanum* l.	Iberia	Corms	[133]
24	*Colchicum ritchii* R. Br	Israel	Corms Whole leafy plant	[141] [142]

Table 11.2 *(continued)* Plants containing colchicine

	Plant species	Occurrence	Organ	References
25	*Colchicum speciosum* Stev.	U.S.S.R.	Corms Leaves Flowers Seeds	[143] [134] [134, 144] [144]
26	*Colchicum steveni* Kunth	Israel	Whole plant in flower	[142]
27	*Colchicum tunicatum* Feinbr	Israel	Whole plant in flower	[142]
28	*Colchicum variegatum* L.	Asia Minor	Corms	[134]
29	*Colchicum vernum* Ker-Gawl	Eastern Europe	Corms	[134]
30	*Dipidax triquetra* Bak	–	Corms	[133]
31	*Gloriosa rothschildiana* O, Brien	Tropical Africa	Corms	[145]
32	*Gloriosa simplex* L.	Africa	Corms seeds	[134] [132]
33	*Gloriosa superba* L.	India and Africa	Corms Seeds	[134] [119] [146]
34	*Gloriosa virescens* Lindl.	Tropical Africa	Corms Seeds	[147] [144]
35	*Iphigenia indica* A. Gray	India	All parts	[141]
36	*Iphigenia pallida* Bak.	India	All parts	[141]
37	*Littonia modesta* Hook	Africa	Corms Leaves Seeds	[134] [132] [132]
38	*Merendera attica* Boiss *et* Sprun	Greece and Bulgaria	Corms	[134]; [144]
39	*Merendera bulbocodium* Ram.	Iberia	Corms	[141]
40	*Merendera caucasia* Spreng	Hungary and Albania	Corms	[134]
41	*Merendera persico* Boiss *et* Kotschy	Abyssinia Pakistan	Corms	[141]; [148]
42	*Merendera robusta* Bge	U.S.S.R.	All parts	[149,150]
43	*Merendera sobolifera* Fisch *et* Mey	Persia	Corms	[134]
44	*Merendera trigina* Stapf	Hungary	Seeds	[144]
45	*Ornithoglossum glaucum* Salisb. var. *grandiflorum*	South and East Africa	Corms	[133]
46	*Ornithoglossum viride* Dyrand	Africa	Corms	[132]
47	*Sandersonia auranticata* Hook	Africa	Corms	[132]

Finnie and Staden [126] reported about the in vitro propagation of *Sandersonia*, another natural source of colchicine, but did not study the colchicine content of the plant in vitro.

Instances where root cultures of *Gloriosa superba* have been established for the production of colchicine are few [127–129]. Callus culture of *G. superba* has been reported to accumulate colchicine [127] and enhancement in colchicine content was obtained by precursor feeding experiments [130].

11.10 Conclusion

Plant cell culture technologies were introduced at the end of the 1960s as a possible tool for both studying and producing plant secondary metabolites. Different strategies, using in vitro systems, have been studied extensively with the objective of improving the production of secondary plant compounds. The evolving commercial importance of secondary metabolites has in recent years resulted in a great interest in secondary metabolism, and particularly in the possibility of altering the production of secondary metabolites by means of genetic engineering. However, progress in this field had been limited; in most cases, very little was known about the biosynthesis of these compounds, and often only theoretical considerations existed about possible biosynthetic routes.

Recently, rapid progress has been made in our understanding of the biochemistry, molecular biology, and cell biology of alkaloid biosynthesis in plants. The data from several different alkaloid-producing plants suggest that their biosynthesis and accumulation involve a highly regulated process that includes cell-, tissue-, development- and environment-specific controls. Our understanding of the biological processes that permit the synthesis and accumulation of alkaloids in plants has advanced considerably over the past decade. Although there are several works elucidating the biosynthetic routes of colchicine, reports on our understanding of the biochemistry, molecular biology, and cell biology of colchicine biosynthesis in plants is meager. Colchicine may be a promising alkaloid for extensive study by plant biotechnologists, leading to exciting opportunities to engineer colchicine metabolism in plants. The socioeconomic importance of the alkaloid is sure to encourage greater interest in the near future.

Acknowledgements

S. Ghosh would like to thank the Council of Scientific and Industrial Research, New Delhi for the award of a Senior Research Fellowship.

References

1. Pichersky E, Gang D R (2000) Trends Plant Sci 5:439–445
2. De Luca V, St-Pierre B (2000) Trends Plant Sci 5:168–173
3. Kossel A (1891) Arch Physiol 181–186
4. Czapek F (1921) Spezielle Biochemie, Biochemie der Pflanzen, vol. 3, G. Fischer, Jena, p 369
5. Bourgaud F, Gravot A, Milesi S, Gontier E (2001) Plant Sci 161:839–851
6. Harborne JB (ed) (1978) Biochemical Aspects of Plant and Animal Coevolution: Proceedings of the Phytochemical Society of Europe, Vol 15. Academic Press, London
7. Li J, Ou-Lee TM, Raba R, Amundson RG, Last RL (1993) Plant Cell 5:171–179
8. Torssell KBG (1997) Chemical ecology. In: Torssell KBG (e.d) Natural Product Chemistry, A Mechanistic Biosynthetic and Ecological Approach. Swedish Pharmaceutical Press, Stockholm, pp 42–79
9. Harborne JB (1999) Classes and functions of secondary products. In: Walton NJ, Brown DE (eds) Chemicals from Plant, Perspectives on Secondary Plant Products. Imperial College Press, London, pp 1–25
10. Verpoorte R, van der Heijden R, Memelink J (2000) Transgenic Res 9:323–343
11. Verpoorte R (2000) J Pharm Pharmacol 52:253–262
12. Verpoorte R, van der Heijden R and Memelink J (1998) Plant biotechnology and the production of alkaloids. Prospects of metabolic engineering. In: GA Cordell (ed) The Alkaloids, Vol. 50. Academic Press, San Diego, pp 453–508
13. Verpoorte R, van der Heijden R, ten hoopen HJG, Memelink J (1999) Biotech Lett 21:467–479
14. Facchini PJ (2001) Annu Rev Plant Physiol Plant Mol Biol 52:29–66
15. Wink M (1999) Plant secondary metabolites: biochemistry, function and biotechnology. In: Wink M (ed) Biochemistry of Plant Secondary Metabolism (Annual Plant Reviews), Vol. 2. Sheffield Academic Press, Sheffield, pp 1–16
16. Rueffer M, Zenk MH (1998) FEBS Lett 438:111–113
17. Budavari S (ed) (1989) The Merck Index: An Encyclopedia of Chemicals, Drugs and Biologicals. Merck, Rahway, NJ, pp 386–387
18. Reynolds JEF (ed) (1993) Martindale, the Extra Pharmacopoeia, 30th edn. The Pharmaceutical Pres, London, pp 335–337
19. Chapman OL, Smith HG (1961) J Am Chem Soc 83:3914–3916
20. Chapman OL, Smith HG, King RW (1963) J Am Chem Soc 85:803–812
21. Loudon JD, Speakman JC (1950) Research 3:583–584
22. Murray SS, Kramlinger KG, McMichan JC (1983) Mayo Clin Proc 58:528–532
23. Stern N, Kupferschmidt H, Meier-Abt PJ (1997) Schweiz Rundsch Med Prax 86:952–956
24. Davies HO, Hyland RH, Morgan CD (1988) Can Med Assoc J 138:335–336
25. Caplan YH, Orloff KG, Thompson BC (1980) J Anal Toxicol 4:153–155
26. Jäger AK, Schottländer B, Smitt UW, Nyman U (1993) Plant Cell Rep 12:517–520
27. Stapczynski JS, Rothstin RJ, Gaye WA (1981) Ann Emerg Med 10:364
28. Watt JM, Breyer-Brandwijk MG (1962) The Medicinal and Poisonous Plants of Southern and Eastern Africa. Edinburgh and Livingstone, London
29. Medical Economics (1995) Physician's Desk Reference. Medical Economics Data Production, Montvale, NJ
30. Darwin C (1875) Insectivorous Plants. John Murray, London
31. Eigsti OJ, Dustin P Jr (1955) Colchicine in Agriculture, Medicine, Biology and Chemistry. Iowa State College Press, Ames, Iowa
32. Pernice B (1889) Silicia Med 265–279
33. Cook J, Loudon J (1951) Colchicine. In: Holmes and Mankse (ed) Alkaloids. Academic Press, New York, pp 261–325

34. Ludford RJ (1936) Arch Exp Zelforsch Mikr Anat 18:411–441
35. Famaey JP (1988) Clin Exp Rheumatol 6:305–317
36. Insel PA (1996) Analgesic-antipyretic and anti inflammatory agents and drugs employed in the treatment of gout: colchicine. In: Goodman LS, Limbird LE, Milinoff PB, Ruddon RW, Goodman Gilman A (eds) Goodman and Gilman's: The Pharmacological Basis of Therapeutics, 9th edn. Mc Graw-Hill, pp 614–658
37. Schiffmann E (1982) Annu Rev Physiol 44:553–568
38. Greenberg MS (2000) Drugs used for connective-tissue disorders and oral mucosal diseases. In: Ciancio SG (ed) ADA Guide to Dental Therapeutics. ADA Publishing, Chicago, pp 438–453
39. Kovacs P, Csaba G (2006) Cell Biochem Funct 24:419–429
40. Sullivan TP, King LE, Boyd AS (1998) J Am Acad Dermatol 39:993–999
41. Yu YS, Youn DH (1987) Korean J Ophthalmol 1:59–71
42. Chang DM, Baptiste P, Schur PH (1990) J Rheumatol 17:1148–1157
43. O'Brien JM Jr, Wewers MD, Moore SA, Allen JN (1995) J Immunol 154:4113–4122
44. Zemer D, Livneh A, Langevitz P (1992) Ann Intern Med 116:426
45. Rosenbaum M, Rosner I (1995) Clin Exp Rheumatol 13:126
46. Callen JP (1985) J Am Acad Dermatol 13:193–200
47. Rodriguez de la Serna A, Guindo Soldevila J, Marti Claramunt V, Bayes de Luna A (1987) Lancet 26:1517
48. Spodick DH (1991) Circulation 83:1830–1831
49. Millaire A, de Groote P, Decoulx E, Goullard L, Ducloux G (1994) Eur Heart J 15:120–124
50. Brucato A, Cimaz R, Balla E (2000) Pediatr Cardiol 21:395–396
51. Voet D, Voet JG (1990) Biochemistry. John Wiley and Sons, New York, pp 758–762
52. Katzung BG (ed) (1995) Basic and Clinical Pharmacology. Apleton and Lange, Norwalk, pp 536–559
53. Laster L, Seegmiller TE, Stetten D, Liddle LV (1957) J Clin Invest 36:908
54. Seegmiller JE, Howell RR, Malawista SE (1962) J Clin Invest 41:1399
55. Singer CJ (1996) A Short History of Scientific Ideas. Barnes and Noble Books, New York
56. Buskila D, Zaks N, Neumann L, Livneh A, Greenberg S, Pras M, Langevitz P (1997) Clin Exp Rheumatol 15:355–360
57. Gökel Y, Canataroğlu A, Satar S, Köseoğlu Z (2000) Turk J Med Sci 30:401–403
58. Levy M, Spino M, Read S (1991) Pharmacotherapy 11:196–211
59. Guttadauria M, Diamond H, Kaplan D (1977) J Rheumatol 4:272–276
60. Wallace SL (1980) Colchicine. In: Kelley WN, Harris ED, Ruddy S, Sledge CB (eds) Textbook of Rheumatology. Saunders, Philadelphia, pp 878–884
61. Folpini A, Furfori P (1995) J Toxicol Clin Toxicol 33:71–77
62. Ruah CB, Stram JR, Chasin WD (1988) Arch Otolaryngol Head Neck Surg 114:671–675
63. Katz J, Langevitz P, Shemer J (1994) J Am Acad Dermatol 31:459–461
64. Andreu JM, Perez-Ramirez B, Gorbunoff MJ, Ayala D, Timasheff SN (1998) Biochemistry 37:8356–8368
65. Berg U, Bladh H, Mpampos K (2004) Org Biomol Chem 2:2125–2130
66. Barnabás B, Pfahler PL, Kovács G (1991) Theor Appl Genet 81:675–678
67. Mathias R, Röbbelen G (1991) Plant Breed 106:82–84
68. Alemanno L, Guiderdoni E (1994) Plant Cell Rep 13:432–436
69. Barcelo P, Cabrera A, Hagle C, Lörz H (1994) Theor Appl Genet 87:741–745
70. Saisingtong S, Schmid JE, Stamp P, Büter B (1996) Theor Appl Genet 92:1017–1023
71. Führer H, Rehbein M (1915) Arch Exp Path Pharmak 79:1–18
72. Rossbach MJ (1876) Arch Ges Physiol 12:308–325
73. Carr GD, King RM, Powell AM, Robinson H (1999) Am J Bot 86:1003–1013
74. Hill RN, Spragg RG, Wedel MK, Moser KM (1975) Ann Intern Med 83:523–524
75. Hobson CH, Rankin AP (1986) Anaesth Intens Care 14:453–455
76. Kontos HA (1962) N Engl J Med 266:238

77. Mouren P, Tatossian A, Poiso Y, Giudicelli S, Jouglard J, Dufour H, Poyen D (1969) Presse Med 77:14:505–508
78. Favarel-Garrigues JC, Bony D, Poisot D (1975) Concours Med 97:5183–5197
79. Bismuth C, Gaultier M, Conso F (1977) Nouv Presse Med 6:1625–1629
80. Kuncl RW (1987) N Engl J Med 316:1562–1568
81. Boruchow IB (1966) Cancer 19:541–543
82. Heaney D, Derghazarian CB, Pineo GF (1976) Am J Mcd Sci 271:233–238
83. Rangam CM, Bhagwat RR (1960) Indian J Med Res 48:549–557
84. Wang D, Greenbaum AL, Harkness RD (1963) Biochem J 86:62–64
85. Ben-Chetrit E, Levy M (1998) Semin Arthritis Rheum 28:48–59
86. The Drug Monitor (2007) Ed:Nasir Anaizi, http://www.thedrugmonitor.com/colchicines
87. Webb DI (1968) N Engl J Med 279:845–850
88. Pelletier PS, Caventou J (1820) J Ann Chim Phy 14:69–83
89. Zeisel S (1888) Monatsch Chem 9:1–30
90. Houde A (1884) C R Acad Sci Paris 98:1442
91. Séris L (1947) La Rev Sci Fas 88:489–493
92. Corrodi H, Hardegger E (1955) Helv Chim Acta 38:2030–2033
93. Oberlin ML (1857) Ann Chim Phys 50:108–114
94. Windaus A (1924) Ann Chem 439:59–75
95. Cohen A, Cook JW, Roe EMF (1940) J Chem Soc 1940:194–197
96. Tarbell DS, Frank HR, Fanta PE (1946) J Am Chem Soc 68:502–506
97. Rapport H, Williams AR, Cisney M (1951) J Am Chem Soc 73:1414–1421
98. Cook JW, Jacj J, Loudon JD, Buchanan GL, MacMillan J (1951) J Chem Soc 1951:1397–1403
99. Koo J (1953) J Am Chem Soc 75:720–723
100. Dewar MJS (1945) Nature 155:50–51, 141–142, 479
101. Doering W von E, Knox LH (1951) J Am Chem Soc 73:828–838
102. Rapport H, Williams AR (1951) J Am Chem Soc 76:1896–1897
103. Rapport H, Williams AR, Campion JE, Pack D (1954) J Am Chem Soc 76:3693–3698
104. Walasek EJ, Kelesey FE, Geiling EMK (1952) Science 116:225–227
105. Battersby AR, Reynolds JJ (1960) Proc Chem Soc 1960:346–347
106. Leete E, Nemeth PE (1961) J Am Chem Soc 83:2192–2194
107. Battersby AR (1965) Angew Chem 4:79–80
108. Battersby AR, Binks R, Yeowell DA (1964) Proc Chem Soc 1964:86
109. Battersby AR, Binks R, Reynolds JJ, Yeowell DA (1964) J Chem Soc 1964:4257–4268
110. Battersby AR, Herbert RB (1964) Proc Chem Soc 1964:260
111. Leete E (1963) J Am Chem Soc 85:3666–3669
112. Leete E, Nemeth PE (1960) J Am Chem Soc 82:6055–6057
113. Hill RD, Unrau AM (1965) Can J Chem 43:709–711
114. Leete E (1965) Tetrahedron Lett 1965:333–336
115. Kuhn R (1993) Molekulare Asymmetrie. In: Fredenberg K (ed) Stereochemie. Franz Deutike, Leipzig-Wien pp 803
116. Eliel EL, Wilen SH (1994) Stereochemistry of Organic Compounds. John Wiley and Sons, New York, pp 1119–1122
117. Wildman WC (1959) Colchicine and related compounds. In: Manske RHF (ed) The Alkaloids VI. Academic Press, New York, pp 247
118. Šantavý F (1956) Oesterr Botan Z 103:300–311
119. Šantavý F, Zajíček DV, Němečková A (1957) Collection Czech Chem Commun 22:1482–1488
120. Šantavý F, Coufalík E (1951) Collect Czech Chem C 16:198–203
121. Linsmaier EM, Skoog F (1965) Physiol Plant 18:100
122. Hunault G (1979) Rev Cytol Biol Végét-Bot 2:103
123. Hayashi T, Yoshida K, Sano K (1988) Phytochem 27:1371–1374
124. Murashige T, Skoog F (1962) Physiol Plant 15:473–497
125. Yoshida K, Hayashi T, Sano K (1988) Phytochem 27:1375–1378

126. Finnie JF, Van Staden J (1989) Plant Cell Tissue Organ Cult 19:151–158
127. Finnie JF, Van Staden J (1994) Gloriosa superba L (Flame Lily): microproapgation and in vitro production of colchicine. In: Bajaj YPS (ed) Biotechnology in Agriculture and Forestry, Vol. 26 Medicinal and Aromatic Plants VI. Springer–Verlag, Berlin, Heidelberg, New York, pp 146–166
128. Ghosh B, Mukherjee M, Jha TB, Jha S (2002) Biotechnol Lett 24:231–234
129. Ghosh S, Ghosh B, Jha S (2006) Biotechnol Lett 28:497–503
130. Sivakumar G, Krishnamurthy KV, Hahn EJ, Paek KY (2004) J Hort Sci Biot 79:602–605
131. Perrot E (1936) Compt Rend 202:1088–1089; Chem Abstr 30:3946
132. Hrbek J, Šantavý F (1962) Collect Czech Chem C 27:255–267
133. Moza BK, Potešilová H, Šantavý F (1962) Planta Med 10:152–159
134. Šantavý F (1957) Collect Czech Chem C 22:652–653
135. Bellet P (1952) Ann Pharm Franç 10:81–88
136. Hamidi A, Fahmy MA (1960) J Chem UAR 3:279
137. Weizmann A (1952) Bull Res Council Israel Sect D 2:21–26
138. Yusupov MK, Sadykov AS (1964) Zh Obsch Khim 34:1672–1676
139. Salo VM (1963) Aptech Delo 12:40–43
140. Podoselov BD (1963) Glasnik Hem Drustva Beograd 28:461–463
141. Kaul JL, Moza BK, Šantavý F, Vrublovský (1964) Collect Czech Chem C 29:1689–1701
142. Fell KR, Ramsden D (1966) J Pharm Pharmacol 18:126S 132S
143. Beer AA, Karapetyan SA, Kolesnikov AI, Snegirev DP (1949) Dokl Akad Nauk SSSR 67:883–884
144. Salo VM (1961) Aptech Delo 10:28–32
145. Bryan JT, Lauter WM (1951) J Am Pharm Assoc 40:253
146. Lang B, Maturova M, Reichstein T, Šantavý F (1959) Planta Med 7:298–309
147. Burden E, Grindley DN, Prowse GA (1955) J Pharm Pharmacol 7:1063–1071
148. Mehra PN, Khoshoo TN (1951) J Pharm Pharmacol 3:486–496
149. Sadykov AS, Yusupov MK (1960) Dokl Akad Nauk Uz SSSR 1960:34–36
150. Sadykov AS, Yusupov MK (1962) Sci Rep Tashkent Univ 203:15–21

Chapter 12
In Vitro Azadirachtin Production

S. Srivastava[1] and A.K. Srivastava[2] (✉)

[1]Department of Biochemical Engineering and Biotechnology,
Indian Institute of Technology, Hauz Khas, New Delhi-110016, India

[2]Department of Biochemical Engineering and Biotechnology,
Indian Institute of Technology, Hauz Khas, New Delhi-110016, India,
e-mail: ashokks@dbeb.iitd.ernet.in

Abstract The secondary metabolite azadirachtin ($C_{35}H_{44}O_{16}$) is a tetranortriterpenoid obtained from the neem tree (*Azadirachta indica*). It has long been investigated for its biopesticidal properties. It is a natural insecticide, known to affect feeding, growth, reproduction and metamorphosis of the insect pests. Because of the broad-spectrum control of insects and the relatively low non-target toxicity it has been widely used in agriculture. To add to its advantage the biopesticidal property of azadirachtin is not only limited to phytophagous insects, but is also known to affect the other pathogenous organisms like nematodes, fungi and micro-organisms. It is a highly oxygenated and complex molecule, which makes its chemical synthesis difficult as well as uneconomical. Studies are still in progress to make its chemical synthesis successful and practically feasible for the mass production of azadirachtin. Currently, azadirachtin is isolated by solvent extraction of the seeds of the *Azadirachta indica* tree. There are various limitations in extracting azadirachtin from plant sources, majorly due to its limited availability/short shelf-life of seeds, degradation during storage and considerable genotypic/environmental variation in its content from different sources. At present, the demand for azadirachtin is greater than the supply. However, due to variability of azadirachtin available in seeds (0.2–0.6%) it is difficult to base its mass production on natural sources. A significantly larger amount of material (seeds) would need to be processed to yield a reasonable amount of azadirachtin. Instead, it would be better to rely on a rather stable parent cell line that can be cultivated in vitro (in bioreactors) with a faster doubling time. A biotechnological approach can be very useful for reaching long-term goals. A deeper understanding of different aspects of large-scale azadirachtin production is therefore very important. In recent years, a considerable amount of information has been obtained on the production of azadirachtin by cell and tissue cultures of *Azadirachta indica*. This approach has an added potential of increasing yield by culture selection and manipulation using elicitors, precursors, permeabilising agents and growth regulators. Azadirachtin is extremely liable to atmospheric degradation in the presence of sunlight. Although few investigations regarding enhancement in its atmospheric stability have been done, more detailed analysis is required to select appropriate stabilisers against the photo-degradation of azadirachtin. A great deal of work with *Azadirachta indica* has been focused on the extraction and quantification of azadirachtin. Purification of azadirachtin is difficult to accomplish, especially on a preparative scale due to its complexity and similarity in structure of the chemicals found in the seeds, foliage and cell culture. Reverse-phase high-performance liquid chromatography is widely used for the qualitative and quantitative estimation of azadirachtin. Use of other methods like super-critical fluid chromatography (SFC) and liquid chromatography-mass spectrometry combined with flash chromatography, thin-layer chromatography and SFC have also been documented for authentic quantification. Even though azadirachtin has long been recognised as a potent biopesticide, no reports are available to date on the commercial production of azadirachtin by plant cell/tissue culture. This chapter provides a deeper insight with respect to the origin, chemical nature, application, mode of action and prevalent technologies for the production of this high-value bioactive molecule.

Keywords Biopesticides, Azadirachtin, Biosynthesis, Chemistry, Mode of action, In vitro, Production, Plant cell/tissue culture, Yield improvement, Scale-up, Stability

Abbreviations

HPLC	High-performance liquid chromatography
IPP	Isopentyl pyrophosphate
DMAPP	Dimethyl allyl pyrophosphate
GPP	Geranyl pyrophosphate
FPP	Farnesyl pyrophosphate
UV	Ultraviolet

12.1 Introduction

In 1959, Heinrich Schmutterer, a German entomologist was witness to a plague of migratory locusts in Sudan. He observed that the only plant survivors amidst the devastation were the neem trees. This led to the investigation of possible reasons for their survival. It was not until 1967 that the compound responsible for this effect was identified as azadirachtin [1]. Out of more than 100 chemicals isolated from various neem tree parts, azadirachtin has been found to have the highest potential to be developed as a bioactive agent for the control of insects and certain harmful micro-organisms [2]. Azadirachtin is biodegradable and shows very low toxicity to mammals. It is less likely to cause environmental damage when compared to some currently used synthetic pesticides. It offers the development of a new class of pesticides as it fulfils many of the criteria needed for a natural insecticide. It is known to affect over 200 species of insects. Its effect as an antifeedant and as an insect growth/reproduction regulator has already been well understood and documented in recent years [3, 4]. Its biochemical effects at the cellular level, however, remain a mystery. The mode of action of azadirachtin lies in the deterrent effects on insect chemoreceptors, disruption in the balance of hormones and direct effects on tissues, resulting in an overall loss of fitness of the insect [2]. The issues of its transformation products, bioactivity, degradation, transport and fate in the environment are yet to be completely characterised.

One essential requirement for the commercial utilisation of azadirachtin in agriculture is its constant availability, with standardised quality. Azadirachtin occurs in all parts of the neem tree (*Azadirachta indica*), but is mainly concentrated in the seed (0.2–0.8% w/w) [5]. Due to environmental and genetic variations, the content of azadirachtin found in seeds varies considerably. Furthermore, the tree does not grow in moderate climates and is not frost-tolerant. Due to the expense of isolating azadirachtin from natural sources, there have been attempts to chemically synthesise the molecule [6–8]. Azadirachtin is a complex and highly oxidised triterpene that has many functional groups. Therefore, despite the recent success in the chemical synthesis of the furan and the decalin moieties, chemists are still faced with the challenge of combining the two fragments for complete synthesis. Neither of the two fragments is biologically active. Production of synthetic azadirachtin might be a reality after some years, but will be more expensive than isolation of the natural product. In nature, synthesis of azadirachtin involves tirucallol, a tetracyclic triterpenoid, and a series of oxidation and rearrangement reactions [4]. As far as quantitative analysis is concerned, high-performance liquid chromatography (HPLC) is currently the most sensitive method of detecting micro-residue levels of azadirachtin-A in plant tissues, soils and water.

Commercial production of chemically complex molecules like azadirachtin can be made feasible via in vitro cultivations. Hence, in order to obtain a constant supply of standardised quality azadirachtin, it seems appropriate to employ a biotechnological approach for its large-scale production. One such approach includes plant cell technology (plant cell/tissue cultivations), which has been suggested as an alternative means for year-round production of azadirachtin. It has an added advantage of increasing metabolite productivity by culture selection and manipulation with various yield-improvement strategies [9, 10]. In vitro cultivation of woody species was not successful until the 1970s, when tissue culture of neem was also among those reported [11, 12]. It was not until the 1990s, however, that efforts were initiated to produce azadirachtin in tissue culture [13]. Some of the successful attempts on the production of azadirachtin in different cell/tissue culture systems are listed in Table 12.1 [14].

The biggest challenge in the azadirachtin production through plant cell/tissue culture has been the low azadirachtin productivities obtained so far as compared to the conventional method of extraction from seeds obtained from the natural resources. Efforts are still needed to make azadirachtin production cost effective via a biotechnological approach. Employing such strategies can resolve the dependence issues on the natural resource for producing increased quality and quantity of high-value bioactive compounds like azadirachtin.

12.2 Chemistry of Azadirachtin

The active ingredient azadirachtin was isolated from the seeds of *Azadirachta indica* by David Morgan [1]. It took almost 17 years after its isolation to determine the structure and complete molecular formula of azadirachtin [22–24] (Fig. 12.1).

The molecule of azadirachtin has eight condensed rings: three carbocyclic and five heterocyclic. It is a complex molecule that boasts a plethora of oxygen functionality, comprising enol ether, acetal, hemi-acetal and tetra-substituted oxirane, as well as a variety of carboxylic esters [26, 27]. Azadirachtin contains 16 stereogenic centres, 7 of which are fully substituted [28]. Chemical synthesis of azadirachtin is a challenge because it is acid and base labile, prone to rearrangement, photosensitive and contains a high density of oxygen atoms.

Table 12.1 Azadirachtin content in tissue culture systems of *Azadirachta indica* [14]. *DCW* Dry cell weight

Culture system	Explant	Azadirachtin	Reference
Callus culture	Leaves	7 $\mu g \cdot g^{-1}$ DCW	[15]
Callus culture	Leaves	5.36 $mg \cdot l^{-1}$	[16]
Callus culture	Leaves Bark	64 $\mu g \cdot g^{-1}$ DCW 44 $\mu g \cdot g^{-1}$ DCW	[13]
Callus culture	Leaves Flowers	0.0268 $g \cdot g^{-1}$ DCW 0.0246 $g \cdot g^{-1}$ DCW	[17]
In vitro roots	–	0.004 $mg \cdot g^{-1}$ DCW	[18]
Hairy root culture	Leaves	27 $\mu g \cdot g^{-1}$ DCW	[19]
Suspension culture	Flower petals	10.11 $mg \cdot l^{-1}$	[20]
Suspension culture	Nodal segment	0.8 $mg \cdot l^{-1}$	[21]

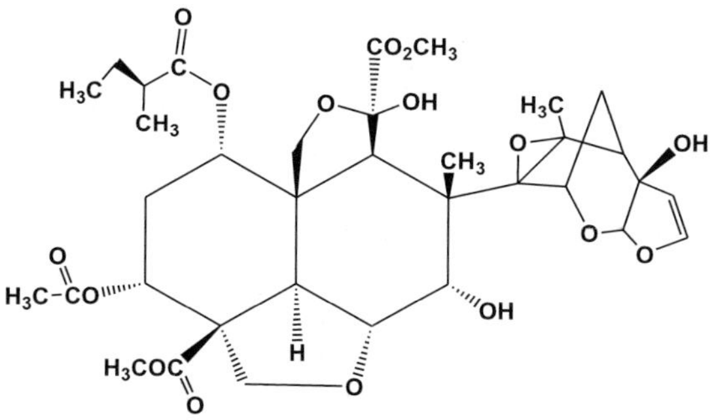

Fig. 12.1 Azadirachtin ($C_{35}H_{44}O_{16}$) [25]

At present, there are nine structural analogues of azadirachtin reported in the literature, of which azadirachtin-A and azadirachtin-B constitute about 99% of the components and are responsible for the bioactivity. The remaining 1% consists of azadirachtins -C, -D, -E, -F, -G and -I, and 22, 23-dihydro-23-a-, b-methoxy-azadirachtin [2, 26].

12.3 Mode of Action of Azadirachtin

Azadirachtin has the potential to be developed into the most potent insecticide of all time due to its multiple mode of action against target pests and micro-organisms. Azadirachtin affects insects as an antifeedant, insect growth regulator and sterilant [4, 13, 26, 29]. The dihydrofuran acetal moiety (Fig. 12.2) of the azadirachtin molecule is said to be mainly responsible for its antifeedant activity, while the decalin fragment (Fig. 12.2) is known to be responsible for the observed insect growth regulation [30, 31]. Its ability to act as a growth inhibitor via hormonal imbalance has been reviewed by Mukherjee et al. [32]. The antifeedant effect of azadirachtin on insects is produced by the stimulation of specific deterrent chemoreceptors on the mouthparts, together with interference in the perception of phagostimulants by other chemoreceptors [33]. The

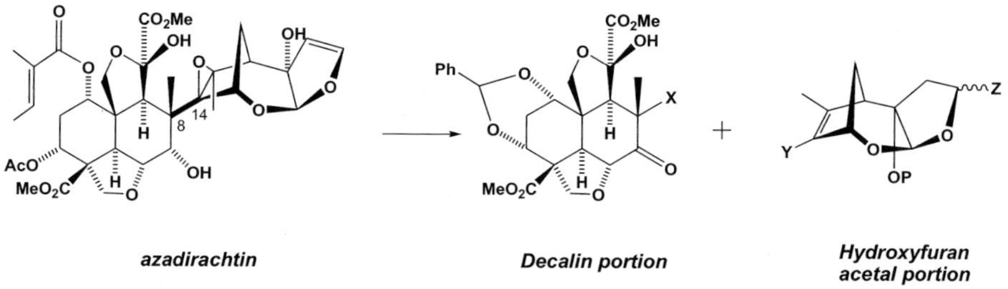

Fig. 12.2 Decalin and hydrofuran acetal fragments of the azadirachtin molecule [28]

developmental and reproductive disruption is caused by the effects of the molecule directly on somatic and reproductive tissues and indirectly through the disruption of endocrine processes. Azadirachtin operates at the cellular level by disrupting protein synthesis and secretion events and, more fundamentally, at the molecular level by altering or preventing the transcription and/or translation of proteins expressed during periods of rapid protein synthesis [29].

Apart from its action against insects, azadirachtin is also known to affect fungi, viruses and protozoa that are harmful to crops [4]. The other pharmacological activities reported in neem seed extracts include antimalarial, anticancer, antioxidant and antifertility [34]. Azadirachtin has been reported to be cytotoxic against human glioblastoma cells [35]. Studies carried out have proved mosquito cells to be highly sensitive to azadirachtin [36].

12.4 Biosynthetic Pathway for Azadirachtin

Azadirachtin (a tetranortriterpenoid) is a member of triterpenoid group of compounds. Triterpenoids in turn belong to the isoprenoid family. Isoprenoids are synthesised through condensation of the five-carbon compound isopentyl pyrophosphate (IPP) and its isomer dimethyl allyl pyrophosphate (DMAPP). Two distinct routes of IPP biosynthesis occur in nature: the mevalonate pathway and the recently discovered deoxyxylulose 5-phosphate pathway [37]. Synthesis of IPP and DMAPP via two pathways is as shown in Fig. 12.3a and b. Alkylation of DMAPP by IPP gives rise to geranyl pyrophosphate (GPP; Fig. 12.3c), which undergoes alkylation with IPP again to give farnesyl pyrophosphate (FPP, Fig. 12.3c). Head-to-head coupling of two FPP units gives rise to squalene catalyzed by squalene synthase (Fig. 12.3d). Squalene is then oxidised to 2,3-oxidosqualene by squalene oxidase (Fig. 21.3e). Lanosterol formation then takes place via cyclase-catalysed conversion of 2,3-oxidosqualene, which establishes the characteristic ring system of all steroids (Fig. 12.3e). The exact biogenesis of azadirachtin is not known, but is surely via steroid modification (steroid ring cleavage). It is likely that it is derived from the steroidal intermediate tirucallol [26] (Fig. 12.3f).

12.5 Qualitative and Quantitative Analysis of Azadirachtin

The extraction procedures of azadirachtin are rather exhaustive involving the partitioning in different solvents. Different organic solvents have been used for extraction like methanol, ethanol, n-hexane, ethyl acetate and dichloromethane [16, 39–41]. However, the partitioning in aqueous and organic layer seems necessary to remove the polar compounds like fatty acids, oil, sugar and proteins. Purification of azadirachtin is difficult to accomplish, especially on

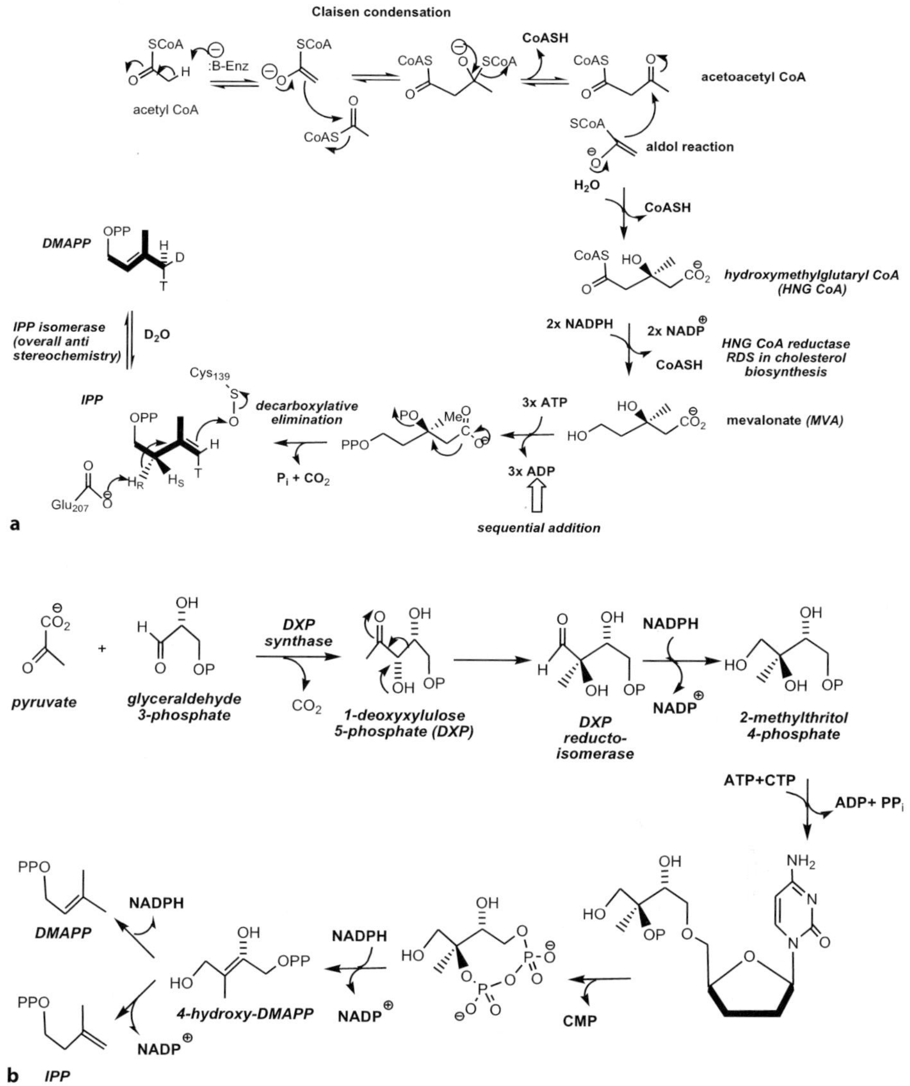

Chapter 12 In Vitro Azadirachtin Production

Fig. 12.3 Biosynthetic pathway for azadirachtin [38]. **a** Biosynthesis of isopentyl diphosphate (*IPP*) and dimethyl allyl diphosphate (*DMAPP*) via mevalonate. **b** Biosynthesis of IPP and DMAPP via 1-deoxyxylulose. **c–f** *see next page*

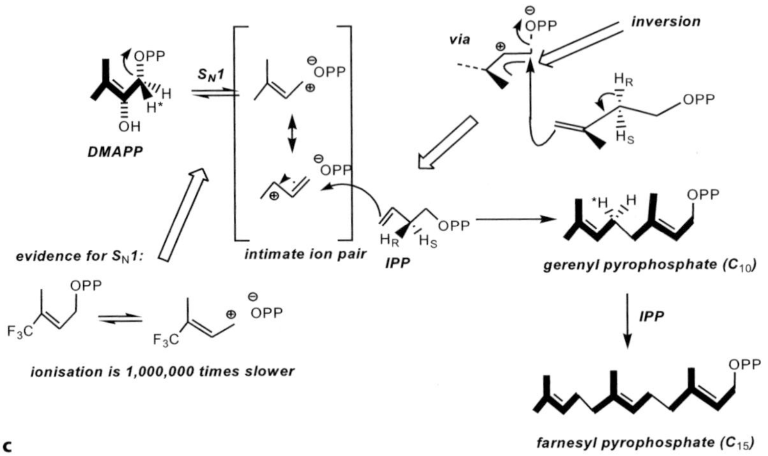

Fig. 12.3 Biosynthetic pathway for azadirachtin [38]. **c–f** *(continued)* **c** Synthesis of geranyl pyrophosphate (*GPP*) and farnesyl pyrophosphate (*FPP*). **d** Formation of squalene from FPP. **e,f** *see next page*

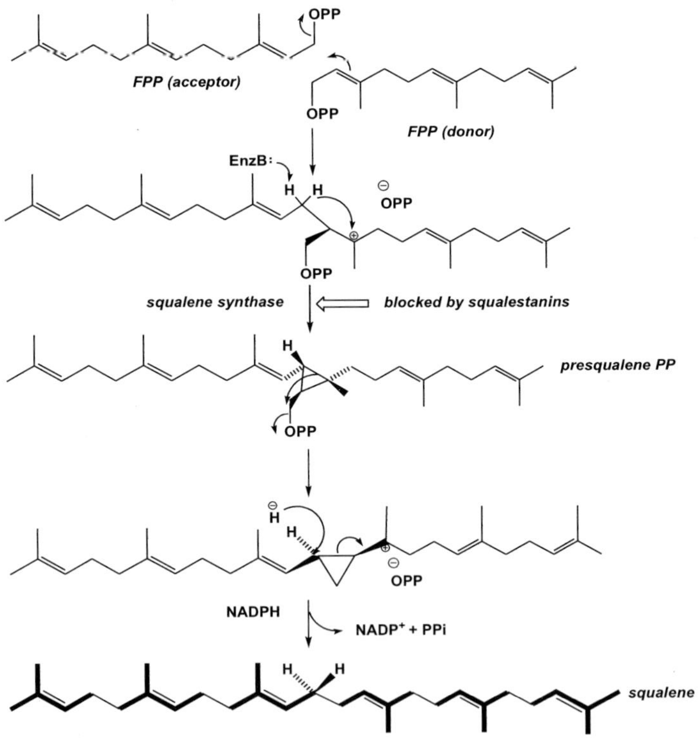

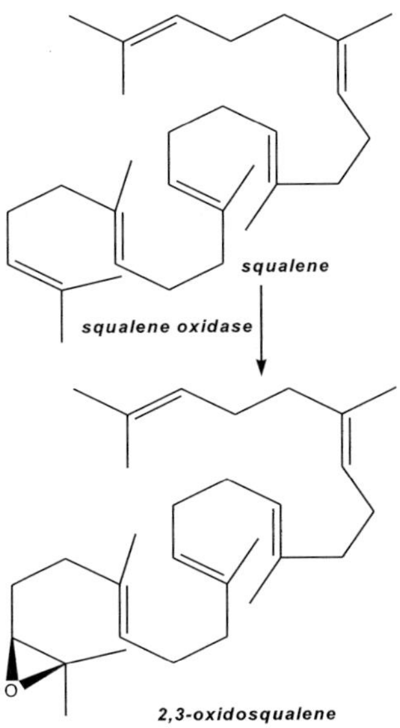

Chapter 12 In Vitro Azadirachtin Production

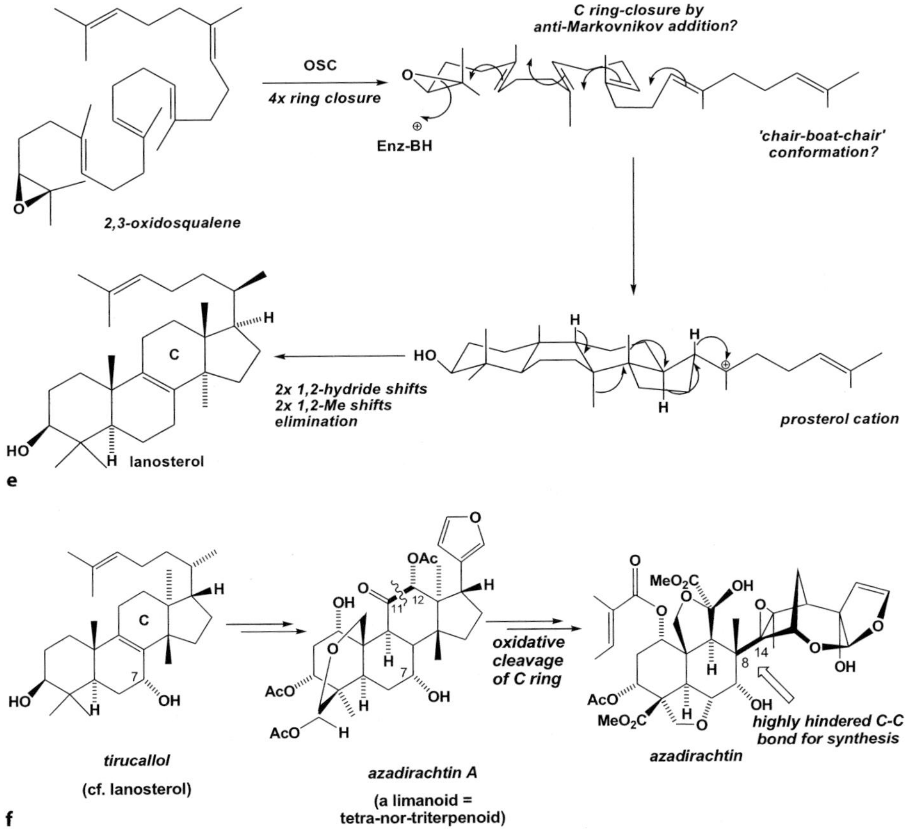

Fig. 12.3 Biosynthetic pathway for azadirachtin [38]. **e,f** *(continued)* **e** Conversion of squalene to lanosterol. **f** Formation of azadirachtin via steroid ring cleavage

the preparative scale, due to the complexity and similarity in structure of the chemicals found in the seeds, foliage and cell culture of the neem tree [42]. Azadirachtin is non-volatile and highly polar, thus it is unsuited for gas chromatography [43]. Reverse phase-HPLC is used for the qualitative and quantitative estimation of azadirachtin because of the polarity of neem compounds [44, 45]. Since the molecule does not carry a strong ultraviolet (UV)-absorbing chromophore, the UV detection that is usually employed for the azadirachtin is not very sensitive. The UV signals have to be recorded at lower wavelength (217 nm) and a UV-transparent solvent system is required for maximising the sensitivity [45]. HPLC analysis of azadirachtin is reported by several workers [39, 46, 47]. Other methods like super-critical fluid chromatography [43, 48] and liquid chromatography-mass spectrometry [49] have also been reported for the analysis of azadirachtin. Colorimetric determination of azadirachtin-related limonoids has also been reported using an acidified vanillin solution [50].

12.6 Availability of Azadirachtin

Extracts of neem seeds containing azadirachtin together with several structurally related molecules have formed the basis of neem usage in insect control [51]. Crude neem extracts have been used as a source of azadirachtin at a local, small-farm level for some time in countries where neem grows indigenously or where plantations have been established. In major Western countries of the world, including the USA, Canada and some of the European countries, only a few commercial neem insecticides have reached the market place to date. Progress has been hampered by lack of supplies of neem seeds (which contain a high amount of azadirachtin as compared to the other parts of the neem tree) of known azadirachtin content, lack of standardisation of formulated products, the cost of the product and the lack of regulatory approval of the complex mixture of compounds found in neem extracts. The first commercial neem insecticide, Margosan-O, was registered by the Environmental Protection Agency in 1985 for use on non-food crop. It contains 3000 ppm azadirachtin [52]. Various other products based on azadirachtin are being formulated and sold by a large number of companies, for example Margosan-O [52], Green Gold Neem Extract, Align, Azatin-R, Turplex, Neemix 90EC (azadirachtin concentration, 90 g·l^{-1}), Neemmzid, Trilogy 90EC, Bio-neem, Turplex and Bollwhip (all Thermo Trilogy), Fortune Aza and Fortune Biotech (Fortune), Neem Surakasha, Proneem, Neem wave and Aza technical (all Karapur Agro), Neem Azal (Trifolio-M), Kayneem (Krishi Rasyan), Neemolin (Rallis), Surfire and Neemachtin (Consep) and Nimbecidine (T stanes) [14].

To date, the only commercially feasible technology to produce azadirachtin is by natural seed extraction, but there are some serious drawbacks to this approach. The reproductive phase in *Azadirachta indica* begins after 5–6 years of plantation and economic yields are obtained only at the age of 10–15 years [14]. The availability of seeds is seasonal. Seeds are known to loose considerable viability and azadirachtin content during storage [53]. Heterogeneity in azadirachtin content in seeds has also been reported by Sidhu and Behl [54]. Moreover, the current supply of azadirachtin is unable to meet the increasing demands. Hence, there is a need for the development of a commercially viable alternative for its enhanced and continuous production.

Keeping in view the drawbacks of the current azadirachtin production methods mentioned herein, plant cell culture technology could be an attractive alternative source to the whole plant for the production of high-value secondary metabolites [55–60].

12.7 Plant Cell/Tissue Culture: an Alternative for Azadirachtin Production

Plant cell and tissue cultures have been suggested as alternative means for continuous (year-round) production of azadirachtin, with the added advantage of increased yield by cell-line selection and manipulation [13, 41, 61].

The basic steps of plant cell/tissue culture employed for the mass production of valuable secondary compounds for the industry include callus culture, cell suspension culture in bioreactors and organ culture (e.g. hairy root cultivation). To make secondary metabolite production economically feasible, proper selection and establishment of a high-yielding (elite) cell or tissue culture system is a necessary prerequisite.

12.7.1 Azadirachtin Production from Plant Cell/Tissue Cultures of Azadirachta indica

The majority of research on azadirachtin production from neem cell/tissue cultures has used undifferentiated cultures [9, 13, 15]. All the earlier studies on in vitro culture of neem have dealt with basic culture techniques of callus induction and plant regeneration [40, 62–67]. All of these studies exhibited variations in the callus induction response depending upon the age and size of explants and cultural conditions employed among the diverse neem genotype [14]. The presence of azadirachtin in callus cultures initiated from leaf and bark explants [13–15, 18], from flower explants [17] and nodal segment explants [21] have been reported in literature. Azadirachtin could not be detected in certain in vitro propagated shoot systems [16, 18, 68]. The formation of transformed callus using *Agrobacterium tumefaciens* has also been reported [69]. However, azadirachtin production from the transformed callus has not been reported to date.

Most of the research is geared toward commercialisation. From an engineering perspective, cell suspension cultures have more immediate potential for industrial application. That is because of well-established expertise in large-scale microbial suspension cultures, which may be considered similar to plant cell suspension cultures in some respects. Therefore, suspension cultures may be used and exploited for the large-scale production of secondary metabolites from plant cells. Allan et al. [15] were the first to report the production of azadirachtin in cell suspension cultures [15]. Since then there have been some successful attempts for azadirachtin production by cell suspension cultures derived from neem leaves and bark, nodal and flower explants [20, 21, 70–73]. Despite several efforts, the amount of azadirachtin production in callus and cell cultures reported to date has been comparatively lower than that found in seeds.

Different research efforts thus far have succeeded in producing a wide range of valuable phytochemicals by unorganised callus or suspension cultures; in some cases, however, production requires more differentiated microplant or organ cultures [10, 74, 75]. Neem roots have been reported to be rich in azadirachtin and related limonoids [76, 77]. Production of azadirachtin has been shown to be higher in in vitro roots and shoots of neem as compared to that of field-grown plants [78]. Seeds obtained from micropropagated neem plants have also been reported to contain azadirachtin [79].

Limitation with root/shoot culture systems of higher plants has been their slow growth; hence, transformed cultures like hairy root cultures have been advocated for the production of plant secondary metabolites. They have higher biochemical and genetic stability than undifferentiated cultures [80], do not require plant growth regulators (hormones) and are equally amenable to scale-up [81]. Hairy roots of *Azadirachta indica* have been reported to contain azadirachtin [61]. Zounos et al. [61] described the development of two cell lines, one derived from the stems of micropropagated shoot cultures and the other from the leaves of aseptically grown whole plants. Both cell lines demonstrated the same growth rate and doubling time, but differed in their azadirachtin concentration [61]. Hairy root cultures were also established from stem and leaf explants, and azadirachtin-related limonoids were detected [19]. The culture had a relatively fast growth rate and exhibited a 100-fold increase in biomass over a 4-week culture period.

12.7.2 Yield Improvement Strategies

In order to obtain products at concentration levels high enough for commercial production, many efforts have been made to stimulate biosynthetic capabilities of cultured plant cells/tissues using different techniques. Typical biotechnological approaches that may increase the productivity of the differentiated/dedifferentiated cultures include media and culture condition optimisation, immobilisation, addition of precursors, elicitors (biotic and abiotic), permeabilising agents and growth regulators. Biotechnological tools like in vitro regeneration and genetic transformation can also be employed for yield enhancement.

12.7.2.1 Strain Improvement and Selection

The biggest drawback with plant cells is their inherent genetic and epigenetic instability. Variability often leads to a reduction in metabolite productivity with subculturing and has been attributed to genetic changes by mutation in the culture, or epigenetic changes, which are due to physiological conditions. The physiological characteristics of individual plant cells are not always uniform; hence, these changes can be reversed by screening for a desired cell population from the heterogenous population. Therefore, strain improvement and selection is the most promising way of increasing the levels of desired metabolites. Several researchers have used cell-cloning methods to enhance the metabolite content [82]. This includes selection of a parent plant with a high content of the desired metabolite to obtain high-producing cell lines. Statistically, high-producing plants give rise to high-producing cell lines [83, 84], but plant cells have also shown variability in yield [85]. Yamada and Sato [86] carried out repeated cell cloning using cell aggregates of *Coptis japonica*, and obtained a strain in which growth was increased by about six-fold and higher amounts of berberine were produced.

As the azadirachtin content is known to vary with genotypic and environment conditions [54], screening and establishment of a productive cell line is necessary for achieving high productivity. A colorimetric procedure has been used as a potential, efficient and fast screening system for the determination of azadirachtin content from different cell lines [49, 14].

12.7.2.2 Media Compositions and Culture Conditions

The productivity of any tissue culture system is greatly influenced by the cultivation conditions [87, 88]. Several-fold enhancements in yield of secondary metabolites have been reported for many cell cultures [89–91]. Several physical and chemical factors influence secondary metabolism in plant cell cultures. Alterations in environmental factors such as nutrient levels, light and temperature have proved to be effective in increasing the product yield and productivity. For example, in *Lythospermum erythorhizon*, the synthesis of shikonin was inhibited by ammonium ions, but these ions promoted cell growth. Hence, for maximum productivity, a medium containing nitrate ions was required at the end of the growth phase [92]. Optimisation of the culture medium therefore is an extremely important aspect for enhancing growth and product concentration.

Wewetzer [13] observed a comparatively higher concentration of azadirachtin on White's medium as compared to that on Murashige and Skoog medium by the same cell line. Two cell lines derived from leaf and bark explants demonstrated variation in azadirachtin content on the different nutrient media and carbohydrate sources used. This observation demonstrated the effect of media composition on azadirachtin production [13]. The effect of major nutrients on the growth and production of azadirachtin-related limonoids in plant cell cultures of *Azadirachta indica* (neem) has been studied and established by Raval et al. [21]. Furthermore, the influence of the culture medium constituents on growth and azadirachtin production in cell suspension culture of *Azadirachta indica* (A. Juss) had been studied and statistical medium optimisation was carried out by Prakash and Srivastava [72, 73]. Satdive et al. [93] have also reported enhanced production of azadirachtin using medium optimisation from hairy root cultures of *Azadirachta indica* A. Juss.

Optimisation of hormone concentrations and their combinations are also found to be highly effective in increasing metabolite yields [84]. Growth regulators mostly used in the literature for cultures producing azadirachtin include indole butyric acid, indole acetic acid, naphthalene acetic acid, 2,4-dichlorophenoxyacetic acid, benzyl adenine and kinetin either alone or in different combinations [13, 15–18, 41]. There is poor knowledge about the effects of environmental conditions like temperature, pH and light intensity on azadirachtin production. In most of the studies, cultures have been grown at 25–28°C under a 16/8 h light/dark regime [14]. Better stability of azadirachtin has been reported under mild acidic conditions [47].

12.7.2.3 Application of Elicitors, Precursors and Permeabilising Agents

Exogenous addition of "elicitor" molecules of biotic and abiotic origin has been one of the most promising strategies to enhance the productivity of secondary metabolites in plant cell technology [94, 95]. Elicitors are defined as molecules that stimulate defence or stress-related responses in plants, which results in improved biosynthesis of the secondary metabolites. Among various elicitors, different classes of jasmonates, chitin or chitosan, salicylic acid, yeast extract and their derived products, heavy metals and fungal/bacterial pathogen cell-wall-derived oligosaccharides have been confirmed as effective elicitors for the production of secondary metabolites in plant cell culture [96]. In recent years, the effect of different biotic and abiotic elicitors has been studied in many important plant secondary metabolites [97–99]. The effect of an elicitor may vary with plant species and there is no universal effect of a particular elicitor on different plant culture systems. Therefore, selection of the right elicitors and appropriate dose optimisation is necessary to identify potentially important elicitors, which may enhance the product yield.

Methyl jasmonate exhibited great influence on the bioproduction of azadirachtin in cell cultures of *Azadirachta indica* over other elicitors (copper sulphate, salicylic acid, pectinase, cellulase, pectolyase, silver nitrate and fungal cell extracts and culture filtrates) employed by Balaji et al. [20]. Enhanced production of azadirachtin by hairy root cultures of *Azadirachta indica* var. A. Juss by elicitation has been reported by Satdive et al. [93].

Any intermediate compound, endogenous or exogenous, that can be converted into a desired secondary metabolite is known as a precursor. They can be classified as natural (non-member of the biosynthetic pathway) or obligatory (member of the biosynthetic pathway leading to the synthesis of a desired secondary metabolite). Metabolic precursors have been commonly applied to the culture medium to enhance the production of secondary metabolites [100]. The addition of a precursor may influence the spatial orientation of enzymes, compartmentation of enzymes and substrate accumulation for secondary metabolite biosynthesis. Azadirachtin is formed via the isoprenoid pathway. Its biosynthesis proceeds through mevalonate, squalene, apo-tirucallol and through a series of oxidation, ring-cleavage, and degradation reactions [22]. A few potential isoprenoid biosynthetic intermediates, like sodium acetate, mevalonic acid lactone, isopentenyl pyrophosphate, geranyl pyrophosphate and squalene, on addition to the culture medium, have resulted in increased azadirachtin production [20].

Permeabilising agents are those compounds that have the ability to increase the pore size in one or more of the membrane systems of the plant cell, enabling the passage of various molecules into and out of the cell. Attempts have been made to reversibly permeabilise the plant cells for a shorter period of time to leach out the bioactive compound and increase mass transfer of the substrate and metabolites across the cell membrane with little or no significant effect on cell viability [84, 101, 102]. Enhanced secretion of azadirachtin was reported after treatment of an *Azadirachta indica* cell suspension culture with different permeabilising agents (Triton X100, dimethyl sulphoxide, cetrimide and chitosan) [41]. Their results also indicated that enhanced release of azadirachtin by permeabilising agents also enhanced the overall productivity of azadirachtin when compared with untreated cultures. In the study conducted by Balaji et al. [20], it was concluded that enhancement of azadirachtin production by permeabilising agents depends on the productivity of the culture. The azadirachtin productivity increased nine-fold (to a value of 90.19 mg·l^{-1}) from the one reported by Kuruvilla et al. [41].

12.7.2.4 Genetic Engineering Approach

Recent advances in plant genetics and recombinant DNA technology to improve secondary metabolite biosynthesis includes identification and manipulation of the enzymes involved in the biosynthetic pathway of the secondary metabolite. Genetic engineering includes isolation, characterisation and reordering of the responsible genetic material and its transfer to foreign organisms [84]. Initial trials of this approach were carried out in *Lupinus polyphyllus*, *Peganum harmala* and *Petunia hybrida* cultures [103]. Despite great success of this technique in herbaceous plants, application in tropical tree species has been challenging [104]. Indeed, successful cases of genetic transformation in tree species have been reported [105–108].

As far as the application of this technology to azadirachtin production is concerned, the genes responsible for the synthesis of azadirachtin can be placed in the engineered yeast or any other rapidly growing micro-organism. The engineered strain can prove to be a more affordable source with higher azadirachtin yield and productivity. The genetic engineering approach enables us to reduce the dependence on the natural resource for phytochemical production by providing an alternative to extraction from plants [109].

12.7.2.5 Somatic Embryogenesis and Regeneration

Somatic embryogenesis is a good example of a potential biotechnology application to conventional tree improvement. Immature embryos selected from the seeds of superior trees under appropriate culture conditions can produce a mass of embryogenic tissues from which several thousands of somatic embryos can be obtained. Thus, allowing the production of a large number of trees from a single seed. Somatic embryogenesis and *Agrobacterium*-mediated genetic transformation have also been used to incorporate additional useful traits of herbicide and/or pest tolerance in tropical trees of *Azadirachta excelsa* [104]. The authors have attempted to develop methods for somatic embryogenesis and gene transfer in *Azadirachta excelsa*.

Methods for *in vitro* propagation of neem have been discussed by Joshi and Thegane [110]. There have been reports on *in vitro* regeneration of *Azadirachta indica* plantlets from various explants like axillary bud [111], immature embryos/seedlings [64, 69, 76], leaves [63, 66], cotyledons and hypocotyls [65], stem tissue [112], anther [40] and through direct and indirect somatic embryogenesis and organogenesis [113, 114]. Further, somatic embryos have also been initiated with mature seeds of neem (*Azadirachta indica* A. Juss.) and regeneration has been carried out via somatic embryogenesis [67, 115].

12.7.2.6 Two-Phase (Stage) Systems

Synthesis and storage of secondary compounds in plant cells often take place in separate phases and compartments. In undifferentiated callus or suspension cultures, these accumulation sites are missing. This is probably the reason for low yields of such compounds reached in these plant cell cultures. The addition of an artificial site for the accumulation of secondary metabolites has been shown to be an effective tool for increased productivity in plant cell cultures. If the formation of a product is subject to feedback inhibition or intracellular degradation, the removal and sequestering of the product in an artificial compartment may increase metabolite production [84].

Some compounds may be non-growth-associated in nature. For them, a two-stage process can prove to be beneficial for enhanced metabolite production. Better volumetric productivity (0.32 mg·l^{-1}·day^{-1}; nearly two-fold) of azadirachtin has been achieved in a two-stage culture system as compared to that obtained in a single-stage process (0.17 mg·l^{-1}·day^{-1}) [21].

12.7.2.7 Immobilisation

Immobilisation of plant cell/tissue cultures has several advantages. In general, cell immobilisation provides a continuous process operation, reuse of biocatalysts, separation of growth and production phases, and a simplified separation of biocatalysts from the culture medium, which allows product-oriented optimisation of the medium and reduction of cultivation periods [116]. It has been observed that immobilisation either through gel entrapment or surface adsorption enhances productivity and prolongs the viability of cultured cells [84]. As far as application of this strategy for enhanced azadirachtin production is concerned, there are no reports available to date. Systematic studies are needed in order to exploit this technique for increased azadirachtin production from plant cell/tissue cultures.

12.8 Stability of Azadirachtin

Azadirachtin is one of the highly oxidised limonoids known. Due to the presence of acid-sensitive groups like tertiary hydroxyls, a ketone group and a dihydrofuran ring azadirachtin is highly unstable under acidic conditions, while the presence of four ester groups makes it equally unstable in alkaline conditions. In addition to these chemical factors, various physical parameters such as temperature, light and humidity are known to affect the molecule, which leads to its rapid decomposition. It has also been reported that interaction between various solvent molecules with active functional groups of azadirachtin could result in chemical changes in the molecule. Azadirachtin is unstable in its native form (i.e. in the neem seed), in extracts and in the pure state. The stability of azadirachtin has been studied extensively under various storage conditions. It is extremely labile in the presence of sunlight. In a study of photostabilisation of azadirachtin in the presence of UV absorbers, photostability has been shown to increase by nearly six-fold [117]. Studies on increasing the half-life of azadirachtin using chemicals, fatty acids, oils, surfactants and organic solvents have been reported in the literature [47, 118, 119].

12.9 Scale-up of In Vitro Azadirachtin Production

Although several studies related to the production of useful compounds by plant tissue cultures have been reported, commercial production is still restricted to very few products like shikonin [120], berberine [121], ginseng saponins [122] and paclitaxel [123]. This is due to the lack of sophisticated technology for the commercial production of plant tissue culture products and the consequent expense of available production methods. The development of large-scale cultivation processes is complicated because of several specific characteristics of plant cells. Bioreactors are considered to be the key step towards the commercial production of secondary metabolites by plant cell technology. They offer optimal conditions for large-scale, plant-derived metabolite production for commercial manufacture [124, 125]. On-line measurement of process parameters such as temperature, pH, dissolved oxygen, carbon dioxide and other gases is possible during bioreactor operations. Hence, efficient control of parameters that affect growth and product synthesis can be achieved using the signals from the sensors mentioned above and various control strategies in the bioreactors. Despite the advantages, transfer from shake flask to pilot plant-scale bioreactor level is a problem because of the slow growth of plant cell/tissue, low shear resistance, and tendency towards cell/tissue aggregation. The choice and design of a bioreactor is therefore determined by factors like shear environment, oxygen transfer capacity, mixing mechanism, foaming, maintenance of aseptic conditions for long fermentation periods and capital investment. Different bioreactor operating strategies can be used in plant cell/tissue cultures to enhance secondary metabolite production. The fed-batch process is one such effective approach to improve the yield and productivity of the bioactive compounds from plant cell cultures [126–128]. Application of continuous culture with and without cell recycling can also be adopted to overcome the limitations of batch and fed-batch processes and for improvement of yield and volumetric productivity [14].

Not many studies have been carried out on azadirachtin production in bioreactors, but among them, a few are suspension culture studies carried out on *Azadirachta indica* for azadirachtin production by Raval et al. [21]; Prakash et al. [71] and Prakash and Srivastava [72]. There has been an attempt in literature to develop a mathematical model for growth and azadirachtin production from suspension cultures of *Azadirachta indica* [73]. This helped in the design of suitable bioreactor strategies (fed-batch and continuous cultivation) for the large-scale production of azadirachtin with an additional advantage of minimisation of the time required for process optimisation. Successful hairy root cultures of *Azadirachta indica* have been established for enhanced azadirachtin production in the literature [19, 93], but bioreactor studies have not yet been reported.

12.10 Conclusion

As awareness towards environmentally friendly and non-toxic pesticide continues to grow, azadirachtin is gaining more and more attention all over the world. It is now well accepted that it is relatively safe for both user and consumer. At present, the demand for azadirachtin is greater than its supply; however, due to the variability of azadirachtin content in seeds it is difficult to depend solely on mass production from natural resource. In addition, the amount of material processed is enormous. Thus, in order to fulfil the increasing demand of biopesticides, other alternatives have been investigated. Together with genetic and biochemical engineering tools, increasing biopesticide (azadirachtin) demand in the market can be met successfully using a process that is continuous, economical and independent of natural resources. Hence, plant cell culture technology has been considered as an attractive alternative source.

Research into azadirachtin production from plant cell/tissue cultures is still in its initial stages and there is a long way to go towards a commercially viable process for azadirachtin production. Knowledge of the biosynthetic pathway of azadirachtin in plants is not yet intricately described and understood. Information is needed at a cellular and molecular level before an efficient alternative for the large-scale commercial production of azadirachtin can be achieved. Despite the various biotechnological advances made in the production technology of azadirachtin to date, efforts are still required in terms of scale up in bioreactors for plant cell/tissue cultivations to economically produce azadirachtin on a large scale.

References

1. Butterworth JH, Morgan ED (1968) J Chem Soc Chem Comm 1:23
2. Wan MT (1994) Environment Canada's Internet resource for weather and environmental information. Chemicals Evaluation, Environmental Protection, Pacific Region, Environment Canada. p 79
3. Schmutterer H (1990) Ann Rev Entomol 35:271
4. Mordue AJ, Blackwell A (1993) J Insect Physiol 39:903
5. Govindchari TR (1992) Curr Sci 63:117
6. Fukuzaki T, Kobayashi S, Hibi T, Ikuma Y, Ishihara J, Kanoh N, Murai A (2002) Org Lett 4:2877
7. Nicolaou KC, Sasmal PK, Koftis TV, Converso A, Loizidou E, Kaiser F, Roecker AJ, Dellios K, Sun XW, Petrovic G (2005) Angew Chem Int 44:3447
8. Nicolaou KC, Sasmal PK, Roecker AJ, Sun XW, Mandal S, Converso A (2005) Angew Chem Int Ed 44:3443
9. Van der Esch SA, Giagnacovo G, Meccioni O, Vitali F (1993) Giorn Bot Ital 127:927
10. Allan EJ, Stuchbury T, Mordue(Luntz) AJ (1999) In: Bajaj YPS (ed) Biotechnology in Agriculture and Forestry Science Series, vol 43. Springer, Berlin Heidelberg New York, pp 11–41
11. Rangaswamy NS, Promila (1972) Z Pflanzenphysiol 67:377
12. Sanyal M, Datta PC (1986) Acta Hort 188:99
13. Wewetzer A (1998) Phytoparasitica 26:47
14. Prakash G, Bhojwani SS, Srivastava AK (2002) Biotechnol Bioprocess Eng 7:185
15. Allan EJ, Eeswara JP, Johnson J, Mordue AJ, Morgan ED, Stuchbury T (1994) Pestic Sci 42:147
16. Kearney ML, Allan EJ, Hooker JE, Mordue AJ (1994) Plant Cell Tiss Org Cult 37:67
17. Veeresham C, Kumar MR, Sowjanya D, Kokate CK, Apte SS (1998) Fitoterapia 69:423
18. Srividya N, Sridevi BP, Satyanarayana P (1998) Ind J Plant Physiol 3:129
19. Allan EJ, Eeswara JP, Johnson J, Jarvis AP, Mordue AJ, Morgan ED, Stuchbury T (2002) Plant Cell Rep 21:374
20. Balaji K, Veeresham C, Srisilam K, Kokate C (2003) J Plant Biotechnol 5:121
21. Raval KN, Hellwing S, Prakash G, Plasencia RA, Srivastava AK, Buchs J (2003) J Biosci Bioeng. 96:16
22. Kraus H, Bokel M, Klank A, Pohnl H (1985) Tetrahedron Lett 26:6435
23. Bilton JN, Broughton, HB, Jones PS, Ley SV, Lidert Z, Morgan ED, Rzepa HS, Sheppard RN, Slawin AMZ, Williams DJ (1987) Tetrahedron 43:2805
24. Turner CJ, Tempesta MS, Taylor RB, Zagorski MG, Termini JS, Schroeder DR, Nakanishi K (1987) Tetrahedron 43:2789
25. Ambrosino P, Fresa R, Fogliano V, Monti SM, Ritieni A (1999) J Agric Food Chem 47:5252
26. Ley SV, Denholm AA, Wood A (1993) Nat Prod Rep 10:109
27. Damarla SR (2001) In: Kelany IM, Reinhard W (eds) Proceedings of the Workshop on Pratice Oriented Results on Use of Plant Extracts and Pheromones, Cairo, Egypt. p 11

28. Ley SV (1994) Pure and Appl Chem 66:2099
29. Mordue (Luntz) AJ, Nisbet AJ (2000) An Soc Entomol Bras 29:615
30. Aldhous P (1992) Science 258:893
31. S. Raina, B.A. Bhanu Prasad, and V.K. Singh (2003) Arkivoc III:16
32. Mukherjee SN, Rawal SK, Ghumare SS, Sharma RN (1993) Experientia 49:557
33. Mordue(Luntz) AJ, Simmonds MSJ, Ley SV, Blaney WM, Mordue W, Nasiruddin M, Nisbet AJ (1998) Pestic Sci 54:277
34. Subapriya R, Nagini S (2005) Curr Med Chem 5:149
35. Akudugu J, Gade G, Bohm L (2001) Life Sciences 68:1153
36. Salehzadeh A, Akhkha A, Cushley W, Adams RLP, Kusel JR, Strang RHC (2003) Insect Biochem Mol Biol 33:681
37. Rohmer M, Knani M, Simonin P, Sutter B, Sahm H (1993) Biochem J 295:517
38. www.ch.ic.ac.uk/spivey/teaching/org4biosynthesis/org405isoprenoids.pdf
39. Govindachari TR, Sandhya G, Ganeshraj SP (1990) J Chromatogr 513:389
40. Gautam VK, Nanda K, Gupta SC (1993) Plant Cell TissOrg Cult 34:13
41. Kuruvilla T, Komaraiah P, Ramakrishna SV (1999) Ind J Exp Bot 37:89
42. Lavie D, Levy EC, Jain MK (1971) Tetrahedron 27:3927
43. Huang HP, Morgan ED (1990) J Chromatogr 519:137
44. Ismann MB, Koul O, Luczynski A, Kaminski J (1990) J AgricFood Chem 38:1406
45. Sundaram KMS, Campbell R, Sloane L, Studens J (1995) Crop Prot 14:415
46. Hull CJ, Dultton WR, Switzen BS (1993) J Chromatogr 633:300
47. Jarvis AP, Johnson S, Morgan ED (1998) Pestic Sci 53:217
48. Johnson S, Morgan ED (1997) Phytochem Anal 8:228
49. Schaaf O, Jarvis AP, Van der Esch SA, Giagnacovo G, Oldham NJ (2000) J Chromatogr 886:89
50. Dai J, Yaylayan VA, Raghavan GS, Pare JR (1999) J Agric Food Chem 47:3738
51. Isman MB (1997) Phytoparasitica 25:339
52. Jacobson M (ed) (1988) The Neem Tree, Vol. 1. CRC, Boca Raton, Florida, p 101
53. Yakkundi SR, Thejavathi R, Ravindranath B (1995) J Agric Food Chem 43:2517
54. Sidhu OP, Behl HM (1996) Curr Sci 70:12
55. Ravishankar GA, Venkataraman LV (1990) Curr Sci 59:914
56. Endress R (ed) (1994) Plant Cell Biotechnology. Springer-Verlag, Berlin, pp 321–330
57. Alfermann AW, Petersen M (1995) Plant Cell Tiss Org Cult 43:199
58. Stockigt J, Obtiz P, Flakenhagen H, Lutterbach R, Endress R (1995) Plant Cell Tiss Org Cult 43:914
59. Dicosmo F, Misawa M (1995) Biotechnol Adv 13:425
60. Scragg AH (1997) Adv Biochem Eng Biotechnol 55:239
61. Zounos AK, Allan EJ, Mordue (Luntz) AJ (1999) Pestic Sci 55:486
62. Van der Esch SA, Giagnacovo G, Maccioni O, Vitali F (1993) Giorn Bot Ital 127:927
63. Ramesh K, Padhya MA (1980) Ind J Exp Biol 28:932
64. Thiagarajan M, Murali PM (1994) Ind Forester 6:500
65. Su WW, Hwang WI, Kin SY, Sagwa Y (1997) Plant Cell Tiss Org Cult 50:91
66. Eeswara JP, Stuchbury T, Allan EJ, Mordue (Luntz) AJ (1998) Plant Cell Rep 17:215
67. Murthy BNS, Saxena PK (1998) Plant Cell Rep 17:469
68. Chaturvedi R (2002) PhD Thesis, Delhi University, Delhi, India
69. Naina NS, Gupta PK, Mascarenhas AF (1989) Curr Sci 58:184
70. Keller H, Pierce L, Irina B, Dennis Ray P (1997) US patent 5698423
71. Prakash G, Emmannuel CSJK, Srivastava AK (2005) Biotechnol Bioprocess Eng 10:198
72. Prakash G, Srivastava AK (2005) Process Biochem 40:3795
73. Prakash G, Srivastava AK (2006) Biochem Eng 29:62
74. Dornenburg H, Knorr D (1997) Food Technol 51:50
75. Hampel D, Mosandl A, Wust M (2005) Phytochemistry 66:305
76. Srividya N, Devi BPS (1998) Indian J Plant Physiol 3:129
77. Siddiqui MA, Alam MM (1985) Neem Newslett 2:43

78. Mishra AS, Rao GP (1988) Phytophylactica 20:93
79. Venkateshwarlu B, Mukhopadhyay K (1999) Curr Sci 76:626
80. Wysokinska H, Chmiel A (1997) Acta Biotechnol 17:131
81. Payne GF, Bringi V, Prince C, Shuler ML (1992) Root cultures. In: Shuler ML (ed) Hanser Series in Biotechnology. Plant Cell and Tissue Culture in Liquid Systems. Hanser, Munich, p 225
82. Misawa M (1985) Adv BioChem Eng/Biotech 31:59
83. Deus B, Zenk MH (1982) Biotechnol Bioeng 24:1965
84. Dornenburg H, Knorr D (1995) Enzyme Microb Tech 17:674
85. Dougall DK (1985) Chemicals from plant cell cultures: yields. and variation. In: Zaitlin M, Day P, Hollaender A (eds) Biotechnology in Plant Science: Relevance of Agriculture in the Eighties. Academic Press, New York, p 627
86. Yamada Y, Sato F (1981) Phytochemistry 20:545
87. Buitelaar RM, Tramper J (1992) J Biotechnol 23:111
88. Rao R, Ravishankar S, Ravishankar GA (2002) Biotechnol Adv 20:101
89. Panda AK, Mishra S, Bisaria VS (1992) Biotechnol Bioeng 39:1043
90. Sakamoto K, Iida K, Sawamura K, Hajiro K, Yoshikawa T, Furuya T (1993) Phytochemistry 33:357
91. Sato K, Nakayama M, Shigeta J (1996) Plant Sci 113:91
92. Tabata M, Fujita Y (1985) Production of shikonin by plant cell cultures. In: Henke RR, Hughes KW, Constantin MP, Hollaender A (eds) Tissue Culture Forestry and Agriculture. Plenum, New York, p 117
93. Satdive RK, Fulzele DP, Eapen S (2007) J Biotechnol 128:281
94. Eilert U, Kurz WGW, Constable F (1985) J Plant Physiol 119:65
95. Bohlmann J, Gibraltarskaya E, Eilert U (1995) Plant Cell Tiss Org Cult 43:155
96. Radman R, Saez T, Bucke C, Keshavarz T (2003) Biotechnol Appl Biochem 37:91
97. Pitta-Alvarez SI, Spollansky TC, Giulietti AM (2000) Enzyme Microb Technol 26:252
98. Zhang CH, Xu HB (2001) Biotech Lett 23:189
99. Zhao J, Hu Q, Zhu WH (2001) Enzyme Microb Technol 28:673
100. Ciddi V, Kokate C (1997) Ind Drugs 34:354
101. Brodelius P, Nilsson K (1983) Eur J Appl Microbiol Biotechnol 17:275
102. Parr AJ, Robins RJ, Rhodes MCJ (1984) Plant Cell Rep 3:262
103. Endre BR (ed) (1994) Plant Biotechnology. Springer-Verlag, Berlin
104. Morimoto M, Nakamura K, Sano H (2006) Plant Biotechnol 23:123
105. MacRae S, Van Staden J (1993) Tree Physiol 12:411
106. Hatanaka T, Choi YE, Kusano T, Sano H (1999) Plant Cell Rep 19:106
107. Bakkali AT, Jaziri M, Foriers A, VanderHeyden Y, Vanhaelen M, Homès J (1997) Plant Cell Tiss Org Cult 51:83
108. Kendurkar SV, Naik VB, Nadgauda RJ (2006) Genetic transformation of some tropical trees, shrubs, and tree-like plants. In: Fladung M, Ewald D (eds) Tree Transgenesis. Springer, Berlin Heidelberg, p 67
109. Paradise EM, Kirby J, Withers ST, Keasling JD (2005) Metabolic engineering of Saccharomyces cerevisiae for the increased production of isoprenoids." In: Tech: Metabolic Engineering of Yeast. Terpnet Conference 2005 (poster presentation)
110. Joshi M, Thengane S (1996) Potential application of in vitro methods for propagation of neem (Azadirachta indica A. Juss). In: Singh RP, Chari MS, Raheja AK, Kraus W (eds) Neem and Enviroment, Vol 2. Science Publishers, Lebanon, New Hampshire, p 967
111. Joarder A, Naderuzzaman R, Islam M, Hossain N, Joarder B, Biswas B (1993) Micropropagation of neem through axillary bud culture. In: Proceedings of the World Neem Conference, 24–28 February 1993, Bangalore, India, p 41
112. Sanyal M, Das A, Banerjee M, Datta PC (1981) Ind J Exp Biol 19:1067
113. Van der Esch SA, Maccioni O, Vitali F, Giagnacovo G, Pasqua G, Monacelli B (2001) Plant Biosyst 135:13
114. Rout GR (2005) J Forest Res 10:263

115. Shrikhande M, Thengane S, Mascarenhas A (1993) In Vitro Cell Dev Biol – Plant 29:38
116. Vorlop KD, Klein J (1987) Method Enzymol 135:259
117. Sundaram KMS, Curry J (1996) Chemosphere 32:649
118. Johnson S, Dureja P (2002) 37:75
119. Johnson S, Dureja P, Dhingra S (2003) J Environ Sci Health 38:451
120. Fujita Y, Hara, Y, Suga C, Morimoto T (1981) Plant Cell Rep 1:51
121. Fujita Y (1988) Shikonin production by plant (Lithospermum erythrorhizon) cell cultures. In: Bajaj YPS (ed) Medicinal and Aromatic Plants, vol. 4. Springer-Verlag, Berlin, Germany, p 225
122. Ushiyama K, Hibino K (1997) National Meeting of the American Chemistry Society 213, San Francisco, CA, USA
123. Zhong JJ (2002) J Biosci Bioeng 94:591
124. Kieran PM, MacLoughlin PF, Malone DM (1997) J Biotechnol 59:39
125. Tripathi L, Tripathi JN (2003) Trop J Pharm Res 2:243
126. Zhang YH, Zhong JJ (1997) Enzyme Microb Technol 21:59
127. Wu J, Ho KP (1999) Appl Biochem Biotechnol 82:17
128. Wang HQ, Yu JT, Zhong JJ (1999) Process Biochem 35:479

Chapter 13
Arabinogalactan Protein and Arabinogalactan: Biomolecules with Biotechnological and Therapeutic Potential

A. Pal

e-mail: amita@bic.boseinst.ernet.in

Abstract Arabinogalactan proteins (AGPs) are high-molecular-weight transmembrane proteoglycans that are implicated in both the vegetative and reproductive stages of plant development. The bulk of an AGP molecule comprises of polysaccharide chains that are mostly anchored to the plasma membrane by glycosyl-phosphatidylinositol. AGP interacts specifically with a class of phenylazoglycoside dyes known as Yariv reagents; these dyes are thus used as histochemical probes to monitor the subcellular localization of AGPs in plant tissues. Arabinogalactans (AGs), a class of polysaccharides, have been reported to be present in a wide range of plant taxa and are prevalent in larch trees. Commercially, larch arabinogalactan (LAG) is used as dietary fiber and prebiotics, as well as in treatment of intestinal disorders. It also has the ability to enhance the activity of the human immune system by stimulating the cytotoxic activity of natural killer (NK) cells via the cytokine network against certain tumor cell lines. Ukonan C, an AG present in the rhizome of *Curcuma longa*, has the ability to activate reticuloendothelial cells; such stimulated cells with enhanced phagocytic activity can protect the human system from pathogenic foreign agents. Thus, AG acts as a strong activator for two cell types involved in the immune system, namely macrophages and NK cells. In addition, LAG has the ability to block the metastasis of liver tumor cells. Because of its numerous activities, LAG is being used as one of the nonconventional therapeutic agents for cancer treatment. On the other hand, AGPs are of considerable current interest due to their involvement in virtually all facets of plant development. In vitro studies have shown that AGPs play an important role in somatic embryogenesis, which suggests that AGP exhibits plant growth-hormone-like activity. The abiotic stress-protective role of AGP has also been demonstrated. Tissue- and organ-specific expression patterns have already been elucidated for this class of biomolecules, indicating that a specific AGP could be used as a marker for cellular differentiation and as a fate determinant of cultured cells during in vitro differentiation. Due to its immense potentiality, it is considered to be a promising biomolecule, which could resolve several of the hurdles that generally delay the process of crop improvement through biotechnological approaches.

Abbreviations

AG	Arabinogalactan
AGP	Arabinogalactan protein
(ß-glc)3Y	ß-Glucosyl Yariv reagent
(ß-D-gal)3Y	ß-D-Galactosyl Yariv reagent
FLA	Fasciclin-like AGP
GPI	Glycosyl-phosphatidylinositol
GPI-PLC	GPI-specific phospholipase C
IFN	Interferon
IL	Interleukin
LAG	Larch arabinogalactan
NK	Natural killer
PCD	Programmed cell death
PGR	Plant growth regulator
WAKs	Wall-associated kinases

13.1 Introduction

Arabinogalactan proteins (AGPs) are a class of high-molecular-weight (~100–200 kD) transmembrane proteoglycans with less than 10% protein moiety. The

▶ **Fig. 13.1** Hypothetical model of a classical arabinogalactan protein (AGP) carrying a glycosyl-phosphatidylinositol (GPI) lipid anchor. The ellipse represents the 15×25-nm size of carrot AGPs. The wavy line represents the core polypeptide, which, for a 141-kD AGP containing 5.6% protein, has a calculated length of 24 nm (essentially the same as the length of the ellipse). The GPI anchor is similarly drawn to approximate scale and is based on the ethanolamine cap found at the truncated C-terminus of pear and *Nicotiana alata* AGP core polypeptides [112] and on the ceramide lipid found in rose AGPs [5]. Biochemical evidence shows that *Arabidopsis* AGPs also contain GPI anchors [113]. The structure of the oligosaccharide linker between the ethanolamine and lipid is as found in animals and microorganisms. The site of cleavage by phosphatidylinositol-specific phospholipase C (PI-PLC) is indicated. The type II arabinogalactan chains typically consist of 30–150 sugar residues and are attached at many Hyp, Ser, and/or Thr residues in the core polypeptide. The site chains shown on the (1→3)-β-D-galactan backbone are based on oligosaccharides characterized from various AGPs, but their placement as shown is hypothetical. No complete structure has been solved for any AGP. Reproduced with permission from Elsevier, copyright year 1999; original artwork of Serpe and Nothnagel [78]

Chapter 13 Arabinogalactan Protein and Arabinogalactan   257

bulk of the molecule is comprised of carbohydrate, with arabinosyl and galactosyl as the major sugar components (Fig. 13.1). It appears that AGPs are almost ubiquitously present in the plant kingdom and are from bryophytes to angiosperms [1]. These proteoglycans are present predominantly in the intercellular spaces and vascular bundles of leaves, stems, and roots; in floral parts and in the cotyledons of seeds. In addition to their role in cell proliferation, the AGPs are implicated in various developmental processes of plants, including cell proliferation, differentiation, cell-cell recognition, somatic embryogenesis, and programmed cell death (PCD) [2]. Most of the AGPs carry a glycosyl-phosphatidylinositol (GPI), a lipid substitute that anchors proteoglycans to the plasma membrane. The GPI anchor is cleaved by phospholipases C and D to release the AGP from the membrane.

AGPs belong to the large family of hydroxyproline-rich glycoproteins that includes extensins, proline/hydroxy-proline rich glycoproteins, and several lectins, such as potato tuber lectin and lectins of other solanaceous plants [3, 4]. The AGPs are classified into two major groups based on their protein backbones, "classical" and "nonclassical" AGPs. Classical AGPs are endowed with at least three distinct domains: an N-terminal signal sequence, a central domain predominantly rich in hydroxyproline and proline, and a C-terminus hydrophobic domain. This hydrophobic domain is replaced by a GPI membrane anchor that helps the proteoglycan to remain attached to the plasma membrane [5]. Nonclassical AGPs, which are poor in hydroxyproline, contain either a cystein- or asparagine-rich C-terminal domain [3]. This class of AGPs neither contains a hydrophobic C-terminal domain nor encodes for GPI modification [5].

Using genomic analysis, Schultz and his associates [6] have extended this class of proteins by identifying 13 classical AGPs, 10 arabinogalactan (AG) peptides, 3 basic AGPs that include a short lysine-rich region, and a fasciclin-like AGP (FLA) from *Arabidopsis*. FLAs are a subclass of classical AGPs bearing a domain similar to *Drosophila melanogaster* cell adhesion molecules, fasciclins [4]. Several genes encoding FLA protein backbones have been identified in loblolly pine and in poplar; these genes are expressed preferentially in the differentiating xylem [7–10].

AGPs interact specifically with certain synthetic phenylazoglycoside dyes, known as Yariv reagents [11, 12]. These reagents are colored multivalent compounds, the most commonly used being ß-glucosyl Yariv reagent [(ß-glc)$_3$Y]. The specific interaction between AGPs and Yariv reagent has been exploited in the use of these dyes: (1) as specific histochemical probes to localize the subcellular distribution of AGPs [13, 14], and (2) as stains to visualize AGPs following electrophoresis [15, 16]. We have also used this reagent in the in situ detection of AGPs in the cotyledonary tissues of *Vigna radiata*. In addition, similar studies have also enabled us to demonstrate the cellular and tissue-specific localization of AGPs in differentiating roots and shoots. Our contention is that the presence of AGP serves as biomarker for differentiation. A series of monoclonal antibodies has also been used as an alternate strategy to isolate and to locate AGPs in plants. Such studies with antibodies have revealed that the carbohydrate epitopes of an AGP are highly regulated within growing vegetative and reproductive systems [17–19]. Other related antibodies, such as ZUM18 and AUM15, have been used for isolation of different types of AGP from total AGPs of carrot seeds [20].

The AGs, a class of polysaccharides, are found in a wide range of plants and found prevalently in the genus *Larix* (Larch tree). Commercially, larch arabinogalactan (LAG) is extracted from the western larch (*Larix occidentalis*). The other common source of LAG is the Mongolian larch (*Larix dahurica*) [21].

AG is also known to be present in several herbaceous plants, such as *Baptisia tinctoria*, *Echinacea purpurea* [22], *Echinacea pallida* [23], *Curcuma longa* [24], and *Angelica acutiloba* [25]. LAG is highly water soluble and produces a low viscous solution. It has a pine-like odor and is sweet in taste [26]. Such physical attributes have facilitated its use as dietary fiber and prebiotics as well as in treatment for intestinal disorders. These highly branched polysaccharides, with molecular weights ranging from 10 to 120 kD, have 3,6-β-D-galactan units [27]. AGs contain either short side chains consisting of a single galactose residue, or longer side chains made up of galactose and arabinose units in 3,4,6-, 2,3,6-, 3,4-, 3,6-, and 3-linkages. These two sugar residues (consisting of β-galactopyranose, β-arabinofuranose, and β-arabinopyranose) are in a molar ratio of ~6:1, and high-grade AG is generally has a total glycosyl content of >98% [21, 24, 26, 27]. These carbohydrate molecules are attached to the protein core via β-D-galactopyranose-hydroxyproline linkage [28, 29].

13.2 Biological Activities of AGP

AGPs are of considerable current interest because of their apparent role in virtually all facets of plant development. A wide range of biological activities has been proposed for AGPs in the last two decades. AGPs have been implicated in cellular proliferation, expansion, and differentiation, also in the regulation of cell shapes and contours. It also influences the hydrodynamic properties of plasma membranes [30]. Several monoclonal antibodies directed to the carbohydrate component of AGPs have been used extensively to elucidate the role of AGP polysaccharides in plant development. That these polysaccharides are indeed the antigenic determinants was confirmed by inhibition of the antibody–AGP interaction by complex oligosaccharides, such as gum arabic and LAG [31, 32]. Such monoclonal antibody-based studies have clearly shown temporal and spatial regulations of AGP expression [16, 33, 34], suggesting a role in plant development.

The hormonal and developmental regulation of two AGPs, PtX3H6 and PtX14A9, which have been implicated in the development of xylem in growing seedlings of loblolly pine (*Pinus taeda*), has been studied by No and Loopstra [35]. The evidences therein suggest that AGPs are tissue specific and probably play important role in xylem development. This conclusion is in line with our observation that a class of AGPs is expressed during cytokinin-induced in vitro differentiation of *Vigna radiata*; the expression of AGP was demonstrated by its in situ colored complex formation with $(\beta\text{-glc})_3Y$ [36].

Several studies have implicated the influence of AGPs in cell elongation as well as in root elongation [37, 38]. Several molecular and biochemical studies also support the idea that AGPs have specific functions during root development and growth [39, 40]; however, the exact mechanism remains to be elucidated. Nonetheless, recent studies apparently suggest the involvement of wall-associated kinases (WAKs) in AGP-induced root elongation: (1) van Hengel and Roberts [41] have put forward evidences to indicate that AGPs are associated with serine-threonine-type receptor kinases; (2) AGP epitopes have been shown to be co-localized with WAKs [42], and (3) antisense WAKs not only decrease significantly the level of WAK proteins, but also impair cell elongation and completely block lateral root development of *Arabidopsis* [43]. This conclusion has further been substantiated by a concurrent investigation in which it was shown that the expression of WAK2, and no other family member, is essential for the cell expansion in primary roots of *Arabidopsis* [44]. All of this evidence put together establishes a crucial role for AGP in cell elongation in general and root elongation in particular. It remains to be elucidated how AGPs and WAKs transduce signals from the plasma membrane to the cell wall. However, it is clear that the direct physical association of these molecules with the plasma membrane and its environment is crucial, since any disruption of the AGP complex formation results in blockage of cell elongation.

Among the plant growth regulators (PGRs), the auxins, gibberellins, and brassinosteroids are known to promote stem elongation. However, the molecular mechanism for regulation of stem elongation by PGRs is still poorly understood. Park et al. [45] have recently shown, by using the fluorescent differential display method, that a gibberellin-responsive gene from cucumber hypocotyls encodes an AGP and is involved in stem elongation. The biological function of the AGP-encoded gene was also investigated by generating transgenic tobacco-overexpressing CsAGP1 sense RNA of *Cucumis sativa*. The transformants with higher AGP contents were taller with longer internodes and were of an early flowering type than the wild tobacco plants, validating the role of AGPs in stem elongation.

Suzuki et al. [46] have shown a role of AGP in gibberellin-induced α-amylase production in barley aleurone layers. Gibberellin is known to promote both the transcriptional activation and secretion of α-amylase. However, in the presence of 20 mM (β-Glc)$_3$Y, transcriptional activation by gibberellin was completely abolished [47]. It was further confirmed that this inhibition by (β-Glc)$_3$Y was specific to gibberellin-induced α-amylase gene expression, and this is not a general effect on promoter activation. Thus, it was inferred that AGP regulates gibberellin during barley seed development.

AGPs also play an important role in plant embryogenesis [47, 48]. Addition of seed AGPs to an old carrot cell line that had lost its embryogenic potential results in the restoration of its ability to produce somatic embryos. This suggests the PGR-like activity of AGPs. In addition to AGP, AG also displays similar activity in a dose-dependant manner in inducing in vitro differentiation in the cotyledonary tissues of *Vigna radiata* [49]. This is further confirmed by the finding that Larcoll, a member of the AG family, increases wheat microspore viability during in vitro culture [47]. Letarte et al. [50] also noted the phytohormone-like, dose-dependant activity of both AG and AGP, a finding that we have also reported [49]. Thus, it is highly likely that like AGP, its precursor, AG, also acts as a PGR.

The arabinose-rich side chains of AGPs are presumably essential for intercellular attachments and tissue organization. This conclusion is supported by the finding that tissues deficient in AGPs exhibited recalcitrance in morphogenesis responses [51]. Van Hengel and Roberts [41] have demonstrated the role of a single AGP of *Arabidopsis*, AtAGP30, in growth-factor signaling. They also reported earlier a similar phenomenon in carrot, where the embryogenic potential of naked protoplasts was restored and even enhanced by adding AGPs to the culture medium. An important question that has been raised relates to the contribution of AGP-derived carbohydrates in embryogenesis. This has recently been resolved by the finding that chitinase-treated AGPs have a higher signaling potential than the intact proteoglycans [52]. Similar results have also been reported in *Picea abies* by Dyachok et al. [53]. All of these results support the conclusions that chitinases, such as embryogenesis-related CH-4 and other chitinases play a major role in the early stages of embryogenesis by generating oligosaccharides, and these carbohydrate components act as signaling molecules. Alternatively, the carbohydrate chains impart a negative constraint on the molecule and this negative constraint is abolished by the chitinase-catalyzed removal of the carbohydrates. Differential accumulation of AGPs has recently been reported in *Araucaria angustifolia* at different stages of seed development [54]. Such spatial-temporal expression of AGPs seems to function in cell positioning and in determining cell fate during embryo development [55]. Therefore, AGPs are not only potential messengers for cell-to-cell communication, but also function as regulators of cell shape and cell contour.

Recently, the involvement of AGPs that are regulated in synchronization with phytohormones in bipolar development processes during microspore embryogenesis of *Brassica napus* L. cv. Topas has been demonstrated by Tang et al. [56]. Yariv reagent was used to perturb AGPs expressed in microspores and microspore-derived embryos, resulting in the termination of normal progress of embryogenesis; it also affects the basic structural pattern of the embryos. The altered developmental fate of embryonic epidermal cells signifies that AGPs play a crucial role in microspore embryogenesis.

13.3 Role of AGPs in Reproductive Organ Development

Several investigators have reported the involvement of AGPs in the process of sexual reproduction of angiosperms. Expression of AGPs has also been detected in many male reproductive organs, such as pollen and pollen tube, and in female reproductive organs, such as stigma, style, and ovary. Five putative AGPs with distinct tissue specificity have been characterized from the styles and stigmas of *Nicotiana alata*. Transcripts of AGPNa1 are expressed in the style and other organs, whereas expression of AGPNa2 is high in cell cultures, and yet is present in low abundance in styles. While expression of AGPNa3 occurs exclusively in pistils, transcripts were found in abundance within the stigma. Two other proline-rich proteoglycans, AGPNa4 and AGPNa5, are highly expressed in the transmitting tract of the style [57, 58]. Recently, Park et al. [59] have isolated and characterized a gene, *BAN102*, from Chinese cabbage that has a high sequence homology with the *AGP23* gene of *Arabidopsis*. It was noted that this gene is expressed specifically in the pollen and pollen tube.

AGPs were also detected during the zygotic and somatic embryo development of many plant species; and they play multiple roles (e.g., providing surface adhesion and transduction of signals for pollen tube growth and emergence) [15, 60–64]. Yet, basic information regarding the mechanism underlying AGP action is lacking.

Pennell and his associates [33, 34] have elegantly demonstrated a critical role of AGP in sporophyte–gametophyte transition in the ovule of pea and oil seed rape. However, they could not determine the involvement of any specific AGP either during ovule development or during early embryogenesis of the plant. Acosta-Garcia and Vielle-Calzada [65] have recently demonstrated that a classical AGP gene, *AGP-18*, is essential for female gametophyte development in *Arabidopsis*.

The distribution and differential expressions of specific AGPs in reproductive organs, evidence of abundance of AGPs during reproductive organ development, and the tissue-specific expression of AGP perhaps suggest their organizational role in reproductive organ formation.

13.4 Signaling Role of AGP

Plants constantly receive different types of chemical and physical signals from the surrounding environment. Chemical signals include hormones, pathogen elicitors, insecticides, fertilizers, and ozone, to name a few, and physical signals include fluctuations in light intensity, temperature, and wind. In order to perceive and suitably respond to protect themselves from harsh conditions, living cells are endowed with some sensor molecules, such as proteins, at the outer surface of the cells. These molecules transduce signals over the plasma membrane into the cell, resulting in the activation of a cascade of defense-related genes involved in specific biotic and abiotic stress responses. Ion channels, receptor kinases, and receptor-activated effector enzymes participate in transmembrane signaling, leading to the generation of intracellular second messengers [66].

It is known that GPI-anchored proteins are capable of transducing signals [67]. Genes encoding analogous proteins have been identified in *Arabidopsis* and most of these genes encode a protein backbone that is predicted to have a GPI anchor domain. Hence, it is highly likely that these GPI-containing AGPs are involved in signaling pathways [68]. In animal cells, the biosynthesis of GPI structures and their attachment to the peptide precursor takes place in the endoplasmic reticulum. The GPI-linked protein is then transported to the cell surface, where it is anchored to specific sites known as lipid rafts. Lipid rafts are often categorized as specialized centers for signaling cascades due to the presence of several signaling molecules within these lipid microdomains in animal cells [69].

Two mechanisms of signal transduction have been proposed for GPI-anchored proteins. By one mechanism, the regulated release of the protein from the membrane allows it to enter the cytoplasm to generate second messengers, such as phosphotidyl-inositol and inositol phosphoglycan [70], without the involvement of a membrane-spanning protein. One way to release the protein from the membrane is by GPI-anchored hydrolyzing phospholipases, which include GPI-specific phospholipase C (GPI-PLC) and phospholipase D. These two enzymes have slightly different substrate specificities [71]. This mechanism appears to play an important role in the functioning of the AGPs of rose cells, as has been demonstrated by Svetek et al. [5]. These AGPs, which are anchored to the plasma membrane through a GPI-lipid, are cleaved by exogenously added phospholipase C, and the released proteins are transported to the cell wall.

Another mechanism by which GPI-anchored proteins transduce signals is by interacting with certain types of transmembrane proteins, and it is these proteins that transmit signals to the cytoplasmic molecules. It appears that such a mechanism is operative in the interaction between AGP and the cytoskeleton, as has been proposed by Sardar et al. [72]. This interaction is mediated either by transmembrane proteins or by an indirect interaction with lipid rafts. The transmembrane protein, with either a transmembrane domain or with an extracellular domain, may interact with AGP to mediate cellular signaling at the cell surface. The other possible mode of interaction involves lipid rafts.

13.5 Abiotic Stress Tolerance Conferred by AGP

It is envisaged that GPI-PLC-mediated release of the membrane-bound AGP is also under genetic control. In this respect, the following observations implicate clearly the role of AGPs, albeit indirect, as sentinels for abiotic stresses, and their participation in stress regulation. Several PLC-related genes, such as *AtPLC1*, *AtPLC4*, and *AtPLC5* in *Arabidopsis*, and *Vr-PLC3* in *Vigna radiata* are under the control of abiotic stresses, such as drought, cold, and high salinity [73–75]. In addition, a FLA-like with 427 amino acids has recently been identified in *Salicomia europeae*. Expression of this protein renders the organism tolerant to salt, moisture, and heat [76]. This proteoglycan is now under Japanese intellectual property rights protection. All this evidences put together assign a role for AGP in conferring abiotic stress tolerance.

13.6 Probable Role in PCD

Li and Showalter [77] demonstrated the occurrence of ultrastructural changes characteristic of apoptosis in the root-tip cells of tomato seedlings treated with Yariv reagent, (β-D-Gal)$_3$. Apoptosis is a type of PCD in which cytoplasmic shrinkage is noted along with nuclear membrane blebbing and chromatin condensation. PCD is also characterized by internuclear DNA damage and changes in cell morphology. The Yariv-reagent-induced cessation of cell growth has been reported by Serpe and Nothnagel [78] as well as by Langan and Nothnagel [79]. Later, it was shown that such cell death in the presence of $(\beta\text{-D-Gal})_3$ is due to apoptosis and not due to necrosis [80]. These investigators also reported a decline in the number of dead cells with increasing time; this phenomenon has been referred to as "societal control of PCD" [81]. Societal control of PCD may be defined as suppression of PCD by signaling molecules or growth factors that are released by the neighboring cells, which subsequently suppress PCD and help the cells to grow. The mechanism by which the Yariv reagent induces PCD is not clear; however, AGP has been implicated in apoptosis [80]. It has been suggested by Showalter et al. [80] that the AGP-mediated interaction between the plasma membrane and cell wall is interrupted by $(\beta\text{-D-Gal})_3$, and such interruption results in activation of apoptosis-inducing genes. Yet another explanation is that AGP exists in two forms – monomeric and oligomeric – and the oligomeric form is the active form whereas the monomeric form is the dormant form. This monomer–oligomer equilibrium shifts to the monomeric form in the presence of $(\beta\text{-D-Gal})_3$, resulting in the loss of growth signals.

13.7 Commercial Uses of Gum Arabic

Gum arabic, an exudate from *Acacia senegal*, is a mixture of polysaccharides and AGPs, and it acts as a low viscosity emulsifier. This makes gum arabic useful in several industries, and especially in the food industry. It is used as a flavor encapsulator and stabilizer of citrus oil emulsion concentrates in soft drinks [82], as food additives, and the candy industry [2]. It is also used as an adhesive in the stamp industry.

13.8 AG as Dietary Fiber and Prebiotics

In addition to its role as a nutrient supplement to symbiotic intestinal bacteria, Larch AG (LAG) also serves as an excellent source of dietary fiber. The amount of LAG absorbed in the intestine following oral administration could not be determined and, as such, its role in pharmacokinetics remains to be established [26]. AG is also used as a prebiotic, which is defined as a nondigestible food ingredient that may selectively and beneficially affect the host by stimulating the growth and/or the activity of a limited number of bacteria in the colon. The unabsorbed LAG in the gastrointestinal tract is fermented by anaerobic bacteria, particularly *Bifidobacterium longum* and *Lactobacillus* [83, 84]. The metabolic products of AG are short-chain fatty acids containing mainly butyrate and to some extent propionate. Butyrate thus produced protects the mucosa against a plethora of intestinal diseases and agents that promote cancer [85]. Prebiotics including AG are now being used to treat intestinal disorders (diverticulosis, leaky-gut, irritable bowel syndrome) and inflammatory bowel diseases, like Crohn's disease and ulcerative colitis, for example [86]. These compounds mainly stimulate the growth of bifidobacteria and are principally oligosaccharides. Effective prebiotics usually escape digestion in the upper gastrointestinal tract and are used by a limited number of the microorganisms comprising the colonic microflora.

It was demonstrated that purified AGs bind in vitro with liver asialoglycoprotein receptors [87] and are transported to the hepatocytes by receptor-mediated endocytosis. Thus, it is highly likely that AGs may especially benefit patients with liver diseases who are unable to detoxify ammonia, since it is known that AG prevents the generation and absorption of ammonia [83, 88].

13.9 AG as Immunomodulators and Immunity Enhancers

The immunomodulators are agents that are capable of modifying immunological responses and activate natural killer (NK) cells. One of the characteristics of the NK cells is their ability to mediate spontaneous cytotoxicity against a variety of tumor cells and virus-infected cells without prior sensitization by antigens and byproducts of the major histocompatible gene complex [27]. LAG has the ability to enhance the activity of the mammalian immune system. Hauer and Anderer [27] have observed the ability of LAG to stimulate cytotoxicity of NK cells via the cytokine network against K562 tumor cells. LAG pretreatment of NK cells induces an increased release of interferon gamma (IFNγ), tumor necrosis factor alpha, interleukin (IL)-1 β and IL-6. Among these effector molecules, the NK cell-mediated cytotoxicity is mostly due to the induction of IFNγ [27]. AG also enhances the phagocytic activity of reticuloendothelial system. Gonda et al. [24] have reported that ukonan C, an AG present in the rhizome of *Curcuma longa*, has the ability to activate reticuloendothelial cells, which are phagocytic cells capable of engulfing and destroying bacteria, viruses, and other foreign substances. They can also ingest worn-out or abnormal body cells. Thus, AG acts as a strong activator for two cell types involved in the immune system: macrophages and NK cells.

The ability of LAG to (1) block the metastasis of tumor cells to liver and (2) stimulate the cytotoxic activity of NK cells, has led to its use as a potential therapeutic agent in nonconventional cancer therapy [89, 90]. As NK cells are the first line of defense in cancer immunosurveillance, any agent that either stimulates these cells or removes any negative constrain on them will definitely be of medicinal value. Incidentally, the occurrence of tumor metastasis to the liver is more common than to any other organ. AG has been shown to reduce tumor cell colonization and increase the survival time of patients suffering from cancers of the spleen, liver, and colon [91–93]. Several chronic diseases, such as chronic fatigue syndrome [93], viral hepatitis [94], HIV/AIDS [95], and autoimmune diseases like multiple sclerosis [96] have been reported to have low NK cell activity. Generally, anti-inflammatory activity is exhibited by low-molecular-weight AGs, whereas high-molecular-weight AGs tend to enhance phagocytosis and NK lymphocyte activity [97]. LAG treatment has already shown promise in improving the clinical prognosis of these patients. Thus, it is highly likely that the ability of LAGs to stimulate NK activity is the primary reason for such significant improvements in clinical reports from various ailments.

13.10 Echinacea-AG as a Nutraceutical

Nutraceuticals, a term first coined by Stephen DeFelice in 1989 [98], are natural plant products that are not only the source of nutrient-rich foods, but also have therapeutic effects with virtually no adverse reactions [99]. Because of their beneficial effects on the prevention and treatment of several diseases, major biotechnology companies and academia have ventured into the screening and large-scale production of the next-generation pharmaceuticals from plants. One of the significant outcomes of such a search is the identification of a medicinal plant, *Echinacea purpurea*, as a rich source of nutraceuticals. The plant extract has been reported to prevent viral infection and reduce tumor progression [100–104]. In addition, daily use of the extract is known to significantly abate leukemia as well as extend the life span of both leukemic and aging mice [105].

Bioactive compounds present in *Echinacea purpurea*, including AG, are of great significance due to their role in immunosurveillance [106]. In vivo study has revealed that *Echinacea* extract has no toxicity when given either prophylactically or therapeutically [105]. Thus, it is highly likely that AG in combination with other ingredients in the extract is the cause of the observed therapeutic effects. This is further supported by the finding that the total extract is much more potent compared to individual compounds in the extract [90, 94]. This is similar to other naturally nutrient-rich or medicinally active compounds that act additively or even synergistically to offer best effects as nutraceuticals/pharmaceuticals under in vivo conditions.

13.11 Other Uses of AG

AG has been used with encouraging results in combating influenza and cold as well as in lung and ear infections. The middle-ear infection, also known as otitis media, is caused mainly by *Escherichia coli* or *Klebsiella* species, and it afflict infants and children, but also affects adults. AG also decreases serum lipids in hyperlipidemic individuals, and modulates serum glucose. One of the unpleasant side effects of AG treatment is loose motions, probably caused by the proliferation of gut microflora, namely *Lactobacillus* and bifidobacteria [84, 95]. Recently, it has been claimed that LAG serves as a protecting agent for maintaining precious metal nanoparticles in colloidal suspension [107].

13.12 Scope of Exploiting the Potentials of AGP and AG in Plant Biotechnology and Therapeutics

The AGPs are wonder molecules, despite their suggested role in plant differentiation and development; their full commercial potential has not yet been explored comprehensively. Tissue- and organ-specific expression profiling of AGPs indicates that specific AGPs could be used as biomarkers for developmental fate determination. The potential of this group of proteoglycans with enormous promise remain underexplored. It is envisaged that use of selected AGPs with phytohormone-like activities may resolve the recalcitrant nature of isolated plant tissues for in vitro regeneration and in raising transgenic plants.

Various characteristics of AG shows promise of a novel fiber as it is a soluble dietary fiber that is easily incorporated into foods and is fermented in the gut with physiological changes that are beneficial to health, and have other immunological functions. Circumstantial research evidence perceives enormous therapeutic efficacy of AG obtained from various sources, especially from herbs like *Echinacea*, which revealed an avenue for exploration of new natural resources of AG and exploitation of the therapeutic efficacy of AG for human welfare.

13.13 Concluding Remarks

AGPs are multifaceted molecules that participate in almost all aspects of plant development. These molecules regulate plant growth, are involved in reproductive organ development, and appear to play a major role in apoptosis. AGPs also act as external stress regulators. It is intriguing how AGPs bring about so many disparate cellular changes. It should be mentioned in this context that the majority of AGPs are GPI-anchored proteins, and GPI-containing proteins are known to play important role in biological signal transduction. Several such proteins of animal origin, such as lymphocyte proteins (TAP and Thy-1) [108], mouse melanoma cell protein [109] have been well characterized and their signaling pathways have been worked out [110]. Although several plant GPI-containing proteins, which are located in *Arabidopsis* plasma membranes [111], have been implicated in signal transduction; their mode of action, however, is yet to be elucidated. Another future area of investigation is to understand how the distribution of signaling molecules across the tissue is regulated by proteoglycans. This distribution cannot simply be mediated by diffusion, as has been suggested by others. Taken together, these findings suggest the possibility that individual AGPs induce specific changes by interacting with a specific set of cellular proteins. WAKs, which are Ser/Thr kinases, are one such group of enzymes that are known to be directly associated with AGPs. Identification of other interacting proteins will eventually aid in the delineation of signal transduction pathways that are mediated by AGPs.

Acknowledgments

The author expresses her deep gratitude to Dr. S. Bishayee for his constant guidance, critical reading, criticism, and editing of the manuscript. She is also grateful to the Department of Science and Technology, Government of India, New-Delhi, for financial support for her work (Sanction No. SR/SO/PS-58/2005).

References

1. Majewska-Sawka A, Nothnagel EA (2000) Plant Physiol 122:3
2. Showalter AM (2001) Cell Mol Life Sci 58:1399
3. Showalter AM (1993) Plant Cell 5:9
4. Gasper Y, Johnson K, McKenna JA, Bacic A, Schultz CJ (2001) Plant Mol Biol 47:161
5. Svetek J, Yadav MP, Nothnagel EA (1999) J Biol Chem 274:14724
6. Schultz C, Rumsewich MP, Johnson KL, Jones BJ, Gasper YM, Bacic A (2002) Plant Physiol 129:1448
7. Loopstra CA, Sederoff RR (1995) Plant Mol Biol 27:277
8. Loopstra CA, Puryear JD, No EG (2000) Planta 210:686
9. Zhang Y, Sederoff RR, Allona I (2000) Tree Physiol 20:457
10. Lafarguette F, Leplé JC, Déjardin A, Laurans F, Costa G, Lesage-Descauses M-C, Pilate G (2004) New Phytol 164:107
11. Yariv JH, Rapport MM, Graf L (1962) Biochem J 85:383
12. Yariv JH, Lis E, Katachalski E (1967) Biochem J 105:1c
13. Clarke AE, Anderson RL, Stone BA (1979) Phytochemistry 18:521
14. Schopfer P (1991) Planta 183:139
15. Du H, Simpson RJ, Moritz RL, Clarke AE, Bacic A (1994) Plant Cell 6:1643
16. Knox JP, Linstead PJ, Peart J, Cooper C, Roberts K (1991) Plant J 1:317
17. Stacey N, Roberts K, Knox JP (1990) Planta 180:185
18. Yates EA, Valdor J-F, Haslam SM, Morris HR, Dell A, Mackie W, Knox JP (1996) Glycobiology 6:131
19. Knox JP (1997) Int Rev Cytol 171:79
20. Toonen MAJ, de Vries SC (1996) Initiation of somatic embryos from single cells. In: Wang TL, Cuming A (eds) Embryogenesis: The Generation of a Plant. Bios Scientific Publishers, UK, pp 173–189
21. Odonmazig P, Ebringerova A, Machova E, Alfoldi J (1994) Carbohydr Res 252:317
22. Egert D, Beuscer N (1992) Planta Med 58:163
23. Thude S, Classen B (2005) Phytochemistry 66:1026
24. Gonda R, Tomoda M, Ohara N, Takada K (1993) Biol Pharm Bull 16:235
25. Kiyohara H, Cyong JC, Yamada H (1989) Carbohydr Res 193:193
26. Kelly GS (1994) Alternative Med Rev 4:96
27. Hauer J, Anderer FA (1993) Cancer Immunol Immunother 36:237
28. Fincher GB, Stone BA, Clarke AE (1983) Ann Rev Plant Physiol 34:47
29. Bacic A, Churms SC, Stephen AM, Cohen PB, Fincher GB (1987) Carbohyd Res 162:85
30. Knox JP (1999) Trends Plant Sci 4:123
31. Anderson RI, Clarke AE, Jermyn MA, Know RB, Stone BA (1977) Aust J Plant Physiol 4:143
32. Pennell RI, Knox JP, Scofield GN, Selvendran RR, Roberts K (1989) J Cell Biol 108:1967

33. Pennell RI, Robert K (1990) Nature 344:547
34. Pennell RI, Janniche L, Kjellbom P, Scofield GN, Peart JM, Roberts K (1991) Plant Cell 3:1317
35. No E-G, Loopstra CA (2000) Physiol Plant 110:524
36. Pal A, Das S (2006) Proc Natl Acad Sci India 76 B-IV:312–320
37. Willats WG, Knox JP (1996) Plant J 9:919
38. Ding I, Zhu JK (1997) Planta 203:289
39. Casero PJ, Casimiro I, Knox JP (1998) Planta 204:252
40. van Hengel AJ, Roberts K (2002) Plant J 32:105
41. van Hengel AJ, Roberts K (2003) Plant J 36:256
42. Gens JS, Fujiki M, Pickard BG (2000) Protoplasma 212:115
43. Lally D, Ingmire P, Tong H-Y, He Z-H (2001) Plant Cell 13:1317
44. Wagner TA, Kohorn BD (2001) Plant Cell 13:303
45. Park MH, Suzuki Y, Chono M, Knox JP, Yamaguchi I (2003) Plant Physiol 131:1450
46. Suzuki Y, Kitagawa M, Knox JP, Yamaguchi I (2002) Plant J 29:733
47. Kreuger M, van Holst G-J (1996) Plant Mol Biol 30:1077
48. Kreuger M, van Holst G-J (1993) Planta 189:243
49. Das S, Pal A (2004) J Plant Biochem Biotech 13:101
50. Letarte J, Simion E, Miner M, Kasha KJ (2006) Plant Cell Rep 24:691
51. Majewska-Sawka A, Munster A (2003) Plant Cell Rep 21:946
52. van Hengel AJ, Tadesse Z, Immerzeel P, Schols H, van Kammen A (2001) Plant Physiol 125:1880
53. Dyachok J, Wiweger M, Kenne L, van Arnold S (2002) Plant Physiol 128:1
54. Wendt dos Santos AL, Wietholter N, Gueddani NEE and Moerschbacher BM (2006) Physiol Plant 127:138
55. Chapman A, Blervacq AS, Vasseur J, Hilbert JL (2000) Planta 211:305
56. Tang X-C, He Y-O, Wang Y, Sun M-X (2006) J Expt Bot 57:2639
57. Sommer-Knudsen J, Clarke AE, Bacic A (1996) Plant J 9:71
58. Sommer-Knudsen J, Clarke AE, Bacic A (1997) Sex Plant Reprod 10:253
59. Park BS, Kim JS, Kim SH, Park YD (2005) Plant Cell Rep 24:663
60. Cheung AY, Wang, H, Wu HM (1995) Cell 82:383
61. Wu H, Wang H, Cheung AY (1995) Cell 82:395
62. Jauh GY, Lord EM (1996) Planta 199:251
63. Roy S, Jauh GY, Hepler PK, Lord EM (1998) Planta 204:450
64. Pereira LG, Coimbra S, Oliveira H, Monteiro L, Sottomayor M (2006) Planta 223:374
65. Acosta-Garcia G, Vielle-Calzada JP (2004) Plant Cell 16:2614
66. Butikofer P, Brodbeck U (1993) J Biol Chem 268:17794
67. Selleck SB (2000) Trends Genet 16:206
68. Schultz C, Gilson P, Oxley D, Youl J, Bacic A (1998) Trends Plant Sci 3:426
69. Zajchowski LD, Robbins SM (2002) Eur J Biochem 269:737
70. Munnick T, Irvine RF, Musgrave A (1998) Biochim Biophys Acta 1389:222
71. Udenfriend S, Kodukula K (1995) Ann Rev Biochem 64:563
72. Sardar HS, Yang J, Showalter AM (2006) Plant Physiol 142:1469
73. Hirayama T, Ohto C, Mizoguchi T, Shinozaki K (1995) Proc Natl Acad Sci U S A 92:3903
74. Hunt L, Mills LN, Pical C, Leckie CP, Aitken FL, Kopka J, Mueller-Roeber B, McAinsh MR, Hetherington AM, Gray JE (2003) Plant J 34:47
75. Kim YJ, Kim JE, Lee JH, Lee MH, Jung HW, Bahk YY, Hwang BK, Hwang I, Kim WT (2004) FEBS Lett 556:127
76. Yamada A, Ozeki Y, Akatsuka S (2006) Japanese Patent- WO/2006/013807
77. Li S-X, Showalter AM (1996) Plant Mol Biol 32:641
78. Serpe MD, Nothnagel EA (1999) Adv Bot Res 30:2027
79. Langan KJ, Nothnagel EA (1997) Protoplasma 196:785
80. Showalter AM, Gao M, Kieliszewski MJ, Lamport DTA (2000) Characterization and localization of a novel tomato arabinogalactan-protein (LeAGP-1) and the involve-

ment of arabinogalactan proteins in programmed cell death. In: Nothnagel EA, Bacic A, Clarke AE (eds) Cell and Developmental Biology of Arabinogalactan-Proteins. Kluwer Academic/Plenum, UK, pp 61–70
81. Mc Cabe PF, Levine A, Meijer P-J, Tapon NA, Pennell RI (1997) Plant J 12:267
82. Yadav MP, Manuel IJ, Yan Y, Nothnagel EA (2007) Food Hydrocolloid 21:297
83. Vince AJ, McNeil NI, Wager JD, Wrong OM (1990) Br J Nutr 63:17
84. Crociani F, Alessandrini A, Mucci MM, Biavati B (1994) Int J Food Microbiol 24:199
85. Tsao D, Shi Z, Wong A, Kim YS (1983) Cancer Res 43:1217
86. Robinson R, Feirtag J, Slavin J (2001) J Am Coll Nutr 20:279
87. Groman EV, Enriquez PM, Jung C, Josephson L (1994) Bioconjug Chem 5:547
88. Englyst HN, Hay S, Macfarlane GT (1987) FEMS Microbiol Ecol 95:163
89. D'Adamo P (1996a) J Naturopath Med 6:33
90. Hagmar B, Ryd W, Skomedal H (1991) Invasion Metastasis 11:348
91. Beuth J, Ko HL, Oette K, Pulverer G, Roszkowski K, Uhlenbruck G (1987) J Cancer Res Clin Oncol 113:51
92. Beuth J, Ko HL, Schirrmacher V, Uhlenbruck G, Pulverer G (1988) Clin Exp Metastasis 6:115
93. Levine PH, Whiteside TL, Friberg D, Bryant J, Colclough G, Herberman RG (1998) Clin Immunol Immunopathol 88:96
94. Corado J, Toro F, Rivera H, Bianco NE, Deibis L, De Sanctis JB (1997) Clin Exp Immunol 109:451
95. D'Adamo P (1996b) Larch arabinogalactan is a novel immune modulator. Townsend Letter for Doctors and Patients 156:42
96. Kastrukoff LF, Morgan NG, Zecchini D, White R, Petkau AJ,. Satoh J, Paty DW (1998) J Neuroimmunology 86:123
97. Kim LS, Waters RF, Burkholder PM (2002) Altern Med Rev 7:149
98. DeFelice SL (2002) FIM Rationale and Proposed Guidelines for the Nutraceutical Research Education Act – NREA, November 10, 2002. Foundation for Innovation in Medicine. Available at:http://www.fimdefelice.org/archives/arc.researchact.html
99. Brewer V (1998) Nat Biotechnol 16:728
100. Melchart D, Linde K, Worku F, Sarkady L, Horzmann M, Jurcic K, Wagner H (1995) J Altern Complem Med 1:145
101. See DM, Broumand N, Sahl L, Tilles JG (1997) Immunopharmacology 35:229
102. Roesler J, Steinmuller C, Kiderlen A, Emmendorffer A, Wagner H, Lohmann-Matthes ML (1991) Int J Immunopharmacol 13:27
103. Roesler J, Emmendorffer A, Steinmuller C, Leuttig B, Wagner H, Lohmann-Matthes ML (1991) Int J Immunopharmacol 13:931
104. Steinmuller C, Roesler J, Grottrup E, Franke G, Wagner H, Lohmann-Matthes ML (1993) Int J Immunopharmacol 15:605
105. Miller SC (2005) Evid-Based Compl Alt Med 2:309
106. Leuttig B, Steinmuller C, Gifford GE, Wagner H, Lohmann-Matthes ML (1989) J Natl Cancer Inst 81:669
107. Mucalo MR, Bullen CR, Manley-Harris M, McIntire TM (2002) J Mater Sci 37:493
108. Bamezai A, Goldmacher V, Reiser H, Rock KL (1989) J Immunol.143:3107
109. Drake SL, Klein DJ, Mickelson DJ, Oegema TR, Furcht LT, McCarthy JB (1992) J Cell Biol 117:1331
110. Robinson PJ (1997) Signal transduction via GPI-anchored membrane proteins. In: F Haag, F Koch-Nolte (eds) ADP-Ribosylation in Animal Tissues. Plenum, New York, pp 365–370
111. Alexandersson E, Saalbach G, Larsson C, Kjellbom P (2004) Plant Cell Physiol 45:1543
112. Youl JJ, Basic A, Oxley D (1998) Proc Natl Acad Sci U S A 95:7921
113. Sherrier DJ, Prime TA, Dupree P (1999) Electrophoresis 20:2027

Chapter 14
Hairy Roots: a Powerful Tool for Plant Biotechnological Advances

S. Guillon[1], J. Trémouillaux-Guiller[1] (✉), P.K. Pati[2], and P. Gantet[3]

[1]UPRES EA 2106 "Biomolécules et Biotechnologies Végétales",
Université François Rabelais, UFR des Sciences Pharmaceutiques – Parc de Grandmont 37200 Tours, France, e-mail: guiller@univ.-tours.fr

[2]Department of Botanical and Environmental Sciences, Guru Nanak Dev University, Amritsar-143 005, India

[3]Université Montpellier2, UMR PIA 1096, Laboratoire de Biochimie et de Physiologie Végétales, Place Eugène Bataillon, Bat 15, CC 002 – 34095 Montpellier cedex 5, France

Abstract Hairy roots in plants are the manifestation of infection caused by *Agrobacterium rhizogenes*, a gram negative soil bacterium. This phytopathogen transfers its large root-inducing (Ri) plasmid carrying a set of genes into plant genome and thereby encoding enzymes capable of modifying the plant hormonal metabolism. Such new hormonal balances induce the formation of proliferating roots, called hairy roots that emerge at the wounding site. Hairy root cultures, owing to their stable and high productivity, have been investigated from several decades to produce the valuable metabolites present in wild-type roots. The emergence of key molecules for overcoming the limiting culture parameters for the regulation of the metabolic pathways has made possible improvements in the production of secondary metabolites by hairy roots. Secretion and harvesting of these metabolites with the aid of trapping systems enhance the interest in such cultures. The use of hairy roots to produce recombinant animal proteins represents an attractive system that may be extrapolated for industrial exploitation. Equally, a good understanding of the underlying molecular mechanism, based on the transfer of the plasmid T-DNA of *A. rhizogenes*, opens a route for developing new strategies in metabolic engineering. Indeed, hairy root systems allow gene gain- or loss-of-function techniques and transcriptome analyses for the discovery of new metabolic genes. Because of the prolific proliferation of the roots, hairy roots could be promising tools for phytoremediation. The hairy root system must be scaled up if they are to be used in industry for the mass production of secondary metabolites.

Ramawat KG, Mérillon JM (eds.), In: *Bioactive Molecules and Medicinal Plants*
Chapter DOI: 10.1007/978-3-540-74603-4_14, © Springer 2008

14.1 Introduction

As a result of their rapid growth and genetic stability, for 25 years hairy roots have been investigated as attractive systems for producing valuable secondary metabolites that are routinely extracted from medicinal plants. The emergence of key molecules for overcoming the limiting culture parameters of the regulation of metabolic pathways has allowed improvements in the production of secondary metabolites by hairy roots. Secretion and harvesting of these metabolites via a trapping system enhances the interest in such cultures. The production of recombinant animal proteins using hairy roots is an attractive system that lends itself to industrial exploitation. Equally, the understanding of the underlying molecular mechanism, based on the transfer of the plasmid T-DNA of *Agrobacterium rhizogenes*, opens the route for developing new strategies for metabolic engineering. Indeed, the hairy root system allows gene gain- or loss-of-function techniques and transcriptome analyses for the discovery of new metabolic genes. Because of the important proliferation of the roots, hairy roots could also be a promising tool for phytoremediation. Considering recent progress, the hairy-root systems must be scaled up in order to meet the industrial demand.

14.2 Hairy Roots Are on the Way to towards an Experimental Model

A large biological diversity of secondary metabolites of interest accumulates in plant roots [1]. However, as harvesting such natural roots may be destructive for the plants, hairy root culture is considered as an alternative source for the production of valuable metabolites. Indeed, owing to their intense development, hairy root cultures have been investigated for several decades in order to produce secondary metabolites that are synthesized naturally in wild-type roots. Moreover hairy root cultures are capable of accumulating these compounds at the same or superior level than the mother plant and for a long period of time. In this way, the hairy roots that are used now as biotechnologically promising tools [2], have been recently established from new medicinal plant species to produce secondary metabolites (Table 14.1).

Hairy roots are derived from the genetic transformation of plant cells by a phytopathogen Gram-negative soil bacterium, *Agrobacterium rhizogenes*. During the infection of previously wounded plant tissues, this bacterium transfers into the host genome T-DNA, a wide part of its Ri plasmid. In hypervirulent bacterial strains, T-DNA possesses two separate segments called TL-DNA, carrying the *rol* genes responsible for the hairy root phenotype, and TR-DNA, which carries a set of genes that encode enzymes controlling auxin biosynthesis and determines a new hormonal balance in the transformed roots. After infection of the tissues with *A. rhizogenes*, emerging roots from wounding sites are excised before their individual transfer onto solid medium and then into agitated liquid medium. The hairy root phenotype is characterized by a fast

Table 14.1 Hairy root cultures established from various plants for the production of secondary metabolites

Family	Genus and species	Major metabolites	Medicinal property	References
Apocynaceae	*Rauvolfia micrantha*	Ajmalicine Ajmaline	Antihypertensive	[3]
Asteraceae	*Saussurea medusa*	Jaceosidin	Antitumorous	[4]
Composeae	*Solidago altissima*	Polyacetylene (cis dehydromatricaria ester)	Unknown	[5]
Cucurbitaceae	*Gynostemma pentaphyllum*	Saponin gypenoside	Several pharmacological activities	[6]
Fabaceae	*Pueraria phaseoloides*	Puerarin	Hypothermic, spasmolytic, hypotensive, antiarrhythmic	[7]
Ginkgoaceae	*Gingko biloba*	Ginkgolide	Against cardiovascular and aging diseases	[8]
Linaceae	*Linum flavum*	Lignans coniferin	Anticancer properties	[9]
Nyssaceae	*Camptotheca acuminata*	Camptothecin	Anticancerous, antiviral	[10]
Papaveraceae	*Papaver somniferum*	Morphine	Sedative, analgesic	[11]
		Sanguinarine		
		Codeine		
Solanaceae	*Solanum chrysotrichum*	Saponin	Antifungal	[12]
Verbenaceae	*Gmelina arborea Roxb*	Verbascoside	Effective against stomach disorders, fevers, and skin diseases	[13]

growth, ageotropism, and exogenous hormone independence as well as a high genetic stability. Recent fundamental research has contributed to a better understanding the transformation processes by *A. rhizogenes* [13]. In addition, the possibility of transferring genes into host plants multiplies the genetic engineering strategy regarding the production of foreign proteins or modification of the limiting metabolic pathways. The hairy root model offers new biotechnological possibilities, which have been underlined in recent reviews [14–16].

Here, we present advances in different studies on hairy root cultures made during the past few years that relate to metabolite production from efficient plants. Possible strategies using hairy roots are multiple, and developments in culture parameters have been reported that improve the production, secretion, and harvesting of secondary metabolites of interest. Based on gene transfer, metabolic engineering applied to hairy roots makes possible the production of

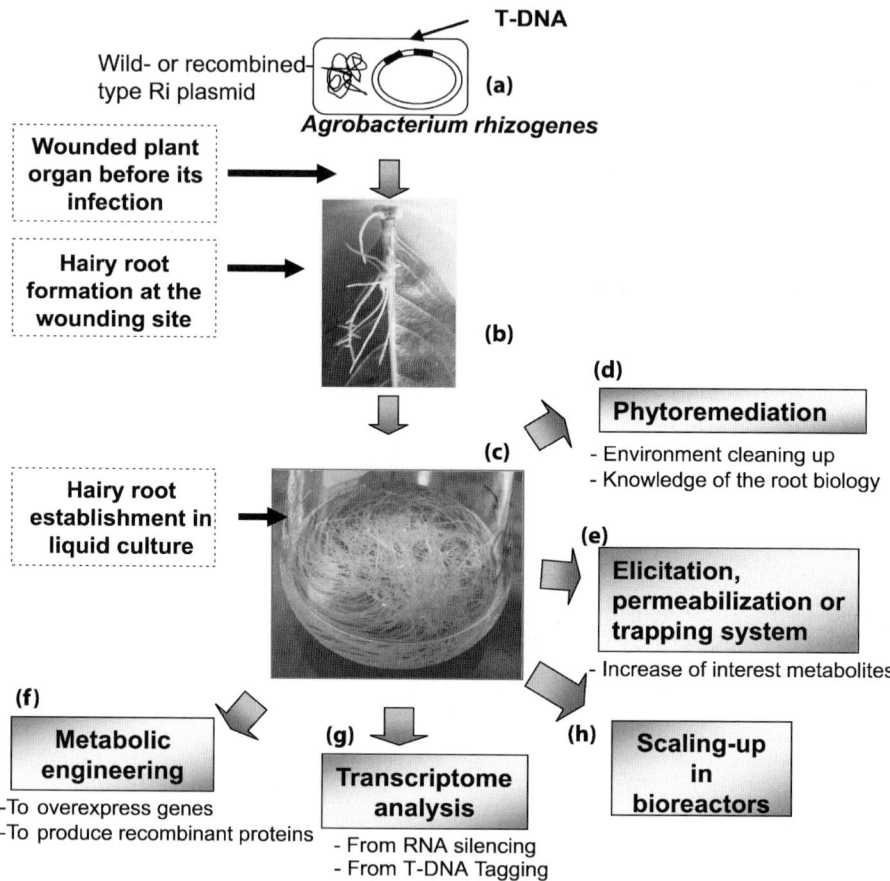

Fig. 14.1 Biotechnological advances from the hairy roots. *Agrobacterium rhizogenes*, a natural pathogen, transfers, from its Ri plasmid, T-DNA into the genomic DNA of a plant species (**a**). Emerging roots from the wounding site (**b**) were individually cultured to establish hairy roots (**c**), which are used as a model for several biotechnological strategies such as phytoremediation (**d**), cleaning up the soil, or understanding the biology of the roots. Likewise, producing metabolites of interest using hairy root cultures can lead, after elicitation treatment, permeability, or trapping processes, to an increase in the amounts of metabolites recovered (**e**). Metabolic engineering research has led to the production of foreign proteins in a confined space or overexpression of a limiting step of a particular metabolic pathway for a given metabolite (**f**). Moreover, fundamental research involving hairy roots has now achieved T-DNA activation tagging or RNA silencing processes followed by transcriptome analyses for the discovery of new genes (**g**). Scaling up in a bioreactor must be mastered for numerous plant species if hairy root cultures are to emerge as a reference model (**h**)

animal protein in a confined space. Use of hairy roots can be envisaged for cleaning up polluted environments via a phytoremediation process (Fig. 14.1). Furthermore, the recent emergence of companies specializing in the large-scale culture of hairy roots emphasizes the importance of such research systems.

14.3 Improvement in the Productivity of Hairy Roots: Biotic and Abiotic Treatments or Metabolite Trapping

Several processes utilizing hairy root cultures such as elicitation, precursor feeding, cellular permeability, and trapping of the molecules released into the culture medium, could have a positive effect on the accumulation and/or secretion of the studied metabolites. As defense against pathogens, plants often synthesize valuable secondary metabolites, and the corresponding biosynthesis pathways are known to be induced by pathogen cell-wall-derived molecules, called elicitors. Several mineral or physical parameters, bacteria, fungi, and yeasts can act as elicitors to successfully increase the production (Table 14.2) and/or secretion of secondary metabolites (Table 14.3) from the hairy roots of different plant species.

Table 14.2 Update on the enhancement of metabolite biosynthesis by elicitation in hairy root cultures, *MeJA* Methyl jasmonate

Family	Genus and species	Metabolites	Treating agent	References
Apiaceae	*Ammi majus*	Coumarine, furocoumarine	Bion	[17]
			Enterobacter sakasaki	
Arialaceae	*Panax ginseng*	Ginsenoside	Vanadyl sulfate	[18]
			MeJA	
			Chitosan	
Labiaceae	*Salvia miltiorrhiza*	Tanshinone	Ag^+	[19]
			Yeast extract	
Rubiaceae	*Rubia tinctorium*	Lucidin antraquinone	MeJA	[20]
Solanaceae	*Solanum tuberosum*	Sesquiterpenes	B cyclodextrin	[21]
			MeJA	
			Rhizoctonia bataticola	

Table 14.3 Update on the secretion of metabolite biosynthesis by biotic or abiotic treatments of hairy root cultures. *CTAB* Cetyl trimethylammonium bromide and MeJA

Family	Genus and species	Metabolites	Treating agent	Reference
Chenopodiaceae	*Beta vulgaris*	Betalaine pigments	Tween 80, CTAB, Triton X100	[22]
			Lactobacillus helveticus	
			Saccharomises cerevisae	
			Candida utilis	
			pH2 sonication, T 50°C oxygen stress in presence of light	[23]
Convolvulaceae	*Pharbitis nil*	Umbelliferone, scopolamine	$CuSO_4$	[24]
		Skimmin	MeJA	
Oxalidaceae	*Oxalis tuberosa*	Harmaline, harmine	*Phytosphtora cinnamoni*	[25]

Likewise, the stress hormone methyl jasmonate (MeJA) is able to increase the production of ginsenosides by *Panax ginseng* hairy roots [18] and the secretion of umbelliferone, scopoletin, and skimmin by *Pharbitis nil* hairy roots [24]. However, most of the time the secretion of metabolites is obtained by addition of permeability agents (e.g., detergents, calcium chelators, pH, sonication, temperature, and oxygen stresses) such as pH or temperature variations for releasing betalaine, a red natural pigment, effluxes from *Beta vulgaris* hairy roots. [22]. The biosynthesis of valuable metabolites by hairy root cultures is sometimes limited by the availability of its precursor. In this case, the precursor feeding process consists of introducing this precursor into the culture medium to increase production of the metabolite of interest. Nevertheless, this process can be costly if the precursor is difficult to synthesize or difficult to obtain from natural sources. In this context, the coculture system has proven to be a judicious alternative, since the production of podophyllotoxin, an antitumor drug, by *Linum flavum* hairy roots is increased when the culture is cultivated with a suspension *Podophyllum hexandrum* cells, which release the coniferin precursor of podophylotoxin [26]. Once the metabolite of interest has accumulated in the medium, a trapping system can be put into that medium to specifically adsorb and harvest the secreted metabolite. Indeed, by introducing a mixture of two adsorbents (alumina and silica; 1:1) into the medium, 97.2% of the betalaine from *Beta vulgaris* hairy roots was recovered [27]. Likewise, the addition of a trapping system contributes to an increase in the productivity of hairy root cultures; adding a hydrophobic polymeric resin, X-5, to *Salvia miltiorrhiza* hairy root cultures makes possible the recovery of 80% of the diterpenoid tanshinones released from them, which are used for the treatment of menstrual disorders and blood circulation diseases [28]. Finally, these strategies could be combined advantageously to enhance hairy root productivity, since multiple elicitors, *in situ* adsorption, and repeated medium renewal with an efficient semicontinuous system increases by 15-fold the production of transhinones, with 76.5% adsorbed by the resin [29].

14.4 Potential Discovery of Metabolic Genes from Transcriptome Analysis of T-DNA Activation Tagging or Elicited Hairy Roots

Hairy roots, as whole plants, can be screened for their metabolic phenotype in order to identify the gain-of-function mutations created by an activation tagging strategy. The dominant mutation is expressed in the plant genome after random insertion of a T-DNA construction that carries constitutive enhancer elements, leading to overexpression of flanking genes and possible findings of hitherto unknown genes. This strategy was adapted to generate tagged hairy roots from *Arabidopsis thaliana, Solanum tuberosum*, and *Nicotiana tabacum* by using *A. rhizogenes* and a binary vector carrying a T-DNA with four tandem repeats of the CaMV 35S (i.e., cauliflower mosaic virus 35S) enhancer promoter elements in its right border [30]. Equally, this technology can be used for regenerating recalcitrant plants such as tree species. Induction of hairy roots has been successfully reported on these woody plant species in order to characterize unknown genes necessary for the root biology [31].

Recently, a strategy based on an inducible system by MeJA elicitation was applied to cell suspensions of *Catharanthus roseus* to find new genes involved in the metabolic pathways of the terpenoid indole alkaloids. Such a strategy allows the definition of a gene-to-metabolite network [32]. Similarly, the treatment of *Panax ginseng* hairy roots with MeJA reveals new genes by generating 3134 expressed sequence tags [33]. In this way, among transcripts, several genes encoding enzymes such as squalene synthase, squalene epoxidase, oxidosqualene cyclase, cytochrome p450, and glycosyltransferase involved in the triterpene glycoside gensenosides could be characterized. These data constitute a gain of information on the secondary metabolite biosynthesis in *Panax ginseng* and on the genes responding to MeJA treatment.

14.5 RNA Silencing via Hairy Root: a Powerful Tool for Loss-of-Function Analyses of Genes

Post-transcriptional gene silencing was successfully achieved in higher plants; there has been no study on RNA silencing in the roots. Kumagai and Kouchi [34] investigated, in *Lotus japonicus* hairy roots, an efficient system for loss-of-function analyses of genes expressed in the roots or root nodules. The seedlings were infected with *A. rhizogenes* harboring the molecular construction for the hpRNAs expression possessing the complementary sequences to the GUS coding region. The GUS activity was decreased or silenced in 60–70% of the hairy roots, demonstrating that transient RNA silencing by hairy root is a powerful process.

14.6 Metabolic Engineering of the Hairy Root System

Metabolic engineering is based on the integration between the T-DNA borders of genes encoding enzymes for transferring these into the plant host via the *A. rhizogenes* molecular machinery. This process has been investigated in *Duboisia hybrid* hairy roots to overexpress the hyoscyamine-6-hydroxylase (H-6-H) encoding enzyme, which catalyzes two consecutive steps of the tropane alkaloid biosynthesis pathway [35]. Introducing the H-6-H gene from *Hyoscyamus niger* placed under control of the CaMV35S promoter increases by threefold the conversion of hyoscyamine into scopolamine, an anticholinergic agent, in *Duboisia hybrid* hairy roots, but also promotes hyoscyamine synthesis. This metabolic engineering strategy was successfully applied to circumvent problems related to precursor availability or negative feedback regulatory loops. Because of a negative feedback by tryptamine and tryptophan on anthranilate synthase (AS) activity, tryptamine and tryptophan are limiting precursors for the biosynthesis of the anticancer drug terpenoid indole alkaloid (TIA) from hairy root cultures of *Catharanthus roseus*, but not in the suspension cell systems [36]. Tryptophan comes from the shikimate pathway, in which chorismate is converted into anthranilate by AS, and then tryptamine is synthesized from tryptophan by a single enzymatic reaction catalyzed by tryptophan decarboxylase (TDC). In *Arabidopsis thaliana*, the gene that encodes the AS α-subunit is resistant to the feedback by tryptophan and tryptamine. Introducing this gene in *Catharanthus roseus* hairy root cultures, simultaneously with the gene encoding the AS β-subunit and the TDC gene, improved dramatically the metabolic flux of the indole precursors, tryptamine and tryptophan, and increased the amount of ajmalicine, an antihypertensive drug [37–41]. In contrast, a reduction in the concentration of lochnericine, horhammericine, and tabersonine TIAs was observed. If no chemical or physical trap is known, the possibility of introducing a transgene encoding a protein capable of trapping the expected metabolite may be possible. In *Solanum khasianum* hairy roots, this strategy has been explored to divert the negative feedback regulatory loop performed by solasodine glycoside, an antineoplastic agent, on its own expression [42]. The binding of solasodine glycoside with its recombinant antibody, expressed from a foreign gene after its integration into the genome of the hairy roots, has been successfully developed. By eliminating the negative feedback, the production of solasodine glycoside was enhanced by two to threefold [43].

14.7 Hairy Roots: A Novel System for Molecular Farming

Compared with field-cultured plants, hairy roots in confined recipients possess several advantages for expression of the functional foreign proteins in controlled conditions of growth that avoid the transgene or pharmacologically active protein dispersion in the environment. Recombinant proteins have been successfully produced as functional antibodies from hairy roots of *Solanum khasianum* [42] or as therapeutically active foreign protein, such as the human secreted alkaline phosphatase (SEAP) from *Tobacco* hairy roots [44]. Complex proteins, expressed in plant organs, must often be extracted from tissues and purified by costly and labourious processes. Therefore, proteins produced by hairy roots are often secreted into the culture medium, such as a non-toxic lectin subunit ricin B, fused to green fluorescent protein (GFP) and expressed in tobacco hairy roots [45]. This fusion protein was tested in mouse as an antigen, showing that proteins fused to ricin B, taken as a mucosal adjuvant in mammalian immune responses, can be efficiently produced by hairy roots. Moreover, the production of this protein can be improved with a system based on culture confinement. In fact, ricin B is sensitive to the proteases present in the culture medium of hairy roots; nevertheless a two-phase extraction process can improve the stability of this protein by increasing its production and facilitating its harvest in the organic phase [46]. A nondestructive rhizosecretion system, coupled to a trapping process, can lead to high amounts of recombinant proteins and facilitate their downstream purification. The transformation mechanism by *A. rhizogenes* can serve to create de novo a new metabolic pathway by introducing genes into, for example, *Beta vulgaris* hairy roots, which produce poly (3-hydroxybutyrate) (PHB) polyester compounds, a raw alternative material for industrial plastic production by introducing three genes, β-ketothiolase, acetoacetyl-CoA reductase, and PHB synthase from *Ralstonia eutropha* bacterium encoding concomitant enzymatic steps [47]. These studies demonstrate that hairy roots are becoming a serious alternative to whole plants for the production of therapeutically functional animal proteins.

14.8 Phytoremediation Process for Cleaning up the Environment and More Knowledge on Root Adsorption

Phytoremediation is based on the use of plant species to remove pollutants from the environment (soil or water) by adsorption of heavy metals, antibiotics, or pesticides. Toxic organic molecules can accumulate in plant organs in an unchanged form (a process called phytoextraction) or converted enzymatically into a harmless form (a process called phytotransformation) [48]. Hairy roots represent a biological study model without interference with other part

Table 14.4 Use of hairy root for phytoremediation. *2,4 DCP* 2,4 Dichlorophenol, *DDT* (1,1,1-trichloro-2,2-bis-(4'-chlorophenyl)ethane)

Pollutant	Method of Phytoremediation	Genus and species	References
2,4 DCP	Phytoextraction	*Brassica napus*	[49]
Cadmium	Phytoextraction	*Thalspi caerulescens*	[50]
DDT	Phytotransformation	*Brassicae juncea*	[51]
		Chitorium intybus	
Nickel	Phytoextraction	*Alyssum bertolinii*	[50, 51]
		Alyssum tenium	[51]
		Alyssum troodi	
Tetracycline, oxytetracycline	Phytoextraction	*Helianthus annuus*	[52]
Uranium	Rhizofiltration	*Brassica juncea*	[53]
		Chenopodium amaranticolor	

of the plant, which makes possible a better understanding of the fundamental mechanisms underlying the absorption, accumulation, tolerant distribution, and detoxification of toxic material by hairy roots. In this way, hairy roots from hyperaccumulator plants able to uptake cadmium, nickel, or uranium have been investigated recently because of their greater penetration, increasing their ability to retrieve contaminants from deeper soils, and enzymatic degradation (Table 14.4).

Moreover, hairy roots from several plant species have been successfully tested to remove 2,4 dichlorophenol (2,4 DCP) from industrial effluent, antibiotics, or pesticides by phytoremediation. In addition to the phytoremediation strategies, knowledge of genetic engineering offers new possibilities by which the environment can be cleaned.

14.9 Scale up and Technological Integration into Industry

With the exciting spin-off of research concerning the engineering of hairy roots and their optimization of production and growth, large-volume bioreactors, until now adapted for use with cell suspension cultures, are suitable for scaling up production from hairy root cultures. Depending on the plant species and the organ culture, bioreactors adapted to liquid, air, or both conditions of cultures have recently been designed and optimized (Table 14.5).

In addition, the airlift system designed for microorganisms or plant cells in a liquid medium, has been reported to work efficiently for the growth of *Artemesia annua* hairy roots [58] and for producing betalaine from *Beta vulgaris* hairy roots [57]. The scale up favored both productivity and tissue growth by using

Chapter 14 Hairy Roots: a Powerful Tool for Plant Biotechnological Advances 281

Table 14.5 Bioreactor types used for growth or secondary metabolite production from hairy roots

Bioreactors	Genus and species	Metabolite or growth	References
Airlift	*Astragalus membranacus*		[54]
Airlift mesh draught	*Solanum chrysotrichum*	Growth	[55]
Basket bubble	*Genista tinctoria*	Phytoestrogen	[56]
Bubble column	*Beta vulgaris*	Betalaine	[57]
Bubble column	*Artemisia annua*	Growth	[58, 59]
Mist	*Tagetes patula*		[60]
Mist trickling	*Stizolobium hassjoo*		[61]
Trickle bed	*Hyoscyamus muticus*	Growth	[62]

a basket bubble reactor with a coculture of shoots and hairy roots of *Genista tinctoria* for the production of phytoestrogens [56]. The difficulty with organ cultures of hairy roots is in achieving good homogenization of the roots in the bioreactor. This problem could be avoided if hairy roots are cultured in an airlift mesh draught reactor, as described by Caspeta et al. [53] for the growth of the *Solanum chrysotrichum* hairy roots in this reactor. Attaching hairy roots to a mesh support allows reduction of the volume of culture and allows the concentration of the secreted metabolite. This system was used in a mist culture system for *Hyoscyamus muticus* [62], *Tagetes patula* [60] and *Campthoteca* hairy roots. These latter cultures were developed by the German company ROOTec for the production of campthotecin, an antitumoral drug, used at the beginning of the 21st century for the treatment of ovarian and colon cancer. The multiplication of reactors, instead of raising the culture volume, enhances the capacity of metabolite production by hairy roots and, if a problem appears, the production of only one reactor is lost. When the expected metabolite is stable and does not require confined conditions of culture, a cheaper hydroponic culture can be designed. Such a hydroponic system was used to produce a recombinant protein rhizosecreted by adventitious roots and hairy roots [44]. Scaling up such cultures for industry can be achieved by optimizing culture medium parameters and overexpression of metabolic genes.

14.10 Perspectives

In the past few years, hairy root technology has been significantly improved in various fields. Improved knowledge in the area of natural plant biodiversity, the possibility of expressing animal proteins in plant systems, and the concept of new bioreactor systems for hairy root cultures has led to a promising technology. The discovery of new genes that participate in the metabolic pathways from hairy root studies increases the tremendous potential of such cultures. It is also predicted that this model of pharmaceutical production is relatively safe and stands as a viable alternative to the whole-plant molecular farming system. This prediction is strengthened by the observation that emerging private companies have converted this technology to allow production at a commercial scale. This is a serious indication that in the near future, hairy roots will become powerful tools for biotechnologists with which to reach the precious underground resources of the plant kingdom.

References

1. Flores HE, Vivanco JM, Loyola-Vargas VM (1999) Trends Plant Sci 4:220
2. Sevon N, Oskman-Caldentey KM (2002) Planta Med 68:859
3. Sudha CG, Obul Reddy B, Ravishankar GA, Seeni S (2003) Biotechnol Lett 25:631
4. Zhao D, Fu C, Chen Y, Ma F (2004) Plant Cell Rep 23:468
5. Inoguchi M, Ogawa S, Furukawa S, Kondo H (2003) Biosci Biotechnol Biochem 67:863
6. Chang CK, Chang KS, Lin YC, Liu SY, Chen CY (2005) Biotechnol Lett 27:1165
7. Shi HP, Kintzios S (2003) Plant Cell Rep 21:1103
8. Ayadi R, Tremouillaux-Guiller J (2003) Tree Physiol 23:713
9. Lin HW, Kwok KH, Doran PM (2003a) Biotechnol Lett 25:521
10. Lorence A, Medina-Bolivar F, Nessler CL (2004) Plant Cell Rep 22:437
11. Le Flem-Bonhomme V, Laurain-Mattar D, Fliniaux MA (2004) Planta 218:890
12. Caspeta L, Nieto I, Zamilpa A, Alvarez L, Quintero R, Villarreal ML (2005) Planta Med 71:1084
13. Tzfira T, Li J, Lacroix B, Citovsky V (2004) Trends Genet 20:375
14. Guillon S, Trémouillaux-Guiller J, Kumar Pati K, Rideau M, Gantet P (2006) Trends Biotechnol 24:403
15. Guillon S, Tremouillaux-Guiller J, Kumar Pati P, Rideau M, Gantet P (2006) Curr Opin Plant Biol 9:341
16. Shanks JV, Morgan J (1999) Curr Opin Biotechnol 10:151
17. Staniszewska I, Krolicka A, Malinski E, Lojkowska E, Szafranek J (2003) Enzyme Microb Technol 33:565
18. Palazon J, Cusido RM, Bonfill M, Mallol A, Moyano E, Morales C, Pinol MT (2003) Plant Physiol Biochem 41:1019
19. Ge X, Wu J (2005) Appl Microbiol Biotechnol 68:183
20. Nakanishi F, Nagasawa Y, Kabaya Y, Sekimoto H, Shimomura K (2005) Plant Physiol Biochem 43:921
21. Komaraiah P, Reddy GV, Reddy PS, Raghavendra AS, Ramakrishna SV, Reddanna P (2003) Biotechnol Lett 25:593
22. Thimmaraju R, Bhagyalakshmi N, Narayan MS, Ravishankar GA (2003) Biotechnol Prog 19:1274
23. Thimmaraju R, Bhagyalakshmi N, Narayan MS, Ravishankar GA (2003) Process Biochem 38:1069
24. Yaoya S, Kanho H, Mikami Y, Itani T, Umehara K, Kuroyanagi M (2004) Biosci Biotechnol Biochem 68:1837
25. Bais HP, Vepachedu R, Vivanco JM (2003) Plant Physiol Biochem 41:345

26. Lin HW, Kwok KH, Doran PM (2003) Biotechnol Prog 19:1417
27. Rudrappa T, Neelwarne B, Aswathanarayana RG (2004) Biotechnol Prog 20:777
28. Ge X, Wu J (2005) Plant Sci 168:487
29. Yan Q, Hu Z, Tan RX, Wu J (2005) J Biotechnol 119:416
30. Seki H, Nishizawa T, Tanaka N, Niwa Y, Yoshida S, Muranaka T (2005) Plant Mol Biol 59:793
31. Damiano C, Monticelli S (1998) Elcctron J Biotechnol 1:3
32. Rischer H, Oresic M, Seppanen-Laakso T, Katajamaa M, Lammertyn F, Ardiles-Diaz W, Van Montagu MC, Inze D, Oskman-Caldentey K, Goosens A (2006) Proc Natl Acad Sci U S A 103:5619
33. Choi DW, Jung J, Ha YI, Park HW, In, DS, Chung HJ, Liu, JR (2005) Plant Cell Rep 23:557
34. Kumagai H, Kouchi H (2003) Mol Plant Microbe Interact 16:663
35. Palazon J, Moyano E, Cusido RM, Bonfill M, Oksman-Caldentey KM, Pinol MT (2003) Plant Sci 165:1289
36. Morgan JA, Shanks JV (2000) J Biotechnol 79:137
37. Li J, Last RL (1996) Plant Physiol 110:51
38. Hughes EH, Hong SB, Gibson SI, Shanks JV, San KY (2004) Biotechnol Bioeng 86:718
39. Hughes EH, Hong SB, Gibson SI, Shanks JV, San KY (2004) Metab Eng 6:268
40. Hong SB, Peebles CA, Shanks JV, San KY, Gibson SI (2005) J Biotechnol 122:28
41. Peebles CA, Hong SB, Gibson SI, Shanks JV, San KY (2005) Biotechnol Prog 21:1572
42. Putalun W, Prasamsiwamai P, Tanaka H, Shoyama Y (2004) Biotechnol Lett 26:545
43. Putalun W, Taura F, Qing W, Matsushita H, Tanaka H, Shoyama Y (2003) Plant Cell Rep 22:344
44. Gaume A, Komarnytsky S, Borisjuk N, Raskin I (2003) Plant Cell Rep 21:1188
45. Medina-Bolivar F, Wright R, Funk V, Sentz D, Barroso L, Wilkins TD, Petri JW, Cramer CL (2003) Vaccine 21:997
46. Zhang C, Medina-Bolivar F, Buswell S, Cramer CL (2005) J Biotechnol 117:39
47. Menzel G, Harloff HJ, Jung C (2003) Appl Microbiol Biotechnol 60:571
48. Suresh B, Ravishankar GA (2004) Crit Rev Biotechnol 24:97
49. Agostini E, Coniglio MS, Milrad SR, Tigier HA, Giulietti AM (2003) Biotechnol Appl Biochem 37:139
50. Boominathan R, Doran PM (2003) J Biotechnol 101:131
51. Suresh B, Sherkhane PD, Kale S, Eapen S, Ravishankar GA (2005) Chemosphere 61:1288
52. Gujarathi NP, Haney BJ, Park HJ, Wickramasinghe SR, Linden JC (2005) Biotechnol Prog 21:775
53. Eapen S, Suseelan KN, Tivarekar S, Kotwal SA, Mitra R (2003) Environ Res 91:127
54. Du M, Wu XJ, Ding J, Hu ZB, White KN, Brandford-White CJ (2003) Biotechnol Lett 25:1853
55. Caspeta L, Quintero R, Villarreal ML (2005) Biotechnol Prog 21:735
56. Luczkiewicz M, Kokotkiewicz A (2005) Plant Sci 169:862
57. Suresh B, Thimmaraju R, Bhagyalakshmi N, Ravishankar GA (2004) Process Biochem 39:2091
58. Kim YJ, Weathers PJ, Wyslouzil BE (2003) Biotechnol Bioeng 83:428
59. Souret FF, Kim Y, Wyslouzil BE, Wobbe KK, Weathers PJ (2003) Biotechnol Bioeng 83:653
60. Suresh B, Bais HP, Raghavarao KSMS, Ravishankar GA, Ghildyal NP (2005) Process Biochem 40:1509
61. Sung LS, Huang SY, Nakanishi F, Nagasawa Y, Kabaya Y, Sekimoto H, Shimomura K (2005) Biotechnol Bioeng 94:441
62. Ramakrishnan D, Curtis WR (2004) Biotechnol Bioeng 88:248

Chapter 15
Hairy Roots of *Catharanthus roseus*: Efficient Routes to Monomeric Indole Alkaloid Production

S. Guillon[1], P. Gantet[2], M. Thiersault[1], M. Rideau[1] and J. Trémouillaux-Guiller[1] (✉)

[1]Université F. Rabelais Tours UPRES EA 2106 "Biomolécules et Biotechnologies Végétales", UFR des Sciences Pharmaceutiques et UFR des Sciences et Techniques – Parc de Grandmont 37200 Tours, France, e-mail: guiller@univ-tours.fr
[2]Université Montpellier2, UMR PIA 1096, Laboratoire de Biochimie et de Physiologie Végétale, Place Eugène Bataillon, Bat 15, CC 002 – 34095 Montpellier cedex 5, France

Abstract Vinblastine, an efficient antineoplastic drug produced in *Catharanthus roseus*, can be semi-synthesized in vitro by coupling catharanthine and vindoline. Highly differentiated hairy root cultures are potentially able to produce all of the precursors found in the natural roots. We have established hairy root clones from *Catharanthus* explants and analysed terpenoid indole alkaloids by thin-layer chromatography, spectrofluorometry and high performance liquid chromatography. Among 441 hairy root clones developing on solid medium, 73 fast-growing clones were transferred into liquid culture, from which 28 well-established clones could be obtained. Six of these hairy root clones, elicited or not elicited by methyl jasmonate, biosynthesized ajmalicine, serpentine, catharanthine, tabersonine and vindolinine.

Keywords Hairy root, *Catharanthus*, *Agrobacterium rhizogenes*, Terpenoid indole alkaloids

Abbreviations

CAS	Ceric ammonium sulphate
HPLC	High-performance liquid chromatography
MeJA	Methyl jasmonate
MSD	Mass spectrometer detection
TIAs	Terpene indole alkaloids
TLC	Thin-layer chromatography

Ramawat KG, Mérillon JM (eds.), In: *Bioactive Molecules and Medicinal Plants*
Chapter DOI: 10.1007/978-3-540-74603-4_15, © Springer 2008

15.1 Introduction

Catharanthus roseus is one of the best-studied medicinal plants because of the therapeutic effectiveness of its secondary metabolites, such as the anti-arrhythmic monomer, ajmalicine, and the anti-neoplastic dimers, vinblastine and vincristine, which are used in clinical cancer chemotherapy [1]. Unfortunately, both of the latter secondary metabolites are accumulated at extremely low levels *in planta* and their chemical synthesis is expensive and difficult because of their complex structures [2]. Nevertheless, vinblastine can be produced semi-synthetically by coupling catharanthine to vindoline extracted from cultured plants [1]. For several decades, in vitro cultures have been considered to be a promising biotechnological approach to enhance the production of secondary metabolites and/or to reveal novel alkaloids, for example those identified in *C. roseus* cell suspensions [3]. However, metabolite production is unstable in most cell suspensions, and becomes even lower during subcultures [4]. It is well known that *in planta*, the biosynthetic capacity is mostly restricted to specialised organs, tissues or cells, as observed in *C. roseus* for the terpene indole alkaloids (TIAs). Therefore, as an alternative biotechnological process, transgenic hairy root cultures were raised to obtain in vitro high accumulation of secondary metabolites.

The Gram-negative soil bacterium *Agrobacterium rhizogenes* induces, after inoculation of wounded plant tissues, the proliferation of adventitious roots called "hairy roots" [4]. Culturing hairy roots is considered as a promising system with which to produce the valuable metabolites present in wild-type roots, or to integrate genes of interest [5]. Moreover, this system can also be used to achieve recombinant animal proteins or to detoxify the polluted environment by the process of phytoremediation [6]. These highly stable transgenic roots show a fast growth in hormone-free medium in comparison with hormone-dependent normal root cultures [2].

Here we report the establishment of numerous *C. roseus* hairy root clones initiated after infecting leaf explants with *A. rhizogenes*. Afterwards, both alkaloid profiles and contents of six hairy root clones, chosen at random, were analysed.

15.2 Materials and Methods

15.2.1 Bacterial Strain

A. rhizogenes strain *Ar*15834 (kindly provided by Dr. Annick Petit, CNRS-France), harbouring an agropine-type Ri plasmid [7,8], was grown on solid yeast extract mannitol medium for 2 days at 28°C in the dark, stored at 4°C and subcultured routinely each month.

15.2.2 Plant Material

Both varieties of *C. roseus*, two plants with pink flowers and two plants with red flowers, were cultured in a greenhouse. In total, 300 leaves were surface-disinfected for 1 min in 70% (v/v) ethanol and for 10 min in 3% (w/v) sodium hypochlorite and rinsed 3 times with sterile distilled water. All solutions contained 0.001% (v/v) sterile Tween-80. Before bacterial inoculation, the leaves were placed upside down on solid B5 medium [9] supplemented with 30 g·l^{-1} sucrose and kept for 1 week at 25±1°C in the dark.

15.2.3 Hairy Root Induction

For each plant, 75 leaves were sterilised. Of these, 65 randomly taken leaves were inoculated with *A. rhizogenes* using a single, tiny, 2-day-old colony, and 10 leaves were taken as controls. After a 3-day co-culture at 28°C, the inoculated leaves were transferred onto solid medium containing half-strength salts and vitamin B5 medium (referred to as B5/2) supplemented with 30 g·l^{-1} sucrose, 8 g·l^{-1} agar (Kalys, France) and 1 g·l^{-1} cefotaxime; the pH was adjusted to 5.7 before autoclaving. Inoculated and control leaves were kept at 25±1°C with a photoperiod of 12 h·day^{-1}. Developed root primordia at the wounding site were excised, separately placed on solid B5/2 medium supplemented with 30 g·l^{-1} sucrose and 1 g·l^{-1} cefotaxime, and then subcultured for two passages at 14-day intervals.

15.2.4 Liquid Hairy Root Culture

Each well-developed hairy root clone was cultured for one passage on B5/2 medium without antibiotic containing only 6 g·l^{-1} agar, then successively subcultured in Erlenmeyer flasks containing 20 ml, then 50 ml of antibiotic-free liquid B5/2 medium supplemented with 30 g·l^{-1} sucrose. Subsequently, 10 root tips (20–25 mm in length) from each clone were subcultured every 2 weeks on an orbital shaker (100 rpm), at 25±1°C in the dark. The hairy root cultures that grew rapidly after four subculture cycles were maintained in liquid B5/2 medium containing 30 g·l^{-1} sucrose.

15.2.5 Methyl Jasmonate Treatment

Nineteen-day-old hairy root cultures, at the late exponential phase, were elicited by adding methyl jasmonate (MeJA) dissolved in ethanol (final concentration of 40 µM) and harvested after 48 h of treatment. Six flasks were run for each sample.

15.2.6 Alkaloid Identification by Ceric Ammonium Sulphate Reagent

Hairy roots were harvested by filtration and immediately frozen at –20°C before being lyophilised. Alkaloid extraction was performed at room temperature from 100 mg of freeze-lyophilised hairy roots with 5 ml methanol [10]. After solvent evaporation, the residue was dissolved in a 28% NH_4OH (v/v) solution and then the alkaloids were extracted three times with 1 ml of CH_2Cl_2. After evaporation at 30°C for a night, the residues were dissolved in 250 µl of methanol. Twenty-five microlitres of the previous extract were spotted onto silica gel plates (25 thin-layer chromatography, TLC, Al. Sheets 20×20 cm, Merk) and developed either with absolute methanol:ethyl acetate (9:1, v/v) or with hexane:ether (1:1, v/v). The alkaloids were visualised by spraying with ceric ammonium sulphate (CAS) reagent, as described by Farnsworth et al. [11]. Alkaloid standards (ajmalicine, catharanthine, tabersonine, vindoline and vindolinine) were spotted on the plate as controls.

15.2.7 Serpentine and Ajmalicine Content Determination by Spectrofluorometry

Alkaloids were separated on TLC plates of silica gel either with methanol:chloroform (6:144, v/v) or acetone:toluene:methanol: NH_4OH (80:60:20:4, v/v/v/v). Serpentine was visualised under ultraviolet light. Ajmalicine (after oxidation by CAS) and serpentine contents were determined by spectrofluorometry (with excitation at 310 nm and emission at 445 nm), as reported previously by Farnsworth et al. [11]. Alkaloid contents in each hairy root clone were determined in triplicate from three hairy root culture flasks.

15.2.8 Catharanthine Content Determination by High-Performance Liquid Chromatography Analysis

One hundred milligrammes of dried hairy roots were ground and extracted with 12 ml of methanol (Hipersolv, BDH):acetic acid (Analar BDH; 96:4, v/v), using an automatic extractor (Dionex, Accelerated Solvent Extractor 200). Alkaloid detection and quantification were effected with an Agilent chain 1100 Liquid Chromatography/Mass Spectrometer Detection (MSD) Trap. The supernatant was filtered through a polytetrafluoroethylene 0.45 µM filter (GHP

Acrodisc from APLL Life Sciences) and 10 µl was injected into an amber glass high-performance liquid chromatography (HPLC) vial. Analyses were carried out, as described by Tikhomiroff et al. [3], at a flow rate of 1 ml·min^{-1} on a Symmetry C_{18}, 4.6×250 mm 5-m column (Waters). The mobile phase consisted of 10 mM Na_2HPO_4, acetonitrile (v/v). The elution was performed for 10 min with a linear gradient from 67:33 to 27:73; for 10–15 min, isocratic elution with 27:73 (v/v; column rinse); for 15–25 min, and isocratic elution with 67:33 (v/v; column equilibration). Stock solutions of catharanthine, vindoline, anhydrovinblastine, vinblastine and vincristine were prepared at a concentration of 1 g·l^{-1} in methanol/acetic acid (96:4; v/v).

15.2.9 Statistical Analysis

The dry mass and alkaloid contents were compared by ANOVA using the Scheffe post hoc test. A probability of less than 0.05% was considered statistically significant. The significant means obtained from ANOVA and Scheffe tests are indicated for each figure by different letters (a–d).

15.3 Results and Discussion

15.3.1 Genetic Transformation of Catharanthus Leaves

We chose the hypervirulent agropine-type *A. rhizogenes* strain Ar 15834, which has been successfully used to genetically transform recalcitrant plant species such as *Ginkgo biloba* [12]. Three hundred leaves of 5–6 cm in length were collected from the plants (Table 15.1).

Transformation was achieved by co-culturing leaves and bacteria during a "window of competence" of 72 h in darkness to avoid bacterial dryness. Primary roots emerged at the wounding site within 2–3 weeks. No root appeared on the control leaves. To initiate hairy root clones, each root reaching about 1–1.5 cm in length was excised and placed separately on solid, hormone-free B5/2 medium supplemented with cefotaxime. These roots come from the genetic transformation of a single plant cell by one bacterium [13]. Finally, 441 hairy root clones were obtained (Table 15.1). After three passages on solid medium, 73 of the clones that showed an obvious phenotype hairy root (fast growth in the hormone-free medium, ageotropism and a high degree of lateral branching) were placed onto a semi-solid medium to facilitate the subsequent passage in liquid medium (Fig. 15.1).

Among these clones, 28 were considered to be well established on liquid medium. Six clones of hairy roots, expressing the *rolB* and *C* genes as well as the *aux* genes (data not shown) and showing a rapid growth in liquid medium, were chosen for TIA analyses.

Table 15.1 Different numbers obtained from leaf transformation and hairy root clones on solid and in liquid medium

Species and varieties	Plants	Transformation numbers		Hairy root clones		
		Inoculated leaves (n)	Transformed leaves (n)	Solid medium (n)	Transferred into liquid medium (n)	Number of clones established[a]
Catharanthus roseus with pink flowers	P1	65	19	78	24	2
	P2	65	26	141	9	2
C. roseus with red flowers	P3	65	19	80	14	11
	P4	65	27	142	26	13
Total		260	91	441	73	28

[a]Well-established hairy root clones after four passages in liquid medium

Fig. 15.1 a,b Two transformed clones cultured on hormone-free B5/2 solid medium exhibited a different root type. **c,d** After root transfer into B5/2 liquid medium the hairy root clones showed a more (**c**) or less (**d**) fast growth and typical phenotype

15.3.2 Alkaloid Profiles

Transformed roots of *C. roseus* are known to accumulate both high content and a large number of monomer alkaloids [14]. In this study, the presence of tabersonine, ajmalicine, vindolinine and catharanthine in six hairy root clones previously elicited by MeJA or not elicited were revealed by TLC with CAS reagent spraying. Although other CAS-positive spots were also visible, their identification was not possible because of the lack of specific standards (Fig. 15.2).

No spot of vindoline appeared, in spite of the detection of its key precursor tabersonine. Presumably, the biosynthetic pathway leading to vindoline bifurcated at the tabersonine level towards two other alternative routes, leading to

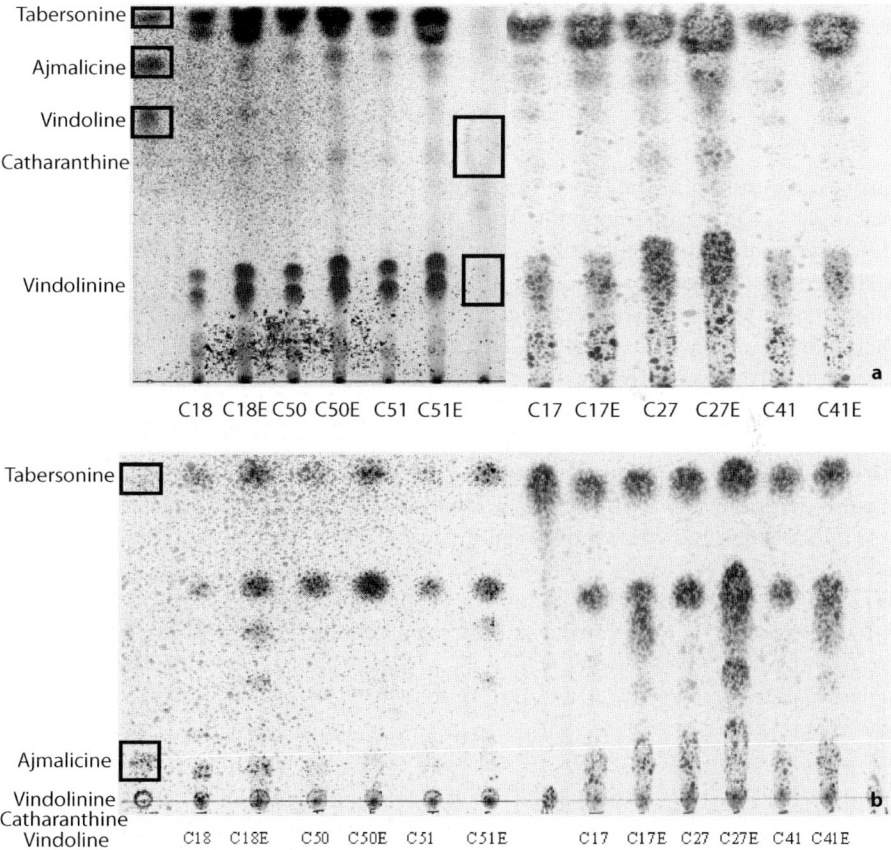

Fig. 15.2 Alkaloid profiles revealed the presence of **a** tabersonine, ajmalicine, catharanthine and vindolinine monomers (separated with absolute methanol:ethyl acetate), and **b** tabersonine and ajmalicine alkaloids separated with hexane:ether after ceric ammonium sulphate (CAS) spraying. Other spots could not be identified because of the lack of standards

either löchnericine or hörhammericine, both of which are devoid of medicinal properties [3]. The tabersonine-to-vindoline conversion, essential to vinblastine biosynthesis, is a complex process that requires at least six enzymatic steps [15]. Similarly, vindoline was not detected in several other studies [16], except in a few *C. roseus* hairy root lines that also produced vinblastine [14].

15.3.3 Alkaloid Contents

Jasmonates are powerful signal molecules triggering the plant defence response through the stimulation of secondary metabolite biosynthesis. They were used to enhance TIA accumulation in *C. roseus* cell suspensions [17] and in the past 5 years to amplify secondary metabolite production in hairy root cultures [18]. In the present experiment, MeJA was added to the culture medium of the six hairy root clones at the 19th day. Treatment with 40 µM MeJA had no deleterious effect on clone growth expressed as dry mass (Fig. 15.3), as found by Rhijwani and Shanks [16] with jasmonic acid. Quantifying alkaloids in C18, C50, C51, C17, C27 and C41 clones by spectrofluorometry and HPLC-MSD trap processes resulted in the following values: 0.1–1.25 mg·g^{-1} DM ajmalicine, 0.5–2.7 mg·g^{-1} DM serpentine and 0.1–0.7 mg·g^{-1} DM catharanthine (Fig. 15.4).

Ajmalicine and catharanthine content in hairy root cultures resembled those detected in the roots of intact plants [14], but serpentine accumulation was higher [19]. HPLC analyses did not detect any traces of vindoline, anhydrovinblastine or vinblastine. Such analytical data are in agreement with the aforementioned alkaloid profiles.

Since cultures of hairy roots are routinely maintained in the dark to palliate their eventual browning, the absence of vindoline may be explained by the lack of expression of the light-dependent genes encoding for the enzymes desacetoxyvindoline-4-hydroxylase and deacetylvindoline 4-O acetyltransferase [20,21] belonging to the vindoline pathway.

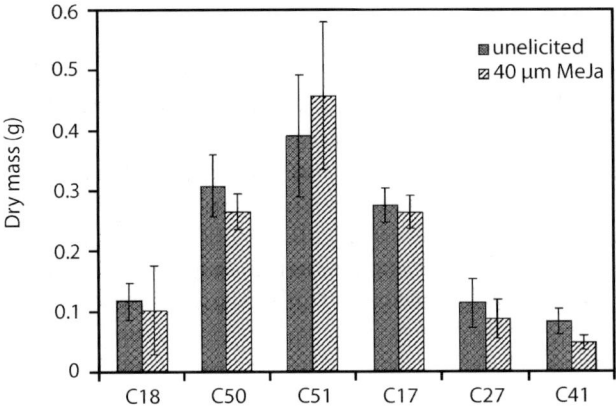

Fig. 15.3 The growth is expressed in dry biomass of roots per gram for the six hairy root clones elicited by methyl jasmonate (MeJA) or non-elicited taken as controls

Chapter 15 Hairy Roots of Catharanthus roseus

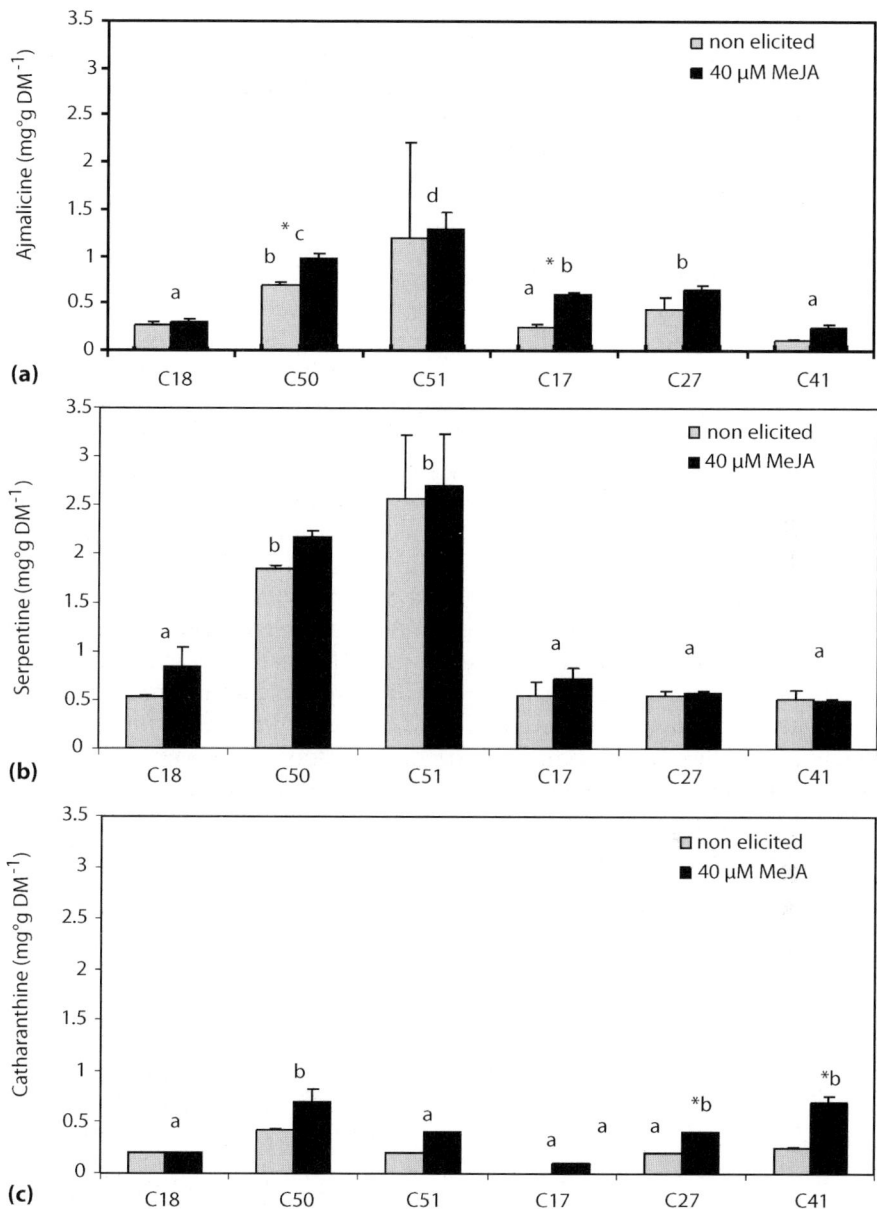

Fig. 15.4 Contents of ajmalicine (**a**), serpentine (**b**) and catharanthine (**c**) were respectively evaluated in the C18, C50 and C51 hairy root clones belonging to *Catharanthus roseus* with pink flowers and in C17, C27 and C41 belonging to *C. roseus* with red flowers elicited by 40 μM MeJA. The same non-elicited clones were taken as controls. Letters *a–d* indicate the results of an ANOVA test. Significantly different means are indicated by two different letters between two clones and by an *asterisk* for the same clone

Ajmalicine contents were significantly increased in the C50 clone elicited by MeJA treatment (Fig. 15.4a). For all clones, serpentine contents were not affected by MeJA elicitation and the highest levels were found in the C51 clone. Catharanthine amounts were significantly increased in C27 and C41 clones after elicitation (Fig. 15.4c). Obviously, MeJA treatment increased the metabolic flux towards the ajmalicine or catharanthine branches of the indole alkaloid pathway in these clones. Similarly, Rhijwani and Shanks [16] considered jasmonate acid as being a good elicitor, able to increase the metabolite flux in several branches of the alkaloid pathway in *C. roseus*.

15.4 Conclusion

Numerous clones of hairy roots could be obtained from leaf tissues of *C. roseus*, demonstrating the efficiency of the transformation process. Such cultures are able to accumulate different TIAs, and some of them, such as ajmalicine, serpentine and catharanthine, were produced at interesting levels after MeJA treatment of the clones. Moreover, the CAS reagent method revealed other alkaloids that could be quantified in the presence of the appropriate standards. Other alkaloids biosynthesised in the wild roots may also be similarly accumulated into hairy root clones. This transformation root system makes possible integration and overexpression of genes encoding either transcription factors [22] or limiting enzymes of the indole alkaloid biosynthesis pathway, such as the 1-deoxy-D-xylulose-5-phosphate synthase (*Crdxs*) and 1-deoxy-D-xylulose-5-phosphate reductioisomerase (*Crdxr*) genes belonging to the methylerythritol 4-phosphate route [23].

References

1. Van Der Heijden R, Jacobs DI, Snoeijer W, Hallard D, Verpoorte R (2004) Curr Med Chem 11:607
2. Verpoorte R, van der Heijden R, ten Hoopen HJG, Memelink J (1999) Biotechnol Lett 21:467
3. Tikhomiroff C, Jolicoeur M (2002) J Chromatogr A 955:87
4. Davioud E, Kan C, Hamon J, Tempe J, Husson H (1989) Phytochemistry 28:2675
5. Guillon S, Trémouillaux-Guiller J, Kumar Pati P, Rideau M, Gantet P (2006) Trends Biotechnol 24:403
6. Guillon S, Tremouillaux-Guiller J, Kumar Pati P, Rideau M, Gantet P (2006) Curr Opin Plant Biol 9:341
7. Bercetche J, Chriqui D, Adam S, David C (1987) Plant Sci 52:195
8. Camilleri C, Jouanin L (1991) Mol Plant Microbe Interact 4:155
9. Gamborg O, Miller R, Ojima K (1968) Exp Cell Res 50:151
10. Mérillon JM, Doireau P, Guillot A, Chenieux JC, Rideau M (1986) Plant Cell Rep 5:23
11. Farnsworth NR, Blomster RN, Damratoski D, Meer WA, Camarato LV (1964) Lloydia 27:302

12. Ayadi R, Tremouillaux-Guiller J (2003) Tree Physiol 23:713
13. Chilton MD, Tepfer DA, Petit A, David C, Casse-Delbart F, Tempe J (1982) Nature 295:432
14. Parr AJ, Peerless ACJ, Hamill JD, Walton NJ, Robins RJ, Rhodes MJC (1988) Plant Cell Rep 7:309
15. Shanks JV, Bhadra R, Morgan J, Rijhwani S, Vani S (1998) Biotechnol Bioeng 58:333
16. Rijhwani SK, Shanks JV (1998) Biotechnol Prog 14:442
17. Gantet P, Imbault N, Thiersault M, Doireau P (1998) Plant Cell Physiol 39:269
18. Shanks JV, Morgan J (1999) Curr Opin Plant Biol 10:151
19. Bhadra R, Vani S, Shanks JV (1993) Biotechnol Bioeng 41:581
20. Aerts RJ, De Luca V (1992) Plant Physiol 100:1029
21. Vazquez-Flota F, De Luca V, Carrillo-Pech M, Canto-Flick A, de Lourdes Miranda-Ham M (2002) Mol Biotechnol 22:1
22. Gantet P, Memelink J (2002) Trends Pharmacol Sci 23:563
23. Veau B, Courtois M, Oudin A, Chenieux JC, Rideau M, Clastre M (2000) Biochim Biophys Acta 1517:159

Chapter 16
Roseroot (*Rhodiola rosea* L.): Effect of Internal and External Factors on Accumulation of Biologically Active Compounds

Z. Węglarz (✉), J.L. Przybył and A. Geszprych

Department of Vegetable and Medicinal Plants, Warsaw University of Life Sciences - SGGW, Nowoursynowska 159, 02-776 Warsaw, Poland, e-mail: zenon_weglarz@sggw.pl

Abstract Roseroot (*Rhodiola rosea* L.) is a perennial that grows wild in the mountains of Siberia, Central Europe and North America. Its underground organs (rhizomes with roots) are used as a medicinal raw material; the plant is considered to be one of the most active adaptogens. The most important biologically active constituents of the raw material are phenolic compounds, including tyrosol and its glycoside salidroside, and *trans*-cinnamic alcohol derivatives (rosavin, rosarin and rosin). The results of several years of study carried out at Warsaw University of Life Sciences - SGGW in Poland indicate a high intraspecific variability concerning accumulation of these compounds. It was also stated that both the weight of the underground organs of roseroot and the content of active compounds changes during plant development. The mean weight of air-dry rhizomes with roots of plants grown in central Poland increased by up to 120 g per plant in the 5th year of plant vegetation. In the 6th year the symptoms of plant aging were observed – the oldest, central part of rhizome decayed and the rhizome divided into many smaller parts characterised by lower content of salidroside and rosavin. The yield and quality of roseroot raw material was also significantly affected by climatic and soil conditions. Plants grown in central Poland were characterised by higher weight of underground organs but lower content of rosavin and salidroside in comparison with those grown in southern Poland (mountain area). Post-harvest treatment of the raw material (stabilisation and extraction method) distinctly affected the quality of the obtained extracts. Both convection drying at 80°C and lyophilisation are good methods of stabilisation of the roseroot raw material. Periodical extraction with ultrasound, and continuous exhaustive extraction using both methanol and 75% ethanol as extraction media allow to get extracts of comparable content of determined phenolic compounds.

Ramawat KG, Mérillon JM (eds.), In: *Bioactive Molecules and Medicinal Plants*
Chapter DOI: 10.1007/978-3-540-74603-4_16, © Springer 2008

Keywords Plant development, Intraspecific variability, Ecological factors, Post-harvest treatment, Phenolic compounds, Salidroside, Rosavin

16.1 Introduction

Roseroot (*Rhodiola rosea* L.) is an alpine perennial that belongs to the *Crassulaceae* family. Its underground organs (i.e. rhizomes with roots) have been used as natural remedies in Siberia, Tibet and the Far East for centuries [1]. The biological activity of this raw material has been proven in contemporary studies. The results of pharmacological investigations indicate that the extracts from roseroot reveal antioxidant activity via inhibition of lipid peroxidation in liver cells and clearing of free radicals. They stimulate the central nervous system, improve learning abilities and prevent stress-induced cardiac damage [2–16]. They also show anti-fatigue, anti-inflammatory, hepatoprotective and anti-tumour activity [17–21]. In clinical studies they have been effective in the treatment of physical weakness, heart diseases, depression, memory and learning problems. Roseroot is regarded as one of the most active adaptogens and it is specially recommended for sportsmen, hard-working people, convalescents and elderly people.

In the severe alpine climate, the growth of roseroot is very slow, so that it may be harvested as a raw material for herbal industry even after several dozen years. The slow development of the plant and growing demand for the raw material has resulted in a rapid diminution of its natural sites and has necessitated legal protection for this species [1, 22, 23]. It seems that the only reasonable way of both preserving wild-growing roseroot and providing for the needs of the phytopharmaceutical industry is to introduce this plant into cultivation. However, it is not easy to obtain the raw material of uniformly high quality from wild plants directly introduced into cultivation [24–32]. This will only be possible after preliminary multi-directional studies. In the present paper we discuss the research concerning the effects of genetic, developmental, ecological and post-harvest factors on the accumulation of active compounds in roseroot cultivated in Poland.

16.2 Plant Characteristics

Roseroot is a heterozygous plant that exhibits high morphological, developmental and chemical variability. Plant height ranges from 5 to 70 cm, the leaves are sessile, elliptic to lance-shaped, wax coated, crenulated or serrulate, 7–35 mm long and 3–18 mm wide. Yellow to red flowers are located in terminal umbel-like clusters (Fig. 16.1). Flowers are male, female or bisexual [32]. Its fruit is 4–6 mm long and 3–5 mm wide. Seeds are 0.5–1 mm long. The weight of 1000 seeds ranges from 185 to 250 mg [33, 34]. The underground part of the plant consists of the fleshy cylindrical rhizome, 2–10 cm in diameter, with sparse roots [1, 23, 35]. The outer part of the rhizome is grey-brown with a golden metallic cork [35]. The inner part of the fresh rhizome is white, and during drying of the sliced raw material those surfaces that have contact with air turn pink (Fig. 16.2).

From the pharmacological point of view, the most important active constituents of the raw material are phenolic compounds, including tyrosol and its glycoside salidroside, and *trans*-cinnamic alcohol derivatives: rosavin, rosarin and rosin (Fig. 16.3) [36–47]. The presence of phenolic acids in roseroot has also been reported [1, 48].

Fig. 16.1 Roseroot plant

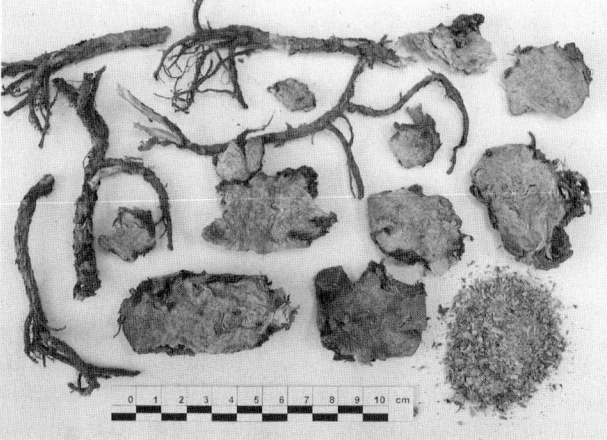

Fig. 16.2 Air-dried rhizomes with roots (raw material of roseroot)

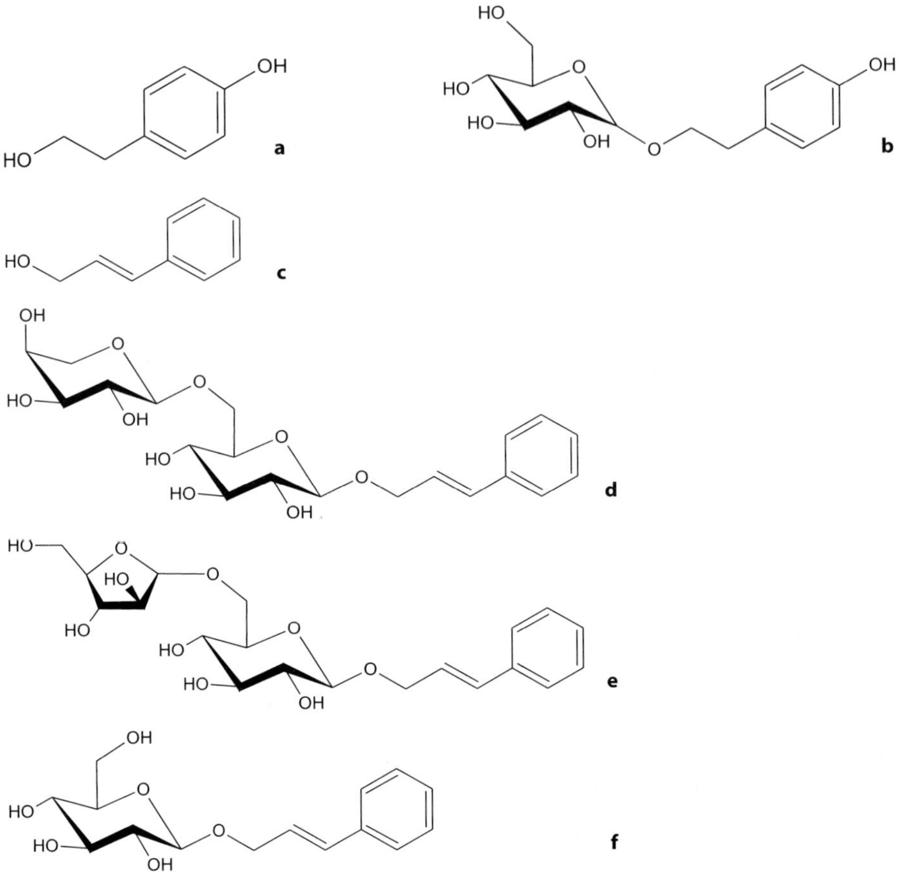

Fig. 16.3 The most important biologically active compounds in the underground organs of *Rhodiola rosea*: **a** tyrosol; **b** salidroside; **c** *trans*-cinnamic alcohol; **d** rosavin; **e** rosarin; **f** rosin

16.3 Intraspecific Variability

16.3.1 Distribution of Phenolic Compounds in Rhizomes and Roots

Phenolic compounds accumulate in the underground organs of *R. rosea* starting from the early stages of plant development. In our investigations they were detected in the roots of 7-week-old seedlings. In the older plants the cells containing phenolics were present mainly in rhizomes. These compounds were located in the parenchymal cells of the secondary conducting tissues of both the rhizomes and roots. In the rhizomes they were also found in the cortical parenchyma cells (Figs. 16.4 and 16.5).

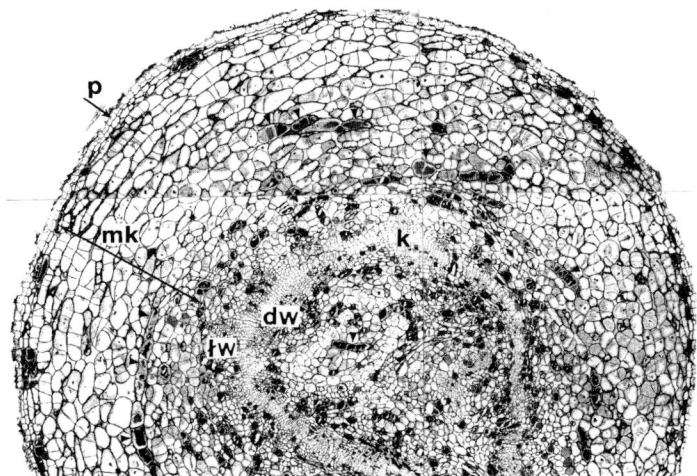

Fig. 16.4 Structure of the rhizome of a 1-year-old plant. Cells containing phenolic compounds are shown with an *arrowhead*. Magnification ×46. *p* Periderm, *mk* cortical parenchyma, *lw* secondary phloem, *dw* secondary xylem, *k* cambium

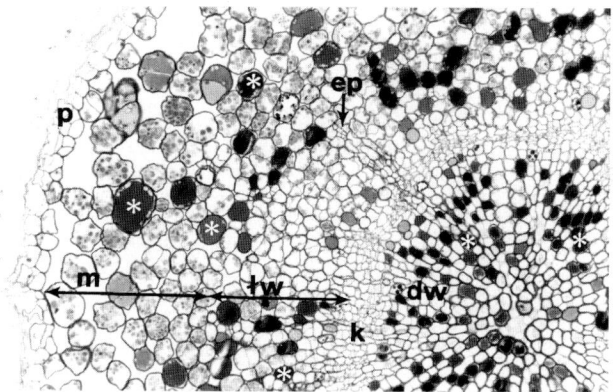

Fig. 16.5 Secondary structure of the root. Cells containing phenolic compounds are shown with an *asterisk*. Magnification ×185. *m* Parenchyma, *ep* conducting elements of secondary phloem

Rhizomes and roots differed significantly with respect to the content of the determined phenolic compounds (Table 16.1). The rhizomes were characterised by higher content of salidroside, rosavin, rosarin and *trans*-cinnamic alcohol, whereas roots by higher content of rosin, tyrosol and phenolic acids.

Table 16.1 Content of phenolic compounds in the rhizomes and roots of 5-year-old plants (mg·100 g^{-1}).

Compound	Rhizomes ($n=13$)	Roots ($n=13$)	Mean
Tyrosol derivatives			
Tyrosol	9.7±3.6*	21.3±6.1	15.5±1.8
Salidroside	675.3±565.5	248.9±222.8*	462.1±242.3
***Trans*-cinnamic alcohol derivatives**			
Trans-cinnamic alcohol	44.6±26.6	21.1±16.0*	32.9±7.5
Rosavin	2961.4±633.5	2270.5±594.3*	2616.0±27.7
Rosarin	335.6±60.0	268.6±41.6*	302.1±13.0
Rosin	616.9±169.5*	774.6±205.7	695.8±25.6
Phenolic acids			
Caffeic acid	4.5±2.6*	6.6±2.8	5.6±0.1
Protocatechuic acid	6.1±2.8	4.8±2.2	5.5±0.4
4-Hydroxybenzoic acid	37.0±12.6*	50.1±21.1	43.6±6.0
Syringic acid	48.1±12.6	37.7±14.3*	42.9±1.2

*$P<0.05$

For each analysis of phenolic compounds, the underground organs of 20 randomly selected plants (dried at 80±5°C) were used. One gramme of air-dried, grounded raw material was extracted with 100 ml of methanol in a Büchi B-811 extraction system. After evaporation of the solvent, the residue was dissolved in 10 ml methanol, filtered through a Supelco IsoDisc polytetrafluoroethylene 25 mm×0.45 µm filters, and subjected to high-performance liquid chromatography (HPLC). The analysis was carried out using a Shimadzu chromatograph with SPD-M10A VP DAD detector equipped with a Luna 5-µm C18 (2) 250 mm×4.6 mm column (Phenomenex). A gradient of 0.2% phosphoric acid in HPLC-grade water (A) and acetonitrile (B) was used as follows: 0 min, 4% B; 10 min, 13% B; 20 min, 15% B; 30 min, 20% B; 33 min 25% B; 38 min, 30% B; held constant for 22 min. The following analysis parameters were used: injection volume: 20 µl, flow rate 1.2 ml·min^{-1}, oven temperature 31°C, time of analysis 60 min, recording wavelength: 190–450 nm, detection wavelength: 275 nm. Peaks were identified by comparison of retention time and spectral data with adequate parameters of standards (*Rhodiola rosea* Standards Kit by ChromaDex). Quantification was based on the peak area. The content of the determined compounds was calculated in mg·100 g^{-1} dry matter. The results were analysed with one-way and multifactor ANOVA Tukey's HSD test at the 0.05 significance level using Statgraphics Plus for Windows v. 4.1

Table 16.2 Content of phenolic compounds in the raw material (rhizomes with roots) of different origins (mg·100 g^{-1})

Compound	Mongolian Altai	Gorkhi Terelj	Russian Altai
Tyrosol derivatives			
Tyrosol	55.1 ± 2.3^a	5.3 ± 0.3^c	9.2 ± 0.4^b
Salidroside	111.4 ± 4.8^b	48.2 ± 4.3^c	141.6 ± 11.9^a
***Trans*-cinnamic alcohol derivatives**			
Trans-cinnamic alcohol	1631.7 ± 40.5^a	60.5 ± 3.4^c	174.7 ± 14.7^b
Rosavin	2250.6 ± 147.9^b	813.9 ± 84.1^c	3140.9 ± 61.0^a
Rosarin	492.7 ± 36.5^a	65.9 ± 4.5^c	315.7 ± 5.9^b
Rosin	275.0 ± 18.2^b	95.1 ± 7.9^c	596.1 ± 33.8^a
Phenolic acids			
Caffeic acid	4.27 ± 1.60^b	4.75 ± 0.68^b	14.22 ± 1.35^a
Protocatechuic acid	5.07 ± 1.54^a	1.78 ± 0.58^b	7.08 ± 0.73^a
4-Hydroxybenzoic acid	4.62 ± 0.75^b	21.11 ± 2.51^a	8.08 ± 0.17^b
Syringic acid	5.03 ± 0.46^c	12.11 ± 1.66^a	8.28 ± 1.04^b

$^{a-c}$Values marked with the same letter do not differ significantly at $\alpha=0.05$

16.3.2 Quality of Raw Material of Different Origin

The results of previous studies [43, 49] indicate that the content of biologically active compounds in the raw material collected from different natural sites of roseroot varies within a wide range. For example, differences in the content of rosavin came up to 60%. One of the most important reasons for such diversity is genetic factors.

In our studies, the raw materials obtained from plants of three different populations originating from distant natural sites – in the area of Russian Altai, Mongolian Altai and Gorkhi Terelj (central Mongolia) – were compared (Table 16.2). The evaluated raw material differed significantly with respect to the content of all determined phenolic compounds. Differences in the content of rosavin were much higher in comparison with those reported by Kir'janov et al. [49] and came up to 400%, and differences in the content of rosin even reached 600%.

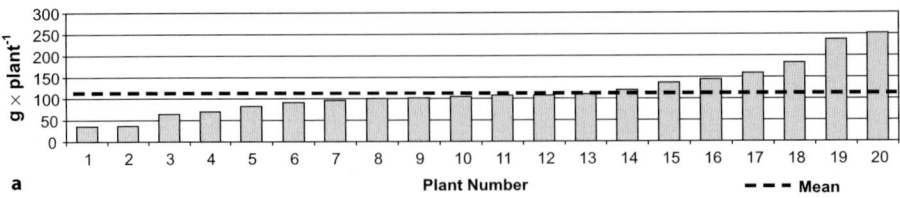

Fig. 16.6 Air-dried weight of the raw material (rhizomes with roots) of individual plants (g·plant^{-1})

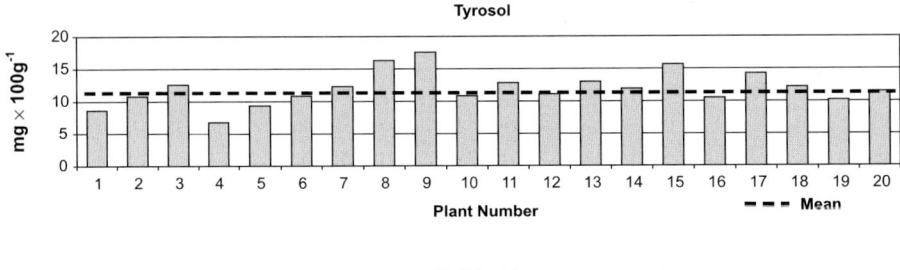

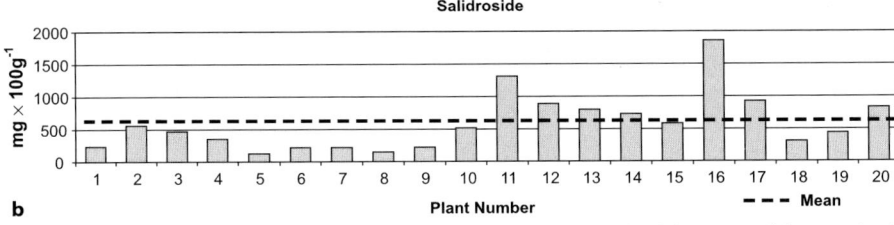

Fig. 16.7 Content of tyrosol derivatives in the raw material (rhizomes with roots) of individual plants (mg·100 g^{-1})

16.3.3 Individual Variation

The chemical variation within the roseroot population originating from the Russian Altai and those cultivated in central Poland was investigated in the 5th year of plant vegetation. High variability concerning both the weight of rhizomes and the content of phenolic compounds was found. The weight of air-dried underground organs ranged from 36 to 250 g (Fig. 16.6). In terms of phenolic compounds, the biggest difference between individual plants concerned the content of salidroside (125–1860 mg·100 g^{-1}; Fig. 16.7) and *trans*-cinnamic alcohol (8.9–79.7 mg·100 g^{-1}; Fig. 16.8). The content of other compounds also varied, but not so remarkably.

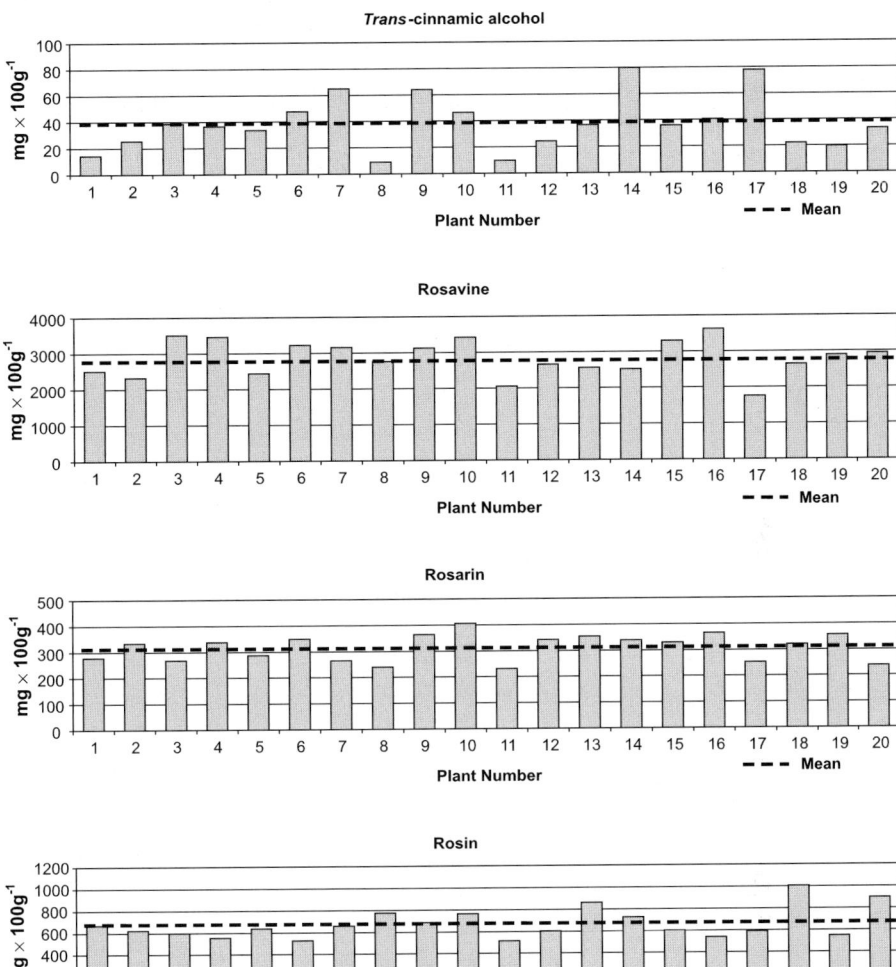

Fig. 16.8 Content of *trans*-cinnamic alcohol derivatives in the raw material (rhizomes with roots) of individual plants (mg·100 g^{-1})

16.4 Accumulation of Biomass and Biologically Active Compounds in the Underground Organs of Roseroot During Plant Development

So far, roseroot is collected mainly from natural sites. The standardisation of such raw material is difficult because it is obtained from the plants of different age. It is easier to control the quality of raw material from cultivation because of the possibility of more precise determination of the dynamics of accumulation of biologically active compounds in such plants.

We studied the growth of the underground organs and the accumulation of phenolic compounds in roseroot grown in central Poland during the period of six vegetation seasons. The mean weight of air-dried rhizomes with roots increased up to 120 g per plant in the 5th year of plant vegetation (Fig. 16.9). Over 60% of 5-year-old plants had underground organs weighing 50–150 g; however, the maximum weight came up to 300 g. Plants collected in the 4th and 5th year of vegetation were characterised by having the highest percentage of rhizome weight in the total weight of the underground part (Table 16.3). In the 6th year of vegetation, symptoms of plant aging were observed. The oldest, central part of rhizome decayed and the rhizome divided into many smaller parts (Fig. 16.10f), so that its mean weight decreased up to 45 g (Fig. 16.9).

Table 16.3 Effect of plant age on the percentage of rhizome weight in the total weight of air-dry raw material (rhizomes with roots; %)

Plant age					
1-year-old	2-year-old	3-year-old	4-year-old	5-year-old	6-year-old
56^c	60^{bc}	76^a	81^a	83^a	69^b

$^{a-c}$Values marked with the same letter do not differ significantly at $\alpha=0.01$

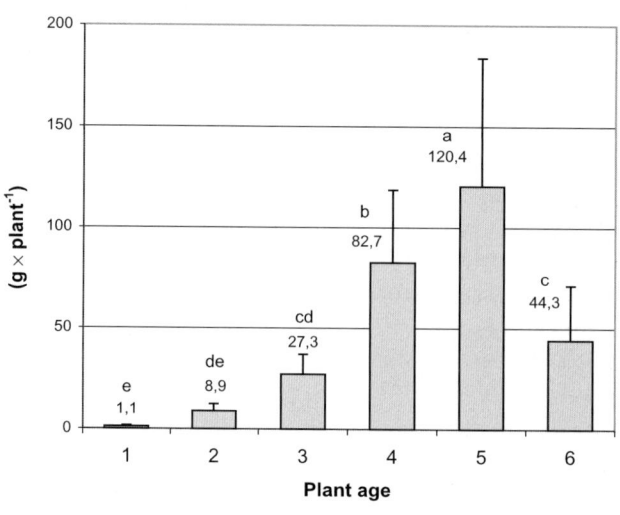

Fig. 16.9 Effect of plant age on the weight of air-dried raw material (rhizomes with roots) (g·plant^{-1}). Columns marked with the same letter (a–e) do not differ significantly at $\alpha = 0{,}05$

Chapter 16 Roseroot (*Rhodiola rosea* L.)

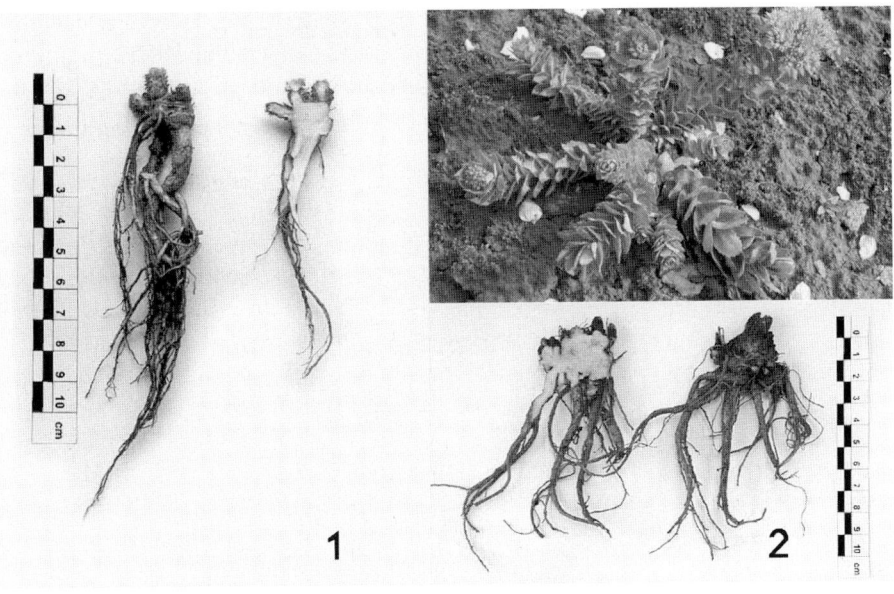

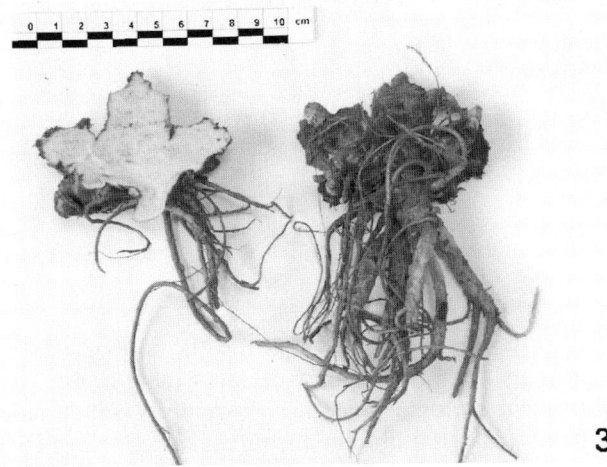

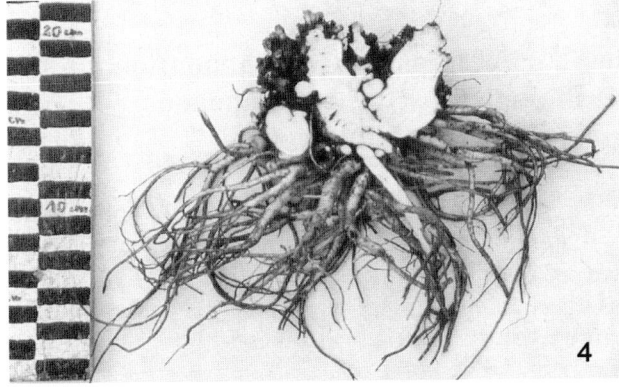

Fig. 16.10 The underground organs:
1 – 1-year-old plant,
2 – 2-year-old plant,
3 – 3-year-old plant,
4 – 4-year-old plant
5–6 see next page

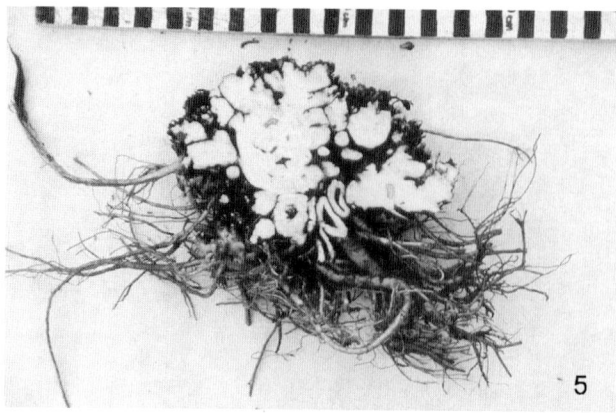

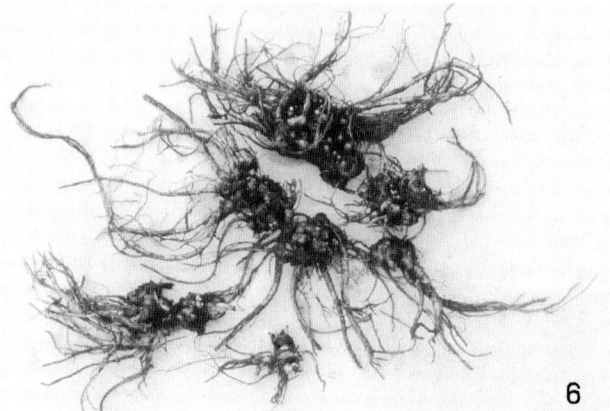

Fig. 16.10 The underground organs: *(continued)* 5 – 5-year-old plant, 6 – 6-year-old plant (the rhizome divided into smaller autonomic parts)

There was no simple relationship between plant age and the content of determined phenolic compounds in the underground organs (Table 16.4). The highest content of the most pharmacologically active compounds (salidroside and rosavin) was found in the raw material obtained from 5-year-old plants.

16.5 Effect of Ecological Factors on the Accumulation of Biomass and Biologically Active Compounds in the Underground Organs of Roseroot

The climatic and soil conditions may significantly affect the yield and quality of the obtained plant raw material. Our studies confirmed the effect of these factors on the development of roseroot, morphology and yield of its underground organs, as well as the content of biologically active compounds in the raw material. The mean weight of air-dried rhizomes with roots of plants

Table 16.4 Effect of plant age on the content of biologically active compounds in the raw material (rhizomes with roots; mg·100 g^{-1})

Compound	Plant age					
	1-year-old	2-year-old	3-year-old	4-year-old	5-year-old	6-year-old
Tyrosol derivatives						
Tyrosol	7.5^c	10.0^{bc}	14.4^{ab}	14.4^{ab}	13.8^{ab}	15.7^a
Salidroside	182.2^c	207.9^{bc}	259.1^{bc}	350.9^{abc}	535.7^a	441.4^b
***Trans*-cinnamic alcohol derivatives**						
Trans-cinnamic alcohol	25.3^{ab}	16.4^b	14.5^b	23.4^{ab}	40.9^a	24.9^{ab}
Rosavin	2415.3^{ab}	2361.9^{ab}	2196.1^{ab}	2186.1^{ab}	2744.4^a	2014.6^b
Rosarin	354.9^a	254.8^{ab}	235.5^b	180.9^b	322.5^{ab}	243.2^{ab}
Rosin	193.3^a	727.3^a	601.6^a	483.6^a	669.2^a	462.2^a
Phenolic acids						
Caffeic acid	3.2^c	7.6^{ab}	7.3^{abc}	4.2^{bc}	5.4^{bc}	11.4^a
Protocatechuic acid	6.6^a	5.8^{ab}	4.0^b	5.2^{ab}	5.9^{ab}	6.1^{ab}
4-Hydroxybenzoic acid	21.8^c	38.6^{ab}	31.5^{abc}	27.2^{bc}	41.9^a	41.1^a
Syringic acid	10.4^c	21.0^{bc}	30.6^{ab}	28.9^{ab}	41.7^a	22.6^{bc}

$^{a-c}$Values marked with the same letter do not differ significantly at $\alpha=0.05$

grown in central Poland (typical temperate climate, 99 m above sea level, vegetation period 216 days, alluvial soil) was twice as high as the weight of underground organs of plants grown in north-eastern Poland (transitional area between continental and Atlantic climates, 164 m above sea level, vegetation period 208 days, sandy soil) and in the mountains (alpine climate, 1000 m above sea level, vegetation period 184 days, clayey soil; Table 16.5). Soil type affected the size and shape of the underground part of a plant. Plants grown on sandy soil formed a highly branched rhizome with few roots, whereas on clayey and alluvial soils they formed a compact rhizome with numerous roots of large diameter (Fig. 16.11).

The content of salidroside and rosavin in the raw material obtained from the plants grown in the mountains was significantly higher in comparison with that of plants grown in the lowlands (central and north-eastern Poland). In the case of other determined compounds, there was no clear relationship between their accumulation in the raw material and the region of plant cultivation (Table 16.6).

Table 16.5 Effect of climatic and soil conditions on the weight of air-dried raw material (rhizomes with roots) of 3-year-old plants (mg·plant^{-1})

Plant organ	Central Poland	North-eastern Poland	Mountains (south Poland)
Rhizome	20.7 ± 7.3^c	12.0 ± 4.8^b	4.3 ± 2.6^a
Roots	6.6 ± 2.3^b	1.6 ± 0.6^a	1.8 ± 1.1^a
Total	27.3 ± 9.6^c	13.6 ± 5.4^b	6.1 ± 3.7^a

$^{a-c}$Values marked with the same letter do not differ significantly at $\alpha=0.05$

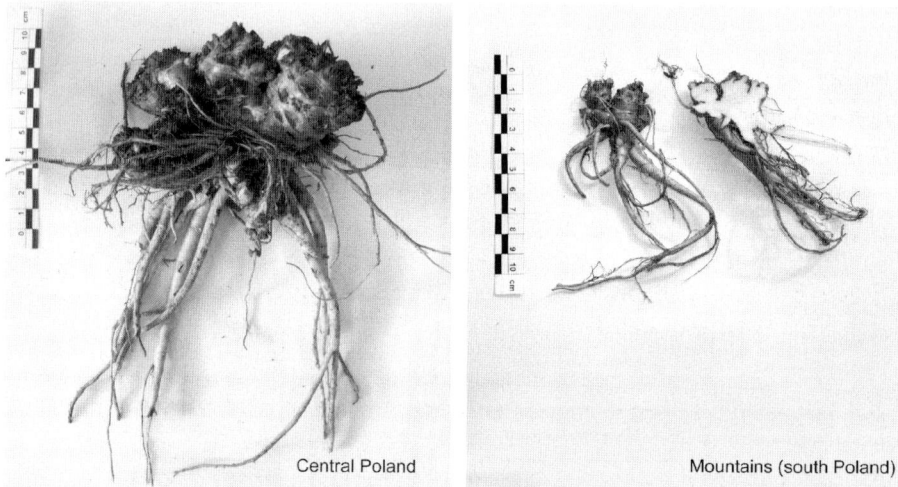

Fig. 16.11 The underground organs of 3-year-old roseroot plants cultivated in different climatic and soil conditions

16.6 Effect of Post-harvest Treatment on the Quality of Raw Material and Extracts

Regarding the high content of water (sometimes over 70%) in fresh rhizomes of roseroot, this plant material is rather difficult to stabilise. Kurkin et al. [42, 45] studied the effect of temperature in the drying chamber on the quality of this raw material. They found that the optimum drying temperature was 80°C or 20°C. Drying at 50–60°C, previously recommended by Syrov [50], resulted in a distinct reduction in salidroside and rosavin content. In our studies, three methods of stabilisation were applied: convection drying at 80°C, freezing and lyophilisation (Table 16.7). It appeared that the content of determined phenolic compounds in dried and lyophilised raw material was comparable and high,

Table 16.6 Effect of different climatic and soil conditions on the content of biologically active compounds in the raw material (rhizomes with roots) of 3-year-old plants (mg·100 g^{-1})

Compound	Central Poland	North-eastern Poland	Mountains (south Poland)
Tyrosol derivatives			
Tyrosol	14.4[ab]	17.1[a]	10.7[b]
Salidroside	259.1[b]	181.0[c]	378.7[a]
Trans-cinnamic alcohol derivatives			
Trans-cinnamic alcohol	14.5 ns	16.1 ns	15.7 ns
Rosavin	2196.1[ab]	1993.0[b]	2420.3[a]
Rosarin	235.5 ns	237.3 ns	239.4 ns
Rosin	601.6[a]	351.8[b]	441.6[b]
Phenolic acids			
Caffeic acid	7.3[b]	7.2[b]	11.1[a]
Protocatechuic acid	4.0[b]	2.7[c]	7.4[a]
4-Hydroxybenzoic acid	31.5[b]	40.9[a]	33.9[b]
Syringic acid	30.6[a]	13.1[c]	25.0[b]

[a-c]Values marked with the same letter do not differ significantly at α=0.05
ns – differences are not significant at α = 0.05

Table 16.7 Effect of the stabilisation method on the content of biologically active compounds in the raw material (rhizomes with roots; mg·100 g^{-1})

Compound	Freezing	Convection drying	Sublimation drying
Tyrosol derivatives			
Tyrosol	14.0 ± 7.9[b]	26.4 ± 10.4[a]	23.8 ± 9.5[ab]
Salidroside	297.3 ± 151.2 ns	443.9 ± 227.6 ns	491.1 ± 180.8 ns
Trans-cinnamic alcohol derivatives			
Trans-cinnamic alcohol	330.0 ± 115.4[a]	66.2 ± 23.5[b]	13.1 ± 2.9[b]
Rosavin	24.7 ± 8.0[b]	3079.9 ± 329.4[a]	3589.6 ± 739.7[a]
Rosarin	224.5 ± 91.1[b]	302.8 ± 83.4[ab]	388.9 ± 104.1[a]
Rosin	450.8 ± 104.4[b]	1029.1 ± 279.7[a]	849.6 ± 390.2[a]
Phenolic acids			
Caffeic acid	19.6 ± 7.3[b]	17.4 ± 3.7[b]	28.7 ± 5.7[a]
Protocatechuic acid	7.5 ± 2.0[ab]	9.7 ± 3.4[a]	6.6 ± 1.3[b]
4-Hydroxybenzoic acid	27.4 ± 7.2[c]	65.4 ± 18.8[b]	87.9 ± 19.1[a]
Syringic acid	26.0 ± 7.2[c]	45.0 ± 6.5[b]	81.1 ± 15.4[a]

[a-c]Values marked with the same letter do not differ significantly at α=0.05
ns – differences are not significant at α = 0.05

Table 16.8 Effect of solvent and extraction method on the content of biologically active compounds in the raw material (rhizomes with roots; mg·100 g^{-1} dry matter). *S* Ultrasonic extraction, *C* continuous exhaustive extraction

Compound	Extraction method	Water	Ethanol	Methanol	Mean
Tyrosol derivatives					
Tyrosol	S	7.5±1.8	10.3±1.1	10.3±0.5	**9.4±1.1**
	C	2.8±1.3	11.8±2.1	12.0±1.9	**8.9±1.8**
	Mean	**5.2±1.6^b**	**11.1±1.6^a**	**11.2±1.2^a**	
Salidroside	S	335.4±7.2	590.0±50.4	664.8±54.2	**530.1±37.3**
	C	486.8±52.1	604.6±27.6	576.5±22.4	**556.0±35.0**
	Mean	**411.1±29.7^b**	**597.3±39.0^a**	**620.7±38.3^a**	
***Trans*-cinnamic alcohol derivatives**					
Trans-cinnamic alcohol	S	21.7±13.4	17.5±3.9	35.9±24.1	**25.0±13.8**
	C	8.2±2.8	46.5±8.8	46.1±15.2	**33.6±8.9**
	Mean	**15.0±8.1^b**	**32.0±6.4^{ab}**	**41.0±19.7^a**	
Rosavin	S	1160.1±55.4	3015.7±107.8	3088.5±82.7	**2421.4±81.9**
	C	1702.6±199.4	2731.4±46.7	2801.6±63.8	**2411.9±103.3**
	Mean	**1431.4±127.4^b**	**2873.6±77.2^a**	**2945.1±73.3^a**	
Rosarin	S	124.5±4.0	426.0±22.4	413.1±30.3	**321.2±18.9**
	C	176.2±24.1	341.8±17.4	342.6±22.3	**286.9±21.3***
	Mean	**150.4±14.1^b**	**383.9±19.9^a**	**377.9±26.3^a**	
Rosin	S	114.4±2.2	471.8±136.7	531.7±26.2	**372.6±55.0**
	C	100.0±10.5	566.9±51.3	612.4±34.5	**426.4±32.1**
	Mean	**107.2±6.4^b**	**519.4±94.0^a**	**572.1±30.4^a**	
Phenolic acids					
Caffeic acid	S	3.4±0.6	2.3±0.5	4.0±0.7	**3.2±0.6**
	C	2.4±0.3	3.3±0.6	2.8±0.4	**2.8±0.4**
	Mean	**2.9±0.5 ns**	**2.8±0.6 ns**	**3.4±0.6 ns**	
Protocatechuic acid	S	1.9±0.1	4.4±0.1	5.3±0.4	**3.9±0.2**
	C	3.5±0.4	4.1±0.2	4.4±0.4	**4.0±0.3**
	Mean	**2.7±0.3^c**	**4.3±0.2^b**	**4.9±0.4^a**	
4-Hydroxybenzoic acid	S	11.0±0.7	39.0±1.2	37.4±2.3	**29.1±1.4**
	C	14.9±1.9	36.9±3.3	37.4±2.0	**29.7±2.4**
	Mean	**13.0±1.3^b**	**38.0±2.3^a**	**37.4±2.2^a**	
Syringic acid	S	12.3±5.6	14.1±0.6	23.6±0.3	**16.7±2.2***
	C	4.3±0.9	33.3±1.5	34.2±3.0	**23.9±1.8**
	Mean	**8.3±3.3^c**	**23.7±1.1^b**	**28.9±1.7^a**	

$^{a-c}$Values marked with the same letter do not differ significantly at α=0.05
ns – differences are not significant at α = 0.05

whereas the frozen rhizome was characterised by a lower content of the majority of these compounds. A remarkable decrease in rosavin content and increase of the content of its aglycone (*trans*-cinnamic alcohol) indicates that freezing was not effective in inactivating hydrolytic enzymes, which is essential for plant material stabilisation.

In order to reliably evaluate the quality of a raw material it is necessary to find the best method for extraction of the main biologically active compounds. Data concerning the recommended solvent and extraction method for standardisation of roseroot is contradictory [39, 40, 49]. Our studies indicate that periodical ultrasonic extraction and continuous exhaustive extraction (in a Soxhlet-like Büchi Universal Extraction System) allowed to get extracts characterised by a similar content of phenolic compounds. Both 70% ethanol and 100% methanol appeared to be better extraction media than water (Table 16.8).

The results of several years studies carried out in the Warsaw Agricultural University indicate that the cultivation of roseroot in the lowlands of the temperate zone is possible. In comparison with the natural mountain habitats of roseroot, the region of central Poland is characterised by a longer vegetation period, which results in a faster increment in the weight of its underground organs, which are used as a medicinal raw material. In such conditions it is possible to obtain a high yield and good quality of the raw material as early as in the 5th year of plant vegetation. In the 6th year, the plants divide into smaller autonomic parts that are characterised by a lower content of salidroside and rosavin, the compounds regarded to be the most important for the pharmacological activity of roseroot preparations.

Taking into consideration the high intraspecific variability of roseroot, it is advisable to undertake research on basic breeding problems, as well as on effective methods of vegetative propagation (e.g. in vitro).

Post-harvest treatment of the medicinal raw materials may distinctly affect their quality (i.e. the content and composition of biologically active compounds). Convection drying is the most common method of roseroot raw material stabilisation. Our studies proved that comparable results might be obtained using lyophilisation.

Regarding the necessity for the fast, cheap and reliable evaluation of a raw material, it seems that the best extraction method for determination of phenolic compounds in roseroot is ultrasonic extraction with methanol as a solvent.

Acknowledgements

The authors would like to express their gratitude to Dr. Joanna Kopcińska from the Department of Botany, Warsaw University of Life Sciences – SGGW for help in anatomical studies.

References

1. Kurkin VA, Zapesochnaya GG (1986) Khim-Farm Zh 20:1231
2. Abidov M, Crendal F, Grachev S, Seifulla R, Ziegenfuss T (2003) Bull Exp Biol Med 136:585
3. Shevtsov VA, Zholus BI, Shervarly VI, Vol'skij VB, Korovin YP, Khristich MP, Roslyakova NA, Wilkman G (2003) Phytomedicine 10:95
4. Spasov AA, Wilkman GK, Mandrikov VB, Mironova IA, Neumoin VV (2000) Phytomedicine 7:85
5. Darbinyan V, Kteyan A, Panossian A, Gabrielian E, Wilkman G, Wagner H (2000) Phytomedicine 7:365
6. Maslov LN, Lishmanov YuB, Maimeskulova LA, Krasnov EA (1998) Bull Exp Biol Med 125:424
7. Afanas'ev SA, Lishmanov YuB, Lasukova TV, Naumova AV (1997) Bull Exp Biol Med 123:447
8. Maslova LV, Kondrat'ev BYu, Maslov LN, Lishmanov YuB (1994) Exp Klin Farmakol 57:61
9. Wagner H, Nörr H, Winterhoff H (1994) Phytomedicine 1:63
10. Afanas'ev SA, Alekseeva ED, Bardamova IB, Maslova LV, Lishmanov YuB (1993) Bull Exp Biol Med 116:480
11. Lishmanov YuB, Maslova LV, Maslov LN, Dan'sina EN (1993) Bull Exp Biol Med 116:175
12. Gossudarstvennaya Farmakopeya SSSR, XI (1991) Meditsina, Moskva
13. Sokolov SYa, Boiko VP, Kurkin VA, Zapesochnaya GG, Rvancova NV (1990) Khim-Farm Zh 24:66
14. Petkov VD, Yonkov D, Mosharoff A, Kamburova T, Alova L, Petkov VV, Todorov I (1986) Acta Physiol Pharm Bulg 12:3
15. Lazarova MB, Petkov VD, Markovska VL, Petkov VV, Mosharoff A (1986) Methods Find Exp Clin Pharmacol 8:547
16. Sokolov SYa, Ivashin VM, Zapesochnaya GG, Kurkin VA, Shchavlinskii AN (1985) Khim-Farm Zh 19:1367
17. Dugan OM, Barilyak IR, Nester TI, Dvornik AS, Kunakh VA (1999) Tsitol Genet 33:19
18. Bocharova OA, Matveev BP, Baryshnikov AYu, Figurin KM, Serebryakova RV, Bodrova NV (1995) Urol Nefrol 2:46
19. Udincev SN, Shakhov VP, Borovskoy IG, Ibragimova SG (1991) Biofizika 36:105
20. Udincev SN, Shakhov VP (1991) Eur J Cancer 27:1182
21. Udincev SN, Shakhov VP (1991) Neoplasma 38:323
22. Polozhii AV, Sakharova NA, Sviridonov GM (1983) Rastit Resur 19:289
23. Nuchimovskii EL (1974) Rastit Resur 10:499
24. Galambosi B (2006) Demand and availability of Rhodiola rosea L. raw material. In: Bogers RJ, Craker LE, Lange D (eds) Medicinal and Aromatic Plants. Springer, the Netherlands, p 223
25. Elsakov GV, Gorelova AP (1999) Agrochimiya 10:58
26. Kirichenko EB, Rudenko SS, Baglay BM, Masikevich YuG (1993) Byulleten' Glavnogo Botanicheskogo Sada 169:50
27. Satsyperova IF, Kurkin VA, Zapesochnaya GG, Pautova IA (1991) Rastit Resur 27:55
28. Kurkin VA, Zapesochnaya GG, Dubichev AG, Voroncov ED (1990) Khim Prir Soedin 26:481
29. Kir'yanov AA, Bondarenko LT, Kurkin VA, Zapesochnaya GG (1989) Khim-Farm Zh 23:449
30. Ishmuratova MM (1998) Rastit Resur 34:13
31. Ishmuratova MM (1998) Rastit Resur 34:72
32. Nukhimovskii EL, Yurtseva NS, Yurtsev BN (1987) Rastit Resur 23:489

33. Ishmuratova MM, Satsyperova IF (1998) Rastit Resur 34:3
34. Tikhonova VL, Kruzhalina TN, Shugaeva EV (1997) Rastit Resur 33:68
35. Khnykina LA, Zotova MI (1966) Aptechnoe Delo 15:34
36. Patov SA, Punegov VV, Kuchin AV (2006) Chem Nat Comp 42:397
37. Kurkin VA (2003) Chem Nat Comp 39:123
38. Tolonen A, Pakonen M, Hohtola A, Jalonen J (2003) Chem Pharm Bull (Tokyo) 51:467
39. Ganzera M, Yayla Y, Khan IA (2001) Chem Pharm Bull (Tokyo) 49:465
40. Linh PT, Kim YH, Hong SP, Jian JJ, Kang JS (2000) Arch Pharm Res 23:349
41. Dubichev AG, Kurkin VA, Zapesochnaya GG, Voroncov ED (1991) Khim Prir Soedin 27:161
42. Kurkin VA, Zapesochnaya GG, Kir'yanov AA, Bondarenko LT, Vandyshev VV, Mainskov AV, Nukhimovskii EL, Klimakhin GI (1989) Khim-Farm Zh 23:1364
43. Kurkin VA, Zapesochnaya GG, Gorbunov JN, Nukhimovskii EL, Sreter AI, Shchavlinskii AN (1986) Rastit Resur 22:310
44. Kurkin VA, Zapesochnaya GG, Shchavlinskii AN (1985) Khim Prir Soedin 21:632
45. Kurkin VA, Zapesochnaya GG, Shchavlinskii AN, Nukhimovskii EL, Vandyshev VV (1985) Khim-Farm Zh 19:185
46. Krasnov EA, Kuvaev VB, Khoruzhaya TG (1978) Rastit Resur 14:153
47. Saratikov AS, Krasnov EA, Khnykina LA, Duvidson LM, Sotova MI, Marina TF, Nechoda MF, Axenova RA, Tscherdinzeff SG (1968) Pharmazie 23:392
48. Kurkin VA, Zapesochnaya GG, Nukhimovskii EL, Klimakhin GI (1988) Khim-Farm Zh 22:324
49. Kir'yanov AA, Bondarenko LT, Kurkin VA, Zapesochnaya GG (1988) Khim-Farm Zh 22:451
50. Syrov YuP (1985) Pravila sbora i suszki lekarstvennych rastenij, Moskwa

Chapter 17
Apoptosis and Plant-Derived Pharmaceuticals

L.F. Brisson

Department of Biochemistry and Microbiology, Research in Heath and Life Science Building, Laval University, Quebec, Quebec, Canada, G1K 7P4,
e-mail: louise.brisson@bcm.ulaval.ca

Abstract The medicinal potential of the plant kingdom is widely exploited for its antitumoral properties. Many plant-derived antitumoral compounds such as paclitaxel, the vinca alkaloids, and catechin derivatives exhibit an extraordinary diversity of chemical structures that might exert their effects either by impairing cell division (mitosis) or by stimulating cells to undergo a cell-death program known as apoptosis. This review focuses on the effects of some plant-derived antitumoral substances on programmed cell death in animals. Several studies have shown a correlation between plant-derived drug-induced cell death and expression of apoptotic regulators or altered signaling pathways. Among the regulators of programmed cell death that have a modified expression is the protein p53 (a protein that effects either arrest of the cell cycle or activation of cell death) and proteins belonging to the Bcl-2 family, including members that activate or inhibit cell death. In several studies, it was shown that interactions with those regulators involved changes in signal transduction, especially with specialized protein kinases. Although some antitumoral substances are used extensively in the clinical setting, their precise mechanism of action on cell death remains to be elucidated. On the other hand, since apoptosis could be initiated through diverse mechanisms, it appears that the molecular targets of various plant molecules that exhibit a wide diversity of chemical structures could also be diverse and may depend on the cellular context.

17.1 Introduction

17.1.1 Molecular Regulation of Apoptosis

The medicinal potential of the plant kingdom is widely known by traditional cultures and by pharmaceutical industries. Accordingly, many of the prescription drugs contain active ingredients derived from plants, including many anti-tumor agents such as paclitaxel, vinca alkaloids, and catechin derivatives, to mention the most studied ones. Many plant-derived antitumoral compounds exhibit an extraordinary diversity of chemical structures that might work either by impairing cell division (mitosis) or by stimulating cells to undergo a program of cell death referred to as apoptosis. The latter process is a genetically encoded cell-elimination program for removing unnecessary or harmful cells during normal development or pathological conditions (for recent reviews see [1–4]). Cells undergoing apoptosis generally display characteristic morphological and biochemical features such as cytoplasmic shrinkage, membrane blebbing, nuclear fragmentation, chromatin condensation, DNA fragmentation, and phosphatidylserine externalization. These features are orchestrated by several proteins. In animal cells, two major pathways are commonly described for apoptosis; the death-receptor pathway, referred to as the extrinsic pathway, and the mitochondrial route. Both pathways lead to caspase (cysteine proteases that their substrates following specific aspartate residues) activation and cleavage of their specific substrates. Of interest, within the mitochondrial pathway of apoptosis, Bcl-2 family members play a key regulatory role either for their proapoptotic (Bax, Bak, Bcl-X_S, Bad, Bid, Bik) or for their antiapoptotic (Bcl-2, Bcl-X_L, Bcl-w, BRAG-1) activities. Therefore, members of Bcl-2 family are regarded as key elements in maintaining a balance between cell growth and cell death. It is believed that in animal cells, apoptosis is controlled by an equilibrium between antiapoptotic and proapoptotic members. For example, a high level of Bcl-2 relative to Bax promotes survival, while an excess of Bax results in cell death. Pro- and antiapoptotic Bcl-2 members are known to interact with each other at the level of protein–protein interactions; such interaction may play a regulatory role in the induction or the suppression of cell death. Among them, the Bax protein can promote apoptosis by interfering with mitochondria features either by forming a channel leading to the release of the electron transfer protein, cytochrome c, or by perturbing the oxidative state. Several other factors, such as post-translational modifications, may also regulate the final cellular outcome. Accordingly, some reports have suggested the importance of Bcl-2 phosphorylation [4]. Since killing cancer cells remain a significant topic, the scope of this paper is to review the known effects of plant antitumoral substances on apoptosis.

17.2 Plant Antitumoral Substances

17.2.1 Plant Substances

Phytochemicals could act either as a chemotherapeutic agent when they exhibited cytotoxic effect or could be used to treat cancer cells. Alternatively, other substances known as chemopreventive agents have low toxicities compared to the chemotherapeutic agents, but they appear promising for preventing, arresting, and reversing cancer [5]. Among the most studied antitumoral drugs figure: (1) the taxanes, (2) the vinca alkaloids, (3) the lapachone, (4) the polyphenol derivatives, and 5) the catechin derivatives from tea plants. Among these substances, the taxanes appears to be the broadest, in terms of antitumor spectrum, of any class of anticancer agents. Taxanes are diterpenes that are produced by plants of the genus *Taxus*, known also under the name of yews. Paclitaxel (commercialized under the name of taxol), initially extracted from the bark of a Pacific yew tree known as *Taxus brevifolia*, and docetaxel (commercialized as taxotere), derived from the needles of the English yew, are the two commercialized taxoid drugs. The vinca alkaloids, principally the two natural compounds vinblastine and vincristine and the semisynthetic derivatives vindesine and vinorelbine, have been used as anticancer drugs for more than 30 years. The lapachone extracted from the bark of lapacho tree is a form of quinone that is quite attractive since its action may be specifically targeted at cancer cells. Besides these chemotherapeutic drugs, several studies have highlighted natural compounds that act as chemopreventive agents, which were discovered following epidemiological observations revealing a low cancer incidence in populations drinking tea or those eating soyfood, fruits, vegetables, flaxseed, peanuts, grapes, and/or red wine, suggesting the presence of cancer protective agents. It has been proposed that other phenolic compounds including the isoflavones (e.g., the genistein, daidzein, and biochanin found in soybean and soy-based food), the lignans (present in most fruits, vegetables and flaxseed), and the stilbenes (found in peanuts, grapes, and red wine) could also act as protective agents against cancer [6].

Several mechanisms to explain the beneficial effects of plant-antitumoral substances have been proposed, among which their effects on target-specific cell-signaling pathways regulating cell growth and proliferation as well as apoptosis have been quite extensively studied. However, the exact mechanism(s) underlying the effects of these compounds on apoptosis remains unclear. The effects of several substances including the taxanes and vinca alkaloids on microtubules have been frequently documented (see [7] and references therein). It is believed that these antimicrotubule agents can induce not only mitotic catastrophe, but could also activate cell death. On the other hand, several studies have shown a correlation between plant-drug-induced cell death and expression of apoptotic regulators or altered signaling pathways.

17.2.2 Chemotherapeutic Drugs

17.2.2.1 Taxanes

Following intracellular uptake, taxanes are known to inhibit cell proliferation. Taxanes bind to B-tubulin and promote their polymerization, interfering then with the function of the mitotic spindle, which leads to cell cycle growth arrest [8]. In addition to their effect on the cell cycle, several studies have documented their action on apoptosis. However, as mention by the group of Huang [9] studying the selective resistance of MCF7 and R3227 cells to paclitaxel, it may be possible that mitotic arrest and apoptotic cell death are two separate events, and that paclitaxel-induced mitotic arrest may not always be followed by apoptotic cell death.

Several studies have documented that taxane-treated cells exhibit morphologic features and DNA fragmentation characteristic of apoptosis [10–14]. The taxane-induced effect on apoptosis could be related to their action on the cytoskeleton. Accordingly, since the cytoskeletal network is important for mitochondrial arrangement, one could speculate that taxanes initiate an apoptotic pathway by interfering with either the extrinsic apoptotic pathway or with mitochondrial function by facilitating the release of cytochrome c or by perturbing oxidative intracellular stress [15–18]. Alternatively, taxane-induced apoptosis could interfere with some members of the regulatory proteins belonging to the Bcl-2 family. Indeed, ectopic expression of Bcl-2 or Bcl-X_L blocked taxol-induced apoptosis in leukemic cells [14]. This might occur through the phosphorylation of Bcl-X_L/Bcl-2 or through the activation of the prodeath regulators such as Bax, Bak, and Bad [15, 19, 20]. Although controversial, the significance of Bcl protein phosphorylation also highlights the involvement of diverse kinases such as protein kinase A, mitogen-activated protein kinase, or Raf kinase, within the biochemical events leading to apoptosis [20–23]. The ectopic expression of the proapoptotic protein Bax or Bad has been shown to sensitize cancer cells to paclitaxel and induce apoptosis [24, 25].

It has been proposed that taxol may increase cellular susceptibility to apoptosis by amplifying the normal downstream events associated with mitotic kinase activation [26]. In addition to the downregulation of Bcl-2, several authors considered that upregulation of p53 and p21 are important for taxane-induced apoptosis, depending on the cellular context [27, 28]. Accordingly, cells lacking p53 or cells from p-53-null mice display increased sensitivity to paclitaxel [29, 30]. However, as demonstrated by the team of Tan [28], low doses of paclitaxel induce apoptosis through the upregulation of p53, while the death induced by higher concentrations occurred in a p53-independent manner, pinpointing the complexity of cellular equilibrium [28].

17.2.2.2 Vinca Alkaloids

While taxanes interact with polymerized tubulin and prevent depolymerization, vinca alkaloids interfere with monomeric tubulin and prevent polymerization [31]. Although the effects of vinca alkaloids on cell growth arrest have been well documented through their interference with the dynamics of microtubules, some reports have shown significant killing activity in a variety of tumor cells, through an apparent induction of apoptosis, as shown by cellular morphology or by DNA fragmentation. As complex as the situation described above for taxoid compounds, alkaloids may act at different levels of the death pathway in numerous cell types, and the mechanisms by which vinca alkaloids induce apoptotic cell death in tumor cells is not clearly defined. However, their effect on death-signaling pathways has been less studied. Similar to taxanes, recent studies have indicated that vinblastine could promote a coordinated cycle of phosphorylation and dephosphorylation of Bcl-X_L and Bcl-2 [31–33]. Bax has been implicated in vinblastine-induced cell death, in which changes in Bax concern its localization, conformation, and oligomerization pattern [34]. According to the study of Longuet et al. [17], Bax activation and the release of cytochrome c are major events in vincristine-induced apoptosis in KB-3-1 cells.

The essential cellular functions associated with microtubules have led to a wide use of microtubule-interfering agents in cancer chemotherapy, with promising results. Considering the complexity of the death pathway, it is quite obvious that the action of microtubule-interfering agents may depend on the cellular context (cell-type specificity, nature of taxane treatment – dose and duration) and may also not be exclusive, but may be the result of several effects leading to an increased sensitivity to death [20]. However, it appears significant to highlight the biochemical events leading to apoptosis under various cellular contexts, since its death action could be selective toward cancer cells. A better knowledge will contribute to the development of efficient strategies for cancer therapy.

17.2.3 Chemopreventive Agents: Catechin and its Derivatives

Many compounds belonging to diverse structural and functional chemical classes have been identified following epidemiologic studies. Among these, special attention has been focused on tea, in which the active compounds are the flavanols and catechins. The major tea catechins include epigallocatechin-3-gallate (EGCG), which is the most abundant, epigallocatechin, epicatechin-3-gallate, and epicatechin [35]. Associated with the action of polyphenolic compounds, it seems that their antioxidative properties could play an important role in chemoprevention. In addition, similar to the chemotherapeutic agents, it appears that tea compounds may act by inducing cell-cycle arrest and apoptosis. Although the precise effects of these compounds on cell proliferation or death remain unclear, it appears that EGCG acts selectively on cancer cells without affecting normal cells [36, 37]. The mechanism under the selectivity was explained by a differential inhibition of nuclear factor-κB (NF-κB) activation [35]. The first evidence linking EGCG to apoptosis showed the typical nuclear condensation of apoptotic cells, caspase-3 activation and poly(ADP) ribose polymerase cleavage [38]. Several recent reviews have outlined the wide range of mechanisms by which EGCG may interfere with cell death [35, 38, 39]. Bode and Dong [35] proposed that their chemoprotective effect is a consequence of their ability to bind to protein kinases in diverse signal transduction pathways. This is further confirmed by proteomic analysis in which several kinases were found to be down- or upregulated following exposure to EGCG; in their study, particular interest was shown in metalloproteinases and to vascular endothelial growth factor, which are upregulated [40]. Of interest, these genes belong to diverse regulatory pathways excluding those involved in the regulation of the cell cycle, such as cyclin-dependent kinase, among others. From several studies, it is clear that EGCG can act on several molecular targets. Action can be effective at the cell membrane (activating receptors), within the cytosol (through diverse transduction pathways by affecting the expression of kinases or death regulators or by activating the translocation of the transcription factor NF-κB), or within the nucleus (following activation of the transcription factors activator protein-1 and NF-κB) [35, 38]. These molecular targets could act on the cell cycle as well as on apoptosis.

With regard to apoptosis, the molecular mechanisms underlying EGCG-induced cell death includes stimulation of diverse protein kinases and modulation of cell survival/cell death genes. Changes such as increase in p53 protein, decreases in the Bcl-2 and Bcl-X_L proteins, an increase in Bax protein, and activation of caspase-9 suggest that they interfere with the mitochondrial pathway, leading to apoptosis [41–43]. It has been shown that EGCG stimulates Bax oligomerization and depolarization of mitochondrial membranes, facilitating the release of cytochrome c into the cytosol [42]. On the other hand, direct interactions of polyphenols to the BH3 pocket of the antiapoptotic Bcl-2 family proteins have been revealed by recent nuclear magnetic resonance studies [38]. Taken together, these events suggest an effect on the mitochondrial pathway of apoptosis. Paradoxically, EGCG could prevent apoptosis induced by some stresses such as nitric oxide and exposure to $CoCl_2$ [44].

17.3 Conclusion

Several studies indicated that plant antitumoral substances could act either by their interaction with the cell cycle or by activating an apoptotic pathway. Killing cancer cells is an important issue in terms of chemotherapy. Since apoptosis could be initiated through multiple mechanisms, the targets of plant substances could be diverse and may be dependent on the cellular context. Several reviews have indicated the diversity and the number of putative molecular targets. Shimizu and collaborators have suggested that the antitumor effects of several phytochemicals are due to binding, with a relative low affinity, to multiple cellular or molecular targets [45]. The wide range of molecular targets could act in synergy to define the cellular threshold for induction of death. This also raises the question of selectivity to choose chemicals that preferentially kill cancer cells without affecting significantly normal cells. Considering that all of these chemicals affect diverse signal transduction pathways, and particularly the activity of diverse kinases, it seems reasonable to speculate that specificity might result from a general cellular state involving the necessity of envisaging the cell as a system. This pinpoints the importance of characterizing the cellular events

Chapter 17 Apoptosis and Plant-Derived Pharmaceuticals 323

leading to apoptosis in order to define the strategic use of these drugs in the treatment of human cancers.

References

1. Fulda S, Debatin KM (2006) Oncogene 25:4798
2. Kim R, Emi M, Tanabe K, Murakami S, Uchida Y, Arihiro K (2006) J Pathol 208:319
3. Edinger AL, Thompson CB (2004) Curr Opin Cell Biol 6:663
4. Mohamad N, Gutierrez A, Nunez M, Cocca C, Martin G, Cricco G, Medina V, Rivera E, Bergoc R. (2005) Biocell 29:149
5. Greenwald P (2002) BMJ 324:714
6. Limer JL, Speirs V (2004) Breast Cancer Res 6:119
7. Wang LG, Liu XM, Kreis W, Budman DR (1999) Cancer Chemother Pharmacol 44:355
8. Rowinsky EK (1997) Annu Rev Med 48:353
9. Huang Y, Fang Y, Dziadyk JM, Norris JS, Fan W (2002) Oncol Res 13:113
10. Bhalla K, Ibrado AM, Bradt JE, Ray S, Huang Y, Tang C, Nawabi A, Hoffman R (1995) Leukemia 9:1851
11. Fang G, Chang BS, Kim CN, Perkins C, Thompson CB, Bhalla KN (1998) Cancer Res 58:3202
12. Ireland CM, Pittman SM (1995) Biochem Pharmacol 49:1491
13. Tang C, Willingham MC, Reed JC, Miyashita T, Ray S, Ponnathpur V, Huang Y, Mahoney ME, Bullock G, Bhalla K (1994) Leukemia 8:1960
14. Yuan SY, Hsu SL, Tsai KJ, Yang CR (2002) Urol Res 30:282
15. Bhalla KN (2003) Oncogene 22:9075
16. Ibrado AM, Kim CN, Bhalla K (1998) Leukemia 12:1930
17. Longuet M, Serduc R, Riva C (2004) Int J Oncol 25:309
18. Wang S, Wang Z, Dent P, Grant S (2003) Blood 101:3648
19. Blagosklonny MV, Schulte T, Nguyen P, Trepel J, Neckers LM (1996) Cancer Res 56:1851
20. Wang TH, Wang HS, Soong YK (2000) Cancer 88:2619
21. Bacus SS, Gudkov AV, Lowe M, Lyass L, Yung Y, Komarov AP, Keyomarsi K, Yarden Y, Seger R (2001) Oncogene 20:147
22. Blagosklonny MV, Giannakakou P, el-Deiry WS, Kingston DG, Higgs PI, Neckers L, Fojo T (1997) Cancer Res 57:130
23. MacKeigan JP, Collins TS, Ting JP (2000) J Biol Chem 275:38953
24. Strobel T, Swanson L, Korsmeyer S, Cannistra SA (1996) Proc Natl Acad Sci U S A 93:14094
25. Strobel T, Kraeft SK, Chen LB, Cannistra SA (1998) Cancer Res 58:4776
26. 26.Scatena CD, Stewart ZA, Mays D, Tang LJ, Keefer CJ, Leach SD, Pietenpol JA (1998) J Biol Chem 273:30777
27. Ganansia-Leymarie V, Bischoff P, Bergerat JP, Holl V. (2003) Curr Med Chem Anticancer Agents 3:291
28. Tan G, Heqing L, Jiangbo C, Ming J, Yanhong M, Xianghe L, Hong S, Li G (2002) Int J Cancer 97:168
29. Blagosklonny MV, Chuman Y, Bergan RC, Fojo T (1999) Leukemia 13:1028
30. Cassinelli G, Supino R, Perego P, Polizzi D, Lanzi C, Pratesi G, Zunino F (2001) Int J Cancer 92:738
31. Wang LG, Liu XM, Kreis W, Budman DR (1999) Cancer Chemother Pharmacol 44:355
32. Du L, Lyle CS, Chambers TC (2005) Oncogene 24:107
33. Fan M, Du L, Stone AA, Gilbert KM, Chambers TC (2000) Cancer Res 60:6403

34. Upreti M, Lyle CS, Skaug B, Du L, Chambers TC (2006) J Biol Chem 281:15941
35. Bode AM, Dong Z (2003) J Biochem Mol Biol 36:66
36. Ahmad N, Feyes DK, Nieminen AL, Agarwal R, Mukhtar H (1997) J Natl Cancer Inst 89:1881
37. Yang CS, Maliakal P, Meng X (2002) Annu Rev Pharmacol Toxicol 42:25
38. Khan N, Afaq F, Saleem M, Ahmad N, Mukhtar H (2006) Cancer Res 66:2500
39. Beltz LA, Bayer DK, Moss AL, Simet IM (2006) Anticancer Agents Med Chem 6:389
40. Adhami VM, Ahmad N, Mukhtar H (2003) J Nutr 133:2417S
41. Hastak K, Agarwal MK, Mukhtar H, Agarwal ML (2005) FASEB J 19:789
42. Qanungo S, Das M, Haldar S, Basu A (2005) Carcinogenesis 26:958
43. Roy M, Chakrabarty S, Sinha D, Bhattacharya RK, Siddiqi M (2003) Mutat Res 523:33
44. Jung JY, Mo HC, Yang KH, Jeong YJ, Yoo HG, Choi NK, Oh WM, Oh HK, Kim SH, Lee JH, Kim HJ, Kim WJ (2006) Life Sci 80:1355
45. Shimizu M, Weinstein IB (2005) Mutat Res 591:147

Chapter 18
The Indian Herbal Drugs Scenario in Global Perspectives

K.G. Ramawat (✉) and S. Goyal

Laboratory of Biomolecular Technology, Department of Botany,
M.L. Sukhadia University, Udaipur 313001, India, e-mail: kg_ramawat@yahoo.com

Abstract Herbal drugs are essential components of traditional medicine in several countries including China and India. India has a well-established system of medicine known as Ayurveda. Ayurveda utilises plants, animal and minerals for the welfare of human beings. India is also a hot-spot of megabiodiversity. There is an urgent need to rationally utilise medicinal plants for curative purposes with proper maintenance of biodiversity. The government of India has taken several initiatives to develop technology for the effective conservation and efficient utilisation of medicinal plants, to coordinate research and developmental activities as well as to prepare databases. Priority of the Department of Biotechnology, Government of India and the Indian Council of Agricultural Research, demand that the top 20 medicinal plants in India, plants be imported and exported from India vis-à-vis world demand. Scientific validation of pharmacological activity of age-old drugs used in Ayurveda reinforces faith in the traditional system, in which plants are selected only on the basis of experience. This review provides information on Indian herbal drug biodiversity, supply and demand, use of herbal drugs in the pharmaceutical industry and quality control methods required for the modern drug industry.

Keywords Herbal drugs, Traditional medicine, Ayurveda, Endangered plants

18.1 Introduction

The world population is likely to touch the 7.5 billion mark at the current growth rate by the year 2020. Mostly this increase is in the developing or under-developed countries, 80% of whose population still relies on a traditional system of medicine based on herbal drugs. These folk or household medicines are readily available in neighbourhood, cheap, and without side effects, having been time tested. The demand for medicinal plants is continuously increasing not only in developing countries, but also in developed countries as drug, food supplements (nutraceuticals) and cosmetics [1]. Tyler defines herbal medicines as "crude drugs of vegetable origin utilised for the treatment of diseased state often of a chronic nature or to attain or maintain a condition of improved health" [2]. If we look at the socio-economic scenario of Asian and African countries, modern medicine is neither affordable nor within the reach of many villagers and tribes inhabiting remote areas and deep forests. There are certain pockets in a country like India where the tribal people have no access to modern amenities like roads, telecommunications or electricity, and therefore, these communities rely only on their traditional knowledge of medicine for day-to-day requirements [3].

It is well established that industrialisation has many direct and indirect effects on the human population. Increased stress is the most evident, although this is offset by increased health awareness among the people and better medical facilities. Nevertheless, increases in the incidence of diseases (mostly in urban populations) such as coronary heart disease, diabetes, hyperlipidaemia, AIDS and cancer cannot be denied [4].

Men learnt to use plants as healers of different ailments with the beginning of civilisation. One of the oldest Pharmacopoeia describing the appearance, properties and use of many plants is by the Greek physician Galen (A.D. 129–200). The great civilisations of India, China and North Africa have developed this science to perfection, and written records are available dating back up to 3000 years B.C. There are many examples where medicines have been obtained from plants known to traditional healers. With the development of modern analytical tools, interest in natural product chemistry has led to the isolation by Serturner of morphine alkaloid from opium, a mixture of plentiful alkaloids. This in turn was obtained from the opium poppy (*Papaver somniferum*) by processes that have been used for over 5000 years. Quinine isolated from the *Cinchona* tree had its origin in the Royal household of South American Incas. Long before the first European explorers arrived, the native people of South America had developed medical systems with complete diagnosis and treatment of various maladies. The leaves of the coca tree have been primarily chewed by Andean people to obtain well-known benefits. In 1860, Carl Koler isolated cocaine from the coca tree, the chemical responsible for its biological activity, and has become infamous as a drug of abuse. As a local anaesthetic, it revolutionised surgical and dental procedures. Similar are stories of the development of many modern drugs such as pilocarpine from the alkaloid-rich oil of the zaborandi tree (*Pilocarpus zaborandi*) used to treat glaucoma, anti-inflammatory agents from the pineapple (*Ananas comosus*), which was used by the American Indians of Guadeloupe. The other botanicals include atropine, hyoscine, digoxin, colchicine and emetine.

India is a very large country (3,280,483 km^2) with diverse geoclimatic zones and biodiversity. Although there are many excellent centres of natural products chemistry, still many plants are yet to be explored for their phytochemical and pharmacological properties, for example *Chlorophytum borivilianum*. Earlier, this plant was collected from forests, but it is now cultivated in several thousand hectares and consumed in tonnes, but nothing is known about its chemistry and pharmacology [5,6]. In India we often wake up to our therapeutic wisdom only after recognition about a plant material or its active ingredient comes from the West [7]. This can be well illustrated by the cases of *Rauwolfia serpentina* [8] and *Withania sonmifera*, both of which are used traditionally in the Indian system of medicine, where Indian reports were ignored and the plants attained importance only after publications by Western scientists [9]. Similarly, guggul, which is obtained from *Commiphora wightii* (syn. *C. mukul*), is used to negate joint pain and to treat arthritis and obesity since time immemorial, but the mechanism of action of guggul is still being worked out by Western scientists [10–13].

In India, it is mainly the central government departments that fund the research on medicinal plants; the contribution of private partnership is almost insignificant. The research and development activity of large traditional Indian pharmaceutical firms is mainly in-house activity. There are several departments under the ministries of Science and Technology (Department of Science and Technology, Department of Biotechnology – DBT, Council of Scientific and Industrial Research), Environment and Forests, Health and Welfare (Medicinal Plant Board) and Agriculture (Indian Council of Agricultural Research, ICAR) working in India on various aspects of medicinal plant research and development.

The plants supported by the DBT (New Delhi; Table 18.1) for research and development are different than those cultivated and prioritised by the Indian ICAR (New Delhi; Table 18.2.) with the exception of three plant spe-

Table 18.1 Plant species supported by the Department of Biotechnology, New Delhi, including endangered species

Aconitum species	*Nardostachys jatamansi*
Acorus calamus	*Panax pseudoginseng*
Amomum species	*Phyllanthus fraternus*
Andrographis paniculata	*Picrorhiza kurroa*
Azadirachta indica	*Podophyllum hexandrum*
Cassia angustifolia	*Pogostemon cavlin*
Commiphora wightii	*Psoralea corylifolia*
Coptis teeta	*Rauwolfia serpentina*
Crataeva nurvuala	*Rheum emodi*
Cymbopogon winteriansis	*Swertia chirata*
Dioscorea deltoidea	*Valeriana jatamansi*
Ephedra species	*Vetiveria zizanoides*
Ferula asafoetida	*Withania somnifera*
Mesua ferrea	

Table 18.2 Plants under cultivation and promoted for agricultural practices by the Indian Council of Agricultural Research, New Delhi, India

Cultivated plants	Area (hectares)	Plants promoted for agro-techniques	
Cassia angustifolia	10,000	Catharanthus roseus	Gloriosa superba
Cephalis ipecacahuna	100	Glycirrhyza glabra	Aloe vera
Cinchona officinalis	6000	Plantago ovata	Solanum laciniatum
Crocus sativus	3000	Hyoscyamus niger	Rosmarinus oficinalis
Cymbopogon flexiosus	20,000	Dioscorea floribunda	Commiphora wightii
Cymbopogon martini	2000	Chrysanthemum cinerariefolium	Silybum marianum
Humulus lupulus	1000	Pogostemon patchouli	Matricaria chamomile
Jasminum officinale	2000	Withania somnifera	Chlorophytum borivilianum
Mentha arvensis	2000	Valeriana jatamansi	Lavandula stoechas
Ocimum basilicum	500	Piper longum	Mucuna pruriens
Papaver somniferum	18,000	Digitalis purpurea	Alpinia officinarum
Pelargonium graveolens	3000	Fornuculum vulgare	Salvia officinalis
Plantago ovata	50,000	Cassia angustifolia	
Rosa demascena	3000	Psoralea corylifolia	
Solanum viarum	3000	Rauwolfia serpentina	
Withania somnifera	4000	Swertia chirata	

cies, namely *Cassia angustifolia*, *Swertia chirata* and *Withania somnifera*. The plants under cultivation are utilised in both the domestic and foreign markets as raw material for drugs and perfumery. The plants promoted by ICAR for agronomic practices are demand driven. These nodal agencies have their own priorities based on their own mandate, and a comprehensive common list of medicinal plants of Indian national priority is yet to be evolved.

18.2 Indian System of Medicine

Ayurveda originated in India way back in the pre-vedic period. "Rigveda" and "Atharva-veda" (5000 years B.C.), the earliest documented ancient Indian treatise, have references on health and diseases. Ayurveda texts like "Charak Samhita" and "Sushruta Samhita" were documented about 1000 years B.C. The term "Ayurveda" means "Science of Life". It deals elaborately with measures for healthful living during the entire span of life and its various phases. In addition, dealing with principles for maintenance of health, it has also developed a wide range of therapeutic measures with which to combat illness. These principles of positive health and therapeutic measures relate to the physical, mental, social and spiritual welfare of human beings. Thus, Ayurveda is one of the oldest systems of health care, dealing with both the preventive and curative aspects of life in a most comprehensive way, and presents a close similarity to the World Health Organization (WHO)'s concept of health propounded in the modern era. In fact, of the 6599 hymns and around 700 prose lines that comprise *Atharva Veda*, a substantial part relates to the human body, its disorders and possible cures, which included recitation of prayers and magical invocations. "Atharva Veda" is considered as the forerunner of Ayurveda. It is because poetic descriptions have different interpretations that it is difficult to draw conclusions from the text. The Indian system of medicine utilises all natural products like plants, animals and minerals for the treatment of human diseases. Modern medicine, or allopathy (a term coined in 1842 by C.F.S. Hahnemann), is not more then 300 years old and gradually developed on the basis of observations recorded about diseases, human anatomy, physiology and the use of natural resources. Much of the information about plants has been derived from traditional medicine and folk medicine. There have been many successful modern drugs developed from botanicals. The most recent examples are guggulsterones from the resin of *Commiphora mukul* and artemisinin from Quinghaosu (*Artemisia annua*), a Chinese medicine. Continuous efforts by chemists, botanists and pharmacologist have resulted in the establishment of modern medicine from Ayurveda-based medicine [14,15]. These includes indole alkaloids for hypertension from *Rauwolfia serpentina*, psoralens for leucoderma from *Psoralea corylifolia*, alkaloids against amoebiasis from *Holarrhena antidysenterica*, guggulsterones as hypolipidaemic agents from *Commiphora wightii* (syn. *Commiphora mukul*), l-Dopa (dihydroxy phenylalanine) from *Muccuna pruriens* for Parkinson's disease, piperidines as bioavailability enhancers, baccosides from *Bacopa monnierri* for memory enhancement, picrosides from *Picrorhiza kurroa*, in hepatic protection, curcumin from *Curcuma longa* as an anti-inflammatory agent and withanolides and many other steroidal lectones as immunomodulators [16].

There are four key concepts in Ayurveda; these concepts collectively guide the preventive, promotive and curative aspects of the Indian system of medicine. These concepts are the Panch bhutas, Tridoshas, Saptdhatus and Malas. The philosophy of Ayurveda is based on the principle of Panch bhutas (five elements: air, sky, water, fire and earth), of which the body is composed. A healthy person is one in whom there is equilibrium of the humours and body tissues, with normal digestive as well as excretory functions, all of which are responses to the gratification of physical sciences and mental as well as spiritual forces. An absence of this equilibrium describes the status of sickness. The Vatta, Pitta and Kapha are known as three humours (tridoshas: related to physiological functioning). In a healthy person, these three humours are in a state of non-functional equilibrium, and loss of this harmony leads to sickness [1]. Dhatus (related to structural components of body) refers to vital body organ or parts. These are Rash (body fluid), Rakta (blood), Mansa (muscular tissue), Meda (adipose tissue), Asthi (bone tissue), Majja (nerve tissue and bone marrow) and Sukra (generative tissue including sperm and ova). Malas deals with production and excretion of waste products by different organs and body. The examination of patient and cause of disease are important in deciding the state of disease and treating a person in Ayurveda. Treatment in the system consists of avoiding the factors responsible for causing change in the equilibrium of body and restoring it by medicine, suitable diet and activity.

In various therapies, Rasayana is an important therapy in Ayurveda. Rasayan preparations are inducers of enzymes and hormones, for example, which the body needs for adaptation and survival during health stress and disease. Some of the plants used in this therapy are *Acorus calamus, Asparagus racemosus, Centella asiatica, Commiphora wightii, Emblica officinalis, Ocimum sanctum, Piper longum, Semecarpus anacardium, Sida cordyfolia, Tinospora cordyfolia* and *Withania somnifera*.

Rasayana therapy deals with promotion of strength and vitality. The integrity of body matrix, promotion of memory, intelligence, immunity against disease, the preservation of youth, lustre, complexion and maintenance of optimum strength of the body and senses are some of the positive benefits credited to this therapy. Prevention of premature wear and tear of body tissues and promotion of total health content of an individual are the roles that Rasayana therapy plays. The procedures of revitalisation and rejuvenation were adopted to increase the power of resistance to disease and these procedures also retarded the progress of aging. Rasayanas are prescribed for a particular period and a strict diet regimen is observed. Thus, these vitalisers were compounds that are closely related to the anti-stress agents of plant origin and may be acting as inducers of interferon (fighting against viral diseases) and succinate dehydrogenase, the enzyme responsible for conservation and utilisation of energy during stress [17]. Ayurveda was fully familiar with concept of vitalisation therapy and the need to keep a disease-free healthy life in its totality of both physical and mental well-being.

India has moved forward in popularising global usefulness of Ayurveda in health care through global networks. As a result, many foreign countries have began looking to India for an understanding of Ayurveda and incorporating it through education, research and practice to meet the overwhelming desire of consumers to access complementary and alternative medicine. Indian Missions in the USA, UK, Russia, Germany, Hungary and South Africa have played an effective role in channelling information regarding Ayurveda and opening up new opportunities for the spread of this Indian medicine in to foreign institutions; general public awareness building about Ayurveda in foreign countries has been identified as an important thrust area.

The world herbal market is growing fast and the Chinese market is projected to increase by US$ 400 billion by 2010 [18]. Therefore, serious efforts are required to make herbal-based economy a major contribution to the Indian economy. To this end, Indian Government initiatives to facilitate research, co-ordination and planning include the establishment in 2000 of the National Medicinal Plant Board under the auspices of the Ministry of Health and Family Welfare, Government of India, and a separate task force on medicinal and aromatic plants by the DBT under the auspices of the Ministry of Science and Technology, New Delhi.

India has a well-recorded and traditionally well-practiced knowledge of herbal medicines. There are very few medicinal herbs of commercial importance that are not found in this country. Two of the largest users of medicinal plants are China and India. Traditional Chinese medicine uses over 5000 plant species, while about 7000 are used in India. However, India's share in the world market is US$ 1 billion, compared to China's share of US$ 6 billion [19]. Indigenous medicinal herbs provide about 75% of the requirement for medicines of the third-world countries [20]. Three of the ten most widely selling herbal medicines in the developed countries, namely preparations of *Allium sativum*, *Aloe barbadensis* and *Panax* species are available in India [21], yet this segment is not fully exploited commercially in India.

In order to prevent grant of patents based on Indian Traditional Knowledge, the Government of India has undertaken an ambitious project of creating a Traditional Knowledge Digital Library. This is a joint venture of the Council of Scientific and Industrial Research and Central Council for Research in Ayurveda and Siddha. This project is intended to cover about 35,000 formulations available in 14 classical texts of Ayurveda to convert the information into patent-compatible format. The work has been initiated with a co-operative set up of 30 Ayurveda experts, 5 information technology experts and 2 patent examiners. The digital library will include all details in digital format regarding international patent classification, traditional research classification, Ayurveda terminology, concepts, definitions, classical formulations, doses, disease conditions and references to documents.

About 90 plants have been described as prime Ayurvedic medicine [22], while demand, production and supply estimates by the Government of India have been prepared for 162 plants [23]. The latter survey enlisted top 20 Ayurvedic plants (Table 18.3) based on the highest market value in terms of their importance in various formulations. The top 20 plants (by volume) account for about 66% of the total demand for 162 medicinal plants and contribute 73% of total value. These plants have all sorts of activities from bioavailability enhancer to immunomodulator and anti-tumour agents. These plants have very diverse active principle and pharmacological activities (Table 18.4).

There are two important points regarding their use: (1) these plants are used in specific combinations, causing a synergistic effect and (2) they are used as prophylactic agents for a longer duration (as compared to allopathic medicine), thus causing fundamental physiological effects and improving the very functioning of the body. Modern tools are validating their established properties and there has been a surge in publications on these materials in recent years; the demand for medicinal plants is growing by 15–16%. Taking this into account, the demand for 162 selected medicinal plants is expected to increase from 120,817 tonnes in 1999–2000 to 272,618 tonnes in 2004–2005. In value terms, demand is expected to increase from Rs. 670 crores (US$ 149 million) in 1999–2000 to Rs. 1453 crore (US$ 323 million) in 2004–2005 [77]. These estimates are

Table 18.3 Top 20 Ayurvedic plants, their annual requirement, active principles and validation of pharmacological activity by modern scientific tools. Chemical constituents are compiled from [17,22,24]

Plant species	Active molecules	Pharmacological action validated	Demand (tonnes)	Reference
Acorus calamus	Essential oils containing sesquiterpene hydrocarbons, ketones and ~80% asarones	Anti-spasmodic, neuroprotective, anti-oxidant	932	[25–28]
Aegle marmelos	Coumarins (marmins), furoquinoline alkaloids (skimmianine) and several others	Anti-hyperglycaemic, anti-diarrhoeal, prevention of myocardial infraction	7084	[28–30]
Azadirachta indica	Di- and tri-terpenoids, limonoids (nimbidinin), flavonoids etc	Inhibitor of carcinoma, chemopreventive in tumorigenesis, inhibit colon cancer	–	[31–33]
Curcuma longa	Curcuminoids (curcumin I-III), essential oils	Anti-stress, anti-proliferative activity	–	[34–36]
Embelia ribes	Embelin, homoembelin, rapanone and vilangin	Anti-proliferative, in lipid disorder	941	[37, 38]
Glycirrhiza glabra	Triterpenoids (glycyrrhizin, glycyrrhizinic acid), flavonoids, pterocarpans, coumarins etc	Anti-angiogenic, anti-proliferative, anti-oxidant, anti-hypercholesterolaemic, prevention of cerebral ischaemia	1328	[39–41]
Hemidesmus indicus	Coumarinolignoids (hemidesminine, hemidesmin-I, II) essential oils etc	Anti-nociceptive, hepatoprotective, anti-bacterial	1614	[42, 43]
Phyllanthus emblica	Several phenols (gallic acid), flavonoids, triterpenoids and tannins	Hepatoprotective, anti-hypercholesterolaemic, anti-atherogenic	34,568	[44–46]
Piper longum	12 amides (piperine) and 10 lactams (alkaloids), lignans	Bioavailability enhancer, chemopreventive	6072	[47, 48]
Piper nigrum	Piperidine, dehydropipernonaline	Anti-carcinogenic, anti-hyperlipidaemic, epilepsy	–	[49, 50]
Plumbago zeylanica	Mono-, di- and tri-napthoquinones (plumbagin), triterpenoids, coumarins etc	Anti-proliferative, anti-oxidant	2530	[51, 52]

Chapter 18 The Indian Herbal Drugs Scenario in Global Perspectives 333

Table 18.3 (continued) Top 20 Ayurvedic plants, their annual requirement, active principles and validation of pharmacological activity by modern scientific tools. Chemical constituents are compiled from [17, 22 ,24]

Plant species	Active molecules	Pharmacological action validated	Demand (tonnes)	Reference
Punica granatum	~20 tannins, alkaloids and anthocyanidins (delphinidin etc)	Anti-tumour, anti-oxidant	–	[53–55]
Riccinus communis	alkaloid (ricinine), lectin (ricin)	Hepatoprotective, anti-oxidant, hypoglycaemic, anti-tumorous etc	–	[56, 57]
Rubia cordifolia	Anthraquinone (rubiadin) and cyclic peptides	Suppresses the activation of mast cells, hepatoprotective	1424	[58, 59]
Santalum album	4–6% essential oils (sesquiterpenoids, cis-α-santalol)	Antiviral (herpes simplex 1,2), chemopreventive-skin cancer	–	[60, 61]
Semecarpus anacardium	Bhilawanol	Immunomodulatory, anti-inflammatory, anti-arthritic, anti-oxidant, hypoglycaemic, anti-hyperglycaemic	–	[62–64]
Terminalia chebula	Tannins, shikimic acid compounds, triterpenoids, ellagic acid etc	Anti-oxidant, anti-diabetic, renoprotective, hepatoprotective	6778	[65–67]
Tinospora cordifolia	Diterpenoid furanolactones (tinosporin), isoquinoline alkaloids	Immunomodulator, chemopreventive, cardioprotective	2932	[68–70]
Withania somnifera	>45 withanolides (withaferin A, etc.) and several alkaloids	Chemopreventive, anti-cancerous, and immunomodulatory	12,120	[71–74]
Zinziber officinalis	Essential oils, mono and sesquiterpenoids, pungent principles (vanilloids: zingerone) and curcuminoids	Cancer preventive, anti-cancerous, hypercholesterolaemic, anti-atherosclerotic	–	[75, 76]

Table 18.4 Major chemical constituents of the top 20 Ayurvedic plants

Acorus calamus	*Aegle marmelos*
α-asarone, β-asarone	skimmianine, marmin
Azadirachta indica	*Curcuma longa*
nimbidinin	curcumin (R=R'=OMe) desmethoxycurcumin (R=OMe; R'=H) bisdesmethoxycurcumin (R=R'=H)
Embelia ribes	*Glycyrrhiza glabra*
homoembelin (R=n-C_9H_{19}) embelin (R=n-$C_{11}H_{23}$) rapanone (R=n-$C_{13}H_{27}$)	glycyrrhetic acid (R=H) glycyrrhizinic acid (R=GlcU-GlcU)
Hemidesmus indicus	*Phyllanthus emblica*
hemidesminine	gallic acid
Piper longum	*Piper nigrum*
piperine	dehydropopernonaline
Plumbago zeylanica	*Punica granatum*
plumbagin	pelargonidin (R=R'=H) delphinidin (R=OH, R'=OH) cyanidin (R=OH, R'=H)

Chapter 18 The Indian Herbal Drugs Scenario in Global Perspectives

Table 18.4 *(continued)* Major chemical constituents of the top 20 Ayurvedic plants

Riccinus communis — ricinine	*Rubia cordifolia* — rubiadin
Santalum album — (+)-cis-α-santalol	*Semecarpus anacardium* — bhilawnol-A
Terminalia chebula — ellagic acid	*Tinospora cordifolia* — tinosporin
Withania somnifera — withaferin A	*Zinziber officinalis* — zingerone

based on data collected from agriculturists, traders and forest official and based on projected 15–16% growth.

A plant extract or a mixture of extracts constitutes a formulation under the Indian system of medicine. The effect of these preparations is the result of a combined effect of active molecules in these extracts. Perhaps this is the reason that the effect of plant-based drugs cannot be reproduced by pure active principles obtained from that plant (e.g. *Ginkgo biloba* extract, GBE 761) [78].

There are about 25,000 plant-based formulations used in folk and traditional medicine in India. There are over 1.5 million practitioners of traditional medicinal system, and over 7000 drug manufacturers consume about 2000 tonnes of herbs annually [79]. Several plants are widely used in different formulations. A few examples are: out of about 75 formulations available in the Indian market for health and vitality, all contain *Withania somnifera* (100%), *Asparagus racemosus* (81.5%), *Asparagus adscendens* (48%) and *Curculigo orchioides* (15%), with other plants in minor quantities [78]. The plants *Glycirrhiza glabra*, *Piper longum*, *Adhatoda vasica*, *Withania somnifera*, *Ciprus rotundus*, *Tinospora cordifolia*, *Berberis aristata*, *Tribulus terrestris*, *Holarrhena antidysenterica* and *Boerhavia diffusa* have been used in 52–141 herbal formulations, and Triphala (*Terminalia chebula*, *Terminalia bellerica* and *Embelica officinalis*) alone has been used in 219 formulations [80]. There is a need for standardisation of individual plant materials (cultural practices, selections, collection periods and germplasm preservation) and final formulations (by thin-layer chromatography –TLC – and high-performance liquid chromatography – HPLC – profiling).

18.3 World-Wide Use of Medicinal Aromatic Plants

India is one of the major raw-material-producing nations of South Asia. Available export statistics indicate that between 1992 and 1995 the country exported about 32,600 tonnes of crude drugs worth US$ 46 million. Commercially, these plant-derived medicines are worth about US$ 14 billion per year in the USA and US$ 40 billion worldwide. Americans paid an estimated US$ 21.2 billion for services provided by alternative medicine practitioners [81]. The nutraceutical market place in Europe is estimated to be worth US$ 9 billion, while the USA market place, estimated to be worth US$ 10–12 billion in 2003, is expanding at a rate of more than 20% per year. The United States congress has fuelled the rapid growth of nutraceuticals with the passage of the Dietary Supplement Health and Education Act (DSHEA) in 1994. Globally, there have been efforts to monitor quality and regulate the growing business of herbal drugs and traditional medicine.

The number of plant species that have at one time or another been used in some culture for medicinal purposes can only be estimated. An enumeration of the WHO from the late 1970s listed 21,000 medicinal species [82]; however, in China alone 4941 of 26,092 native species are used as drugs in Chinese traditional medicine [83], an astonishing 18.9%. If this proportion is calculated for other well-known medicinal florae and then applied to the global total of 422,000 flowering plant species [84,85], it can be estimated that the number of plant species used for medicinal purposes is more than 50,000 (Table 18.5).

These medicinal plants are not evenly distributed in different florae and regions. It is known that certain plant families have higher proportions of medicinal plants than others. Families like Apocynaceae, Araliaceae, Apiaceae, Asclepiadaceae, Canellaceae, Solanaceae, Leguminaceae, Rubiaceae, Composi-

Chapter 18 The Indian Herbal Drugs Scenario in Global Perspectives

Table 18.5 Worldwide utilisation of medicinal plant species

Country	Plant species (*n*)	Medicinal plant species	%
China	26,092	4941	18.9
India	15,000	3000	20.0
Indonesia	22,500	1000	4.4
Malaysia	15,500	1200	7.7
Nepal	6973	700	10.0
Pakistan	4950	300	6.1
Philippines	8931	850	9.5
Sri Lanka	3314	550	16.6
Thailand	11,625	1800	15.5
USA	21,641	2564	11.8
Vietnam	10,500	1800	17.1
Average	13,366	1700	12.5
World	**422,000**	**52,885**	

tae, Guttiferae and Menispermaceae are rich in plants with secondary metabolites. In addition, these families are not distributed uniformly across the world. As a consequence, not only do some florae have higher proportions of medicinal plants than others, but also certain plant families have a higher proportion of threatened species than others. Some of the prominent commercial plant-derived medicinal compounds include colchicine, betulinic acid, camptothecine (CPT), Topotecan (Hycmptin), CPT-11 (Irinotecan, camptosar), 9-aminocamtothecin, α-tetrahydrocannabinol (dronabinol, marinol), b-lapachone, lapachol, podophyllotoxin, etoposide, podophyllinic acid, vinblastine (Velban), vincristine (Leurocristine, oncovin), vindicine (eldisine, fildesine), vinorelbine (Navelbine), docetaxel (Taxotere), paclitaxel (Taxol), tubocuranine, pilocarpine and scopolamine.

18.4 Supply and Demand of Medicinal Plants

It is difficult to assess how many medicinal aromatic plants (MAP) are traded commercially, either on a national or even on an international level. The bulk of the plant material is exported from developing countries, while major markets are in the developed countries. An analysis of United Nations Conference on Trade and Development trade figures for 1981–1998 reflects this almost universal feature of the MAP trade (Table 18.6). If the volumes for the five European countries in this list are added together (94,300 tonnes), it becomes

Table 18.6 The 12 leading countries of import and export of medicinal and aromatic plant material from 1991 to 1998. Source: United Nations Conference on Trade and Development COMTRADE database, United Nations Statistics Division, New York

Country of import	Volume [tonnes]	Value [US$1000]	Country of export	Volume [tonnes]	Value [US$1000]
Hong Kong	73,650	314,000	China	139,750	298,650
Japan	56,750	146,650	India	36,750	57,400
USA	56,000	133,350	Germany	15,050	72,400
Germany	45,850	113,900	USA	11,950	114,450
Republic of Korea	31,400	52,550	Chile	11,850	29,100
France	20,800	50,400	Egypt	11,350	13,700
China	12,400	41,750	Singapore	11,250	59,850
Italy	11,450	42,250	Mexico	10,600	10,050
Pakistan	11,350	11,850	Bulgaria	10,150	14,850
Spain	8600	27,450	Pakistan	8100	5300
UK	7600	25,550	Albania	7350	14,050
Singapore	6550	55,500	Morocco	7250	13,200
Total	342,550	1,015,200	Total	281,550	643,200

clear that Europe dominates as an import region. Germany ranks fourth and third as importer and exporter, respectively, expressing the country's major role as a turntable for medicinal plant raw materials worldwide.

Iqbal [86] estimates that about "4000 to 6000 botanicals are of commercial importance", and the Secretariat of the Convention on Biological Diversity in 2001 referred to 5–6000 "botanicals entering the world market". A thorough investigation of the German medicinal plant trade identified a total of 1543 MAP being traded or offered on the German market [87]. An extension of this survey to Europe as a whole arrived at 2000 species in trade for medicinal purposes [88]. Recognising the role of Europe as a sink for MAP traded from all regions of the world, it is a qualified guess that the total number of MAP in international trade will be around 2500 species worldwide.

18.5 Medicinal Plant Biodiversity

India is one of the 12 mega biodiversity centres, having over 45,000 plant species (17,500 flowering plants, of which 5725 are endemic to India), 8000 of which are medicinal [89]. The florae of India is rich is biodiversity, being a subtropical country, and in Himalaya alone, over 8000 angiosperms, 44 gymnosperms, 600 pteridophytes, 1737 bryophytes and 1159 lichens have been a source of medicine for millions of people in the country and elsewhere in the world [90]. Some important species that have become endangered and need immediate attention for conservation in India are *Acquilaria malaccensis, Dioscorea deltoidea, Podophyllum hexandrum, Pterocarpus santalinus, Rauwolfia serpentina, Saussurea lappa* and *Taxus wallichiana* [89].

To satisfy the regional and international markets, the plant sources for expanding local, regional and international markets are harvested in increasing volumes and largely from wild populations [88, 91]. In developing countries, besides tribals, who are authorised to collect minor forest produce for their livelihood, traders collect plant products illegally. Supplies of wild plants in general are increasingly limited by deforestation from logging and conversion to plantations, pasture and agriculture [1, 92].

In many cases, the impact through direct off-take goes hand-in-hand with decline owing to changes in land use. Species favoured by extensive agricultural management like *Arnica montana* in central Europe go into decline with changes in farming practices towards higher nutrient input on the meadows. This requires habitat management as the key factor in managing species populations [93]. One of the goals of the International Union for the Conservation of Nature and Natural Resources Medicinal Plant Specialist Group is to identify the species that have become threatened by non-sustainable harvest and other factors. The enormity of this task is illustrated by the following estimate: according to Walter and Gillett [94], 34,000 species or 8% of the world's florae are threatened with extinction. If this is applied to our earlier estimate that 52,000 plant species are used medicinally, it leads us to estimate that 4160 MAP species are threatened.

Table 18.7 Medicinal plants being exported from India

Botanical name	Part of the plant	Botanical name	Part of the plant
Aconitum species	Root	*Juniperus macropoda*	Fruit
Acorus calamus	Rhizome	*Picrorhiza kurroa*	Root
Adhatoda vasica	Whole plant	*Plantago ovata*	Seed and husk
Berberis aristata	Root	*Podophyllum hexandrum*	Rhizome
Cassia angustifolia	Leaf and pod	*Punica granatum*	Flower, root and bark
Colchicum luteum	Rhizome and seed	*Rauvolfia serpentina*	Root
Hedychium spicatum	Rhizome	*Rheum emodi*	Rhizome
Heraleum candicans	Rhizome	*Saussurea lappa*	Rhizome
Inula racemosa	Rhizome	*Swertia chirata*	Whole plant
Juglans regia	Bark	*Valeriana jatamansi*	Rhizome
Juniperus communis	Fruit	*Zingiber officinale*	Rhizome

Table 18.8 Medicinal plants being imported into India

Botanical name	Native name
Cuscuta epithymum	Aftimum vilaiyti
Glycyrrhiza glabra	Mullathi
Lavandula stoechas	Ustukhudus
Operculina turpethum	Turbud
Pimpinella anisum	Anise fruit
Smilax china	Chobchini
Smilax ornata	Ushba
Thymus vulgaris	Hasha

18.6 Traditional Medicine in Healthcare

Traditional medicine is the synthesis of therapeutic experience of generations of practicing physicians of an indigenous system of medicine. While traditional preparations utilise medicinal and aromatic plants, minerals and other organic matter, herbal drugs constitute only those traditional medicines that use primarily medicinal plant preparations for therapy. Traditional medicine has been defined as the sum total of the knowledge, skills and practices based on the theories, beliefs and experiences indigenous to different cultures, whether applicable or not, used in the maintenance of health as well as in the prevention, diagnosis, improvement or treatment of physical and mental illness [23]. According to a 1983 WHO estimate, the majority of population in developing countries depend upon traditional and herbal medicines as their primary source of health care.

It is estimated that 70–80% of people worldwide rely chiefly on traditional, largely herbal, medicine to meet their primary health-care needs. The global demand for herbal medicine is not only large, but also growing. The market for Ayurvedic medicine is estimated to be expanding at 20% annually in India, while the quantity of medicinal plants obtained from just 1 province of China has grown by 10 times in the last 10 years [95]. Factors contributing to the growth in demand for traditional medicine include the increasing human population and the frequently inadequate provision of Western (allopathic) medicine in developing countries.

In developed countries, non-conventional medical modalities, also designated as complementary and alternative medicines (CAM), are often used concomitantly with conventional medicine in medical treatment, including cancer therapy. The popularity of CAM in the USA is reflected in a survey, which showed its use increased from 34% in 1990 to 42% in 1997. The same survey showed that American consumers spent US$ 27 billion on alternative treatments and an estimated US$ 5.1 billion on herbal medicines in 1997 [81]. A large percentage with life-threatening disorders use alternative medical therapies. This may be because of the poor prognosis that many of these patients face despite the use of the full spectrum of conventional medical approaches. In developing countries, patients are brought to hospitals at a very late stage when treatment cannot cure the disease. At this juncture, these patients turn to alternative therapies and paranormal treatments. Worsening physical symptoms, troubling side effects from prescription drugs and diminishing hope may further add to the allure of less orthodox approaches. There are several examples where patients with chronic diseases like cancer and HIV have tried one or other form of alternative medicine [96]. In South-eastern Rajasthan (India), 400 medicinal plants belonging to 97 families are currently used in ethnomedicine [3]. Folklore claims about several natural drugs were verified on modern scientific grounds and WHO recommended more efforts in this direction [97, 98].

18.7 Indian Pharmaceutical Industries

The Indian economy is one of the fastest growing economies, and the current growth is expected to exceed 8% for the fiscal year 2006–2007 (9.1% on 1st December, 2006). The Indian pharmaceutical market is growing at a compounded average growth rate (CAGR) of 15%, compared to a world industry rate of 8% [99]. According to a report by Mckinsey, the industry will grow further at a CAGR of 19% to reach US$ 25 billion in revenue by 2010. The market capitalisation of Indian pharmaceutical companies is projected to grow dramatically to US$ 150 billion from the present US$ 15–20 billion [23].

There are about 15,000 licensed manufacturing units to manufacture traditional and allopathic medicines; about 300 are in the organised sector, of which multinationals account for 40%. Together, the top ten industry players account for only 30% of the market share. At the turn of the new millennium, the top five multinationals grew at a rate of 7.2%, while the top five domestic companies achieved a growth rate of 14% [100]. The turnover of herbal medicines in India is about US$ 1 billion, with an export of about 80 million [80]. Most of the export products are crude drugs, herbs, extracts and unprocessed low-value materials. *Psyllium* seeds and husk, castor oil and opium extract alone account for 60% of the export. In addition to these 3 plant products, 20 other plants are exported as crude drugs worth US$ 8 million (Table 18.7), while 8 plants are imported in significant quantities (Table 18.8). The major traditional sector pharmaceuticals like Himalaya, Zandu, Dabur, Hamdard and Maharshi and allopathic manufacturers like Ranbaxy, Lupin and Allembic are standardising their herbal formulations by TLC and HPLC fingerprinting [80].

18.8 Quality of Herbal Drugs

Much of the medicinal plants and their products/extracts are collected and prepared in the developing countries like India and China. The quality of these products is a major hindrance to the use and integration of these materials into modern medicine. Poor quality control parameters or not following these regulations associated by inappropriate technical tools affects both the safety and efficacy of the materials. It was demonstrated that the majority of the preparations prepared using guggulipid, an extract of *Commiphora wightii*, for hypercholesterolaemia, did not contain guggulsterones in the amounts mentioned on the labels [101, 102]. Similarly, *Panax ginseng*, *Panax quinquifolius* and *Eleuthrococcus senticosus*, marketed as a botanical supplement in North America, showed that the ginsenoside contents of 232 *Panax ginseng* and 81 *Panax quinquifolius* products range from 0.00 to 13.54% and 0.009 to 8.00%, respectively, and that approximately 26% of these products did not meet label claims [103]. In another study, silymarin from milk thistle (*Silybum marianum*) was detected at 58–116% of the label claim [104]. Studies on St. John's wort (*Hypericum perforatum*) products showed the hypericin content ranging from 22 to 140% of the label claim when analysed using an official spectrophotometric procedures [105], and from 47 to 165% employing HPLC methods [106].

Recently, many international authorities and agencies including the WHO, European Agency for the Evaluation of Medicinal Products and European Scientific Cooperation of Phytomedicine, US Agency for Health Care Policy and Research, European Pharmacopeia Commission and the Department of Indian System of Medicine have started creating new strategy for inducing and regulating quality control and standardisation of botanical medicine. The term "nutraceutical" is of recent origin and is used for nutritionally or medicinally enhanced foods with health benefits. Nutraceuticals include engineered grain, cereals supplemented with vitamins and minerals, genetically manipulated or enriched soya food and canola oil without trans-fatty acids.

There has been a flood of vitaliser and aphrodisiac formulations in the market. This has become a huge market that does not require the approval of the drug controllers, and hence many pharmaceutical and biotechnology companies have extended the term nutraceutical to include pure compounds of natural origin like lovastatin (a lipid-lowering agent from *Monascus ruber* and *Aspergillus terreus*) [107] and curcumin (*Curcuma longa*). Since herbal drugs/ formulations are based on traditional knowledge, the United States Food and Drug Administration banned the dietary supplement cholestin (i.e. lovastatin). Nutraceuticals are in great demand in the developed world, particularly in countries like the USA (US$ 80–250 billion) and Japan. A surge in the consumption of nutraceuticals took place because of the DSHEA, passed by the USA in 1994. Many of these nutraceuticals have anti-oxidant and chemopreventive properties; therefore, they have a direct bearing on disease prevention and consequently less burden on the health-care system.

The major drawback with Indian herbal manufacturers, particularly small-scale industries, is that their products are not standardised. Adulterations are caused mostly at collection points, sometimes at trader level and rarely at the manufacturer level, thus affecting the efficacy of the formulation, and as a result, faith in indigenous drugs has declined [108]. Illiterate tribal peoples and villagers collect raw materials and they do not understand the importance of quality and standards. There are several examples of substitution of highly priced material with a cheap product for example, bark of *Holarrhena antidysenterica* with *Wrightia tinctoria*, *Saraca indica* with *Trema orientalis* [109], roots of *Cholorophytum borivilianum* with *Asparagus racemosus* [78], and gum resin of *Commiphora wightii* with gum of *Acacia arabica* and *Boswellia serrata*.

Identification of active molecules in a medicinal plant is an essential requirement towards developing methods for quality controls. A serious problem in the country is that authentic compounds are usually not available for comparison on various chromatographic techniques. Isolation and identification of compounds using nuclear magnetic resonance or mass spectrophotometry is not available to many small industries or universities, while a few national laboratories (in India) have their hands full with institutional work. The USA and a few European countries have recognised about 20 herbal drugs and strict quality control is required for such materials. It has been emphasised in ancient Ayurvedic literature that the season of harvest and the age of the plant affect the quality of herb. The amount and nature of secondary metabolites are not constant throughout the year. The age of the plant also affects the quantity and relative proportions of different constituents. The drying conditions, storage and processing of raw material need standardisation and control to maintain uniform quality. It has been shown that drugs such as Indian hemp and sarsaparilla deteriorate even when carefully stored.

18.9 Concluding Remarks

Herbal-based traditional medicine has become popular in developed countries in recent years and its use is likely to be increased in the coming years. This system has advantages over the allopathic system, being prophylactic. This increased utilisation of herbs has direct repercussions on the collection of raw materials and consequently requires sustainable utilisation of these plants along with methods of conservation, and studies of reproductive biology, phytochemistry and pharmacological validation. Standardisation of chemical fingerprinting (TLC, HPLC) towards quality control is another major requirement in developing countries. Although herbal drugs have been used in the Indian system of medicine for last several hundred years, and they are prepared by a procedure prescribed in Ayurvedic text, their toxicity/safety must be evaluated on modern models for universal acceptance. Most of the herbal industries are in the small sector, which need improvements regarding the processing of raw material, packaging, quality control (most have no research and development or quality control system) and technical know-how regarding global demand and marketing.

Acknowledgements

Research on medicinal plants in the laboratory is supported by funds from UGC under the DRS programme and DST under the FIST programme. SG thanks UGC for a fellowship.

References

1. Ramawat KG, Sonie KC, Sharma MC (2004) Biotechnology of medicinal plants: vitalizer and therapeutic. In: Ramawat KG (ed) Biotechnology: Medicinal Plants. Science Publishers, USA, p1
2. Tyler VE (1994) Herbs of Choice: The Therapeutic Use of Phytomedicals, Pharmaceutical Products, Binghampton, New York
3. Katewa SS, Jain A (2006) Traditional Folk Herbal Medicine. Apex, Udaipur, India
4. Mérillon JM, Ramawat KG (2007) Biotechnology for medicinal plants. In: Ramawat KG, Mérillon JM (eds) Biotechnology, Secondary Metabolites, Scientific Publishers, Enfield, USA, p 1
5. Arora DK, Jain AK, Ramawat KG, Mérillon JM (2004) Chlorophytum borivilianum: an endangered aphrodisiac herb. In: Ramawat KG (ed) Biotechnology of Medicinal Plants: Vitalizer and Therapeutic. Scientific Publishers, Enfield, USA, p 111
6. Arora DK, Suri SS, Ramawat KG (2006) Indian J Biotech 5:527
7. Vaidya ADB (1996) In: Handa SS, Kaul MK (eds) Supplement to Cultivation and Utilization of Medicinal Plants. CSIR, RRL, Jammu-Tawi, p 1
8. Sen G, Bose KC (1931) Indian Med World 2:194
9. Nittala S, Velde VV, Frolow F, Lavie D (1981) Phytochemistry 20:2547
10. Urizar NL, Liverman AB, Dodds DT, Silva FV, Ordentlich P, Yan Y, Gonzalez FZ, Heyman RA, Mangelsdorf DF, Moore DD (2002) Science 269:1703
11. Cui J, Huang L, Zhao A, Lew JL, Yu J, Sahoo S, Meinke PT, Royo I, Pelaz F, Wright ST (2003) J Biol Chem 278:10214
12. Wang X, Greilberger J, Ledinski G, Kager G, Paigen B, Jurgens G (2004) Atherosclerosis 172:239
13. Burris TP, Montrose C, Houck KA, Osborne HE, Bocchinfuso WP, Yaden BC, Cheng Cc, Zink RW, Barr RJ, Hepler CD, Krishnan V, Bullock HA, Burris LL, Galvin RJ, Bramlett K, Stayrook KR (2005) Mol Pharmacol 67:948
14. Dev S (1999) Environ Health Persp 107:783
15. Dahanukar SA, Kulkarni RA, Rege NN 9 (2000) Indian J Pharmacol 32:81
16. Patwardhan B (2000) Indian Drugs 37:213
17. Handa SS, Kaul MK (1996) Supplement to Cultivation and Utilization of Medicinal Plants. CSIR, RRL, Jammu-Tawi, p 64
18. Wang Z-G, Ren J (2002) Trend Pharmacol Sci 23:347
19. Rawat RBS (2002) Medicinal plants sector in India with reference to traditional knowledge and IPR issues (online). Paper presented at the International Seminar for the

Protection of Traditional Knowledge, New Delhi. Available from the Internet: http://r0.unctad.org/trade_env/test1/meetings/delhi/India/mik-094.doc.
20. Rajshekharan PE (2002) Herbal medicine. In: World of Science, Employment News. Ministry of Information and Broadcasting, New Delhi, India, p 3
21. Dubey NK, Kumar R, Tripathi P (2004) Curr Sci 86:37
22. Dev S (2006) A Selection of Prime Ayurvedic Plant Drugs, Ancient-Modern Concordance. Anamaya, New Delhi
23. Anonymous (2001) Demand Study for Selected Medicinal Plants. Centre for Research, Planning and Action, New Delhi, India
24. Rastogi RP, Mehrotra BN (1990, 1991, 1993, 1995, 1998) Compendium of Indian Medicinal Plants, Vol 1 (1990), Vol 2 (1991), Vol 3 (1993), Vol 4 (1995), Vol 5 (1998). Central Drug Research Institute, Lucknow, NISCAIR, New Delhi, India
25. Gilani AU, Shah AJ, Ahmad M, Shaheen F (2006) Phytother Res 12:1080
26. Shukla PK, Khanna VK, Ali MM, Maurya R, Khan MY, Srimal RC (2006) Hum Exp Toxicol 25:187
27. Manikandan S, Devi RS (2005) Pharmacol Res 52:467
28. Kesari AN, Gupta RK, Singh SK, Diwakar S, Watal G (2006) J Ethnopharmacol 107:374
29. Mazumder R, Bhattacharya S, Mazumder A, Pattnaik AK, Tiwary PM, Chaudhary S (2006) Phytother Res 20:82
30. Prince PS, Rajadurai M (2005) J Pharm Pharmacol 57:1353
31. Haque E, Baral R (2006) Immunobiology 211:721
32. Gangar SC, Sandhir R, Rai DV, Koul A (2006) Phytother Res 20:889
33. Roy MK, Kobori M, Takenaka M, Nakahara K, Shinmoto H, Tsushida T (2006) Planta Med 72:917
34. Xu Y, Ku B, Tie L, Yao H, Jiang W, Ma X, Li X (2006) Brain Res 1122:56
35. Karmakar S, Banik NL, Patel SJ, Ray SK (2006) Neurosci Lett 407:53
36. Cui SX, Qu XJ, Xie YY, Zhou L, Nakata M, Makuuchi M, Tang W (2006) Int J Mol Med 18:227
37. Ahn KS, Sethi G, Aggarwal BB (2006) Mol Pharmacol 71:209
38. Bhandari U, Kanojia R, Pillai KK (2002) Int J Exp Diabetes Res 3:159
39. Visavadiya NP, Narasimhacharya AV (2006) Mol Nutr Food Res 11:1080
40. Zhan C, Yang J (2006) Pharmacol Res 53:303
41. Sheela ML, Ramakrishna MK, Salimath BP (2006) Int Immunopharmacol 6:494
42. Baheti JR, Goyal RK, Shah GB (2006) Indian J Exp Biol 44:399
43. Verma PR, Joharapurkar AA, Chatpalliwar VA, Asnani AJ (2005) J Ethnopharmacol 102:298
44. Duan W, Yu Y, Zhang L (2005) Yakugaku Zasshi 125:587
45. Kim HJ, Yokozawa T, Kim HY, Tohda C, Rao TP, Juneja LR (2005) J Nutr Sci Vitaminol 51:413
46. Pramyothin P, Samosorn P, Poungshompoo S, Chaichantipyuth C (2006) J Ethnopharmacol 107:361
47. Selvendiran K, Thirunavkkarasu C, Singh JP, Padmavathi R, Sakthisekaran D (2005) Mol Cell Biochem 271:101
48. Pattanaik S, Hota D, Prabhakar S, Kharbanda P, Pandhi P (2006) Phytother Res 20:683
49. Vijayakumar RS, Nalini N (2006) J Basic Clin Physiol Pharmacol 17:71
50. Selvendiran K, Singh JP, Sakthisekaran D (2006) Pulm Pharmacol Ther 19:107
51. Tilak JC, Adhikari S, Devasagayam TP (2004) Redox Rep 9:219
52. Zhao YL, Lu DP (2006) Zhongguo Shi Yan Xue Ye Xue Za Zhi 14:208
53. Malik A, Mukhtar H (2006) Cell Cycle 5:371
54. Ricci D, Giamperi L, Bucchini A, Fraternale D (2006) Fitoterapia 77:310
55. Jung KH, Kim MJ, Ha E, Uhm YK, Kim HK, Chung JH, Yim SV (2006) Biol Pharm Bull 29:1258

56. Ross IA (2001) Medicinal plants of the world, Humana Press, Totowa, New Jersey, p 375
57. Khare CP (2004) Encyclopedia of Indian Medicinal plants, Springer-Verlag Berlin Heidelberg, New York, p 404
58. Lee JH, Kim NW, Her E, Kim BK, Hwang KH, Choi DK, Lim BO, Han JW, Kim YM, Choi WS (2006) J Pharm Pharmacol 58:503
59. Rao GM, Rao CV, Pushpangadan P, Shirwaikar A (2006) J Ethnopharmacol 103:484
60. Benencia F, Courreges MC (1999) Phytomedicine 6:119
61. Dwivedi C, Abu-Ghazaleh A (1997) Eur J Cancer Prev 6:399
62. Ramprasath VR, Shanthi P, Sachdanandam P (2006) Biol Pharm Bull 29:693
63. Ramprasath VR, Shanthi P, Sachdanandam P (2006) Chem Biol Interact 160:183
64. Kothai R, Arul B, Kumar KS, Christina AJ (2005) J Herb Pharmacother 5:49
65. Lee HS, Won NH, Kim KH, Lee H, Jun W, Lee KW (2005) Biol Pharma Bull 28:1639
66. Rao NK, Nammi S (2006) BMC Complement Altern Med 6:17
67. Tasduq SA, Singh K, Satti NK, Gupta DK, Suri KA, Johri RK (2006) Hum Exp Toxicol 25:111
68. Rao PR, Kumar VK, Vishwanath RK, Subbaraju GV (2005) Biol Pharm Bull 28:2319
69. Singh RP, Banerjee S, Kumar PV, Raveesha KA, Rao AR (2006) Phytomedicine 13:74
70. Nair PK, Melnick SJ, Ramachandran R, Escalon E, Ramachandran C (2006) Int Immunopharmacol 6:1815
71. Senthilnathan P, Padmavathi R, Magesh V, Sakthisekaran D (2006) Cancer Sci 97:658
72. Ichikawa H, Takada Y, Shishodia S, Jayaprakasam B, Nair MG, Aggarwal BB (2006) Mol Cancer Ther 5:1434
73. Rasool M, Varalakshmi P (2006) Vascul Pharmacol 44:406
74. Visavadiya NP, Narasimhacharya AV (2007) Phytomedicine 14:136
75. Kaul PN, Joshi BS (2001) Prog Drug Res 57:43
76. Fuhrman B, Rosenblat M, Hayek T, Coleman R, Aviram M (2000) J Nutr 130:1124
77. Anonymous (2003) Medicinal Plants in India – Report and Directory. Institute of Economic and Market Research, New Delhi, 110001
78. Ramawat KG, Jain S, Suri SS, Arora DK (1998) Aphrodisiac plants of Aravalli Hills, with special reference to safed musli. In: Khan IK, Khanum A (eds) Role of Biotechnology in Medicinal and Aromatic plants-1. Ukaaz, Hyderabad, India, p 210
79. Ramakrishnappa K (2002) Impact of cultivation and gathering of medicinal plants on biodiversity: case studies from India. In: Biodiversity and the Ecosystem Approach in Agriculture, Forestry and Fisheries online. FAO. Available from the Internet: http://www.fao.org/DOCREP/005/AA021E/AA021E00.htm
80. Kamboj VP (2000) Curr Sci 78:35
81. Einsberg DM, Davis RB, Ettner SL, et al (1998) JAMA 280:1569
82. Penso G (1980) WHO Inventory of Medicinal Plants used in Different Countries. WHO, Geneva, Switzerland
83. Duke JA, Ayensu ES (1985) Medicinal Plants of China. Reference Publications, Algonac, USA
84. Bramwell D (2002) Plant Talk 28:32
85. Govaerts S (2001) Taxon 50:1085
86. Iqbal M (1993) International Trade in Non-Wood Forest Products, An overview. FAO, Rome
87. Lange D, Schippmann U (1997) Trade Survey of Medicinal Plants in Germany. Bundesamt fur Naturschutz, Bonn, Germany
88. Lange D (1998) Europe's Medicinal and Aromatic Plants: Their Use, Trade and Conservation. TRAFFIC International, Cambridge, UK
89. Rao RR (2006) Proceedings of the 29th All India Botany Conference, Oct 9–11, Department of Botany, ML Sukhadia University, Udaipur, India, Abstract 1
90. Singh N, Squier C, Sivek C, Nguyen MH, Wagener M, Yu VL (1996) Arch Intern Med 156:197

91. Kuipers SE (1997) Trade in medicinal plants. In: Bodeker G, Bhat KKS, Burley J, Vantomme P (eds) Medicinal Plants for Forest Conservation and Healthcare. FAO (Non-Wood Forest Products 11) p 45
92. Ahmad B (1998) Plant exploration and documentation in view of land clearing in Sabah. In: Nair MNB, Ganapathi N (eds) Medicinal Plants, Cure for the 21st century, Biodiversity Conservation and Utilization of Medicinal Plants. Proceedings of a seminar, 15–16 October 1998, Serdang, Malaysia, Faculty of Forestry, Universiti Putra Malaysia, p 161
93. Ellenberger A (1999) Assuming responsibility for a protected plant. Weleda's endeavour to secure the firm's supply of Arnica montana. In: Traffic Europe (ed) Medicinal Plant Trade in Europe. Proceedings of the First Symposium on the Conservation of Medicinal Plants in Trade in Europe. TRAFFIC Europe, Brussels, Belgium, p 127
94. Walter KS, Gillett (1998) 1997 IUCN Red List of Threatened Plants. Gland, Switzerland
95. Pei S (2002) Paper presented at a workshop on Wise Practices and Experimental Learning in the Conservation and Management of Himalayan Medicinal Plants, Kathmandu, Nepal, the WWF, Nepal Program
96. Crone CC, Wise TN (1998) Psychosomatics 39:3
97. WHO (1978) Traditional Medicine. WHO, Geneva
98. Zhang X (1997) The International Symposium on Herbal Medicine, 25–27 March, King Fahad Hospital, Jeddah, Saudi Arabia
99. Bhat S, Sharma N, Maheshwari KK (1998) Status and issues affecting the drugs and pharmaceuticals industry in India. Chemical Industry News p 153
100. Viswanathan H, Salmon JW (2002) J Manag Care Pharmacy 8:211
101. Mesrob B, Nesbitt C, Misra R, Pandey RC (1998) J Chromatography B 720:189
102. Tanwar YS (2006) PhD thesis, ML Sukhadia University, Udaipur, India
103. Fitzloff J, Yat P, Lu Z, Awang DVC, Arnason JT, van Breeman RB, Hall T, Blumethal M, Fong HHS (1998) In: Huh H, Choi KJ, Kim YC (eds) Advances in Ginseng Research – Proceedings of the 7th International Symposium on Ginseng, 22–25 September, Seoul, Korea. The Korean Society of Ginseng, p 138
104. Schulz V, Hubner W-D, Ploch M (1997) Phytomedicine 4:379
105. Monmaney T (1998) The Los Angeles Times, 9 September, A1
106. Constantine GH, Karchesy J (1998) Pharm Biol 36:365
107. Demain AL, Zhang L (2005) Natural products and drug discovery. In: Demain AL, Zhang L (eds) Natural Products:Drug Discovery and Therapeutic Medicine. Humana Press, Totowa, New Jersey, p 3
108. Gupta AK, Vats SK, Lal B (1998) Curr Sci 74:565
109. Prajapati ND, Purohit SS, Sharma AK, Kumar TA (2003) A Handbook of Medicinal Plants. Agrobios (India), Jodhpur

Chapter 19
Phytochemical Standardization of Herbal Drugs and Polyherbal Formulations

M. Rajani (✉) and N.S. Kanaki

Pharmacognosy and Phytochemistry Department, B.V. Patel Pharmaceutical Education and Research Development (PERD) Centre, Thaltej-Gandhinagar Highway, Thaltej, Ahmedabad – 380054; Gujarat, India, e-mail: rajanivenkat@hotmail.com

Abstract The recent global resurgence of interest in herbal medicines has led to an increase in the demand for them. Commercialization of the manufacture of these medicines to meet this increasing demand has resulted in a decline in their quality, primarily due to a lack of adequate regulations pertaining to this sector of medicine. The need of the hour is to evolve a systematic approach and to develop well-designed methodologies for the standardization of herbal raw materials and herbal formulations. In this chapter, various methods of phytochemical standardization, such as preliminary phytochemical screening, fingerprint profiling, and quantification of marker compound(s) with reference to herbal raw materials and polyherbal formulations, are discussed in detail and suitable examples are given.

19.1 Introduction

19.1.1 Herbal drugs

Traditional systems of medicine are have been in vogue for centuries all over the world. According to one estimate, 80% of the world population still depends on herbal products for their primary healthcare needs. The toxic side effects of the drugs of modern medicine and the lack of medicines for many chronic ailments has led to the reemergence of the herbal medicine, with possible treatments for many health problems. Consequently, the use of plant-based medicine has been increasing all over the world [1], especially for conditions like cancer, high blood pressure, allergies, and for general well being [2–4]. According to an estimate 20,000 plant species out of 250,000 species are in use as medicines all over the world.

Most diseases, like diabetes, heart diseases, cancer, and psychiatric disorders, are multifactorial and hence need therapeutic intervention at more than one level. Plants with complex phytochemical mixtures have advantage over single molecules in treating such diseases, with an added advantage of being devoid of toxic side effects. The World Health Organization (WHO) encourages the use of plant-based medicine, especially in developing countries, even if the rationale is to reduce the financial burden on the respective governments. In view of the increasing demand for herbal products in Western countries, well-defined herbal products of proven efficacy and safety have been introduced in the last two decades. In the draft of the National Policy on the Indian Systems of Medicine [5], priority is being given to research on standardization, pharmacology, toxicology and clinical trials of herbal drugs.

19.1.2 Trade Scenario

With the growing interest in plant-based medicine over the last 20 years, herbal medicine has been enjoying a renaissance throughout the world. The global market of herbal medicinal products was estimated at approximately US$ 60 billion in the year 2000. Demand for herbal products has been growing at the rate of 7% per year and is expected to reach US$ 5 trillion by 2050. More than 50 medicinal plants are traded extensively on the international market.

Export of herbal products and essential oils from India is more than Rs. 2 billion; 15 herbal drugs and essential oils are regularly exported from India. Within the country, the turnover of herbal drugs was estimated to be ~Rs. 2000 crore, which includes classical formulations of Ayurveda, Unani, Siddha, Homeo, proprietary medicines and over-the-counter products.

19.1.3 Bottlenecks and Steps to be Taken

Despite the promise that plant-based medicine exhibited, the one major obstacle in using plant-based drugs has been the reproducibility of the activity. In olden times the traditional medicine used to be a personalized one, with the healers preparing the medicines on an individual basis, where the quality of the medicine and hence the safety and efficacy were taken care of completely. Large-scale production of herbal drugs has only started in the last 100 years or so. Now that the commercialization of the herbal medicine has happened, the onus of maintaining their quality falls to a large extent on the scientists and to a certain extent on the manufacturers. In many countries, the herbal market is poorly regulated. In this scenario, the assurance of safety, quality and efficacy of medicinal plants and herbal products has become an important issue.

The herbal raw material is prone to a lot of variation due to several factors, the important ones being the identity of the plants and seasonal variation (which has a bearing on the time of collection), the ecotypic, genotypic and chemotypic variations, drying and storage conditions and the presence of xenobiotics. The National Center for Complementary and Alternative Medicine and the WHO stress the importance of the qualitative and quantitative methods for characterizing the samples, quantification of the biomarkers and/ or chemical markers and the fingerprint profiles. It is indeed a challenging task to develop suitable standards for herbal drugs. The advancements in modern methods of analysis and the development of their application have made it possible to solve many of these problems. Extremely valuable are techniques like high-performance thin-layer chromatography (HPTLC), gas chromatography (GC), mass spectrometry (MS) high-performance liquid chromatography (HPLC), LC-MS, and GC-MS.

As mentioned above, development of standards for plant-based drugs is a challenging task and it needs innovative and creative approaches, different from the routine methods [6]. Starting from sourcing of the raw material, standardization, preparation of the extracts, to formulation of the extracts into suitable dosage form, the problems vary with each plant species and part of the plant that is being used. At each and every step, phytochemical profiles have to be generated and a multiple-marker-based standardization strategy needs to be adopted to minimize batch-to-batch variation and to maintain quality and ensure safety and efficacy.

19.1.4 Strategy

Methods of standardization should take into consideration all aspects that contribute to the quality of the herbal drugs, namely correct identity of the sample, organoleptic evaluation, pharmacognostic evaluation, volatile matter, quantitative evaluation (ash values, extractive values), phytochemical evaluation, test for the presence of xenobiotics, microbial load testing, toxicity testing, and biological activity. Of these, the phytochemical profile is of special significance since it has a direct bearing on the activity of the herbal drugs.

19.1.5 Status of Herbal Drugs in Pharmacopoeias

Although there is much activity all over the world with regard to establishing standards for medicinal plants, there is no consensus as to how these should be standardized. This is generally because there are a myriad of factors mentioned above that affect the quality of the herbal raw material. Several publications, United States Pharmacopoeia [8], British Herbal Compendium [9], British Herbal Pharmacopoeia [10], Chinese Pharmacopoeia [11], and Physician's Desk Reference (PDR) for Herbal medicines [12], carry monographs for herbal raw material. The government of India also brought out the Ayurvedic Pharmacopoeia of India [13].

Pharmacopoeias carry monographs for herbs and herbal products to maintain their quality. Several pharmacopoeias including Indian Pharmacopoeia, European Pharmacopoeia, and British Pharmacopoeia do cover monographs and quality control tests for a few medicinal plants.

19.1.5.1 Official

1. British Herbal Pharmacopoeia (1983) [10]: has 232 monographs, and in the 1990 edition 70 more monographs are included for herbs commonly used in the UK; 169 monographs were revised and updated in 1996.
2. Japanese Standards for Herbal Medicines (1993) [7]: 248 monographs.
3. European Scientific Cooperation for Phytotherapy [14]: 20 monographs (1996).
4. Chinese Pharmacopoeia [11]: more than 1000 monographs.
5. Ayurvedic Pharmacopoeia of India, Part I, vols. I–V [13]: 418 monographs.
6. Pharmacopoeia of India, 2005 – Addendum [15]: Monographs on 12 plants.

19.1.5.2 Others

1. PDR for herbal drugs [12]: USA.
2. American Herbal Pharmacopoeia [16]: around 17 monographs.
3. Quality Standards of Indian Medicinal Plants, vols. 1–4; Indian Council of Medical Research [17]: 136 monographs.
4. Indian Herbal Pharmacopoeia [18]: 52 monographs.
5. WHO monographs, vols. 1 and 2 [19]: 59 monographs.

Over the past 50 years, many changes are being incorporated into the pharmacopoeias, as reflected by the availability, development, and subsequent refinement of analytical techniques. Although initially most of the aforementioned pharmacopoeias did not describe methods for the phytochemical evaluation of the raw material, more recently, assays of chemical compounds are recommended and are being included (e.g., 12 monographs in the Pharmacopoeia of India – Addendum, 2005 [15], in the monographs published by the Indian Council of Medical Research [17] and Indian Drug Manufacturer's Association [18], and in the American Herbal Pharmacopoeia [16]).

19.2 Phytochemical Standardization

The analysis of plants and herbal formulations presents several problems arising from their complex nature and the inherent variability of their constituents. Plants are complex mixtures of varied chemicals, which pose a problem in standardization and quality control, but that very fact is responsible for imparting them with the feature of being therapeutically effective with the advantage of

Chapter 19 Phytochemical Standardization

synergistic and additive effects and at the same time having less side effects [20]. Consequently, the herbal drug preparation itself as a whole is regarded as the active substance. Hence, the reproducibility of the total configuration of herbal drug constituents is important. To meet this requirement it is essential to establish the chemoprofiles of the samples encompassing the following:

1. Thin-layer chromatography (TLC)/gas-liquid chromatography (GLC)/high-performance liquid chromatography (HPLC) fingerprint profiles.
2. Fingerprint profiles with marker compounds.
3. Quantification of marker compound/s (active principles, chemical markers).

Of these, the fingerprint profiles serve as guideline to the phytochemical profile of the drug in ensuring the quality, while quantification of the marker compound/s would serve as an additional parameter in assessing the quality of the sample.

Recent advances in chromatographic techniques have enabled reproducible, rapid, and efficient semiquantitative and quantitative analysis of the chemical constituents in complex mixtures [15, 18, 21–25].

Most of the herbal formulations, especially the classical formulations of Ayurveda and Unani, are polyherbal. Furthermore, the unique processing methods followed for the manufacturing of these drugs turn the herbal ingredients into very complex mixtures, from which the separation, identification, and estimation of chemical components is very difficult. As in the case of the raw material, parameters can also be defined for polyherbal formulations, where it includes comparison of the phytochemical profile of the formulations with that of individual ingredients (e.g., churnas and simple mixtures) by cochromatography of herbal extracts and formulations [26–30]. However, in case of many other preparations, it is not very easy since the chemical constituents of the herbs undergo complex changes during the unique processing steps involved in traditional preparations (e.g., Avaleha, Asava, Arishta, Sharbats). In such cases, sample preparation for phytochemical analysis should take into consideration the process of preparation of the formulation.

Phytochemical standardization encompasses all possible information generated with regard to the chemical constituents present in a herbal drug. Hence, the phytochemical evaluation for standardization purpose includes the following:

1. Preliminary testing for the presence of different chemical groups.
2. Quantification of chemical groups of interest (e.g., total alkaloids, total phenolics, total triterpenic acids, total tannins).
3. Establishment of fingerprint profiles.
4. Multiple marker-based fingerprint profiles.
5. Quantification of important chemical constituents.

19.2.1 Sample Preparation

The method of extraction of the drug should be such that all or most of the chemical constituents that contribute to the therapeutic efficacy of the drug are extracted in an unaltered form and remain stable in the extractive. This is rather difficult since plants are known to contain many chemical constituents that either directly or indirectly participate in their therapeutic outcome. For sample preparation, some guidance can be taken from the type of extracts used in traditional medicine for each drug. Reports on the pharmacological activity of the specific plant can also be taken into consideration.

19.2.2 Preliminary Screening for Chemical Groups and Quantification of Chemical Groups

Since the herbal drugs are known to contain chemical constituents belonging to different chemical groups, it is important to ascertain the major chemical groups employing simple chemical tests for the presence of, for example, alkaloids, steroids, terpenoids, flavonoids, coumarins, tannins, and anthraquinones [31].

The next logical step would be to quantify the major chemical group/s, since this will have a bearing upon the efficacy of the drug. Some examples include total alkaloids [32–35], total phenolics [36], total triterpenic acids [37], and total tannins [38].

19.2.3 Phytochemical Profiles – Fingerprinting

To establish fingerprint profiles, good resolution of the components of different polarity is achieved by devising a mobile phase based on the chemical nature of the compounds present therein, and optimized. Furthermore, pre- or postchromatographic derivatization of the sample with suitable chemical reagents may be employed for detection of the sample constituents. The potential of the phytochemical profiling in quality control can be well appreciated in the reported detailed fingerprint profiles generated with or without chemical markers [17, 18, 25, 39]. In the phytochemical evaluation of herbal drugs, TLC is being employed extensively for the following reasons: (1) it enables rapid analysis of herbal extracts with minimum sample clean-up requirement, (2) it provides qualitative and semiquantitative information of the resolved compounds, and (3) it enables the quantification of chemical constituents. Fingerprinting using HPLC and GLC is also carried out in specific cases.

In TLC fingerprinting, the data that can be recorded using a high-performance TLC (HPTLC) scanner includes the chromatogram, retardation factor (R_f) values, the color of the separated bands, their absorption spectra, λ_{max} and shoulder inflection/s of all the resolved bands. All of these, together with the profiles on derivatization with different reagents, represent the TLC fingerprint profile of the sample. In this way a lot of information can be generated with regard to even the unknown chemical constituents of the drug. The information so generated has a potential application in the identification of an authentic drug, in excluding the adulterants and in maintaining the quality and consistency of the drug [25].

Some examples of reported work include fingerprint profiles of oleogum resin of *Dorema ammoniacum* [25], *Rosa damascena* [40], *Citrus aurantium* spp. *aurantium* [41], oleogum resin from *Boswellia* species [42], Triphala [43], *Allium sativum* [44].

HPLC fingerprinting includes recording of the chromatograms, retention time of individual peaks and the absorption spectra (recorded with a photodiode array detector) with different mobile phases. Similarly, GLC is used for generating the fingerprint profiles of volatile oils and fixed oils of herbal drugs [17, 31, 45].

Such fingerprint profiles are usually distinctive and would form a benchmark for the drug, especially when the identity of the active principles is not known or when chemical markers are not available. In this process, it is important to take into account all of the information available in the fingerprint analysis *in toto* to ascertain the quality of the sample. This phytochemical profile can form an important component of quality-control criteria for herbal drugs.

Apart from serving the purpose of standardization and quality control, the fingerprint profiles, especially of TLC, aid in the experiments for bioassay-guided fractionation leading to the isolation of active compounds. The fingerprint profiles are also useful for characterization of extracts showing specific activity.

19.2.3.1 Multiple Marker-Based Fingerprinting

Marker compounds are of three kinds:
1. Active principles: one or a few of the compounds specific to the drug that are proved to be responsible for the claimed activity of the respective herbal drug [e.g., vasicine, curcumin, E- and Z-guggulsterones (Fig. 19.1), quinine, withaferin, andrographolides, ginsenosides, bacosides, capscaicinoids, l-dopa, silymarin, piperine, sennosides, digitoxin, berberine, strychnine, brucine, and boswellic acids] (Table 19.1).
2. Chemical markers: compounds reported from the respective drugs, although not specific to the drug, and activity specific to the drug has also not been proven (e.g., hederagenin, lapachol, cucurbitacins, rutin, quercetin).
3. General markers: compounds widely present in many plants (e.g., gallic acid, lupeol, stigmasterol, β-sitosterol) for which some activity may or may not be reported.
4. For those plants for which active principles are known, their presence in the sample can be ascertained by cochromatography and comparison of the R_f and absorption spectra with that of the standards of the marker compounds. Wherever active principles are not known, fingerprint profiles can include the general marker compounds. In addition, to have a complete picture of phytochemical profile, in the former case the fingerprint profiles should also include the other marker compounds [31, 46].

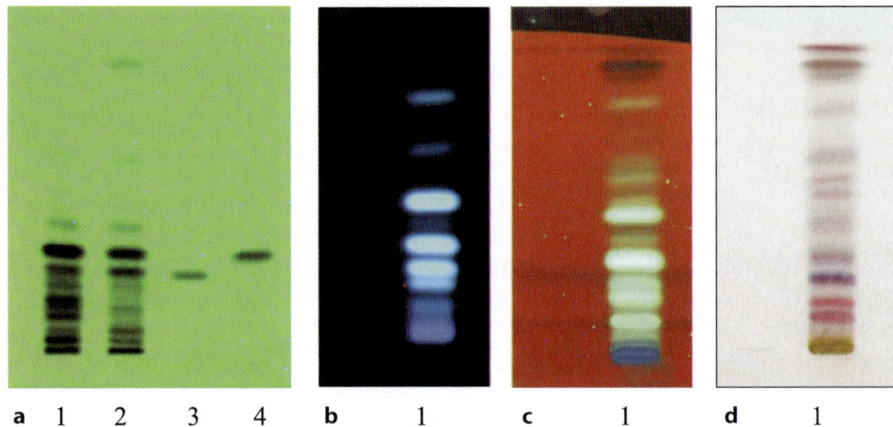

Fig. 19.1 Thin-layer chromatography (TLC) fingerprint profiles of guggul (oleogum resin of *Commiphora wightii*). **a** Under ultraviolet (UV) 254 nm; **b** under UV 366 nm; after derivatization with anisaldehyde-sulphuric acid reagent - **c** under UV 366 nm; **d** in natural light. *Tracks 1* and *2*, sample; *track 3*, E-guggulsterone standard; *track 4* Z-guggulsterone standard

Table 19.1 Some examples of active principles as marker compounds

Active principles	Common name	Latin name
Withanolides	Aswagandha	*Withania somnifera*
E-, Z-guggulsterones	Guggul	*Commiphora wightii*
Azardirachtin, Nimbidine	Neem	*Azadirachta indica*
Quinine alkaloids		*Cinchona* sp
Curcuminoids	Turmeric	*Curcuma longa*
Ginsenosides	Ginseng	*Panax ginseng*
Valepotriates	Valerian	*Valeriana officinalis*
Bacosides	Jal brahmi	*Bacopa monnieri*
Andrographolides	Kalmegh	*Andrographis paniculata*
Aegelin	Bilva	*Aegle marmelos*
Shatavarine	Shatavari	*Asparagus racemosus*

19.2.4 Marker Compound Analysis

19.2.4.1 Marker Compounds

Ideally, a marker compound is the one chemical compound specific to the plant material. For example, E- and Z-guggulsterones in the oleogum resin of *Commiphora wightii*. However, in most of the cases it is difficult to find such specific marker compounds. For example, vasicine, the active principle of *Adhatoda vasica* leaf, is the major alkaloid of the drug, and by virtue of it, it is an important biomarker of *Adhatoda vasica* leaf. However, it cannot be considered as "marker compound" of *Adhatoda vasica* since it is present in other plants as well (e.g., other species of *Adhatoda, Sida* sp., *Paganum harmala*). Hyoscyamine is present in several species including *Datura* sp., *Atropa belladonna, Hyoscyamus niger, Dubosia myoporoides*. Paclitaxel is found in several species of *Taxus* and also a fungal species. Secondary metabolites are usually found to be present in many members of the same family and sometimes across several families [47], so much so that in chemotaxonomy some of the secondary metabolites are used to settle disputes regarding the classification of certain taxa. Can we call such important secondary metabolites markers or not? Considering the above, is it possible to find a specific marker for each of the herbal drugs? Can we call the other compounds widely present, like gallic acid, quercetin, and rutin markers, albeit general markers? The question remains whether such chemical constituents can be called "marker compounds" at all. By definition, the marker should be specific to the drug in question. If a compound is identified that is specific to a drug, but is not an active principle and is a minor constituent, can it still be called a marker by virtue of being specific to the drug? It may best be left to the discretion of the scientist working with specific herbal samples to decide which important chemical constituents to use in chemoprofiling, without passing the judgment as to the chemical compound selected is a marker or not. In this scenario, the existing categorization of marker compounds as active principles, chemical markers, and general markers as described in the previous section may be taken into consideration.

19.2.4.2 Quantification of Marker Compounds

For those plants for which active principles are known, quantification of the active principles is carried out (examples given above; Fig. 19.2). In addition, some chemical markers can also be quantified [23, 24]. In the absence of known active principles any other compound/s that are predominantly present in the herb can be used as chemical markers. There are many such examples in the literature in which the active principles and the chemical markers are quantified by HPLC, GLC, and HPTLC methods (eugenol, gallic acid, ursolic acid, oleanolic acid – *Ocimum sanctum* leaf, [48]; phyllanthin, hypophyllanthin, gallic acid, ellagic acid – *Phyllanthus amarus* whole plant, [49]; boswellic acids in the

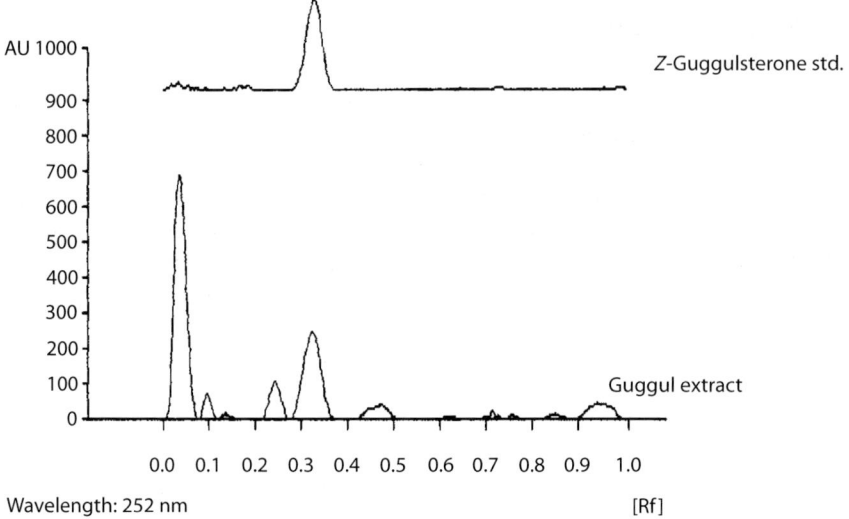

Fig. 19.2 TLC densitometric chromatogram of guggul (oleogum resin of *Commiphora wightii*) extract scanned at 252 nm

oleogum resin of *Boswellia serrata* [17, 18] eugenol and gallic acid in clove [50], alliin in *Allium sativum* [51], hecogenin in *Agave americana* leaf [52], luteolin [53], gallic acid and ellagic acid [54], rutin [55], sennosides [56], swertiamarine in *Enicostemma littorale* and *Swertia chirata* [57]).

Several examples of the methods for the quantification of phytochemicals using HPLC have been reported, which can form an important component of the quality parameters. For example: Piperine from *Piper* sp. [58], glycyrrhizin from *Glycyrrhiza* [59], allicin from *Allium sativum* [60], alliin from *Allium sativum* [51], guggulsterones from *Commiphora wightii* [61]; GLC of several herbal extracts [18, 62], GC-MS analysis of essential oil of *Chrysanthemum maximum* [45].

19.2.5 Multiple Marker-Based Evaluation

Since plants have complex mixtures of chemical compounds, multiple-marker-based analysis has recently been gaining importance. The multiple markers of a drug are usually a mixture of active principles and chemical markers. A few examples include quantification of four alkaloids of *Cinchona officinalis* stem bark [23, 24], four markers in *Ocimum sanctum* [48], phyllanthin, hypophyllanthin, gallic acid and ellagic acid in *Phyllanthus amarus* [49], four triterpenic acids of *Terminalia arjuna* by HPLC [63], triterpenic acids of the oleogum resin of *Boswellia serrata* [17], guggulsterones of the oleogum resin of *Commiphora wightii* [17].

19.2.6 Polyherbal Formulations

All the parameters described above – testing for different chemical groups and their quantification, the fingerprint profiles (Fig. 19.3), multiple-marker-based phytochemical profiling, and marker compound analysis (Fig. 19.4) – form important parameters for the polyherbal formulations as well. The different ingredients of a formulation contain important marker compounds, which can be active principles and/or chemical markers. Suitable extraction procedures are adapted to effect complete extraction of the compounds from the samples. The presence of the markers in a sample extract is ascertained by cochromatography of the sample extracts and standards of marker compounds and comparison of the R_f/retention time and absorption spectra with that of the standards of the marker compounds.

The mobile phases used for the TLC fingerprint profiles can be different from the mobile phases used for cochromatography with marker compounds. This is imperative because to resolve the maximum number of compounds for fingerprinting purposes, suitable mobile phases needed to be evolved. The mobile phase requirement for resolving specific markers, however, can be different. In TLC, while trying to obtain a desired band of maximum purity for the purpose of quantification, sometimes it becomes imperative to compromise on resolution for the rest of the compounds.

The methods so developed help in identifying the presence of these important markers and indicate the presence of the respective raw materials in the formulation. Spectral comparison is a further confirmation. Although multiple-marker-based evaluation ensures the quality with respect to the ingredients containing these marker compounds, it is practically impossible to have marker compounds specific to each of the ingredients of the formulation. Hence, TLC fingerprints are established to represent chemical profile of the formulations, along with cochromatography with important biomarkers. Furthermore, quantification of important marker compounds of the formulation forms an additional parameter in maintaining the quality of the product [26–31, 64–67].

In cases where neither the active principles nor the chemical markers are available, TLC fingerprint profile can be established, which can serve as a guideline for ensuring the quality.

In fact, complete fingerprint profiles should be established for all of the drugs and formulations, even in cases where active principles are known, keeping in view the possible synergistic activity of several chemical components of the herbal drugs. The profile so evolved by taking in to account all of the information *in toto*, is distinctive for the drug and would form its benchmark, especially when the active principles are not known or when chemical markers are not available for analysis.

A comprehensive specification for the drug would thus include fingerprint profiles along with chemical marker/active principle analysis that would establish identity and purity and, to an extent, ensure efficacy.

The estimation of individual compounds from a particular raw material has two applications:

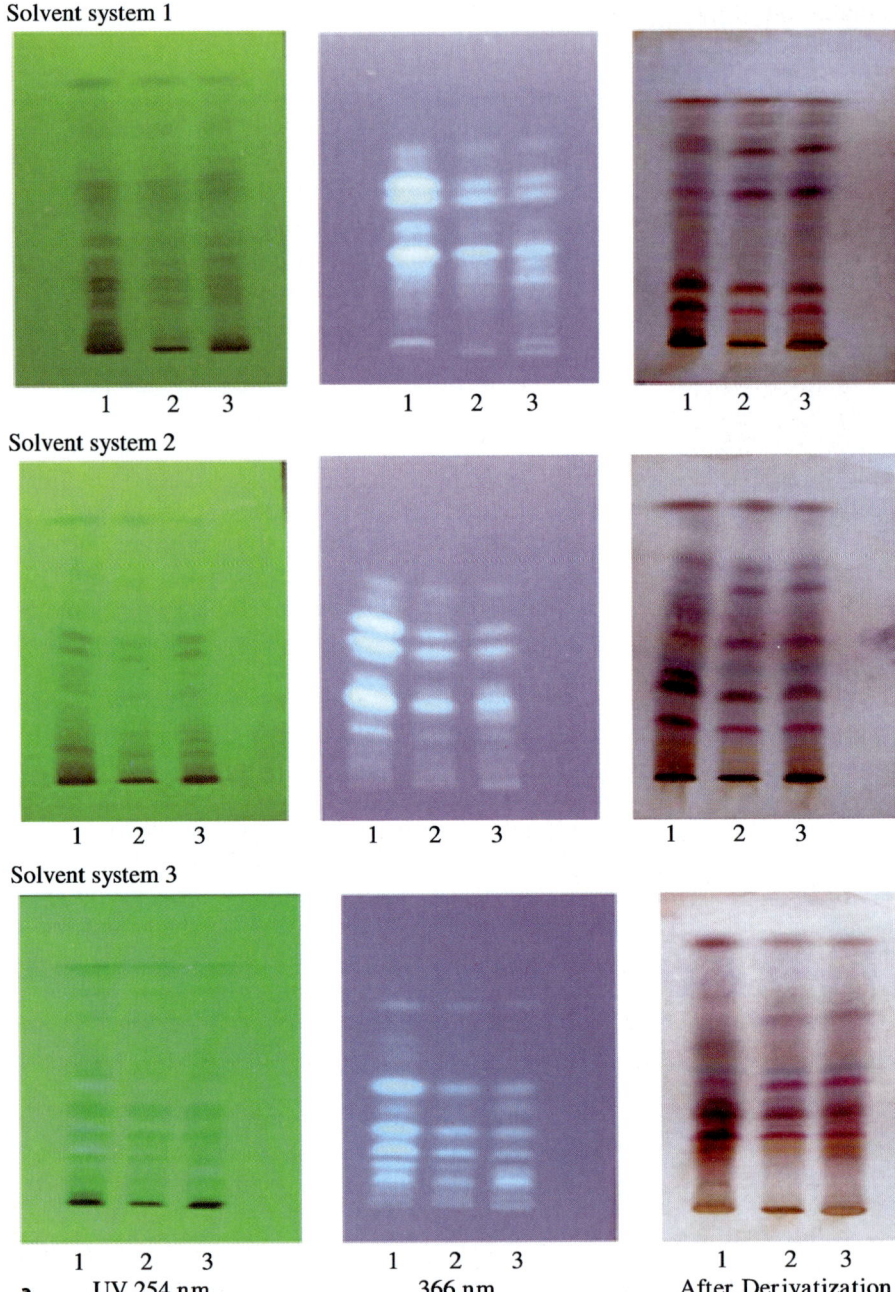

Fig. 19.3 a TLC fingerprint profiles of Chandraprabhavati, a polyherbal formulation [46]. *Track 1*: authentic sample; *Tracks 2 and 3*: market samples **b** *see next page*

Volatile oil of the formulation

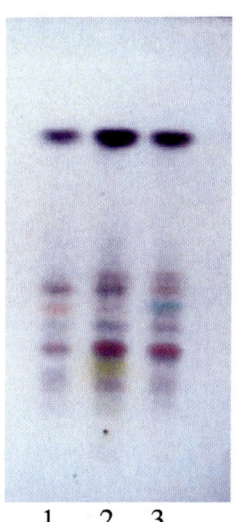

Fig. 19.3 *(continued)* **b** TLC fingerprint profiles of Chandraprabhavati, a polyherbal formulation [46].
Track 1: authentic sample; *Tracks 2 and 3*: market samples

1. If the compound happens to be one of the active principles, the analysis would ensure the quality of the raw material and the possible efficacy of the formulation in which it becomes a part (e.g., andrographolide, curcumin, azadirachtin, withanolides).
2. It will ensure the quality of the raw material when one is interested in extraction of that compound as a bulk drug phytopharmaceutical (e.g., quinine, artemisinine, paclitaxel, vincristine, vinblastine, l-dopa).

However, very little work has been reported on developing complete fingerprint profiles for herbal drugs, especially using HPTLC [40–42].

19.2.7 Hyphenated Techniques

The potential of the hyphenated techniques like LC-MS, LC-MS-MS, LC-MS coupled to NMR and MS and GC-MS are largely unexplored in phytochemical analysis of herbal drugs for standardization purposes, although there are some reports of the application of these techniques in the analysis [68, 69]. The potential of these methods should be exploited in deriving parameters as per the requirement.

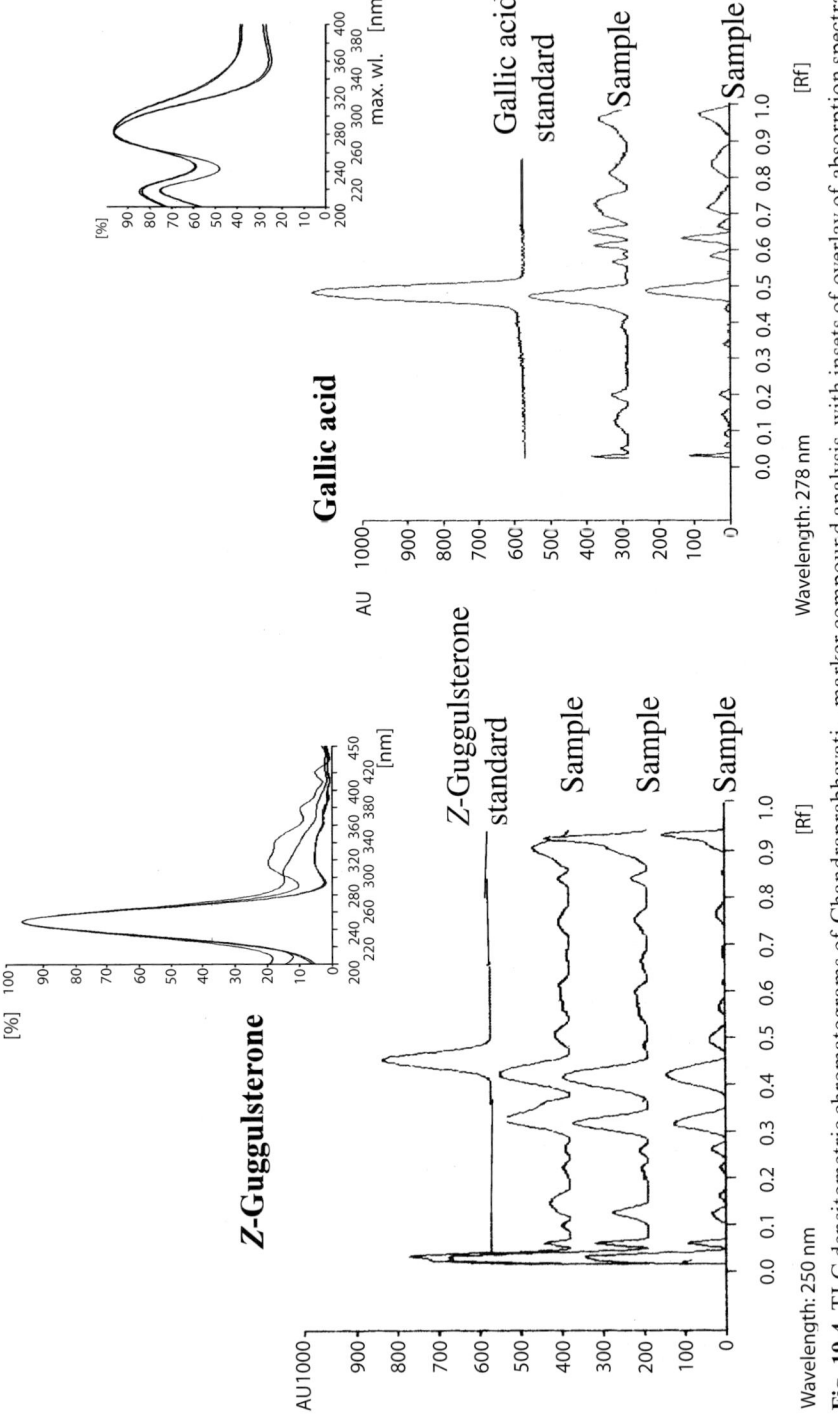

Fig. 19.4 TLC densitometric chromatograms of Chandraprabhavati – marker compound analysis, with insets of overlay of absorption spectra of respective standards and the corresponding band in the sample tracks [46]

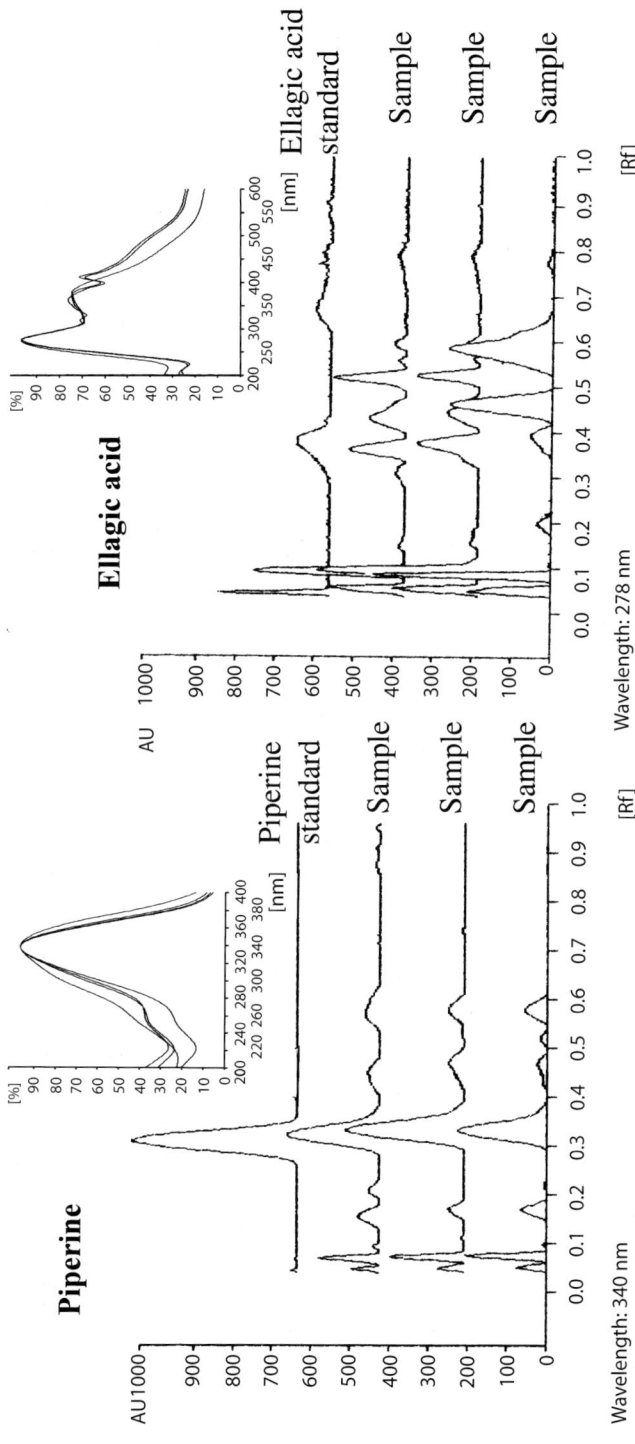

Fig. 19.4 (*continued*) TLC densitometric chromatograms of Chandraprabhavati – marker compound analysis, with insets of overlay of absorption spectra of respective standards and the corresponding band in the sample tracks [46]

19.2.8 Reference Compounds

When it comes to marker compound/active principle analysis, the availability of the reference compounds or even working standards is difficult. A few of these are available on the market, but they are very expensive. Hence, there is a need to make them available from a central facility by the Government before including the marker compound analysis in regulatory requirements.

19.3 Some Examples

19.3.1 Raw Material

19.3.1.1 Ammoniacum Gum

Ammoniacum gum [25] is an oleogum resin obtained as an exudate from the stem of the flowering and fruiting plants of *Dorema ammoniacum* (family Umbelliferae). The plant is native to Central and Eastern Iran and the oleogum is imported into India for its domestic market. The drug is described in British Herbal Pharmacopoeia (1983) [10]. In Unani medicine it is being used as an expectorant, stimulant and antispasmodic. It is also used in the treatment of catarrh, asthma, chronic bronchitis, and enlargement of liver and spleen. It contains volatile oil, resin, and gum. Some of the chemical constituents include free salicylic acid, ammoresinol, doremone A, doremin, and ammodoremin [25].

The phytochemical fingerprint profile of ammoniacum gum (from *Dorema ammoniacum*) was defined by subjecting different extracts of the ammoniacum gum to qualitative and semiquantitative analysis using HPTLC. The different compounds from alcoholic extract, *n*-hexane extract, and volatile oil of the gum were resolved by TLC and analyzed by scanning at 254 and 366 nm and by postchromatographic derivatization. The resolved bands were evaluated for their spectral details, the relative concentrations by densitometry and also for parameters like color of the bands, their fluorescent/nonfluorescent nature, R_f, and λ_{max} [25].

Therefore, to ensure efficacy it is necessary to evolve a phytochemical fingerprint profile of genuine drugs with the help of the following parameters:
1. The chromatogram.
2. The migration distances of the separated compounds (R_f).
3. The bands as observed with the naked eye, as examined under UV (254 and 366 nm).
4. The UV absorption spectra of the resolved compounds.
5. Densitometric and fluorimetric measurements of the resolved compounds and calculation of their relative percentages.
6. Response to several reagents during derivatization.

This profile would be distinctive and would form a benchmark for the drug, especially when the active principles are not known or when the chemical markers are not available. For this, it is important to take into account all the information available in the TLC fingerprint analysis of different extracts of the drug *in toto*, to ascertain the quality of the sample [25].

This work provides an insight into the potential of the method of standardization by TLC fingerprinting using HPTLC, in the absence of the availability of reference standards of either chemical markers or active principles for analysis of herbal drugs. The work further emphasizes that TLC fingerprint profiles thus generated have the potential for: (1) applicability in authentication of herbal drugs and (2) forming an important component of quality control criteria for herbal drugs [25].

In the case of this sample, this is only the beginning. A lot of work is yet to be done by working out several samples from different places and evaluating them before setting the parameters. If a correlation can be made in terms of activity, the fingerprint profile has the potential to ensure the efficacy apart from ensuring the authenticity and purity of the sample.

19.3.1.2 *Cinchona officinalis* Stem Bark

Stem bark of *Cinchona* was analyzed for alkaloids by the following:
1. Fingerprinting of the alkaloid fraction [17].
2. Colorimetric method – for total alkaloids [33].
3. Spectrophotometric method – for quinine-type alkaloids and cinchonine-type alkaloids [70].
4. Estimation of the four important alkaloids by HPLC [23].
5. TLC densitometric analysis using HPTLC – for the four important individual alkaloids [24].

In all of these, one important step is the sample preparation. It is known that in *Cinchona* stem bark, the alkaloids are present combined with the tannins (cinchotannic acid and quinic acid). European Pharmacopoeia (1997) [70] gives a method of sample preparation that involves adsorption of the tannins using tragacanth, which facilitates extraction of alkaloids by organic solvents. We replaced tragacanth with carboxymethyl cellulose.

While the colorimetric method enabled the estimation of total alkaloids and the spectrophotometric method the quinine-type and cinchonine-type alkaloids, the HPTLC method could be used to analyze all four alkaloids individually. Quinidine is reported to be present in very small quantities in *Cinchona* [71]; this was estimated based on fluorescence enhancement and detection, and quantification of quinidine by HPTLC. The plates were scanned and quantified at 226 nm for quinine, cinchonine, and cinchonidine, and at 366 nm quinidine in fluorescence and reflectance modes [24].

In the case of *Cinchona*, the method of analysis of the raw material could be applied directly to the formulations. However, it is not so easy and direct for many other formulations and it can involve complex procedures for eliminating interfering substances like sugars and tannins from the sample [72].

19.3.2 Formulation

19.3.2.1 Chandraprabhavati

This is an Ayurvedic formulation containing 54 ingredients. We developed fingerprint profiles and also quantified the marker compounds in Chandraprabhavati [64]. Of the markers we evaluated, Z-guggulsterone is one of the important active principles of oleogum resin of *Commiphora wightii* [73], which is the major ingredient of the formulation. Piperine is from *Piper nigrum* and *Piper longum*, berberine is from *Berberis aristata*, gallic acid and ellagic acid are from the fruit pulp of *Embelica officinalis*, *Terminalia chebula*, and *Terminalia bellirica*. Z-Guggulsterone, piperine, and berberine are specific markers, the presence of which in the sample confirms the presence of the respective raw materials in the formulation. The presence and quantity of Z-guggulsterone is important, since guggul is the major ingredient of the formulation. Gallic acid and ellagic acid are general marker compounds since they are present in many ingredients of the formulation. However, the amount of these general marker compounds in the formulation can serve as an extra parameter in quality control.

19.4 Conclusion

With the global increase in the demand for plant-derived medicine as an alternative to synthetic medicine, there is a need to ensure the quality of the herbal drugs using modern analytical techniques, for therapeutic efficacy and safety. Various methods of standardization and testing are needed immediately in the interest of both the manufacturer and the consumer. In the present business and industrial scenario, and considering the interest and faith that people have in herbal drugs, the need for the development of the aforementioned methods of standardization can not be over emphasized. Scientists all over the world have a very important contribution to make in this area.

Acknowledgments

We are grateful to Professor Harish Padh, Director of the B.V. Patel PERD Centre, for providing the facilities. We thank Dr. Ravishankara for his valuable comments and suggestions.

References

1. British Medical Association (1993) Complementary Medicine – New Approaches to Good Practice. Oxford University Press, London, p 9
2. Eisenberg DM, Ronald CK, Foster C, et al (1993) New Engl J Med 328:246
3. Sawyer MG, Gannoni AF, Toogood IR, et al (1994) Med J Aust 160:320
4. MacLennan AK, Wilson DH, Taylor AW (1996) Lancet 347:569
5. Government of India (2001) The Draft of National Policy on Indian Systems of Medicine (2001) Department of Indian Systems of Medicine and Homeopathy, Ministry of Health and Family Welfare, Government of India, New Delhi
6. Raskin I, Ribnicky DM, Komarnytsky S, et al (2002) Trends in Biotechnol 20:522
7. The Japanese Standards for Herbal Medicines (JSHM) (1993)
8. The United States Pharmacopoeia (2002) US Pharmacopoeia, Rockville, USA
9. Bradley P (2006) British Herbal Compendium, vol 2. British Herbal Medicine Association, Dorset, UK
10. The British Herbal Medicine Association (1996) British Herbal Pharmacopoeia, Dorset, UK
11. The State Pharmacopoeia Commission (2001) Pharmacopoeia of the People's Republic of China. State Food and Drug Administration, Beijing, China
12. Gruenwald J, Brendler T, Jaenicke C (2000) PDR for Herbal Medicines, Medical Economics, Montvale
13. The Ayurvedic Pharmacopoeia of India, part I, vol I–V (1990–2006) Government of India, New Delhi
14. The European Scientific Cooperative on Phytotherapy (ESCOP) (1999) ESCOP Monographs on Medicinal Uses of Plant Drugs, ESCOP Secretariat, UK
15. Indian Pharmacopoeia (1996) Ministry of Health and Family Welfare, Government of India, New Delhi
16. Upton R (ed) (1999) American Herbal Pharmacopoeia. American Herbal Pharmacopoeia, California, USA
17. Gupta AK, Tondon N, Sharma M (eds) (2003–06) Quality Standards of Indian Medicinal Plants, vol I–IV. Indian Council of Medical Research, New Delhi
18. Indian Herbal Pharmacopoeia (2002). Regional Research Laboratory, Jammu-Tawi and IDMA, Mumbai
19. World Health Organization (1999) WHO monographs on selected medicinal plants. vol 1 and 2. World Health Organization, Geneva
20. Sharma M (1997) Industry Highlights Feb 1
21. Houghton PJ (1999) Pharm News 6:21
22. Ravishankara MN, Shrivastava N, Jayathirtha MG, Padh H, Rajani M (2000) J Chromatogr B 744:257
23. Ravishankara MN, Shrivastava N, Padh H, Rajani M (2001) Indian Drugs 38:248
24. Ravishankara MN, Shrivastava N, Padh H, Rajani M (2001) Planta Med 67:294
25. Rajani M, Ravishankara MN, Shrivastava N, Padh H (2001) J Planar Chromatogr 14:34
26. Neeta S, Niranjan K, Ravishankar MN, Rajani M (2006) Standardization of Tila-e-Mubahhi: A Unani formulation. In: Abdin MZ, Abrol YP (eds) Traditional Systems of Medicine. Narosa Publishing House, New Delhi, India, p 107
27. Pathak SB, Bagul MS, Rajani M (2006) Multiple marker based Standardization of a polyherbal Unani Formulation Majoon-E-Jograj Gugal. In: Abdin MZ, Abrol YP (eds) Traditional Systems of Medicine. Narosa Publishing House, New Delhi, India, p 113
28. Veerapur VP, Bagul MS, Srinivasa H, Padh H, Rajani M (2006) Phytochemical evaluation of polyherbal Unani formulations – A Case Study of Itrifal Mulaiyin and Sharbate-E- Dinar. In: Abdin MZ, Abrol YP (eds) Traditional Systems of Medicine. Narosa Publishing House, New Delhi, India, p 123

29. Srinivasa H, Bagul MS, Ravishankara MN, Rajani M (2006) Phytochemical Standadization of Unani formulations – Sharbat-E-Nilofer and Majoon Zabeeb. In: Abdin MZ, Abrol YP (eds) Traditional Systems of Medicine. Narosa Publishing House, New Delhi, India, p 139
30. Bagul MS, Pathak SB, Ravishankara MN, Rajani M (2006) Phytochemical standardization of polyherbal Unani formulation Sharbat-E-Ejaz. In: Abdin MZ, Abrol YP (eds) Traditional Systems of Medicine. Narosa Publishing House, New Delhi, India, p 131
31. Bagul MS, Rajani M (2006) Phytochemical evaluation of Chyavanprash. In: Govil N, Singh VK, Arunachalam C (eds) Recent Progress in Medicinal Plants – Search for Natural Drugs, vol 13. Stadium, Houston, p 251
32. Rajani M, Pundarikakshudu K (1996) Int J Pharmacog 34:308
33. Ravishankara MN, Shrivastava N, Mahendru N, Padh H, Rajani M (2001) Indian J Pharmaceut Sci 63:76
34. Ravishankara MN, Shrivastava N, Padh H, Rajani M (2002) Indian Drugs 39:494
35. Niranjan K, Ravishankara MN, Padh H, Rajani M (2002) J Nat Remed 2:168.
36. Bagul M, Srinivasa H, Sheetal A, Rajani M (2006) Indian Drugs 43:665
37. Kalola J, Rajani M (2006) Chromatographia 63:475
38. Anandjiwala S, Srinivasa H, Kalola J, Rajani M (2006) J Nat Med 61:59
39. Ravishankara MN (2002) PhD thesis, North Gujarat University, Gujarat, India
40. Pachaly P (1999) Dtsch Apoth-Ztg 139:52
41. Pachaly P (1999) Dtsch Apoth-Ztg 139:57
42. Hahn-Deinstrop E, Koch A, Miller M (1998) J Planar Chromatogr 11:404
43. Lalla JK, Hamrapurkar PD, Mamani HM (2000) J Planar Chromatogr 13:390
44. Keusqen M (1997) Planta Med 63:93
45. Józefczyk A, Marlowski W, Mardarowicz G (1999) Pharmaceut Biol 37:8
46. Bagul MS (2004) PhD thesis, North Gujarat University, Gujarat, India
47. Evans WC (ed) (2002) Pharmacognosy 15th edn. Saunders, Edinburgh, UK
48. Anandjiwala S, Kalola J, Rajani M (2006) J AOAC Int 89:1467
49. Dhalwal K, Biradar YS, Rajani M (2006) J AOAC Int 27:619
50. Pathak S, Niranjan K, Padh H, Rajani M (2004) Chromatographia 60:241
51. Kanaki NS, Rajani M (2005) J AOAC Int 88:1568
52. Ghoghari AM, Rajani M (2006) Chromatographia 64:113
53. Srinivasa H, Bagul MS, Padh H, Rajani M (2004) Chromatographia 60:131
54. Bagul MS, Srinivasa H, Padh H, Rajani M (2005) J Separation Sci 28:581
55. Prashanthkumar V, Ravishankara MN, Bagul M, Padh H, Rajani M (2003) J Planar Chromatogr 16:386
56. Shah SA, Ravishankara MN, Nirmal A, Shishoo CJ, Rathod IS, Suhagia BN (2000) J Pharm Pharmacol 52:445
57. Vishwakarma SL, Bagul MS, Rajani M, Goyal RK (2004) J Planar Chromatogr 17:128
58. Szabo B, Botz L, Lakatos A, Koszegi T (2001) J Planar Chromatogr 14:181
59. Chauhan SK, Singh BP, Agrawal S (1999) Indian Drugs 36:521
60. Jansen H, Muller B, Knabloch K (1982) Planta Med 46:559
61. Mesrob B, Nesbitt C, Misra R, Pandey RC (1998) J Chromatogr B 720:189
62. Martinelli EM (1980) Fitoterapia LI:35
63. Singh DV, Verma RK, Singh SC, Gupta MM (2002) J Pharm Biomed Anal 28:447.
64. Bagul MS, Rajani M (2007) Phytochemical Evaluation of Polyherbal Formulations Using HPTLC. In Natural Products – Essential Resources for Human Survival, edited by Yi-Zhun Zhu, Benny K-H Tan, Boon-Huat Bay (National University of Singapore, Singapore) & Chang-Hong Liu (Nanjing University, China) World Scientific Publishing Co. Pte. Ltd. Singapore (3rd International Conference on Natural Products 2004, Nanjing, China); p 149
65. Trivedi PD, Batel BC, Rathnam S, Pundarikakshudu K (2006) J AOAC Int 89:1519
66. Bagul MS, Rajani M (2005) Indian Drugs 42:15

67. Bagul MS, Rajani M (2006) J Nat Remed 6:53
68. Renukappa T, Roos G, Klaiber B, Kraus WV (1999) J Chromatogr A 847:109
69. Cech NB, Eleazer MS, Shoffner LT, Crosswhite MR, Davis AC, Mortenson AM (2006) J Chromatogr A 1103:219
70. European Pharmacopoeia (1997) European Pharmacopoeia. Council of Europe, Strasbourg, p 634
71. Wallis TE (1985) Textbook of Pharmacognosy, 5th edn. CBS, Delhi, p 98
72. Forni GP (1980) Fitoterapia 51:13
73. Pandey VN, Malhotra SC (eds) (1992) Pharmacological and Clinical studies of Guggulu (*Commiphora wightii*) in Hyperlipidaemia/Lipid Metabolism. Central Council for Research in Ayurveda and Siddha, New Delhi

Subject Index

A

Acacia catechu 16
Acacia senegal, gum arabic, AGPs 264
Accelerated discovery techniques 1
Acetaminophen 133
Acetylcholinesterase (AchE) 10, 167
– inhibitors 14
Acitus sulcatus, guggulsterol 108
Acquilaria malaccensis 339
Actinic keratoses 14
Adaptogens, roseroot 298
Adhatoda vasica 357
Adhyperforin 150
Aegelin 356
AG 256
– dietary fiber/prebiotics 264
Agave americana 358
AGP 256, 259
– abiotic stress tolerance 263
– reproductive organ development 261
– signaling 262
Agrobacterium rhizogenes 170, 243, 272, 274
Ailanthus grandis, guggulsterol 108
Ajmalicine 273
– spectrofluorometry 288
Ajmaline 273
Alkaloid identification 288
Alkaloids, plant cell/tissue cultures 165

Alliin 358
Alzheimer's disease (AD) 9
– galanthamine hydrobromide 168
– ginkgo 17
– huperzine 11
– ZT-1 14
Amanita phalloides, silymarin 131
Amanitin, silymarin 131
Ammi majus, coumarine/ furocoumarine 275
Ammoniacum gum 364
Amyloidosis, colchicine 220
Anaplastic thyroid cancer 13
Andrographolides 356
Angelica acutiloba, arabinogalactans 259
Anthranilate synthase (AS) 278
Anthraquinones, light-activated 150
Antibiotics, phytoremediation 279
Anticarcinogenic effects, silymarin 134
Anticholinesterase activity, bacosides 184
Antidepressant, St. John's wort 152
Antidiarrheals 8
Antihypertensive 273
Anti-inflammatory action, silymarin 132, 136
Antimitotic agents, colchicine 220
Antimuscarinic agent 8
Antioxidant activity 39
Antipsychotic agents 8
Antiretroviral drugs 12

Antithrombotic properties   41
Antivirals   11
Anxiety neurosis,
    Bacopa monnieri   188
Apokyn   10
Apomorphine   10
Apoptosis   17, 317
Arabidopsis thaliana,
    tagged hairy roots   277
Arabinogalactan   256
Arachidonic acid   16
Araucaria angustifolia, AGPs   261
Aromatic plants   336
Artemisia annua   329
Artharvaveda   3
Asparagus racemosus   336, 343, 356
Asthma   8
Astringin   45
Atropa belladonna   2, 7, 166, 170, 357
Atropine   7
– analog   10
Ayurvedic medicine   3
Azadirachta excelsa   248
Azadirachta indica   234, 243
Azadirachtin,
    in vitro production   234
– marker compound   356
– plant cell/tissue culture   243
– stability   249

B

Bacopa monnieri   175
– saponins   184
– tissue culture   190
Bacosides   176, 356
Bacterial strains, selection   286
Baicalin   16
Balsam (guggul)   103
Banisteriopsis caapi   2
Baptisia tinctoria,
    arabinogalactans   259
Bcl-2   318
Behcet's syndrome, colchicine   220
Benzylisoquinoline alkaloids   167
Berberine   366

Bergenin   45
Beta vulgaris, betalaine   276
Betulinic acid   11, 12
Bevirimat   11, 12
Bilobalide   17
Biochanin A   59
Boswellic acids   357
Bottlebrush plant   10
Bradycardia (low heart rate)   8
Brahmine   183
Bronchodilators   8
Bronchospasm   10

C

CA4P   13
Cadmium, phytoextraction   280
Calabar bean
Callistemon citrinus   10
Camellia sinensis   2
Camptotheca acuminata   199, 273
Camptothecin (CPT)   7, 11, 197, 273, 337
Cancer chemoprevention   25, 43
Cannabidiol   17
Cannabinoids   17
Cannabis sativa   2, 17
Capsaicin   11, 12
Carcinogenesis promotion,
    models   9
Cardiovascular disease   11
Cardiovascular protection   25, 39
Castanospermine   13
Castanospermum austral   13
Castor oil   341
Catechins   318, 321
Catharanthine, HPLC   288
Catharanthus roseus, alkaloids   19
– globular cell cultures   159
– hairy roots   286
– leaves, genetic
    transformation   289
– tagged hairy roots   277, 278
Cathechin   16
Ceflatonine   11, 13
Celgosivir   11, 13
Cell factories   86

Cephalotaxus harringtonia 13
Ceric ammonium sulphate 288
Chandraprabhavati 366
Chantix 11
Chemopreventive effects,
   silymarin 134
Chlorophytum borivilianum 326
Cholinesterase inhibitory
   activity 167
Chronic obstructive pulmonary
   disease (COPD) 10
CHS 61
Cicer arietinum, isoflavonoid 77
Cigarette smoking 11
Cinchona officinalis 326, 365
Cinchotannic acid 365
Clinical trials 1, 11
Cocaine 2
Cocoa 2
Codeine 167, 273
Coffea arabica 2
Coffee 2
Colchicine 4, 192, 216, 326, 337
– gout pain 220
– production 225
– toxicity 218
Coleus blumei, rosmarinic acid 86
Combretastatin A4 11, 13
Combretum caffrum 13
Commiphora wightii 101, 327, 343,
   356–358, 366
Complementary and alternative
   medicines (CAM) 340
Coniferin 273
Cortisone 5
Coumestrol 61
CPT 7, 11, 197, 273, 337
Croton oil 9
Croton tiglium 9
Curcuma longa,
   arabinogalactans 259
Cybister tripuncatus, prothoracic
   defensive gland secretion 108
Cyclic nucleotide-gated (CNG)
   channels 8
Cyclooxygenase (COX-1/COX-2) 16

Cyclopseudohypericin 151
Cytisine 11
Cytisus laburnum 11
Cytochrome P450 45

D
Daidzein 15, 58, 64, 71, 80, 319
Datura spp. 2
DDT, hairy root
   phytoremediation 280
10-Deacetylbaccatin 5
Deadly nightshade 2
Dehydromatricaria ester 273
1-Deoxyxylulose 239
Depression, St. John's wort 152
Devil's claw 16
Diabetes 7
Dianthrones, St. John's wort 149
2,4-Dichlorophenol,
   hairy root phytoremediation 280
Dimethylbiguanide 7
N,N-Dimethyltryptamine 2
Dioscorea deltoidea 339
Dioscorea spp 5, 19
Diosgenin 5, 19
Disodium phosphate prodrug 13
Diterpene ester 14
Diterpenoid, tetracyclic 9
Docetaxel 7
Dopamine receptor agonist 10
Dorema ammoniacum,
   ammoniacum gum 364
Droperidol 8
Drug discovery/development 1
– plant-derived compounds 4
Drug precursors 1, 4
Drug prototypes 1, 6
Duboisia hybrid hairy roots 278
Duboisia myoporoides 357
Dysplastic melanocytic nevi 12
Dytiscus marginalis,
   guggulsterol 108

E
Echinacea pallida,
   arabinogalactans 259

Echinacea purpurea,
  arabinogalactans 259, 266
Echinacea-AG, nutraceutical 266
Echinatin 64
EGb 761 17
Eleuthrococcus senticosus 342
Embelica officinalis,
  gallic acid/ellagic acid 336, 366
Emodinanthrone 154
Enterobacter sakasaki 275
Epicatechin 321
Epigallocatechin-3-gallate (EGCG),
  apoptosis 321
Eravatamia heyneana, CPT 199
Erythroxylum coca 2
Ethanol, silymarin 134
Ethephon 103
Etoposide 7
Euphorbia peplus 14

F
Farnesenoid X receptor (FXR)
  agonists 101, 110
Flavocoxid 16
Flavone glycosides 17
Flavonoids 16
Flavonolignans 123
Formononetin 59, 77
Fumaryl acetoacetate hydrolase
  (FAH) 10

G
Galactosamine 133
Galantamine/galanthamine 9, 10
Galanthamine 167
Galanthus woronowii 10
Galega officinalis 7
Gemcitabine 17
Genista tinctoria, isoflavonoid 77
Genistein 8, 77
Gibberellin regulation, AGP 260
Gingko biloba 17, 273
Ginkgo extracts 17
Ginkgolides 17, 273
Ginsenosides 19
α-Glucosidase I inhibitor 13
Glyceollin 64

Glycine max 8
- isoflavonoid 77
Gmelina arborea 273
Gout/urate metabolism, pain 220
Grapevine stilbenes 27
Guanidine 7
Guggul (balsam) 103
Guggulsterones 102, 329
- marker compound 357
- production 115
- toxicity 113
Gum arabic 264
Gum resin 102
Gynostemma pentaphyllum 273
Gypenoside 273

H
Hairy root induction 287
Hairy roots 170, 244, 272
Harmine/harmaline 276
Harpagophytum procumbens 16, 19
Harpagoside 16
HCV 13
Heavy metals,
  phytoremediation 279
Hecogenin 358
Hepatic fibrosis, silymarin 132
Hepatitis C (HCV) 11
Hepatoprotection, silymarin 132
Herbal drugs, pharmacopoeias 351
- phytochemical
  standardization 349
- quality 342
Herbicide 10
Herpestine 183
HIV-1 maturation 12
Homoharringtonine 11, 13
Hopeaphenol 45
Human immunodeficiency viral
  (HIV) 11
Huperzia serrata 14
Huperzine 11, 14
Hycamtin 7
4-Hydroxyphenyl lactic acid 10
4-Hydroxyphenyl pyruvate
  dioxygenase (HPPD) 10
4-Hydroxyphenyl pyruvic acid 10

Subject Index

11α-Hydroxyprogesterone 5
Hyoscyamine 7, 166
Hyoscyamus niger 2
– H-6-H gene 278
Hypericin/hyperforin 150, 152
Hypericum perforatum 150
Hypocholesterolaemic action, silymarin 134
Hypoglycemic activity 7
Hypolipidemic agent 102

I

Illicium verum 6
Ilybius fenestratus, prothoracic defensive gland secretion 108
Immunity enhancers 265
Immunomodulation 17
Immunomodulators, AG 265
Indian herbal drugs 325
Indian pharmaceutical industries 341
Indian system of medicine 328
Indole alkaloids 286
Influenza 6
Ingenol 3-angelate 11, 14
Interferons 13
Interleukins 17
Ipratropium bromide 8
Irinotecan (CPT11) 7, 197, 199, 337
Iscador 17
Isobutyrophenone synthase (BUS) 155
Isoflavone reductase (IFR) 61
Isoflavones 8, 57
– in vitro cultures 57
Isoflavonoid phyto-oestrogens, biosynthesis precursors 70
– biotransformation 71
– elicitation 61
– genetic modifications 73
– plant cell cultures 57
Isohypericin 151
Isoliquiritigenin 80
Isoquinoline alkaloids 167
Isorhamnetin 17

J

Jaceosidin 273
Jujubogenin 176

K

Kaempferol 17
Karyokinesis, colchicine 220
Kava 3
Khaya grandifolia, guggulsterol 108
Kievetone (5-hydroxyflavone) 59
Knee osteoarthritis 16
Korean ginseng 15

L

Lactofen 64
Lapacho tree/lapachone 319
Larch arabinogalactan (LAG) 255, 264
Lectin 17
Legume plants 59
– phyto-oestrogens 77
Leptogorgia sarmantosa, guggulsterol 108
Leptospermone 10
Leucojum aestivum 166
Licodione 64
Lignans 273
Limbrel 16
Linum flavum 273
Lipopolysaccharide-induced inducible nitric oxide synthase 16
Lipoxygenase (5-LOX) 16
Loperamide 8
Lophophora williamsii 2
Lotus japonicus, hairy roots 277
Lupins 63
Lupinus spp., isoflavonoid 77
Luteolin 358

M

Maackia amurensis, isoflavonoid 77
Maackiain 77
Mandragora officinarum 2
Mandrake 2
Margosan-O 242
Marker compounds 357
Medicarpin 61, 77

Medicinal aromatic plants (MAP)   337
Medicinal plants, biodiversity   338
– supply/demand   337
Mediterranean fever, colchicine   220
Merendera spp., colchicine   227
Merriliodendron megacarpum, CPT   199
Mescaline   2
Metabolite trapping   275
Metformin   7
Methadone   8
Methyl jasmonate   61, 89, 94, 123, 246, 275, 285, 287
– azadirachtin   246
Micropropagation   101
Microtubules, colchicine   221
– vinca alkaloids   320
Milk thistle (*Silybum marianum*)   125
Mistletoe   17
Molecular farming, hairy roots   279
Monnierasides   183
Morphinan, accumulation, hairy roots   170
– alkaloids   165, 170
Morphine   2, 10, 14, 167, 171, 186, 217, 273, 326
– substitutes   8
Morphine-3-glucuronide   14
Mostuea brunonis   199
Multiple sclerosis   17
MX-3253   13
Mydriatic agent   8
Myeloid leukemia   13
Myopic macular degeneration   13

N

Naphthochinone   63
Natural products   1
Neem tree (*Azadirachta indica*)   233, 356
Neuraminidase inhibitor   6
Neurodegenerative diseases   46
Neuronal nicotinic receptors   11
Neuroprotection   25

New therapeutic agents   1
Nickel, phytoextraction   280
Nicotiana alata, AGPs   261
Nicotiana tabacum, tagged hairy roots   277
Nicotinic acetylcholine receptor   10, 11
Nitisinone   10
Nivalin   9
NK cells, AGPs   265
Nootropic drug   175
Nothapodytes nimmoniana   198, 203
Nuclear factor-κB activation   16

O

Oestrogen β receptors   59
Oleogum-resin   101
Ophirrohiza mugos/O. pumila, CPT   199
Opiate analgesic alkaloid   10, 14
Oral contraceptive   5
Orfadin   10
Oseltamivir   6
Osteoarthritis   16
Oxalis tuberosa   276
Oxytetracycline, phytoextraction   280

P

PA-457   12
Paclitaxel   5, 19, 92, 93, 318, 319, 357
Pain reliever   16
PAL   61
Panax ginseng   15, 19, 275–277, 331, 342, 356
– ginsenoside, vanadyl sulfatehitosan   275
– hairy roots, MeJA   277
– methyl jasmonate   276
Papaver somniferum   14, 166, 273, 326
Parkinson's disease   10
PCD   263
Peginterferon α2b   13
PEP005   14
Peripheral claudication   17

Subject Index

Pesticides, phytoremediation 279
Pethidine 8
Phalloidin, silymarin 131
Pharbitis nil, umbelliferone 276
Pharmacological probes 1
Phaseolin (5-deoxyisoflavone) 59
PHB synthase,
   Ralstonia eutropha 279
Phenoxodiol 11, 15
Phenylpropanoids, esterified 19
Phloroglucinol, St. John's wort 149
Phorbol 9
Photodynamic therapy (PDT) 153
Phyllanthin 357
Physostigma venenosum 2
Phytoalexins 59
Phytoanticipins 59, 63
Phytochemical profiles,
   fingerprinting 354
Phytochemical standardization 352
Phytoextraction 279
Phytoremediation 272, 280
Phytotransformation 279
Piceatannol 45
Piper methysticum 3
Piperine 366
PKC 14
Plant-derived drugs 1
Plastoquinone 10
Podophyllotoxin 7, 276
Podophyllum hexandrum 276, 339
Poly(3-hydroxybutyrate) (PHB)
   polyester 279
Polyacetylene 273
Polyherbal formulations,
   phytochemical
   standardization 349, 359
Polyketide synthases (PKSs) 154
Polyphenols 25, 42
Post-harvest treatment 310
Postoperative pain 11
Pregnancy-related complications 11
Progesterone 5
Programmed cell death (PCD) 317
– AGPs 258, 263
Protein kinase C (PKC) activators 9

Protein tyrosine kinases (PTK) 8
Protopanaxadiol 11, 15
Pseudohypericin 151
Pseudojujubogenin 176
Psoralea spp., isoflavonoid 77
Psoralens 329
Psyllium seeds/husks 341
Pterocarpus santalinus 339
Pueraria phaseoloides 273
Puerarin 273
Pyrenacantha klaineana, CPT 200

Q
Quercetin 17
Quinine/quinidine/quinic acid 326, 365
Quinolizidine alkaloid 11

R
Rauvolfia micrantha 273
Rauvolfia serpentina 339
Razadyne 9
Reminyl 9
Resveratrol 25, 45
Retrochalcones 63
Retusin 77
Rheumatoid arthritis 16
Rhodiola rosea 298
Rhodiola sachalinensis,
   globular cell cultures 159
Ribavirin 13
RNA silencing 277
Rosavin 298
Roseroot 298
Rosin 300
Rosmarinic acid 85, 87
Rubia tinctorium, antraquinone 275

S
Salidroside 298
Salvia divinorum 2
Salvia miltiorrhiza 275
Salvinorin A 2
Sandersonia auranticata,
   colchicine 227
Sanguinarine 273

Saponins 2, 15, 176, 184, 249, 273
– triterpenoidal,
  dammarane type 176
Sarcoidosis, colchicine 220
Sativan 61
Sativex 17
Saussurea lappa 339
Saussurea medusa 273
Scleroderma, colchicine 220
Scopolamine 2, 166, 276
Scutellaria baicalensis 16
Secoadhyperforin 151
Secohyperforin 151
Secreted alkaline phosphatase
  (SEAP), tobacco hairy roots 279
Sennosides 358
Serpentine, spectrofluorometry 288
Sesquiterpene lactones 19
Shikimate kinase 6
Shikimi tree 6
Shikimic acid 6
Silybin 128
Silybum marianum 122
Silychristin 123, 128
Silydianin 123, 128
Silymarin 120
– antioxidant activity 131
– hepatocyte membranes 131
– pharmacology 130
Skimmin 276
Smoking cessation 11
Snowdrop 10
Solanum chrysotrichum 273
Solanum khasianum hairy roots 278
Solanum tuberosum,
  sesquiterpenes 275
– tagged hairy roots 277
Solidago altissima 273
Soybean 8
Spiriva 10
Star anise 6
Steroidal sapogenin 5
Stilbenes 25, 27
– dimers 29
– monomers 27
– oligomers 29

– trimers/tetramers 30
Summer snowflake 167
Swertiamarine 358

T
Tamiflu 6
Tanshinone 275
Taxanes 7, 319
Taxol 5, 85, 92
– accumulation, in vitro 94
– production in immobilized
  cell cultures 97
– scale up production 96
Taxus baccata 92
Taxus brevifolia 5, 92, 319
Taxus wallichiana 339
Tea catechins 321
Terminalia chebula/bellirica,
  gallic acid/ellagic acid 336, 366
Terpenoid indole alkaloid
  (TIA) 278
Terpenoid lactones 17
Tetracycline, phytoextraction 280
12-*O*-Tetradecanoylphorbol-13-
  acetate (TPA) 9
Δ^9-*trans*-Tetrahydrocannabinol 2, 17
Thapsia garganica 19
Theobroma cacao 2
Thioacetamide, silymarin 134
Tinnitus 17
Tiotropium bromide 10
Tobacco dependence 11
Tocopherol biosynthesis 10
Topotecan (TPT) 7, 199
Toxic organic molecules,
  phytoremediation 279
Traditional medicine 340
Transcriptome analysis,
  T-DNA activation 277
Triphala 336
Triterpenes 15
– lupane-type 12
Tropane alkaloids 7, 166
– accumulation, hairy roots 170
Tryptophan decarboxylase
  (TDC) 278

Tubulin, colchicine 221
Tumor promoters 9
Tyrosine catabolism, FAH deficiency 10
Tyrosinemia type 1 (HT-1) 10

U
Uranium, hairy root rhizofiltration 280

V
Valepotriates 356
Varenicline 11
Vascular dementia 17
Vasicine 357
Vasoprotective Properties 41
Vaticanol 45
Vepesid 7
Verbascoside 273
Vigna radiata, AGPs 258
Vinblastine/vincristine 319
Vinca alkaloids 318, 320
Viniferin 45

Viscotoxins 17
Viscum album 17
Viscumin 17
Vitis vinifera 25
– stilbenes 33
Vitisifuran 45
Vitisin 45

W
Warts 14
Water hyssop 175
Wine, stilbenes 33
Withaferin-A 207
Withania somnifera 207, 327, 336
Withanolides 207, 356

Y
Yariv reagents 258, 261

Z
Zaborandi tree (*Pilocarpus zaborandi*) 326
ZT-1 14

Printing: Krips bv, Meppel, The Netherlands
Binding: Stürtz, Würzburg, Germany